Structural Steel Design

Structural Steel Design

Edited by

Patrick J. Dowling BE PhD DIC FEng FICE FIStructE FRINA
British Steel Corporation Professor of Steel Structures
Head of Civil Engineering Department, Imperial College, London

Peter R. Knowles MA MPhil CEng MICE FIHT
Consultant Engineer

Graham W. Owens BSc(Eng) PhD DIC CEng MICE MWeldI
Assistant Director, The Steel Construction Institute

Butterworths
London Boston Singapore Sydney Toronto Wellington

First published, 1988

British Library Cataloguing in Publication Data

Structural steel design.
1. Steel structures. Design.
I. Dowling, Patrick. II. Knowles, P.R.
(Peter Reginald) 1928- III. Owens,
Graham
624.1′821
ISBN 0-408-03705-9 (cased edition)
ISBN 0-408-03717-2 (paperback edition)

Library of Congress Cataloging-in-Publication Data

Structural steel design.
Bibliographies.
Includes index.
1. Building, Iron and steel. 2. Structural design.
3. Steel, Structural. I. Knowles, Peter Reginald.
II. Owens, Graham (Graham Wynford) III. Dowling, P.J. (Patrick J.)
TA684.S7892 1988 624.1′821 88-16726
ISBN 0-408-03705-9
ISBN 0-408-03717-2 (pbk.)

Typeset at The Alden Press Oxford London and Northampton
Printed in Great Britain at the University Press, Cambridge

Preface

This textbook is intended for students of structural steel design and is the outcome of a cooperative effort involving some one hundred engineers in universities, polytechnics and industry. It builds on the highly successful Structural Steel Design teaching project for lecturers which was sponsored by the British Steel Corporation at the writer's suggestion and produced under his general direction with Graham Owens as project manager. We were joined by Peter Knowles in editing this book, and his experience of a lifetime in teaching steel design has been of great benefit to us.

We believe this text to be a major advance in many respects over other existing ones. It is based on the modern limit state approach to design and covers areas which are traditionally rather neglected or treated inappropriately in other texts. Such areas include a treatment of metallurgy at a level which should cover a structural designer's needs, treatments of fabrication, erection and important areas of structural steel technology such as the protection of steel against both fire and corrosion. A section on the approach to design itself, notoriously difficult to cover properly in texts, is presented in a form which we believe from trial use of the material to be successful. As might be expected, the design of elements and structures is covered in depth in a modern and, we believe, authoritative fashion, and numerous worked examples are provided to illustrate the methods covered in the text.

One of the most exciting aspects of our involvement in the teaching project and subsequently in editing this text has been to discover the large number of people within Britain with a genuine enthusiasm for steel design. This enthusiasm has been infectious and the interest which has already been stimulated in the proper and efficient use of steel both within and outside our universities has been gratifying.

A list of those many steel enthusiasts who have contributed to the source material is given at the beginning of this book. For those who devoted their time so generously there could be no greater reward for their efforts than to see the benefits accruing to the next generation of structural steel designers, the undergraduates of today.

P.J. Dowling
Imperial College, June 1988

List of contributors

B.O. Allwood, Bolton Institute of Higher Education

Dr D. Anderson, University of Warwick

H. Arch, Consultant

D. Beckett, Thames Polytechnic

C.W. Brown, Freeman Fox and Partners

Professor E.R. Bryan, University of Salford

N.J. Cavagham, British Steel Corporation

Dr S. Cullimore, University of Bristol

Dr G.J. Davies, University of Liverpool

Professor H.R. Evans, University College Cardiff

Dr A.R. Gent, Imperial College of Science and Technology

M. Glover, Ove Arup Partnership

L. Gourd, The Welding Institute

Professor J.E. Harding, University of Surrey

Dr W. Harvey, University of Dundee

J. Hicks, Consultant

Dr R.E. Hobbs, Imperial College of Science and Technology

Dr B. Hayman, Det Norske Veritas

Professor M.R. Horne, University of Manchester

Professor R.P. Johnson, University of Warwick

J.P. Le Good, Portsmouth Polytechnic

J. Kinsella, Dorman Long Bridge and Engineering Ltd

Dr R.M. Lawson, The Steel Construction Institute

Dr W. Manners, University of Leicester

Dr R. McConnel, University of Cambridge

T.J. MacGinley, Nanyang Technological Institute

Dr J.R. Moon, University of Nottingham

Dr L.J. Morris, University of Manchester

F. Needham, The Institution of Structural Engineers

Dr D.A. Nethercott, University of Sheffield

Dr M. Ogle, The Welding Institute

Dr G.W. Owens, The Steel Construction Institute

Dr J. Rhodes, University of Strathclyde

J.T. Robinson, British Steel Corporation

R. Stainsby, Consultant

Mr J. Surtees, University of Leeds

J.C. Taylor, The Steel Construction Institute

Dr D. Tordoff, British Constructional Steelwork Association

Dr L.P. Walpole, British Steel Corporation

Dr J. Whitbread, Conder Group Services

Dr T.A. Wyatt, Imperial College of Science and Technology

Contents

1

Historical development and modern usage of steel

Objective To survey the development of the use of steel in the construction of buildings and bridges.

Summary The chapter outlines the early history of the introduction of steel into the construction of bridges and tall buildings. It describes the present-day use of steel in medium-rise, high-rise and industrial buildings, refers briefly to the employment of cold-formed sections and ends with some discussion of long-span structures.

1.1 Introduction

Following the completion of the Iron Bridge in Coalbrookdale, Shropshire, in 1779 (Figure 1.1) there was a rapid development in the use of metal in the form of cast and wrought iron for bridges and buildings. However, it was not until the latter half of the nineteenth century that steel was manufactured in sufficient quantity and at a price cheap enough to be a serious competitor to wrought iron. The chief difference in the chemical composition of the above metals is the proportion of carbon contained in each. Table 1.1 gives typical percentages of carbon and other impurities in the metals.

Figure 1.1 Iron Bridge elevation

Table 1.1 Typical percentages of carbon and other impurities in metals

	Cast iron (%)	*Wrought iron (%)*	*Steel (%)*
Carbon	2.0–6.0	0–0.25	0.15–1.8
Silicon	0.2–2.0	0.032	0.015
Phosphorus	0.038	0.004	0.041
Manganese	0.013	Trace	0.683
Sulphur	0.014	0.114	0.035

Opposition in England to the use of steel as a replacement for wrought iron extended until the end of the nineteenth century. The situation was different elsewhere, and the St Louis bridge built over the Mississippi River and completed in 1874 was the first major bridge in which steel was used (Figure 1.2). It was designed by James Eads (1820–87) with three spans of 153, 159 and 153 m. The arch ribs are braced tubes formed of chromium steel. John Roebling's Brooklyn Bridge, built over the East River, New York City, with the world's largest span of 486 m when completed in 1883, was the first to use hard-drawn steel wire for the suspension cables (Figure 1.3).

In America, between 1890 and 1900 a large number of steel-framed buildings was constructed with riveted steel frameworks and connections designed to resist wind moments. Typical examples are Jenney's Fair Store, Chicago (1892) (Figure 1.4) and Bruce Price's American Surety Building, New York (1894–5). The latter included a complete steel frame to support the 20 storeys with a total height of 93 m.

The first major steel structure to be built in England was the monumental Forth Rail Bridge, completed in 1889 with a main span of 521 m (Figure 1.5). The bridge has immense physical

Figure 1.2 James Eads's St Louis Bridge

Figure 1.3 Brooklyn Bridge

presence, which is still admired by engineers throughout the world. The principle of its cantilever construction is elegantly expressed by the designer Benjamin Baker, who devised a living model to illustrate the equilibrium of the structure (Figure 1.6).

The Ritz Hotel was the first significant structure in London to utilize a steel skeleton to support the full weight of the building (Figure 1.7). The design contravened the London Building Act of 1894 in a number of ways and the construction technique followed American practice, where multi-storey framed buildings were commonplace. The majority of the site connections were riveted, the columns were spliced above floor level and account was taken of the eccentricity of the beam/column connections. For fire protection the steel framework was cased in 50 mm of concrete. In America, developments in the use of steel frameworks for high-rise buildings proceeded at a much greater rate, and the term 'skyscraper' is synonymous with the architecture of numerous American cities, particularly New York and Chicago.

1.2 High-rise buildings

In the first decade of the twentieth century tower blocks in America were constructed up to a height of 215 m, as typified by the 53-storey Metropolitan Tower, New York (1909). The Woolworth Building, New York, erected between 1911 and 1913, rose to a height of 230 m and consisted of a 35-storey tower on a 25-storey base. The doyen of all tall buildings is the Empire

Figure 1.4 William Le Baron Jenney's Fair Store

State Building, New York (Figure 1.8). The 381 m high, 102-storey frame which required over 50 000 tons of steel was erected in 1930. It remained the world's tallest building for over 20 years.

For several decades, tall buildings were designed using the technique of considering the floor framing as simple beams for vertical load with no allowance for continuity. Frame action was

Figure 1.5 The Forth Rail Bridge

Figure 1.6 The Forth Rail Bridge: living model

Figure 1.7 The Ritz Hotel

Figure 1.8 The Empire State Building

Figure 1.11 The Sears Tower, Chicago

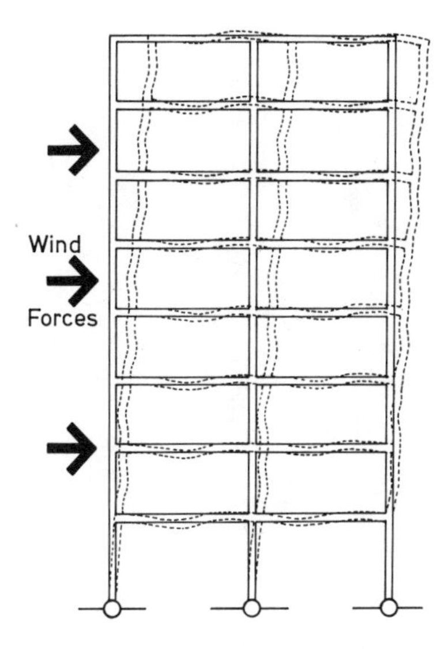

Figure 1.9 Sway of skeletal frames

assumed for wind loading, sufficient strength being provided in the beam/column connections to accommodate the wind moments.

By assuming that points of contraflexure occur at mid-height of the columns and mid-span of the beams it is possible to carry out a rapid hand analysis of the frame for lateral loads. This design approach is referred to as 'simple' or 'wind-moment' design, but for buildings with more than 20 storeys the weight of steelwork required becomes uneconomic and sway under lateral loads may become excessive (Figure 1.9). After the Second World War attention was paid to developing more efficient structural systems to resist wind loading, and significant reductions in the weight of steel required have been achieved (see Table 1.2).

The sway limitations of semi-rigid frames led to the introduction of internal X or double-diagonal bracing in buildings. Further efficiency can be achieved by shortening the diagonals into K-bracing – the K brace does not attract vertical load (Figure 1.10). Another geometrical form is the use of closely spaced columns on the exterior face of the building to which spandrel (deep) beams are attached by moment-resisting connections. The resulting pierced tube (framed tube) is extremely rigid, and additional stiffness can be provided by the addition of diagonal bracing. This tube or hull concept has been extended by providing an interior tubular core which increases the stiffness by sharing the loads with the facade tube. The Sears Tower, Chicago, 442 m high (excluding the television antennae), consists of nine 23 m square tubes which are nested together to form a bundled tube, the tubes rising to different heights (Figure 1.11). There are numerous structural systems combining the concepts described above which have been applied in high-rise construction (Figure 1.12), but as a general guide it should be

Table 1.2 Comparative weights of steel in high-rise buildings. (Source: *The Guinness Book of Structures*)

Building	*Date*	*Height*		*Ratio height/width*	*Gross area*		*Steel weight*		*Number of storeys*
		ft	*m*		$ft^2 (\times 10^6)$	$m^2 (\times 10^5)$	lb/ft^2	kg/m^2	
Empire State	1931	1250	381	9.3	2.75	2.6	42.2	203.6	102
Chrysler	1930	1046	319	8.5	1.1	1.04	37.0	177.6	77
World Trade	1973	1350	411.5	6.9	9.0	8.52	37.0	177.6	110
Sears	1974	1450	442	6.4	4.4	4.16	33.0	158.4	109
John Hancock	1968	1127	343.5	7.9	2.8	2.65	29.7	142.6	100

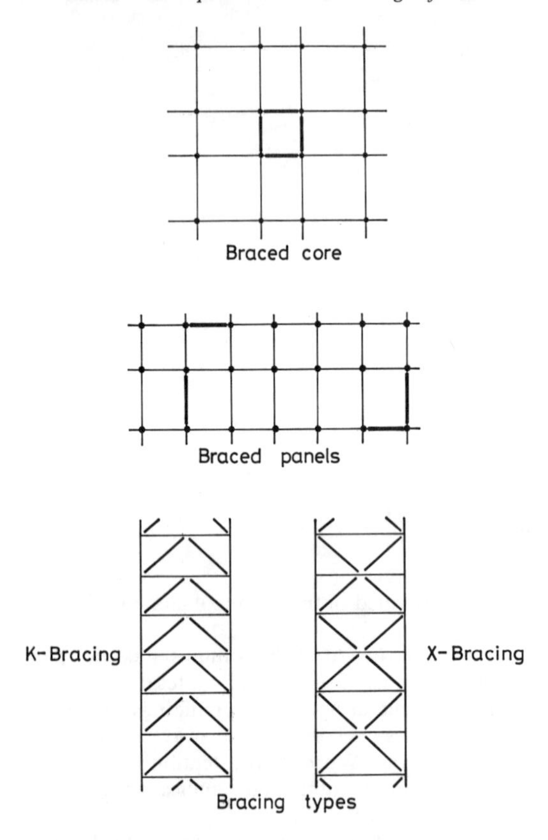

Figure 1.10 X-bracing and K-bracing

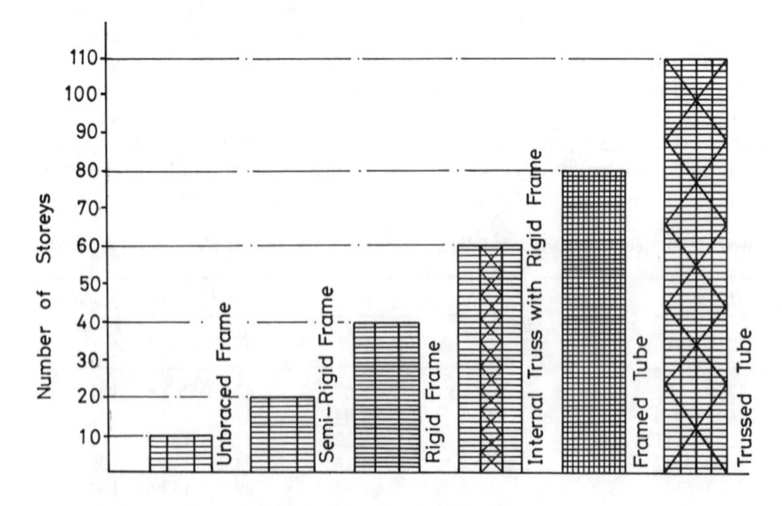

Figure 1.12 Structural systems for high-rise buildings

noted that the weight of framing per floor remains approximately constant with increase in building height, the column weight increases linearly and the weight of wind bracing increases exponentially with increase in building height. By sloping the external columns and combining the hull structure with diagonal bracing it was possible to achieve in the design of the 343.3 m high John Hancock Centre, Chicago, a 30% reduction in the weight of steelwork per unit area compared with the Empire State Building (see Table 1.2).

The Shell Building, South Bank, is one of the few high-rise buildings to be constructed in London with a steel framework. A diagonal bracing system is incorporated to resist wind loading and the framework is clad in stone. The use of heavy cladding is typical of the early American skyscraper, but the current need to decrease building weight and cost necessitates the use of lightweight cladding; various glazed curtain wall systems have been developed. This in turn poses the problem of controlling the temperature gradient across the building and excessive differential vertical movement may occur.

Social factors and relatively poor ground conditions have militated against the widespread construction of high-rise buildings in London, with the notable exception of the National Westminster Bank's 49-storey headquarters tower of height 183 m, constructed of steel and concrete. However, for medium-rise buildings there has been a substantial increase in the use of a steel framework as an alternative to reinforced concrete; steel structures having the advantage of reduced weight and faster construction. In 1987 steel overtook reinforced concrete as the more popular material.

1.3 Medium-rise buildings

It is not possible to draw a clear dividing line between medium- and high-rise buildings, but a medium-rise building could arguably be considered as one in which the structural configuration is not significantly influenced by wind loading. Building techniques in steel have naturally been influenced by developments in America, where overall economy is strongly linked to construction times. There are increasingly demanding requirements for value engineering, utilizing high technology to provide high-quality office space with efficient environmental control. Two typical examples of high-technology office space utilizing steel framing are Cutlers Court and Lloyds Chambers. The former is within the City of London and the latter straddles the boundary between the City and the neighbouring Borough of Tower Hamlets. The buildings collectively embrace a number of features of high-technology steel construction, including:

1. Planning flexibility;
2. Dry construction;
3. Rationalized fire engineering;
4. Flexibility of service runs;
5. Site welding;
6. Composite floor construction including through-deck stud welding.

Cutlers Court is a five-floor steel frame with a roof and plant room over, with a steel weight of 300 tons (Figure 1.13). The bolted connections are designed to resist vertical load only and lateral stability is provided by means of a series of diagonally braced frames. The floor framing consists of primary and secondary universal beams. The floor slab consists of a galvanized, profiled, cold-rolled steel deck onto which a layer of mesh reinforcement was placed prior to lightweight concrete being poured to form a composite deck of 125 mm overall thickness. The span of the secondary beams was limited to 3.0 m to avoid the necessity for temporary propping (Figure 1.14). Composite action between the secondary beams and the concrete slab is achieved by welding 20 mm diameter studs through the profiled steel deck onto the beams.

Lloyds Chambers is a ten-storey structure incorporating an atrium (a term originally used to describe the central courtyard of a Roman house (Figure 1.15)). The structure was designed in a way similar to that adopted for early high-rise steel frames. The beams were designed as simply supported for vertical load without composite floor action. The beam/column connections were adequate to resist wind loads and wind bracing was only used adjacent to the

Figure 1.13 Cutlers Court

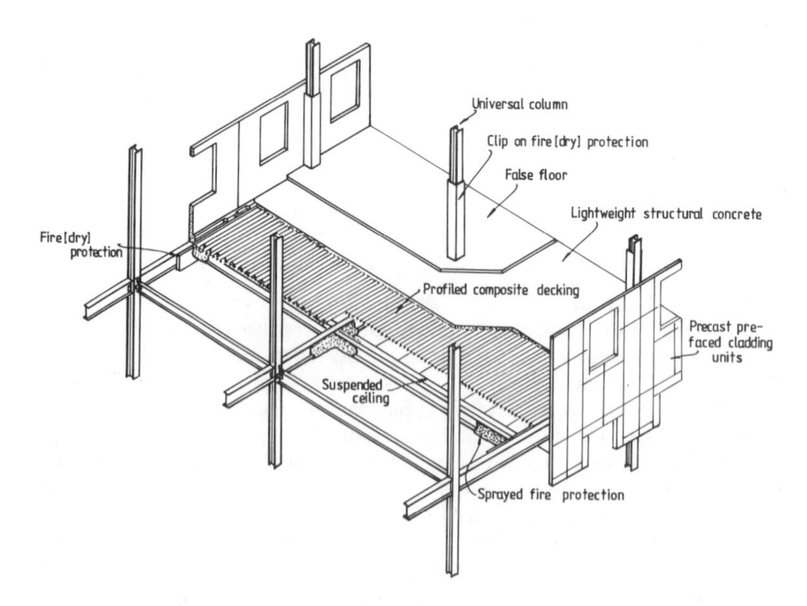

Figure 1.14 Cutlers Court

Figure 1.15 Lloyds Chambers

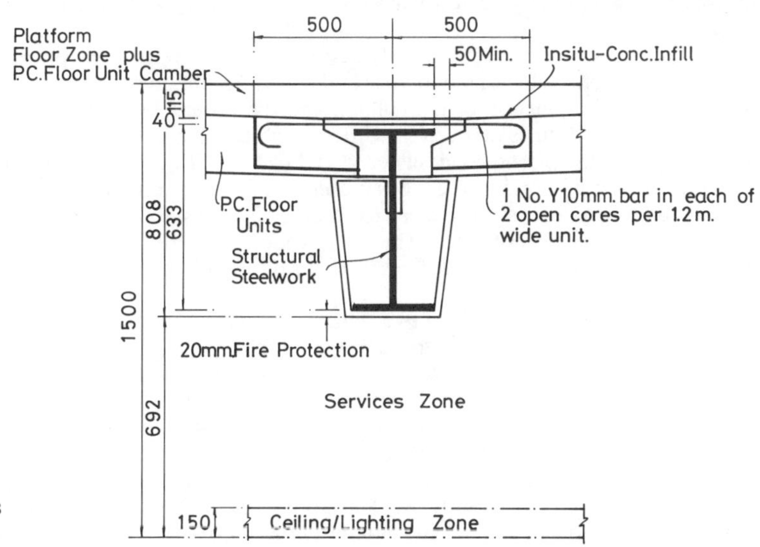

Figure 1.16 Lloyds Chambers

vertical glazed wall of the atrium. All the site connections of the main steel frame were bolted.

It was required to minimize the number of internal columns, and this led to the adoption of a 14.4 × 7.2 m column grid. The 7.2 m span pre-cast concrete floor units are supported on shelf angles attached to the beam webs. This minimizes the overall floor depth. The floor loading is transmitted to the pre-cast floor units via a raised platform flooring system to allow maximum flexibility of service runs (Figure 1.16).

The servicing of modern buildings, including air conditioning, has become increasingly complex and should be considered at the conceptual design stage. One solution is to provide holes in a solid web beam which, in effect, converts it into a Vierendeel section, named after the Belgian engineer who first introduced it at the end of the nineteenth century as an alternative to the lattice beam with diagonal members. With the development of welded connections some elegant designs have been produced, in particular for footbridges. Web holes for services should be located at the mid-depth of the beam and the height of the hole should not be greater than half the overall beam depth. Any web hole should be located at a distance equal to at least half the beam depth from an end-reaction or interior point load. A castellated beam, in which a solid beam is increased in depth by 50% by cutting the web into a half-hexagonal profile and welding the two sections together, is an alternative, but the capacity of the web to resist concentrated loads is limited. Another approach is to run the services below the beam, but this increases the overall depth of the floor zone.

Recent developments have helped to improve the integration of services within the structural zone. These include the use of stub girders and the parallel beam approach, where continuous secondary beams pass above continuous spine beams which are arranged in pairs to pass on side of the columns.

1.4 Industrial buildings

Traditionally, industrial roofs were supported by triangulated assemblies of pin-jointed members of various configurations. Following the work of the Steel Structures Research Committee, a completely new approach to design was developed, attention being turned to the behaviour of steel structures when they were loaded beyond the elastic range into the plastic range, and then to collapse. A steel section of suitable profile (a compact section) is capable of developing its full plastic moment of resistance, and the basis of the design method is to determine the minimum load at which sufficient plastic hinges are formed to turn the structure into a mechanism.

Over the past two decades plastically designed portal frames using prismatic sections have become the norm for industrial buildings in the UK. An optimum design requires the provision of a haunch at the junction of the column and the rafter. Recently, based on an American precedent, tapered frames manufactured from high-tensile steel plate have been introduced as an alternative to the plastically designed portal frames. The cost of fabrication of the tapered members can be minimized by the use of automated welding techniques. With this type of frame the web depth-to-thickness ratio can exceed 200, and the stability of both the flanges and the webs has to be carefully examined. The beam section is described as slender and the full plastic moment of resistance cannot be achieved. Spans of up to 50 m can be achieved with both types of frame.

The portal frame is essentially a two-dimensional system; for covering large roof areas with an approximately square grid the space frame can be used to advantage. This is a three-dimensional assembly of structural members, of which a typical example is the Nodus system. The design of the joints is the key to the successful application of space frame systems.

Industrial organizations are paying increasing attention to the architectural and engineering quality of their buildings and this has proved to be a challenge to architects and engineers. The Renault Parts Building at Swindon is an example of high technology applied to a large single-storey building (Figure 1.17). The grid is 24 × 24 m and the headroom 9 m. The structural system consists of braced columns extending above the roof level and the perforated

Figure 1.17 The Renault Parts Building

purpose-made beams are braced by steel rods. It is now common practice to express the form of the structure both externally and internally.

1.5 The use of cold-formed sections

By means of cold forming it is possible to produce more efficient structural forms, and these have been used extensively in aeroplanes and car bodies. The construction industry has been slow to follow, but, over the past 40 years, extensive studies of the design problems of thin-walled sections have led to the increasing use of cold-formed steel in building structures. In Britain current applications include purlins, roof and floor decking and cladding. Until recently, cold-formed sections were normally used structurally in situations where the total load intensity did not exceed 2.0 kN/m^2. However, cold-formed sections are now being employed as an alternative to hot-rolled sections for primary structural elements. Twin-lipped channels as floor and roof beams, in conjunction with hot-rolled rectangular hollow section columns, are being used for institutional, commercial, residential and light factory buildings.

1.6 Long-span bridges and roofs

Because the tensile strength-to-weight ratio of steel is high, its potential for long spans when used in tension is immediately apparent. For long-span bridges the self-weight will represent up to 90% of the total load to be carried, and thus the appropriate structural configuration is that which has minimal weight, i.e. the suspension form.

Above a span of about 550 m this form dominates, and the nine largest spans, in metres, are given below:

Figure 1.18 The Humber Bridge

1. Humber Bridge	1410
2. Verrazano Narrows, New York	1298
3. Golden Gate, San Francisco	1280
4. Mackinac, Michigan	1158
5. Bosphorus, Istanbul, Turkey	1074
6. George Washington, New York	1067
7. Tagus, Lisbon, Portugal	1013
8. Forth Road Bridge	1006
9. Severn Bridge	988

Traditionally, the deck structures of suspension bridges (2, 3, 4, 6, 7 and 8) were constructed of deep lattice girders, but a recent British innovation is the welded aerofoil box section deck (1, 5 and 9). The Humber Bridge, the world's largest span, has a box section deck made of welded stiffened steel plate, each section of which weighs 140 tons (Figure 1.18). This form of deck is much lighter than the deep lattice girder. Using current materials and design standards, a clear span of 3000 m is technically feasible.

The provision of large roof spans, uninterrupted by columns, for aircraft hangars, sports arenas, exhibition buildings, etc. presents a similar challenge to that of long spans in bridge building. For long-span roofs, both strength and stiffness (resistance to deformation) are

Figure 1.19 The Patcenter Project, Princeton

required, and it should be noted that deformation is proportional to the cube of the span. Thus to minimize the deflection the flexural rigidity (*EI*) has to be extremely large if long spans are to be achieved. The Boeing 747 hangar erected in 1970 at London Airport required a clear door opening 138 m wide by 23 m high. The solution was to erect a deep, and therefore stiff, three-dimensional lattice girder over the door opening. The four main parts of the roof are the fascia girder, a spine girder and high- and low-level space decks. The 1500-ton roof is supported on eight columns and is capable of supporting up to 700 tons of equipment.

An alternative structural solution to providing a long, clear roof span is to support the structure from above. An elegant example of this is the Patcenter project in Princeton, New Jersey (Figure 1.19). The roof deck between the external and internal columns is supported at two intermediate points by steel rods, which in turn are connected to tubular steel masts which rise from the internal columns. The structure has imaginative articulation and is typical of the recent trend to express the structure externally.

1.7 Concluding summary

1. The structural potential of steel for long-span structures is immediately apparent, but for the majority of building types – domestic, office, commercial and leisure – there are several

options available in terms of choice of material. In a number of situations the most economic solution is composite construction, typically the use of composite steel and concrete floors and beams. The cost and energy usage in the manufacture of steel is high compared with other materials of construction, and thus attention must be given to minimizing the material, fabrication and erection costs of steel structures. A further consideration is durability; no building, whatever the materials of construction, can be considered as maintenance-free. The current trend, of expressing the structural form of buildings, requires careful attention to the protection of exposed steel from corrosion. Developments in galvanizing and paint formulations can extend the life to first maintenance of steel structures to an economic level.

2. There is great scope for the imaginative development of the structural potential of steel, taking advantage of its high strength and stiffness, automated fabrication, speed of erection and improved means of protection against corrosion and fire, to produce elegant, economic and durable structures.

Background reading

BECKETT, D. (1980) *Brunel's Britain*, David and Charles, Newton Abbot
BECKETT, D. (1984) *Stephenson's Britain*, David and Charles, Newton Abbot
CONDIT, C.W. (1960) *American Building Art, the 19th Century*, Oxford University Press, Oxford
Constrado Series, *Framed in Steel*, Constrado, London
LIN, T.Y. and STOTESBURY, S.D. (1981) *Structural Concepts and Systems for Architects and Engineers*, Wiley, Chichester
SINGER, C., HOLMYARD, E.J., HALL, A.R. and WILLIAMS, T.J. (1975) *A History of Technology*: Vol. IV, *The Industrial Revolution* (*c. 1750 to c. 1850*), Oxford University Press, Oxford
STEPHENS, J.H. (1976) *The Guinness Book of Structures*, Guinness Superlatives, London
The Engineers (1982) The Architectural Association, London

2

Process of design

Objective To introduce the challenge of creative design and explain approaches by which it may be achieved.

Prior reading A general knowledge of basic applied mechanics.

Summary The section begins by considering a definition of design and some design objectives. It discusses how a designer can approach a new problem in general and how a structural designer can develop a structural system. It concludes by considering differences of emphasis in design approach for different classes of structure.

2.1 Design objectives

We can all see and use the results of successful design in structural engineering (see Figure 2.1). The question is: how can we become professional designers and eventually produce better designs than those previously encountered, to benefit and enhance the performance of our human activities? In particular, how can we utilize steel effectively in our structures for:

1. Travelling more easily over awkward terrain, requiring bridges;
2. Enabling basic industrial processes to function, requiring, for example, machinery supports, docks and oil rig installations;
3. Aiding communications, requiring masts;

Figure 2.1 Comparisons of scale: buildings, bridges and offshore platforms

Figure 2.2 Finished garage

4. Enclosing space within buildings, as in Figure 2.2.

Design is 'the process of defining the means of manufacturing a product to satisfy a required need'.

'Designers'? All people are capable of creative design ideas. They are continuously processing information and making conscious imaginative choices – for example, of the clothes they wear, their activities and the development of ideas they pursue, causing changes.

In structural design, our prime objectives are to ensure the best possible:

1. Unhindered functioning of the designed object over a desired lifespan;
2. Safe construction system, completed on time and to the original budget;
3. Imaginative and delightful solution for both users and casual observers.

These points could be satisfied by either:

1. Simply making an exact copy of a previous object; or
2. 'Re-inventing the wheel', by designing every system and component afresh.

Both these extreme approaches are unlikely to be entirely satisfactory! In the former case the problem may well be slightly different (for example, the previous bridge may have stimulated more traffic flow than predicted or vehicle weights may have increased). Economic and material conditions may have changed (for example, the cost of labour to fabricate small built-up steel elements and joints has increased compared with the production cost of large rolled or continuously welded elements; also, corrosion-resistant steels have reduced maintenance costs relative to mild steel). Functioning deficiencies may have been discovered with time (for example, vibrations may have caused fatigue failures around joints). Finally, too much repetition of a visual solution may have induced boredom and adverse cultural response.

With the latter, 'life is often just too short' to achieve all the optimal solutions while the client frets . . . . Civil engineering projects are usually large and occur infrequently, so a disenchanted client will not make a second invitation. Realization of new theoretical ideas and innovations invariably takes much time; history shows this repeatedly. Thus methodical analysis of potential risks and errors must temper the pioneering enthusiast's flair.

Positive creative responses must be achieved for all aspects of every new problem. These will incorporate components from the above, both of fundamental principles and recent developments.

2.2 How does the designer approach a new problem?

1. Recognize that a problem exists and clearly define the overall objectives for a design.
2. Research around the problem and investigate likely relevant information (analysis).
3. Evolve possible solutions to the problem (synthesis).

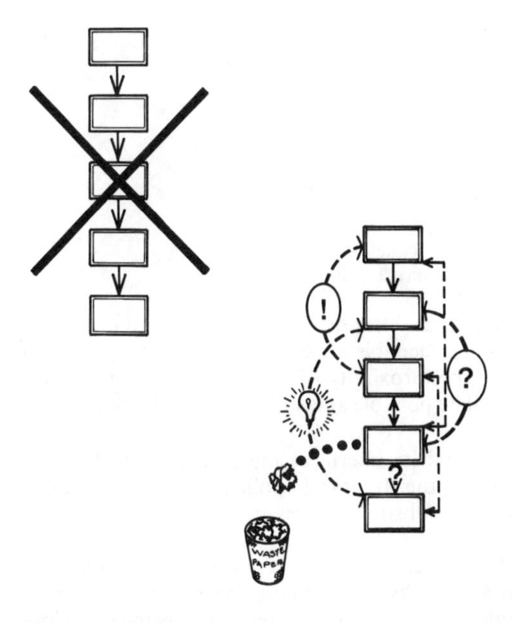

Figure 2.3 Design iterations

4. Decide on, and refine, the best solution (evaluation), establishing clear priorities for action (in terms of manufacture, construction, operation and maintenance).
5. Communicate decisions to others involved in the problem.

At the outset these five stages appear as a simple linear chain; in fact the design process is highly complex, as all factors in the design are interdependent to a greater or lesser degree. Hence there will be many steps and loops within and between the stages, as seen in Figure 2.3. The first rapid passage through stages 1, 2 and 3 will decide if there is 'any problem' (for example, is the likely traffic flow adequate to justify a convenient but high-cost bridge?).

All factors and combinations must be explored comprehensively, from idea to detail, with many compromises having to be finely balanced. Ideas may be developed: verbally (for example, 'brainstorming' or Edward de Bono's 'lateral thinking' approach), graphically, numerically or physically. Qualitative assessment should always precede quantitative evaluation.

The starting point for analysis may thus be the designer's current preconceived notion or visual imagination, but the synthesis will reveal the flexibility of his mind to assimilate new ideas critically, free of preconception!

2.3 How does the designer develop a structural system?

An example of structural design, and the various decision stages, will be briefly considered for a simple two-lorry garage building with an office, toilet and tea room, shown completed in Figure 2.2. (*NB*: It is assumed in this hypothetical case that an initial decision has been made by the client to have this set of requirements designed and built.)

2.3.1 Pose an initial concept that may well satisfy the functions

It is invariably a good idea to start by looking at the functions required and their relationships. Make a list of individual functions; then generate a 'bubble' (or flow) diagram of relationships between different functional areas to decide possible interconnections and locations (Figure

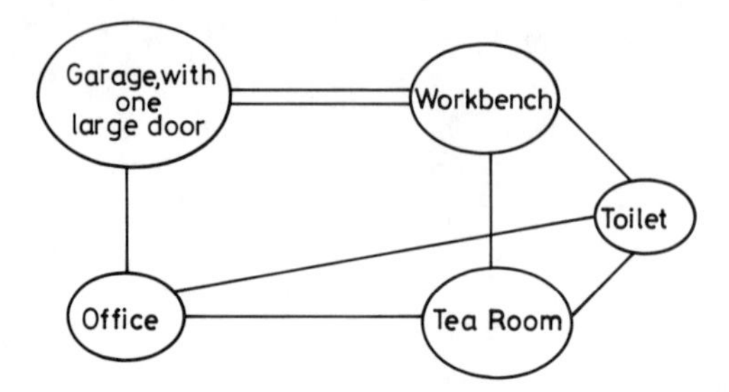

Figure 2.4 Bubble diagram

2.4). Find, or assume, suitable plan areas and minimum clear heights of each three-dimensional 'volume of space'. A possible plan layout may then be indicated, noting any particular complications of the site (for example, plan shape, proximity of old buildings, slope or soil consistency). Many other plan arrangements will be possible and should be considered quickly at this stage.

The requirements of each 'volume of space' and its interfaces must be examined for all functional, cost and aesthetic criteria (for example, the structural applied live loads that must be resisted; the heating, ventilating, lighting and acoustic requirements that are likely to be desired) (Figure 2.5).

The main criteria can easily be recognized and then followed up and tested by numerical assessment. Incompatibilities may be 'designed out' by re-arranging the planned spaces or making other compromises. For example, would you accept an office telephone being very close to the workshop drill or lorry engine, without any acoustic insulation?

Prepare a set of initial assumptions about possible materials and the structural 'frame', 'planar' or 'membrane' load-bearing systems that might be compatible with the 'volumes of space', as shown in Figure 2.6. These assumptions will be based on a previous knowledge and understanding of actual constructions or structural theory, as well as the current availability of materials and skills. Initial consultation may be needed with suppliers and fabricators (for example, for large quantities or special qualities of steel).

Steelwork, with its properties of strength, isotropy and stiffness, and its straight and compact linear elements, lends itself to 'frame' systems, which gather and transfer the major structural loads as directly as possible to the foundations, as a tree gathers loads from its leaves through branches and main trunk to the roots.

Now (and continuously) elucidate and test your ideas by making quick three-dimensional sketches, or simple physical models, to explore the likely compatibility and aesthetic impact.

Note that all principal specialists (architects, engineers for structure and environmental services, and also major suppliers and contractors) must collaborate freely with each other –

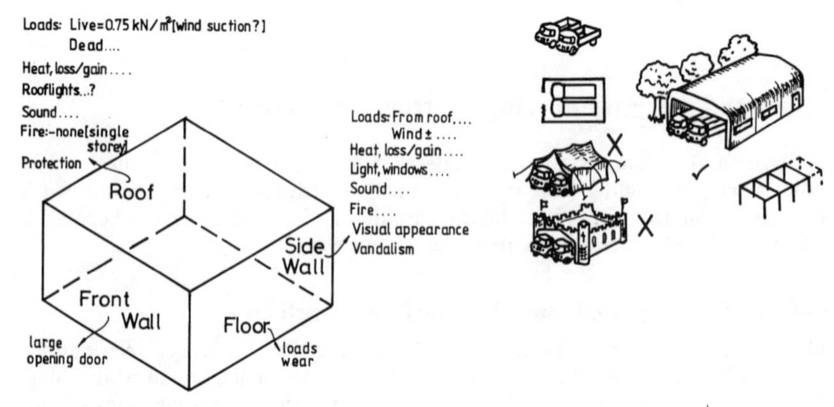

Figure 2.5 Volumes of space

Figure 2.6 Initial concepts

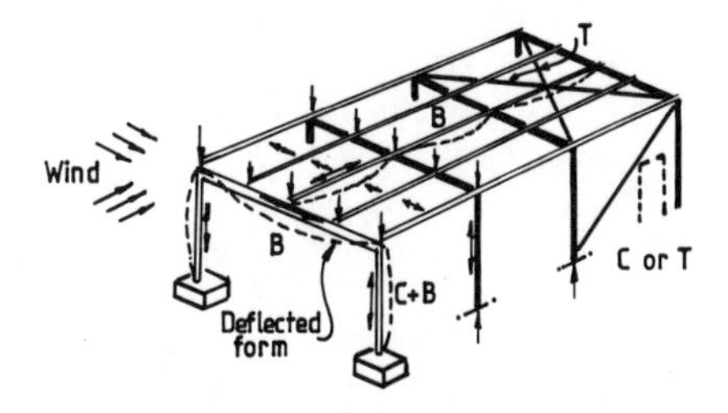

Figure 2.7 Main structural system

and also with the client – at this conceptual design stage. Bad initial decisions cannot subsequently be easily and cheaply rectified when carrying out more detailed design.

Be prepared to modify the concept readily (use 4B pencils) and work quickly (timescale for initial structural design: seconds/minutes).

2.3.2 Recognize the main structural systems and contemplate the necessary strength and stiffness

Consider the applied live loads from roofs, floors or walls, and trace the 'load paths' through the integral three-dimensional array of elements to the foundations (Figure 2.7). If the roof is assumed to be profiled steel decking the rainwater should run to the sides, and a manufacturers' data table will indicate both the slope angle to be provided (4–6° minimum) and the secondary beam (purlin) spacing required (commonly 1.4–2.6 m). The purlins must be supported (commonly 3–8 m) by a sloped main beam or truss, usually spanning the shorter direction in plan, and supported by columns stabilized in three dimensions.

Wind loads on the longer side of the building can be resisted by cladding that spans directly to the main columns, or onto sidewall purlins spanning between columns. The columns could resist overturning by:

1. Cross-bracing (in this case the large entry door would be impeded);
2. Rigidly fixing the columns to the foundation bases ('linked cantilevers'); can the soil resist the extra overturning effect of this base? or
3. Rigidly fixing the tops of the columns to the main beams (creating 'portals') and giving smaller, cheaper 'pin' base foundations.

Wind loads on the open short side of the building can be resisted by the opening door spanning top to bottom or side to side. At the closed short side the wind loads can be resisted by cladding that either spans directly between secondary end wall columns or onto purlins fixed to these columns.

Then, at both ends of the building, longitudinal forces are likely to be induced at the tops of the columns. Trussed bracing can be introduced, usually at both ends of the roof plane, to transfer these loads to the tops of a long side-column bay, which must then be braced to the ground.

Identify the prime force actions (compression *C*; tension *T*; bending *B*) in the elements and the likely overall and elemental deflected forms for all applied loadings both separately and when combined.

It is always useful to have the elements drawn to an approximate scale, which can be done using manufacturers' data tables for decking and cladding, from observations of existing similar buildings or using 'rules of thumb' (for example, the span/depth ratio for a simply supported beam equals about 20 for uniform light roof loading). At this stage the design becomes more definite (use a B pencil) and takes longer (timescale: minutes).

2.3.3 Assess loads accurately and estimate sizes of main elements

Establish the dead load of the construction and, with the live loads, calculate the following (Figure 2.8):

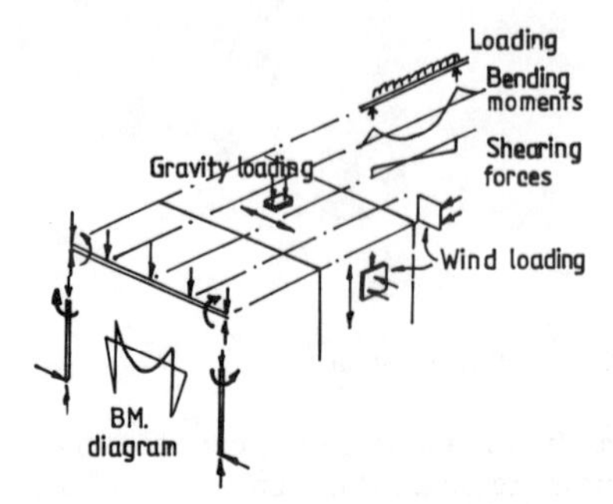

Figure 2.8 Approximate element sizes

1. Beam reactions and column loads (taking half the span to either side of an internal column);
2. Maximum bending moments (for example, $WL^2/8$ for a simply supported beam under uniform load, $WL^2/12$ for a fully fixed beam and $WL^2/10$ for most continuous beams);
3. Maximum shearing forces in beams;
4. Deflection values (for example, $(5/384) \times (WL^4/EI)$ for a simply supported beam, $(1/384) \times (WL^4/EI)$ for a fully fixed beam with uniform load).

The size of columns carrying little moment can be estimated from safe load tables by using a suitable effective length (significant bending moments should be allowed for by a suitable increase – i.e. twice or more – in section modulus for the axis of bending). Beam sizes should be estimated by checking bending strength and stiffness under limiting deflections. Likely jointing methods must be considered carefully: is the beam to be simply supported or fully continuous, and what are the fabrication, erection and cost implications?

Calculations are now being performed (use HB pencil with slide rule or simple calculator) and the time involved is more significant (timescale: minutes/hours).

2.3.4 Full structural analysis, using estimated element sizes with suitable modelling of joints, related to actual details

Carry out a full structural analysis of the framework, either elastically or plastically. A computer may be used, though some established 'hand' techniques will often prove adequate; the former is appropriate when accurate deflections are required (Figure 2.9).

Note that for the analysis of statically indeterminate structures an initial estimate of element stiffnesses (I) and joint rigidity must be determined by stage 3 above before it is possible to find the disposition of bending moments and deflections. If subsequent checking of element design leads to significant changes in element stiffness the analysis will have to be repeated. The role of the individual element flanges and web in resisting local forces within connections must also be considered very carefully when determining final element sizes. Excessive stiffening to light sections can be prohibitively expensive. The analysis cannot be completed without careful structural integration and consideration of the compatibility of the entire construction system, including its fabrication details.

Element joints will normally be prepared in the factory using welding, with bolts completing joints of large untransportable elements at site. Bracings, deckings and claddings will usually

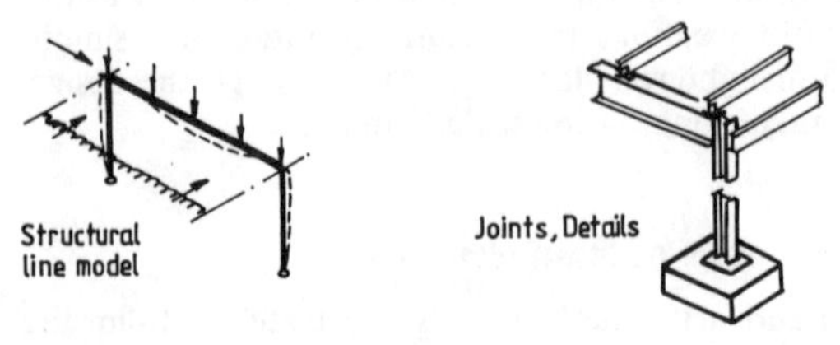

Figure 2.9 Accurate analysis and details

be fixed on site with bolts or self-tapping screws. Failures frequently arise from poor jointing, details and their integration!

The calculations and details are progressing (use HB pencil with slide rule, calculators and computers) (timescale: hours/days).

Iteration of stages 1–4 above will undoubtedly be required, not least to ensure that the structural decisions are compatible with further investigations concerning the functional, environmental, cost and aesthetic aspects. The effect of any change must be considered throughout the complete design, usually necessitating a partial 'redesign'.

2.3.5 Communicate design intentions through drawings and specifications

Prepare detail drawings and specifications for contractors' tenders. Iteration of the design may again be necessary, due to variations in constructors' prices or preferred methods (for example, welding equipment available, difficulties in handling steelwork in the fabricating shop or for transportation and erection). Innovations in the design must be communicated and specified very carefully and explicitly.

The design is being finalized (use 2–4H pencils and pens, or computers) (timescale: days/weeks).

2.3.6 Supervise the construction operation

Stability of the structure must be ensured at all stages of the construction. High-quality components and skilled erectors must be available at the right place and time, calling for very careful organization. If 'all goes to plan' every piece will fit into the complete jigsaw!

The design ideas are now being put into operation (use gumboots) (timescale: weeks/months).

2.3.7 Conduct regular maintenance

Hopefully, only regular maintenance already planned into the design will be needed, with occasional change and renovation if necessary. Correction of design faults due to innovation and errors should never be needed!

This is the operation phase. (Use a serene outlook on life! Timescale: years/decades.)

2.4 Differences of emphasis in design approach compared with that of a building

1. *For a bridge*: The magnitudes of gravity loading are often greater, with particular load patterns to be assessed; also, trains of moving wheel loads will occur, giving marked dynamic effects. Dynamic effects of wind loading are significant in long-span structures. Accessibility of site, constructability of massive foundations, type of deck structure and regular maintenance costs will govern the system adopted. Aesthetics for users and other observers are important; long-distance scale should be appropriately slender but psychologically strong; careful attention is needed for fairly close viewing of abutments and deck undersides.
2. *For an offshore oil rig*: The scale of the whole operation is very many times that of an onshore building. Gravity loading, wind speeds, wave heights and depth of water are the significant design parameters for structure size and stability (here larger elements cause larger wind and wave loads). The scale of the structure also poses special problems for fabrication control, floating out, anchorage at depth by divers and, not least, cost (see Figure 2.1).

2.5 Concluding summary

1. This chapter introduces the challenge of creative design and suggests a comprehensive strategy for designing structural steelwork. It seeks to answer questions about what a

designer is trying to achieve and how he can start putting pen to paper, and illustrates how a successful design is iterated, through qualitative ideas to quantitative testing and, finally, construction.

2. Creative and imaginative designing of structures is most challenging and fun – now try it and gain confidence for yourself. Do not be afraid of making mistakes but make sure that you are right, before it is built, using your own developed in-built checking mechanisms.

Background reading

DE BONO E., e.g. (1977) *Lateral Thinking*, Penguin, Harmondsworth, or (1976) *Practical Thinking*, Cape, London or (1967) *The Use of Lateral Thinking*, Cape, London

FRANCIS, A.J. (1980) *Introducing Structures*, Pergamon, Oxford

GORDON, J.E. (1976) *The New Science of Strong Materials*, Pelican, Harmondsworth

GORDON, J.E. (1978) *Structures*, Pelican, Harmondsworth

JONES, J.C. (1981) *Design Methods*, 2nd edn, Wiley, Chichester

LEGOOD, J.P. (1983) *Principles of Structural Steelwork for Architectural Students*, Constrado, London

LIN, T.Y. and STOTESBURY, S.D. (1981) *Structural Concepts and Systems for Architects and Engineers*, Wiley, Chichester

MAINSTONE, R.J. (1975) *Developments in Structural Form*, Allen Lane, London

TORROJA, E. (1962) *Philosophy of Structures*, University of California Press

3

Design philosophies

Objective

To outline the uncertainties that affect the design process and the resulting need for a consistent and rational approach; to describe the philosophies commonly used ('allowable stress' and 'limit state' design).

Prior reading

Basic concepts of statical determinacy and redundancy, elastic and plastic behaviour.

Summary

The chapter begins by considering the uncertainties of civil engineering design. It examines society's demand for the standardization of design procedures. The development of regulations for building and construction is reviewed with sections on design by geometric ratio, allowable stress design, limit state design and partial factor format.

3.1 Introduction

A civil engineering designer has to design structures that are both safe and economic. However, it is difficult to assess at the design stage how safe and economic a proposed design will actually be in practice. This is because there is much uncertainty about many of the factors which govern safety and economy.

The uncertainties that affect the safety of a structure fall roughly into three groups: loading, material strength and structural behaviour. These uncertainties arise from both the natural variability of the physical world and from our lack of complete knowledge about it and future events. Together, they mean that a designer cannot guarantee that a structure built to his design will be absolutely safe but only that the risk of failure is extremely small.

An illustration of the statistical meaning of safety is given in Figure 3.1. Consider, for example, the case of a typical structural element such as a beam or a column designed to carry a certain nominal design load. The statistical distribution on the right of Figure 3.1 illustrates the variation which might occur between different elements, all nominally the same, due to variations in material strength, fabrication tolerances, etc. The distribution on the left illustrates the variation in the maximum load which different samples of this element might encounter during their life in service. The uncertainty here is not only due to the variability of the loads applied to the structure but also includes the uncertainty about the distribution of these loads through the structure. Clearly, if a particularly weak element were subjected to a very heavy load, the load might exceed the element's strength, and failure would occur. If enough data are available to define probability distributions for the load and strength then it is possible to calculate a probability of failure for the design in question.

Calculated probabilities of failure are usually found to be extremely small (one in one million would not be unusual). In practice, the amount of useful data is often small, and this means that the calculated probability of failure is sensitive to the way in which the data are selected

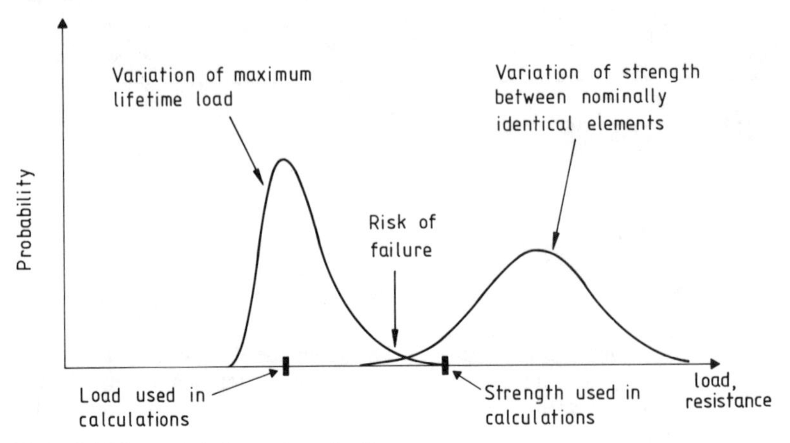

Figure 3.1 The statistical meaning of safety

and analysed. Such calculations are therefore only suitable for careful use in research and are not an acceptable basis for ordinary design work.

Instead, normal design calculations are made, using a single value for each load, each material property and all other structural parameters. This can be called the design, nominal, specified or characteristic value, or almost any combination of these terms, depending on which set of design rules are referred to. The requirement for safety is met by introducing safety factors into the design calculations, as will be discussed later.

In view of the high degree of safety that structural designs appear to possess it is not surprising that most structural failures can be attributed to people making mistakes through carelessness or ignorance during design or construction. The possibility of this occurring is another important source of uncertainty in the design process. An effect of these uncertainties is a tendency for design assumptions and calculation methods to become standardized.

3.2 Standardization of design procedures

A designer faced with a structure similar to those designed by other people will want to know how they did their work. In particular, if a similar design has been built he will want to follow aspects of the design that have proved successful. Standard designs and methods will tend to arise initially within individual firms and then, more gradually, throughout the industry. In addition, because the consequences of bad design can be catastrophic, society needs assurance that structures are safe, and it expects designers to be able to explain and justify their decisions. Thus it is clearly to the designer's advantage to use methods and assumptions that have proved safe in the past. Another advantage in having standardized design methods is that it is then possible to compare alternative designs and organize competitions while minimizing the risk that the cheapest design will be significantly less safe than the others.

As a result of these pressures most governments attempt to ensure structural safety through laws and regulations. Designers then seek to achieve maximum economy within the range of designs that are permitted by the regulations, except, of course, when they feel that these regulations are in some way inadequate. In particular, the avoidance of certain causes of failure, especially fatigue and corrosion, depends heavily on good detailing, and it is very difficult to draft regulations that deal adequately with these aspects of design.

Laws and regulations also apply to partially completed structures. Especially in the case of larger structures, this is often the time when the structure has the least reserves of strength and stability to withstand unforeseen loadings.

3.3 Regulation of building and construction

3.3.1 General

There are a number of different ways in which the law can operate. It can simply place the designer under a general obligation to design safe structures and prescribe punishments to be applied if a structure fails. The earliest known building laws were published by King Hammurabi of Babylon, in about 2200 BC. They are preserved as a cuneiform inscription on a clay tablet and include such provisions as:

> If a builder builds a house for a man and does not make its construction firm and if the house which he has built collapses and causes the death of the owner of the house, then that builder shall be put to death. If it causes the death of the son of the owner of the house, then a son of the builder shall be put to death. If it causes the death of a slave of the owner of the house, then the builder shall give the owner a slave of equal value.

In Britain, government regulations of some sort have existed since the early Middles Ages, and their development has been strongly influenced by events such as the Great Fire of London in 1666 and the collapse of the Ronan Point flats in 1968. Nowadays, structural safety is controlled through a wide range of government regulations made under various Acts of Parliament, such as the Public Health Acts of 1936 and 1961. Most of these regulations do not contain detailed requirements about structural design. Instead, reference is made to codes of practice published by the British Standards Institution, and these are the main source of detailed design requirements in the UK. Most other developed countries have similar codes of practice.

The British Standards Institution also plays a major role in the production of economic steel structures by producing standards for steel sections and plates, the classification of grades of steel and types of bolts, to give just a few examples. Before standards such as these were written each manufacturer produced his own range of products, and an early example of successful standardization was the reduction of the number of different designs of tramway rail being produced from 75 to 5.

The major problem with codes of practice is that they cover only certain types of structure and, as time goes on, this range is reduced by new developments until the code is updated. Designers need to be constantly aware of the limitations of the codes they are using.

How, then, does a code of practice set about regulating structural safety? What basic approaches can be used to give coherence to what must often be a fairly complicated document?

3.3.2 Design by geometrical ratio

Before mathematics and science were applied to building work design rules were largely a matter of experience and tradition. However, one general approach was widely used to express these rules, and that was the use of geometrical ratios giving limits on what had been safely built (for example, keeping the ratio of the height of a wall to its width to less than a certain value). This approach was all that the medieval masons used to construct the great cathedrals of western Europe, which demonstrates its potential. Their rules were established by trial and error, and the not-infrequent collapses of parts of cathedrals while under construction showed the errors.

This geometrical ratio approach is by no means extinct. It is, of course, still used for many small-scale buildings, and by designers of all structures as a way of making a first attempt at a design that is then proved and elaborated by more sophisticated techniques. It also makes frequent appearances in codes that are mainly arranged on more mathematical principles. For example, the easiest way to deal with the design of wide plates in compression is to say that if the width-to-thickness ratio does not exceed a specified ratio then the plate can be considered as 'fully effective'. If it is wider than this, then the excess width is ignored in calculations. Rules of this kind are generally attempts to find simple approximations to results that could only be obtained accurately by a much more complex analysis.

3.3.3 Allowable stress design

The other major design philosophies arose with the application of science and mathematics to engineering. There are, of course, some fairly simple applications of mathematics. For example, calculating whether a given horizontal force, such as a wind load, is sufficient to overturn a structure or estimating the distribution of forces in a statically determinate structure only required the application of the concept of equilibrium. The real advance came in the nineteenth century, with the development of the linear elastic theories which accurately represented the behaviour of the new structural materials such as wrought iron and mild steel up to a yield stress which was taken to be the onset of failure. This enabled indeterminate structures to be analysed and the distribution of bending stresses and shear stresses to be worked out in detail.

The natural way to present such calculations was the allowable stress format. The governing principle of design is that the stresses caused by the nominal or characteristic design loads should not exceed an allowable stress which is the failure stress reduced by a safety factor. This factor is intended to cover the uncertainties discussed earlier in this chapter with an acceptable margin of safety.

In general, each member or element in a structure is checked for a number of different combinations of loading. This is because many of the loadings vary with time, and it would be unnecessarily severe to consider them all acting simultaneously with their full design value while maintaining the same safety factor.

In practice, there are severe limitations on this approach. In many complex circumstances (for example, at connections) the elastic stress distribution can only be worked out using very elaborate calculations that have only become possible with the use of computers. In addition, when buckling takes place, although the material still behaves in a linear elastic way the member as a whole behaves non-linearly. Also, one of the advantages of steel as a structural material is its ability to tolerate high theoretical elastic stresses by yielding locally and redistributing the loads.

The strength of elements such as connections and struts is determined from advanced theory and the results of tests, and the allowable stress format has to be modified before it can be used in such cases. Here, the calculated stress and the allowable stress are rather artificial concepts: they are not the elastic stresses that actually occur.

However, although the allowable stress format could be modified in this way to allow for redistribution of stresses within a member it could not readily allow for redistribution of load from one member to another in a redundant framework. The search for ways to calculate the true strength of redundant frames led to the development of the plastic theory, and the success of this development provided an incentive to find a philosophy which could provide a framework for the full range of structural calculations.

3.3.4 Limit state design

The new philosophy which was developed is known as 'limit state design', but this is a somewhat misleading title because the 'limit state' idea is only one of three which were introduced simultaneously into the new generation of codes of practice written over the last 20 years or so. The three ideas are:

1. Explicit reference to 'limit states';
2. The definition of the nominal loads and stresses used in calculations in terms of statistical concepts; and
3. The use of the partial safety factor format.

'Limit states' are the various conditions in which a structure would be considered to have failed to fulfil the purposes for which it was built. The description of these varies from code to code, but a typical list is given in Figure 3.2. It will be seen that there is a general division into ultimate and serviceability limit states. The former are those catastrophic states which require a large safety factor in order to reduce their risk of occurrence to a very low level and the latter are the limits on acceptable behaviour in normal service.

ULTIMATE LIMIT STATES	SERVICEABILITY LIMIT STATES
1. Strength (yield, buckling, collapse as a mechanism)	5. Deflection
2. Stability against overturning and sway	6. Vibration
3. Fracture due to fatigue	7. Repairable damage due to fatigue
4. Brittle fracture	8. Corrosion

Figure 3.2 Limit states

Note that not all these limit states require structural calculations. For example, failure by brittle fracture is normally handled by writing material specifications which ensure that the steel is sufficiently ductile, so that failure always occurs by yield. Corrosion is also normally covered by specifying painting requirements or other forms of protection, although good detailing remains very important for avoiding corrosion.

Limit states were not, of course, absent from the older, allowable stress format codes. All that has changed is the explicit reference to them as such and the resulting effect on the organization of the code. Similarly, the introduction of statistical concepts into codes of practice has not meant any fundamental change. In general, the values used for loads and yield stress are exactly the same as they were before, but the explanations and definitions given now recognize the uncertainties discussed earlier, and unrealistic terms like 'minimum yield stress' are no longer used. Instead, terms such as 'characteristic strength' are employed, and a typical definition would be 'that strength which is expected to be exceeded by 95% of cases'.

3.3.5 Partial factor format

The major innovation in these new codes is the introduction of the partial factor format. This aims to provide a framework suitable for most of the structural calculations, and it can cover collapse, overturning and deflection. Fatigue and dynamic behaviour generally require a different approach because they depend on aspects of the loading different from those of the other limit states.

In a typical, general, partial factor format produced by an international committee as a basis for comparing and harmonizing national codes seven different factors are defined; five to cover various aspects of the uncertainties discussed earlier and two to enable the overall safety margin, and hence the risk of failure, to vary with the severity of the consequences of failure. Unfortunately, the lack of data about these uncertainties and consequences makes it difficult to provide a rational basis for determining values for these factors, and hence it is also difficult to justify the inconvenience of using so many factors in every calculation. In most practical codes, which deal with only one specific type of structure, the number of factors is reduced by combining factors together. In serviceability calculations load factors of 1.0 are often used. Some codes also check the possibility of permanent deformation occurring due to the yield stress being exceeded in some area of high stress concentration. In these calculations the load factors are also 1.0 or thereabouts. This calculation may govern the design of elements whose collapse calculations allow for a considerable amount of load redistribution to occur between first yield and total failure.

3.4 Concluding summary

1. The process of design in civil engineering has to be carried out in spite of a lack of complete knowledge about loads, materials and structural behaviour.
2. Society has a great interest in the safety of structures, and governments exercise control over design through regulations and codes of practice.
3. The traditional basis for codifying structural design rules is the use of geometric ratios.
4. In the nineteenth century the development of linear elastic theory led to the allowable stress format being used generally for structural design.
5. Recently the 'limit state' philosophy has been introduced, including explicit reference to 'limit states' and to statistical concepts, and using the partial factor format for design calculations.

Background reading

BAKER, J.F. (1960). *The Steel Skeleton*, Vol. I: Baker, J.F., Horne, M.R. and Heyman, J. (1965) *The Steel Skeleton*, Vol. II, Cambridge University Press, Cambridge

CORNES, D.L. (1983) *Design Liability in the Construction Industry*, Granada, St Albans

PUGSLEY, SIR A. (1966) *The Safety of Structures*, Edward Arnold, London

THOFT-CHRISTENSEN and BAKER, M.J. (1982) *Structural Reliability Theory and its Applications*, Springer-Verlag, New York

4

Design of industrial buildings

Objective To outline the factors which may need to be considered in the design of industrial buildings.

Prior reading Chapter 2 (Process of design); Chapter 3 (Design philosophies).

Summary The different categories of industrial building are described; purpose-made, advance and standard units. The anatomy of an industrial building is reviewed with brief discussion of cladding, purlins and sheeting rails, main frames in portal frame construction, bracing and main frames. Some alternative arrangements for main frames are outlined.

4.1 Introduction

In simple terms the primary structure of an industrial building is a series of columns supporting roof members. Common examples are beams or trusses simply supported on columns, plane frames or space frames. Excluding load-bearing brick construction (which is only practicable for short spans of, say, up to 12 m), the choice of material is between steel and concrete. Steel is probably more competitive than concrete for buildings with clear spans up to 18–21 m and is the only economic solution for longer spans, due to the high strength/weight ratio and speed of erection of steel. The primary structural steelwork for an industrial building represents about 15% of the total cost of the building, while the cladding system costs approximately 30% of the total.

Essentially, industrial buildings have to provide sheltered, fully serviced, fully adaptable spaces. Adaptability is an important factor in industrial building design: achieving adaptability and accommodating services is the job of the engineer. However, one cannot emphasize too strongly the importance of collaboration between the designer/engineer, architect and client at an early stage in order to produce an economic solution which is best suited to the client's current and potential requirements. As the fabrication of structural steelwork forms an early part of the building process it becomes necessary for the client to define his requirements at a very early stage. Later changes will generally incur penalties in terms of both time and money.

4.2 The function of industrial buildings

There are three fundamentally different categories of industrial buildings: purpose-made units, advance or nursery units and standard units. Many industrial processes require a special layout for the building in which they are to be carried out, or special services, handling facilities, etc. Often, the only way in which these can be provided is for a building to be specially designed to suit the process (for example, car production, or the manufacture of electronic components). To accommodate possible changes or improvements in the process steps can be taken at the design stage to introduce a measure of flexibility.

On the other hand, it sometimes occurs, particularly in the development of a new industrial estate, that facilities must be available on demand in order to attract industry. This is done by constructing units, usually of small or medium size, in advance of any knowledge of the client's requirements or the processes which will ultimately be carried out. The design of such advance units requires very great care. To allow for any eventuality in the way of flexibility of layout, handling facilities, etc. could prove too costly for the accommodation to be let at an economic rental or sold at a realistic price. Therefore it becomes necessary to assess carefully the type of industry likely to be attracted to the area and to provide limited facilities to suit.

A number of fabricators advertise ranges of standard buildings, or standard components of buildings, which can be provided from stock or in a very short time after receipt of order. Such buildings can be very economical to construct because the supplier has geared his manufacturing processes to produce the components as cheaply as possible. The disadvantage of standard buildings lies in the lack of flexibility. Slight variations in the basic dimensions, perhaps due to restrictions on the site or the provision of special facilities not allowed for in the basic design, immediately invalidate the 'standardization', and the structure reverts to a purpose-built unit. That is, it is not often that a client really requires a 'standard' building.

4.3 The anatomy of an industrial building

The most common structural arrangement for single-storey buildings found on any modern industrial estate is that of portal frame construction. Therefore a portal-framed building will be used to illustrate the basic concepts of framing an industrial building and then other structural forms will be considered. For economic reasons, most industrial buildings have a rectangular floor plan with the main frames spanning the shorter distance. The external loading (due to wind and snow) is resisted by the cladding system, which is supported by the secondary members (purlins and sheeting rails). Then the load is transferred back to the main (or plane) frames and hence to the ground via the foundations.

The various structural components that make up a building will now be examined in detail (Figure 4.1).

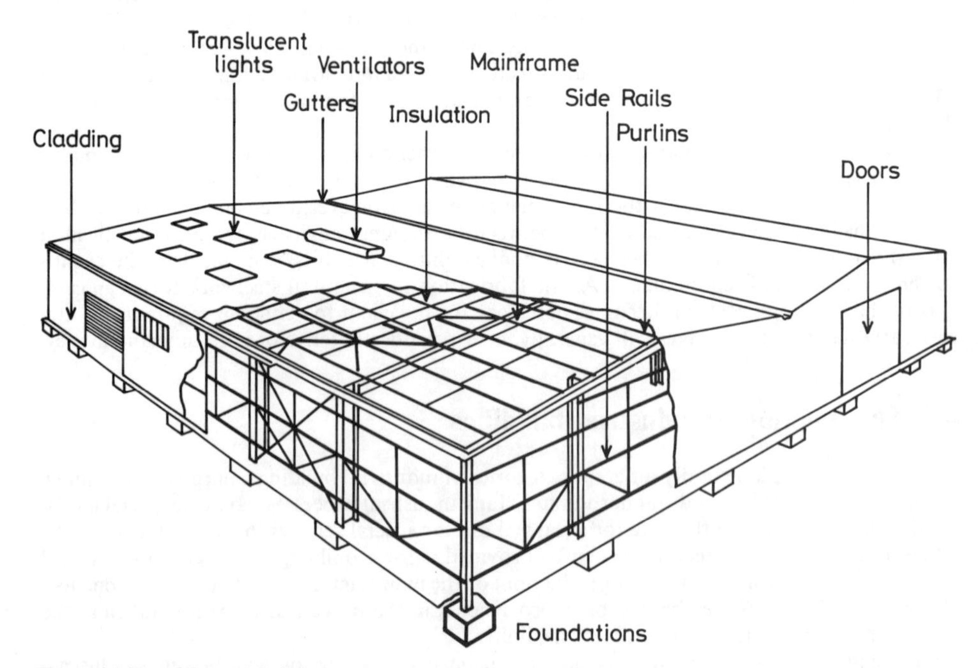

Figure 4.1 Portal-framed building and its elements

4.3.1 Cladding

The cladding system is required to create a protected working environment by providing an external envelope while satisfying an increasing demand for aesthetically pleasing buildings. Cladding normally carries only its own dead weight plus the loads imposed by snow, wind and maintenance. It is the term used for roof and side-wall systems when the steel sheets are exposed to the elements. However, in the case of roofs if the steel sheet supports insulation and waterproofing in addition to the normal loading such sheeting is termed 'roof decking' (Figure 4.2).

A great variety of profiles, colours and textures is available in coated steel sheets which can be produced in lengths up to about 13 m (depending on the profile adopted), thereby minimizing the number of end-laps. However, long lengths of sheeting (> 4 m) can cause handling problems on site. When used on shallow slopes, end-laps need to be bedded in a mastic sealant to prevent rain penetration by capillary action. Glazing can be readily incorporated into the cladding system but it is expensive. A more economical way of providing natural lighting is by the use of translucent sheets, moulded to the profile of the cladding.

The spacing of cladding supports is dependent on the cladding profile and thickness and can vary between 1.00 and 3.75 m. This information is normally obtainable direct from the manufacturers' cladding catalogues. Therefore one of the initial decisions in the design of an industrial building is to choose the type of cladding system for the building, as this controls the spacing and therefore the design of the secondary members (purlins and sheeting rails).

4.3.2 Purlins and sheeting rails

Secondary members will usually be cold-formed sections. Though there is a British Standard covering the design of cold-formed members the manufacturers tend to develop new profiles based on the results of extensive research. Note that if the same profile is selected for both purlins and sheeting rails then the spacing of the rails can be increased (compared with purlin centres) because the rails are not subjected to snow loading. The purlin/rail spacing may also be affected by the need to provide adequate lateral restraint to the portal frames. Though the design span of purlins/rails is taken as being the distance between main frames (commonly 6 m) these members are usually supplied in two-span lengths for the normal frame centres and the resulting continuity is exploited in design.

The purlin/rail joints are usually staggered around the main frame, thereby ensuring that each frame will receive approximately the same loading. In the end-bays this results in alternate purlins/rails spanning only one length, i.e. being simply supported. The purlin/rail can be increased in size (either by selecting a stronger section with the same depth or by changing the gauge thickness) or, alternatively, the centres between the end-bay frames can be reduced so that the purlin/rail size remains constant along the length of the building.

4.3.3 Main frames: portal frame construction

The portal frame is the most common form of construction for industrial buildings, distinguished by its simplicity, clean lines and economy. These frames can provide large clear floor

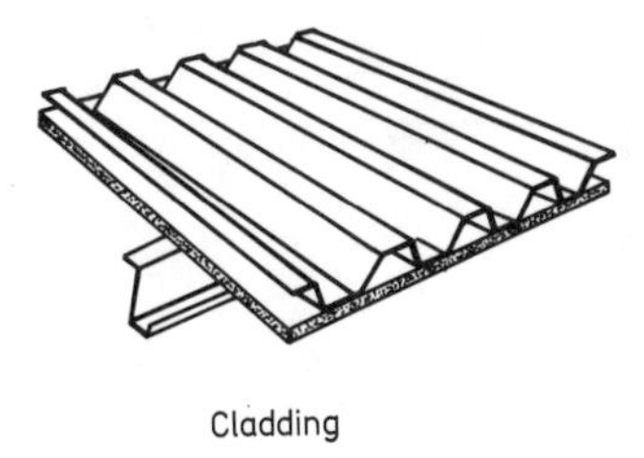

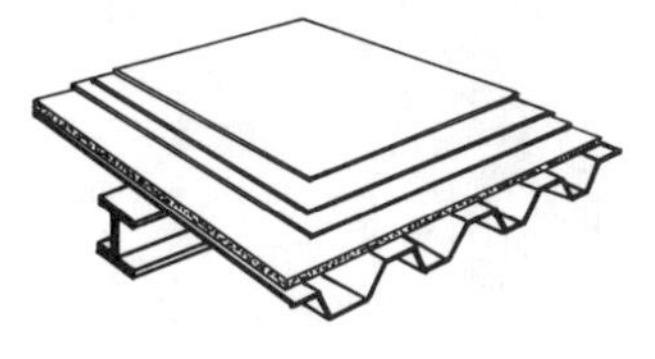

Figure 4.2 Roof cladding and decking

areas, offering maximum adaptability of the space inside the building. Such large-span buildings require fewer foundations, eliminate internal columns, valley gutters and internal drainage. A portal framed building can be extended at any time in the future.

Basically, the portal frame is a rigid-jointed plane frame made from hot-rolled sections, supporting the roofing and side cladding via cold-formed purlins and sheeting rails. The typical span of the portal frame is in the range of 30–40 m, though spans can range from 15 to 80 m. The common spacing between frames is about 6 m, but can vary from 4.5 to 10.0 m. The height to eaves in a normal industrial building is usually 4.5–6.0 m (which is about the maximum effective height for one level of sprinklers).

Since the mid-1950s portal frame construction in the UK has been widely based on the principles of plastic design, developed by Baker and his team at Cambridge. By taking advantage of the ductility of steel, plastic design produces lighter and more slender structural proportions than similar rigid frames previously designed by elastic theory.

Recently, tapered frames designed by elastic methods have been introduced to the market. Their efficiency depends very much on modern fabrication techniques and computer-aided design, backed up by full-scale tests. A frame can now be designed so that member profiles and plate thicknesses can be carefully proportioned to optimize the use of high-tensile steel plate.

4.3.4 Gables

The end-walls or gables of an industrial building can be framed in two different ways. They can be formed by using a frame identical to the intermediate portals and then framing gable posts up to the underside of the portal rafter. Suitable arrangements must be made at the post/rafter connections to accommodate vertical movement of the portal rafter. This method of gable framing can prove economic if the building might need to be extended in the future, because it requires minimum removal of steelwork prior to the extension.

Alternatively, if there is no possibility of an extension the gable rafter can be designed as a continuous member, spanning between the gable posts. This particular form of gable framing has poor sway resistance (unlike the portal frame) and it will need in-plane bracing.

Gable posts are generally designed as propped cantilevers. They support the gable sheeting rails and sometimes other secondary steelwork, for doors, openings, etc.

4.3.5 Bracing – overall stability of building

Every structure standing above the ground is affected by wind and may also be subject to earthquakes (Figure 4.3). Buildings with travelling cranes are also subjected to longitudinal and transverse horizontal forces resulting from the operation of the cranes. Various bracing systems have to be used to provide paths for these horizontal forces from their point of application down to the foundation. In addition, the change in a building's shape may have to be kept

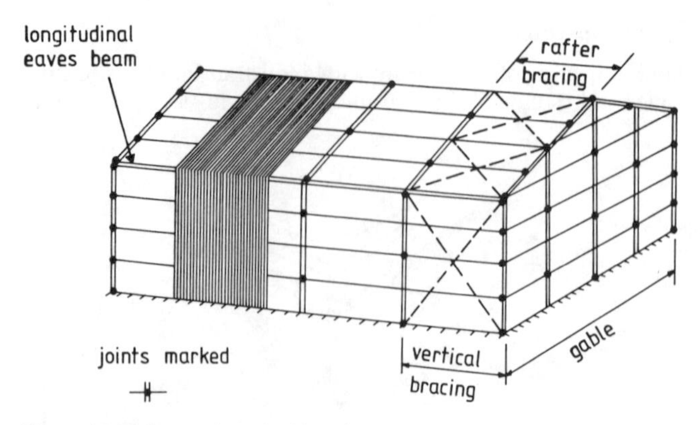

Figure 4.3 Rafter and vertical bracing

within acceptable limits. For example, longitudinal crane runway beams must be kept as straight as possible if the crane is to operate safely and without excessive wear on the rails.

The bracing members that transfer these forces to the ground may do so by their resistance to either bending moments or axial forces. Bracing designed to resist flexural action is more expensive and is usually confined to cases where axially loaded bracing members would interfere with the use of space. For example, when the normal diagonal bracing in an end-bay cannot be used portal framing can be placed in the plane of the side walls.

Wind acting normal to the longitudinal direction of the building is resisted by the flexural rigidity of the portal frames, and wind on the gables is transferred by the gable posts partly directly to the ground and partly into the rafter bracing. The latter takes the form of an open web girder located in the plane of the roof system. Wind forces from the rafter bracing and/or crane surge loading are transferred to the foundations by means of vertical bracing in the plane of the column members in at least one bay, usually the end-bay(s). The tops of the portal legs are tied back to the bracing by means of longitudinal eaves beams. This ensures the longitudinal stability of the building. It is customary to provide vertical side-bracing in the same bay as the rafter bracing, though if a large opening is required at that location the designer may have to offset the vertical bracing by one bay or resort to using portal framing. Bracing is usually made from round bar, angle or tubular sections.

4.4 Alternatives for main frames

Although portal frame construction has dominated the single-storey market in the past there is a growing demand from the 'hi-tech' industries for higher-quality, flexible-use buildings. Such buildings have to accommodate changes during their life to a mix of offices, production and storage space. The structures of these buildings are frequently flat roofed with steel frames of solid beam, castellated or lattice girder construction. Internally, provision is usually made for future construction of mezzanine floors. Several alternative forms of structural arrangement for industrial buildings are now examined.

4.4.1 Plane frames

Continuous beam construction is essentially 'flat roof' and can take the form of solid web sections, or castellated beams, supported by columns (Figure 4.4). Solid web beams may prove to be too shallow to allow the passage of services through the web. Though castellated sections increase roof volume they do allow some services to be accommodated within the castellations.

There is a number of different arrangements using both roof truss and lattice girder construction. These structural arrangements are similar to the portal framed building previously discussed, except that the portal frame is replaced by some form of open web frame supported by vertical column members. Typical roof truss arrangements are shown in Figure 4.5. The Fink truss usually proves to be economic for small spans (< 15 m), while the bowstring truss is frequently used to support a curved roof. Both the bowstring and the mansard trusses are used for longer spans.

For longer spans, lattice girders are often used instead of continuous frames (Figure 4.6). Because of their greater depth, these usually provide a greater stiffness against deflection. Note that the Pratt and Warren trusses have a top chord which is not necessarily parallel to the bottom one. Such an arrangement is used to provide a slope for the drainage of 'flat' roofs. In some buildings which require the provision of a free passage within the trusses and where the appearance of exposed steelwork is acceptable the Vierendeel girder might be a better solution.

In these alternative forms of structural arrangement the beams/girders/trusses are usually designed as 'simply supported' to carry the vertical dead load (plus imposed load) and as 'fixed' when considering the horizontal and inclined wind loads, i.e. portal action. This portal action is provided by using knee braces in the case of truss construction, and means that the columns

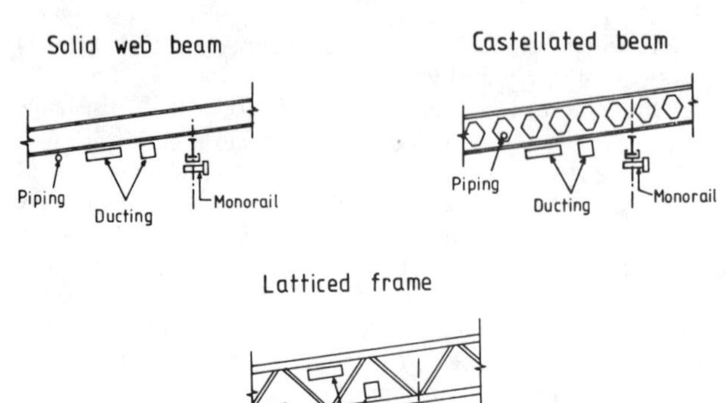

Figure 4.4 Solid web, castellated and lattice rafters

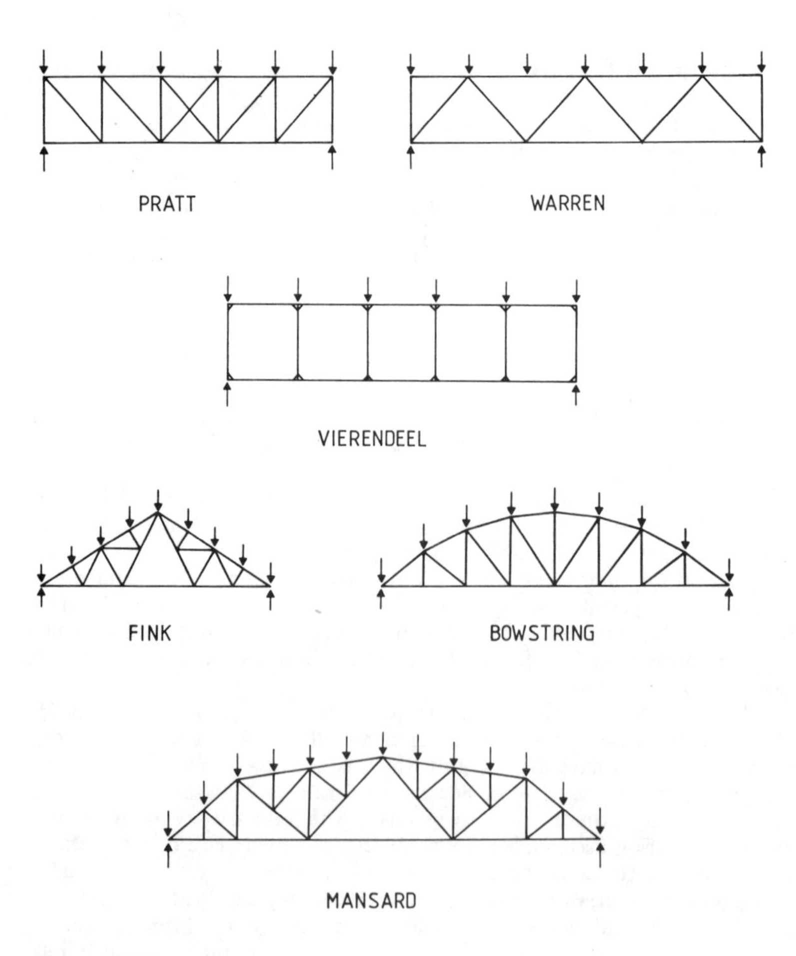

Figure 4.5 Types of trusses

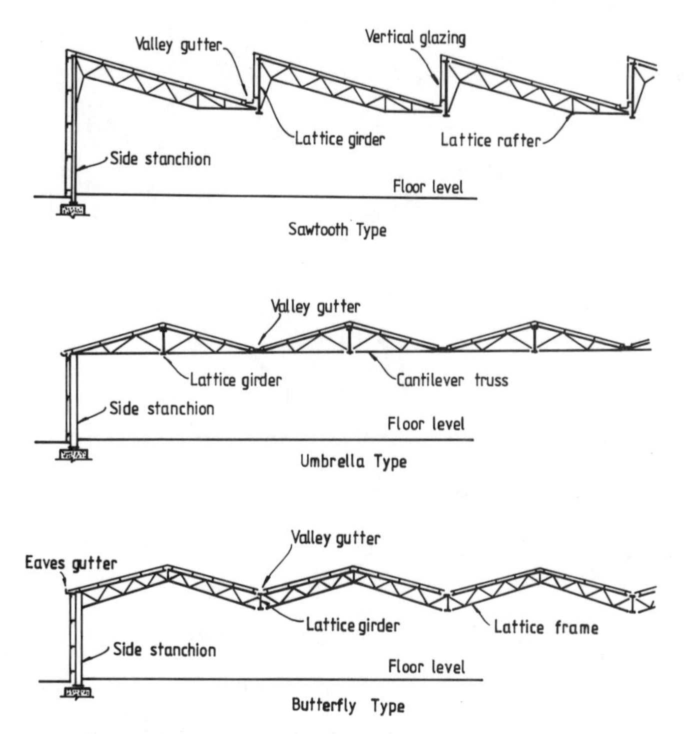

Figure 4.6 Sawtooth, umbrella and butterfly roofs

can be designed to have pinned bases, leading to simple foundations. the longitudinal stability of the framework is ensured by means of bracing (Figure 4.7).

When wind acts in the absence of imposed loading, i.e. with dead load alone, the forces in the members are, more often than not, reversed, with the result that the bottom boom of a girder, instead of being in tension, becomes a compression member. Lateral support may be necessary and this can be provided by introducing a bracing system to the bottom boom or, more economically, by braces to the purlins.

4.4.2 Space frames

The last decade or so has witnessed a growing worldwide interest in space frame structures. They owe this measure of interest to several characteristics, being aesthetically pleasing and able to provide an answer to many design requirements – large uninterrupted spans, unimpeded future extensions in any direction and the capacity to carry extensive overhead services. The space frame spans two ways, and therefore the square grid is most efficient, although rectangular plan shapes can be accommodated. The relatively small size of components permits speedy construction in the fabrication shop, easy transportation to site and quick erection. Where overseas markets are concerned, the small size of components has an extra advantage, as freight charges will be based on weight and not on volume. These considerations, and the ease with which space frames can be dismantled and re-assembled elsewhere, make them an obvious choice for exhibition structures. The British Steel Corporation's Nodus system is a typical example of space construction. However, space frames are often more expensive than other forms of structural arrangement.

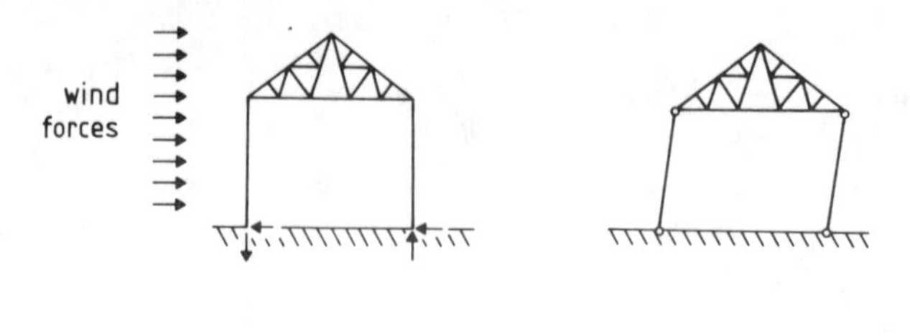

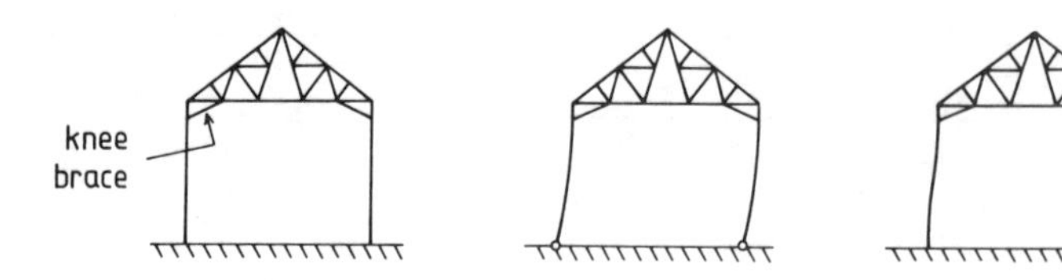

Figure 4.7 Requirements for knee braces for roof trusses

Background reading

BAKER, J.F., HORNE, M.R. and HEYMAN, J. (1965) *The Steel Skeleton*, Vol. 2 – *Plastic Behaviour and Design*, Cambridge University Press, Cambridge

BRETT, P. (1984) Light industrial buildings. Institution of Structural Engineers/IABSE Conference on 'The Art and Practice of Structural Design', London

DOWLING, P.J., MEARS, T.F., OWENS, G.W. and RAVEN, G.K. (1982) A development in the automated design and fabrication of portal framed industrial buildings. *The Structural Engineer*, **60A**, October, 311–319

Holding Down Systems for Steel Stanchions (1980) Concrete Society/BCSA/Constrado, London

MORRIS, L.J. A commentary on portal frame design. *The Structural Engineer*, **59A**, Dec. 1981, 394–404; **61A**, June 1983, 181–189; **61A**, July 1983, 212–221

Nodus Space Frames – a Design Guide for Architects and Engineers (1981) British Steel Corporation, Tubes Division

Protection of Steel from Corrosion (1983) British Steel Corporation

RHS 1 Lattice Beams (1983) British Steel Corporation, Tube Division

Single-storey Steel Framed Buildings – a Design Manual (1984) British Steel Corporation/Conder

The Behaviour of Steel Portal Frames in Boundary Conditions (1980) Constrado. Also supplement, Part D (1982) *Design Recommendations for Portal Frames with Tapered Rafters Subject to Boundary Conditions*, Constrado, London

5

Design of bridges

Objective To outline the factors which have to be considered when deciding the general form of a steel bridge.

Prior reading Chapter 2 (Process of design); Chapter 3 (Design philosophies).

Summary The principal classes of bridge structure are surveyed in ascending order of span. A discussion of simple single-span bridge is followed by a commentary on continuous beam bridges and cable-stayed structures. Suspension, arch and truss bridges are also reviewed.

5.1 Fundamentals

The purpose of a bridge is to carry a service (which may be a highway, a railway, a public utility, etc.) over an 'obstacle' (which may be another road or railway, a river, a valley, etc.). There may be restrictions on the clearance over the 'obstacle' or on the blockage of it (i.e. limitations on the number of intermediate supports or piers). Major considerations in choosing a type of bridge include the type and magnitude of the loading to be carried, the clear span requirements, the topography and geology of the site, and local constructional skills.

5.2 Loading

Loading on bridges may be largely of known magnitude (deterministic) or of a random nature (stochastic). Examples of the former are the dead load of the structure and certain forms of applied load. For instance, on a short-span railway bridge it might be known that the maximum load that could possibly be carried would be two locomotives passing at mid-span. Examples of the latter are traffic loading on a highway bridge where the mix and positioning of different vehicles can never be exactly predicted and also most natural phenomena (wind, earthquake, etc). Statistical analysis is then used to predict an intensity of loading, with an acceptably low probability of occurrence during the life of the bridge, for use in design.

Wind loading is seldom of significance in short-span bridges, while it normally affects only the design of the piers on medium-span structures. It becomes a highly specialized subject for long-span bridges where aerodynamic effects become very important.

The design philosophy adopted in limit state design of steel bridges is generally that all elements should be designed for the ultimate limit state. A few (very few) elements which have a large reserve of strength even after yielding may require checking for the serviceability limit state in order to avoid the development of large permanent deformations. Steel bridges also have to be checked for fatigue, since by their nature fluctuating and repetitive stresses will be induced.

It is always simpler to design a structure for deterministic loading since the magnitude and positions of the main load-bearing elements are usually obvious. The remainder of this chapter will thus be concerned primarily with highway bridges where no such clear indication is usually apparent. The first and obvious requirement of such a structure is a deck on which the traffic loads run, and in most cases there is a considerable variability in the position of such loads.

5.3 Simple single-span bridges

In the case of a very short-span bridge the deck itself may form the complete structure, but such spans are of little relevance in steel bridge designs since this would normally be a reinforced concrete slab. For all but the shortest spans the deck needs additional support, and in simple bridges this normally takes the form of beams spanning between the bridge piers. Two arrangements are commonly used (see Figure 5.1):

1. The deck spans transversely between longitudinal beams which have to be at comparatively close centres;
2. A smaller number of longitudinal beams is used (perhaps only two may be provided) but, because the distance between them is now too great for the slab to span, there are additional transverse 'cross-beams' to support the slab which now spans longitudinally between them.

The beams (both cross-beams and longitudinals) may be either standard rolled sections or fabricated. Up to about 25 m span it is common to use rolled sections, but over this limit it is more usual to adopt fabricated sections.

Modern practice is to make all elements participate in carrying the load (for example, a concrete slab will be made to act compositely with the beams, in both directions if provided, by connecting it to the beams with shear connectors). While a true analysis of such a system can be complex, using finite element or grillage techniques, it cannot be emphasized too strongly that this is a final check analysis and not a design procedure. For preliminary design, the method of static load distribution:

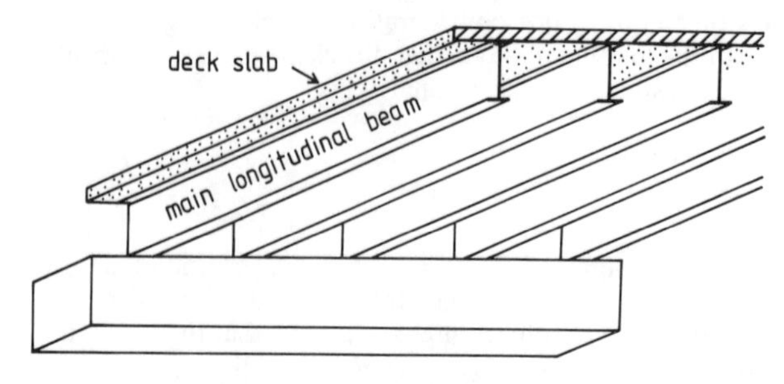

(a) Main longitudinal beams only

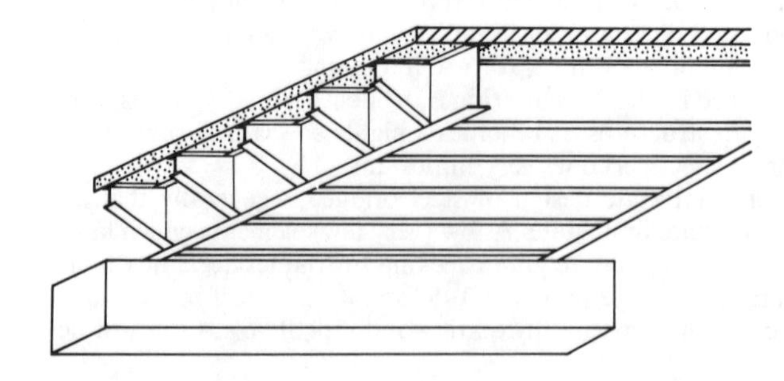

(b) Cross beams and longitudinal beams

Figure 5.1 Typical beam and slab construction (views from below)

'Deck slab' to 'cross-beams if any' to 'main beams'

will still produce an acceptable solution.

5.4 Increasing spans

While the simple single-span bridge of the previous section could be made to span 100 m or more it becomes a very doubtful proposition above 40–50 m because of economic considerations. The obvious first suggestion is to consider intermediate piers. For example, it would probably be acceptable in a bridge crossing a motorway to place a pier in the central reserve, thus reducing it from a single span of perhaps 50 m to two spans of 25 m. Normally, further savings can be effected by making the two spans continuous, thus taking advantage of the resultant reduction in maximum moment at the expense of slight additional calculation. The only situation in which continuity would not normally be used is in areas where a statically determinate structure is essential – for instance, in areas of mining subsidence. Except for the shortest continuous spans, maximum economy of material is normally achieved by using fabricated plate girders for the main longitudinal beams, with the webs and flanges varying along the length and proportioned to resist the shears and moments generated. It is difficult to 'tailor-make' standard rolled joists since added material (flange thickening, etc.) introduces problems such as a tendency to fatigue.

This form of construction (continuous beam and slab) can be extended to substantial span lengths – well over 100 m is frequently economic – although considerable variation in web and flange thickness will occur on such spans. Furthermore, it can be advantageous to deepen (haunch) the main beams at the piers. This not only increases their capacity for carrying bending moment but also enables some of the shear force at the piers to be carried by the compression in the sloping bottom flange (see Figure 5.2).

As spans increase beyond 100 m further expedients can be adopted to stretch the beam and slab type of construction up to about 250 m. One problem is that of the stability of the beams. This can be improved greatly (at the expense of more complex fabrication) by using main beams of fabricated steel box cross-section, which also has the advantage of allowing more steel to be concentrated in the flanges of the beams without recourse to very thick plates, with all the metallurgical problems that these imply. As spans increase above 100 m the dead weight of a concrete deck becomes a large part of the total loading, and two alternatives have been adopted:

1. Lightweight concrete may be used, possibly acting compositely with the top flanges of the box girders.
2. A stiffened steel plate deck is employed, surfaced with mastic asphalt on which the traffic runs direct. Examples generally occur in box girder bridges with spans in excess of 150 m (see Figure 5.3).

While these long spans are normally fully continuous, in certain circumstances little loss of economy occurs if cantilever and suspended span construction is adopted. This can, on occasion, offer some advantages:

1. The structure can be made statically determinate without the cost penalty of simple spans, thus allowing for settlement.

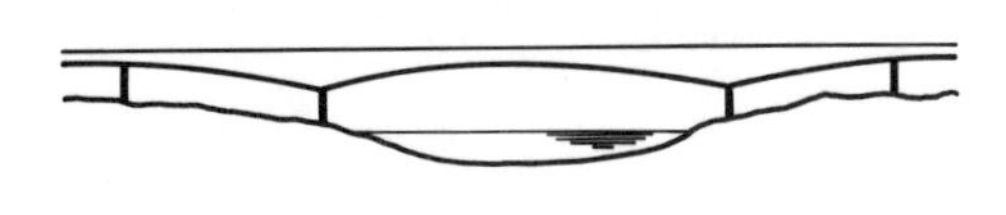

Figure 5.2 Haunched beam bridge

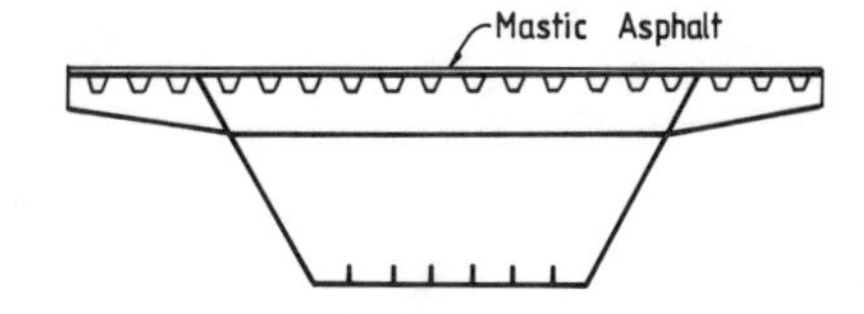

Figure 5.3 Cross-section of steel box girder

2. The problem of accommodating thermal movement in very long continuous spans may be eased.
3. Construction can sometimes be simplified (the suspended span may be lifted as a whole).

5.5 Away from beams

All the bridge types so far described rely ultimately on the bending action of the main longitudinal beam to transmit the applied loadings to the piers and thence to the ground. There is a limit of span for which such an arrangement is feasible; above 250 m alternative means of carrying the load must normally be sought, and on occasion such means may be used to advantage on shorter spans.

Perhaps the commonest form of extending a span to 550 m (or sometimes more) is by the use of intermediate stays of high-tensile steel wire formed into strands. Many detailed layouts are possible, but an elevation of a typical scheme is shown in Figure 5.4. In this a number of stays pass over saddles on top of towers and are anchored into the main longitudinal beams of the bridge. At least one 'backstay' is anchored at a point which is directly above an abutment or bearing, and thus the vertical component of the force in it is transferred directly to the ground. By careful adjustment of the stays the main beams can be levelled, so that the dead-load moments and shears are those that would be generated in a continuous beam with spans between the stay points. Under live load the main beam behaves as a continuous beam on elastic supports, whose stiffness is determined by the stay characteristics. All loading conditions cause tension in the stays, the horizontal components of which have to be resisted by compression in the main beams. If the tower saddles are free to move longitudinally (either by placing them on rollers or by pinning the base of the towers) the horizontal components are internally self-equilibrating.

It is normal for the backstay reaction to be one of uplift; if problems are encountered in designing a bearing to withstand uplift it is frequently possible, particularly if the main beam is a box, to balance this by means of kentledge.

The stayed girder bridge is normally very elegant and has frequently been used for comparatively short spans either for its appearance or if the depth of the main beam has had to be kept to an absolute minimum.

As spans increase still further to 1000 m and more, the economic solution is normally the stiffened suspension bridge (Figure 5.5). In this design the ultimate load path to ground is via a high-tensile steel cable which stretches between anchorages at each end of the bridge and is slung over tall towers within the overall length. The longitudinal beam is hung from this cable by comparatively closely spaced hangers, as a result of which dead-load stresses in the beam are negligible; live loads are distributed by the beam to a few hangers and thus to the main cable. Because of the very high forces induced in it the main cable possesses considerable stiffness against vertical loading. The cable tension is resisted at the ends by anchorages which may (depending on ground conditions) be of the gravity type, relying for their stability purely on mass, or of the tunnelled type, where a concrete plug in suitable rock relies on the strength of the rock in addition to its own mass.

A third type of structure which may be used for long spans (> 200 m) is the arch. In its

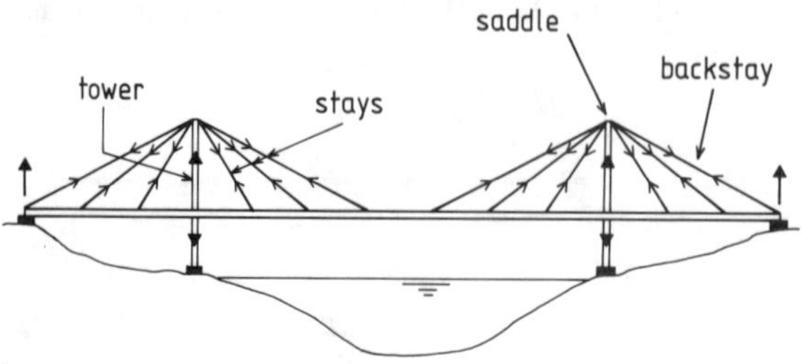

Figure 5.4 Stayed girder bridge

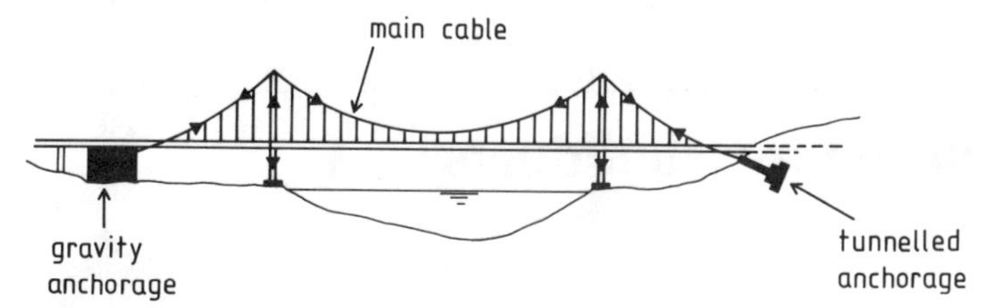

Figure 5.5 Suspension bridge

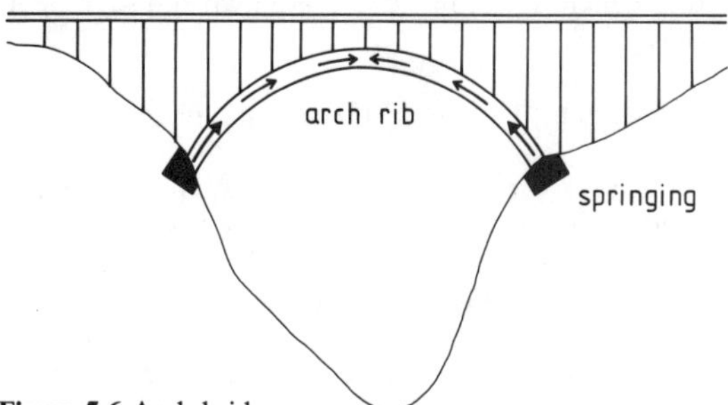

Figure 5.6 Arch bridge

Figure 5.7 Truss bridge: simple spans

simplest form, shown in Figure 5.6, this is essentially an inverted suspension bridge in that the longitudinal beam is suspended or supported from a rib which acts in compression. As there is a large outward horizontal component of this compression at the springings, arches are ideally suited to rocky gorges which can resist such forces without movement. Any spreading of the springings can have serious implications for bridge stresses. Two expedients to alleviate this (which may be implemented separately or in combination) are:

1. If the ground is capable of withstanding the horizontal forces at the springings but may undergo some limited movement the rib is made statically determinate and tolerant of such movements by inserting pin joints in the arch rib at the springings and the crown.
2. If the deck is suspended from the arch rib at the level of the springings it may be made to act as a tie, thus imposing only vertical loads on the abutments.

5.6 Other types

Minor or major variations are possible in all types of bridge. For example, particularly in short single-span bridges, it may be advantageous to make the longitudinal beams monolithic with the piers, thus forming portal frames. One major type of bridge not discussed so far is the truss. In effect, trusses may be used in place of beams in almost any type of bridge, and in terms of initial design may be thought of as open-web beams. They are seldom used nowadays in the UK since they are considered to be aesthetically displeasing. However, they are frequently a very economic form of construction and are still much used abroad, particularly for railway bridges (Figure 5.7).

6

Design of multi-storey buildings

Objective To discuss the factors which have to be considered when designing a multi-storey building.

Prior reading Chapter 2 (Process of design); Chapter 3 (Design philosophies).

Summary This chapter commences with a description of the primary components of a building frame, i.e. columns, beams, floors, bracing and connections. It discusses common structural arrangements to resist vertical and horizontal loadings and the importance of minimizing structural floor depth. It concludes with a review of the most important considerations for practical and economic design.

6.1 Building framework

Figure 6.1 shows the vertical load-carrying structure of a building; the rectilinear framework is composed of vertical column elements interconnected by horizontal beams supporting floor element assemblies. The resistance to horizontal loads is provided by diagonal bracing elements, or wall elements, introduced into the vertical rectangular panels bounded by the columns and beams to form vertical trusses or walls. Alternatively, horizontal resistance may be provided by developing the 'portal' action between the beams and columns. The resistance to horizontal loads in the horizontal plane is provided by the floor element assemblies.

The components of a building frame may therefore be summarized as columns, beams, floors and bracing.

6.1.1 Columns

These are generally standard, universal column, rolled sections which provide a compact, efficient section for normal-building storey heights. Also, because of the section shape they give unobstructed access for beam connections to either the flange or web. For a given overall width and depth of section there is variety of weights to enable the overall dimensions of structural components to be maintained for a range of loading intensities.

Where the loading requirements exceed the capacity of the standard sections the required capacity may be achieved either by welding additional plates to the section to form plated columns or by forming a purpose-designed element by welding plates together to form a plate-column. The use of circular or rectangular tubular elements marginally improves the load-carrying efficiency of components as a result of their higher stiffness-to-weight ratio. However, connections to beams are more complicated.

6.1.2 Beams

The most common section is a hot-rolled universal beam. A range of section capacities for each depth enables a constant depth of construction to be maintained for a range of spans and

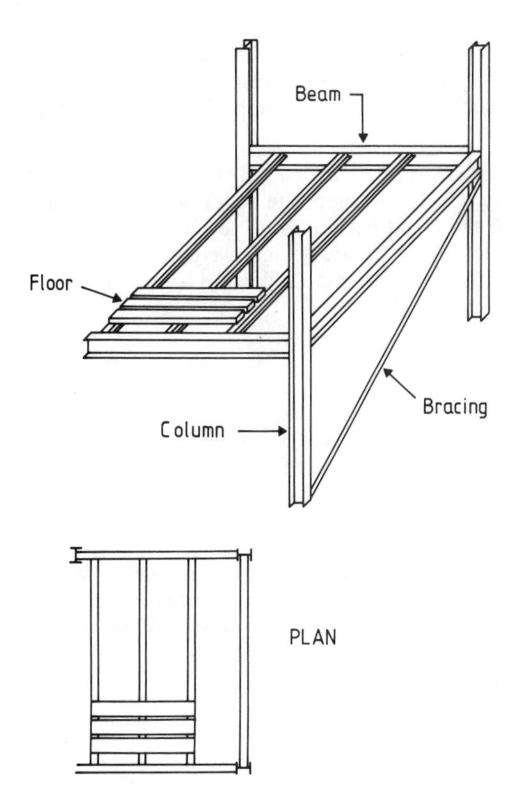

Figure 6.1 The building frame

loading. As with column components, plated-beams and fabricated plate beams or girders may be used where heavier loading or longer spans are required and overall depth is limited. For medium to lightly loaded floors and long spans, beams may also take the form of castellated beams fabricated from standard sections.

6.1.3 Floors

These take the form of concrete slabs of various construction spanning between floor beams. The types generally used are shown in Figure 6.2:

1. Pre-cast concrete slabs acting non-compositely with the floor beams;
2. *In situ* concrete slab with conventional removable shuttering acting compositely with the floor beams;

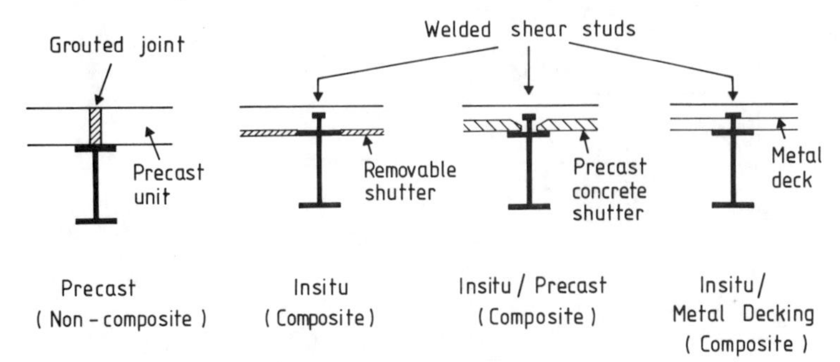

Figure 6.2 Flooring systems

3. *In situ* concrete slab cast over thin permanent formwork pre-cast concrete slabs to form a composite slab, which in turn acts compositely with the floor beams;
4. *In situ* concrete slab cast onto permanent metal decking acting compositely with the floor beams.

Pre-cast concrete systems have been used extensively in the UK. The most widely used construction internationally is metal decking in which the composite action with the beam is provided by shear connectors welded through the metal decking onto the beam flange. The reasons for the emergence of this form of floor construction are its light weight, enabling it to be manhandled easily on site, and its high stiffness and strength, allowing it to support the weight of the wet *in situ* concrete without propping.

6.1.4 Bracings

These are generally flat plate, angle or channel sections, often designed on a 'tension only' basis and arranged between columns as cross-bracing. However, there is a wide range of bracing structures which is described in greater detail later.

6.1.5 Connections

Perhaps the most important aspect of structural steelwork for buildings is the design of the connections between individual frame components. The selection of a component must be governed not only by its capability to support the applied load but also by its ease of connection to other components. Basically, as shown in Figure 6.3, there are three types each defined by its structural behaviour; simple, rigid and semi-rigid:

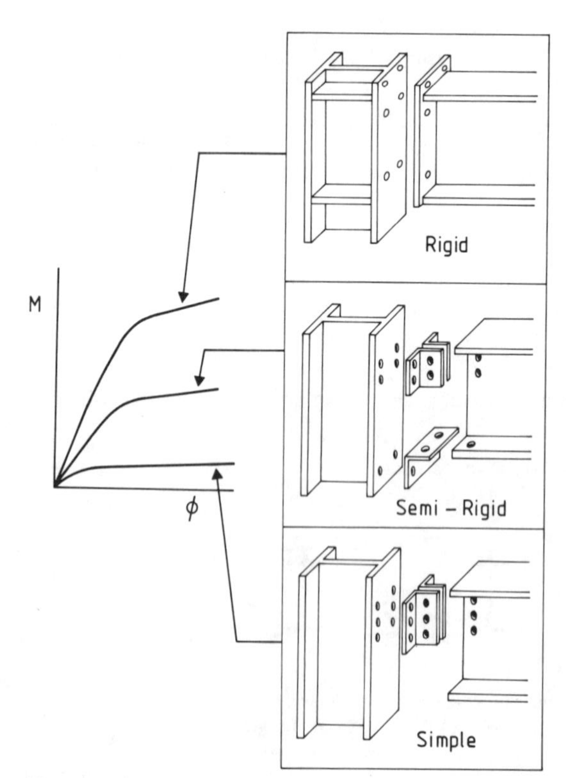

Figure 6.3 Types of connections

1. Simple connections transmit negligible bending moment across the joint; the connection is detailed to allow the beam-end to rotate. The beam behaves as a simply supported one.
2. Rigid connections are designed to transmit shear force and bending moment across the joint. The connection is detailed to ensure a monolithic joint. The beam behaves as a continuous beam within the rectangular moment-resisting framework formed by the columns and beams.
3. Semi-rigid connections are designed to transmit the shear force and a proportion of the bending moment across the joint. The principle of these connections is to provide a partial restraint to beam end rotation without introducing complicated fabrication to the joint.

6.2 Structural arrangement

The intended use of the building dictates the planning module, the intensity of floor loading and the level of mechanical and electrical servicing required. The structural arrangement, and depth selected, must satisfy these requirements.

6.2.1 Floor arrangement

Steel components are uni-directional and, consequently, orthogonal structural column and beam grids have been found to be the most efficient, as shown in Figure 6.4.

The most efficient floor plan is rectangular, not square, in which main beams span the shorter distance between columns and closely spaced floor beams span the longer distance between main beams. The spacing of the floor beams is controlled by the spanning capability of the concrete floor construction.

The spanning capability of the concrete floor construction can be extended by increasing the slab depth, but this increases the weight of construction, the depth of the floor beams and the overall depth of the structural floor arrangement. The best arrangement of beam spacing and concrete depth is therefore a balance of factors. Experience has shown that the most efficient floor arrangements are those using metal decking as permanent shuttering spanning 2.4–3 m

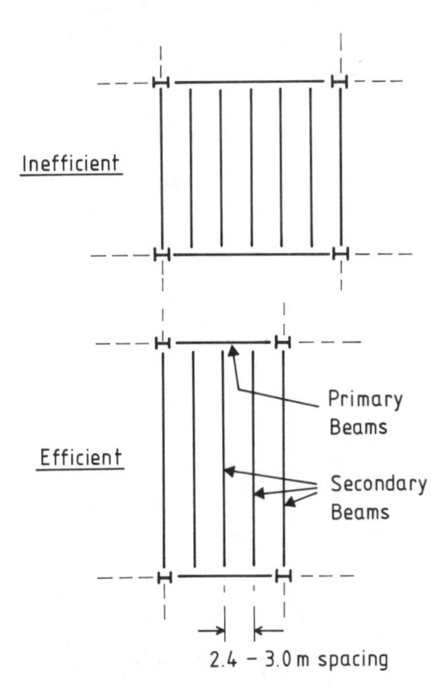

Figure 6.4 Floor arrangement

between floor beams. For these spans the metal decking does not normally require propping during concreting and the concrete thicknesses are near the practical minimum.

6.2.2 Bracing structures

As illustrated in Figure 6.5, building frames resist horizontal loading in one of three ways: rigidly jointed frames, reinforced concrete walls and braced bay frames. Combinations of these systems are frequently used and details of each system are as follows:

1. Rigidly jointed frames are those with full moment-resisting connections between beam and columns. It is not necessary that all connections in a building are detailed in this way; only enough frames are required to satisfy the performance requirements of the building. The advantage of this system in comparison to the others is the increased flexibility of the plan because there are no bracings or walls to obstruct circulation. The main disadvantages are the increased fabrication required for the connections and the increased site connection work, particularly if the connections are welded.
2. Reinforced concrete walls constructed to enclose lift, stair and service cores are generally sufficiently strong and stiff to resist the horizontal loading. The beam-to-column connections throughout the frame are simple, easily fabricated and rapidly erected. The disadvantages are that the construction of walls, particularly in low- and medium-rise buildings, is slow and less accurate than steelwork. Furthermore, the walls are difficult to modify if alterations to the building are required in the future.
3. Braced bay frames are positioned in similar locations to reinforced concrete walls, so they

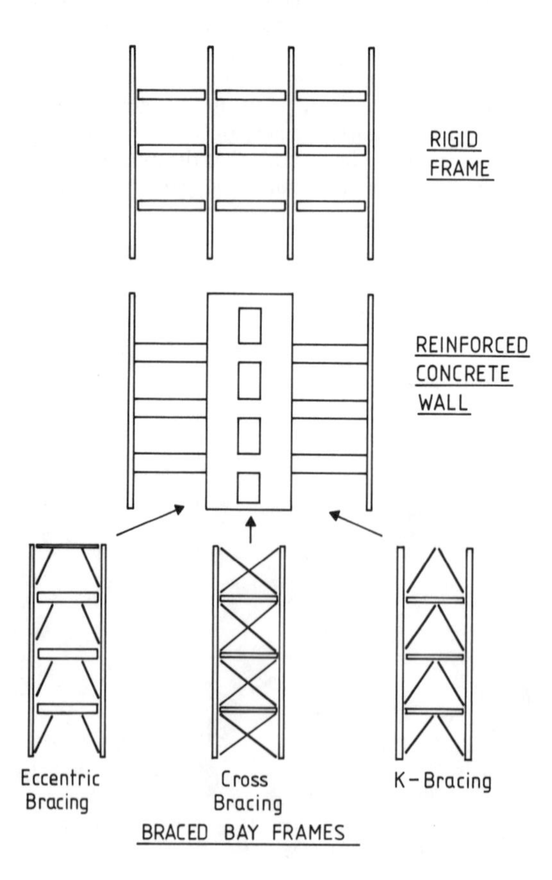

Figure 6.5 Bracing structures

have minimal impact upon the planning of the building. The main advantages of this bracing system are that all beam-to-column connections are simple, the braced bays are concentrated in location on plan, the bracing configuration may be adjusted to suit planning requirements (eccentric bracing) and the system is adjustable if building modifications are required in the future.

Cores should be located to avoid eccentricity between the line of action of the horizontal load and the centre of stiffness of the core arrangement. However, the core locations are not always ideal, because they may be irregularly shaped, located at one end of the building or too small. In these circumstances additional braced bays or rigidly jointed frames will be required at other locations.

6.2.3 Structural depth

The maintenance of overall structural floor depth within a building is an important objective in simplifying the coordination of the mechanical and electrical services and the building finishes. For the range of structural grids used in conventional buildings traditional steel floor construction is generally deeper than the equivalent reinforced concrete flat slab: the difference is generally 100–200 mm for floor structures which utilize composite action and greater for non-composite floors. The increased depth is only at the beam position: elsewhere, between beams, the depth is much less and the space between them is usable for services, particularly if the beams may be penetrated. The greater depth of steel construction does not necessarily result in an increase in building height if the services are integrated within the building zone occupied by the structure. However, integrated systems requiring numerous and irregular penetrations through the beams will increase the cost of fabrication. Figure 6.6 shows possible solutions for integrated systems which overcome this difficulty.

Depth may be reduced, however, by utilizing rigid or semi-rigid rather than simple connections at the end of the beams: this allows the beam-bending moments to be more uniformly distributed and the beam deflection to be reduced.

The depth may also be reduced by using higher-strength steel, but this is only of advantage where the element design is controlled by strength. The stiffness characteristics of both steels are the same, hence where deflection or vibration govern no advantage will be gained by using the stronger steel.

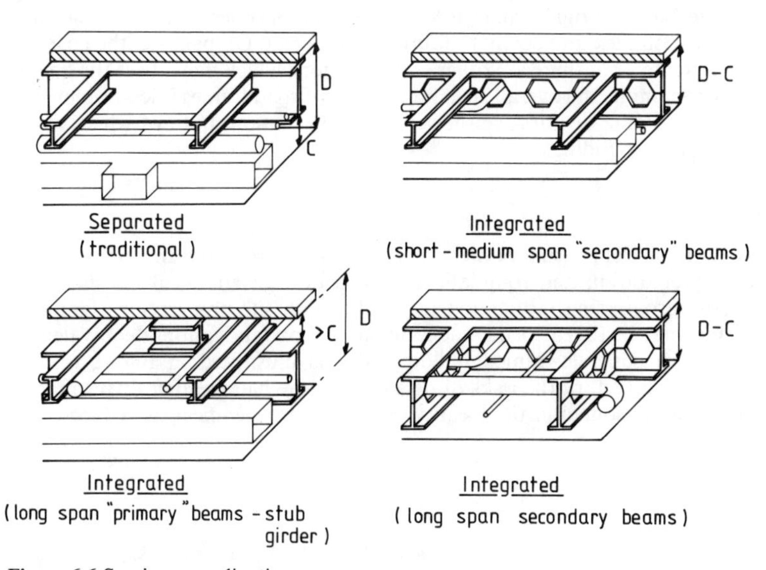

Figure 6.6 Services coordination

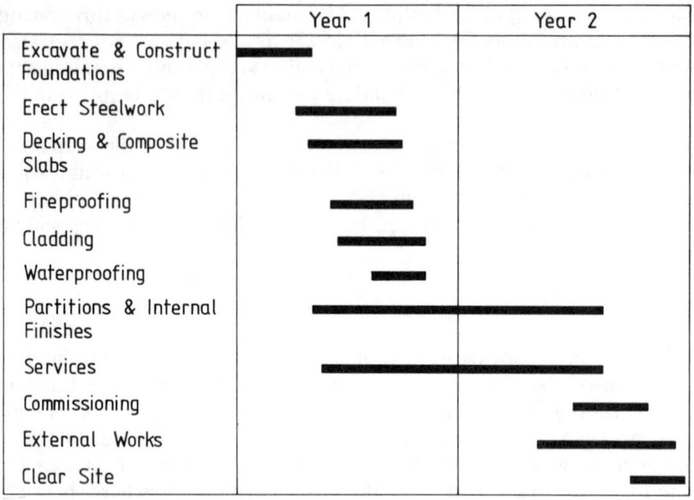

	Year 1	Year 2
Excavate & Construct Foundations		
Erect Steelwork		
Decking & Composite Slabs		
Fireproofing		
Cladding		
Waterproofing		
Partitions & Internal Finishes		
Services		
Commissioning		
External Works		
Clear Site		

Figure 6.7 Typical construction programme

6.3 Practical fabrication and construction

6.3.1 Fabrication

Because steel is prefabricated from components in a factory, repetition of dimensions, shapes and details streamlines the industrial process and is a major factor in economic design. The selected structural form and connection detailing of the building also have a significant effect upon the cost and speed of fabrication and erection. Simple braced frameworks with bolted connections are considered the most economic and the fastest to build for low- to medium-rise buildings. Site-welded connections require special access, weather protection, inspection and temporary erection supports. By comparison, on-site bolted connections enable the components to be erected rapidly and simply into the frame and require no further handling.

The total weight of steel used in rigid frames is less than in simple ones, but the connections for rigid frames are more complex and costly to fabricate and erect. On balance, the cost of a rigid frame structure is greater, but there are other considerations which may offset the greater cost of rigid frames. For example, the overall structural depth of rigid frames is less, which may reduce the height of the building or improve the services' distribution, both of which would reduce the overall cost of the buildings.

6.3.2 Construction

A period of around 12–16 weeks is usual between placing a steel order and the arrival of the first steel on site. Site preparation and foundation construction generally take a similar or longer period. Hence by progressing fabrication in parallel with site preparation significant on-site construction time may be saved, as commencement of shop fabrication is equivalent to start on site for an *in situ* concrete-framed building; this is illustrated in Figure 6.7. By manufacturing the frame in a factory the risks of delay caused by bad weather or insufficient or inadequate construction personnel in the locality of the site are significantly reduced.

Background reading

HART, F., HENN, W. and SONTAG H. (1985) *Multistorey Buildings in Steel*, Crosby Lockwood Staples, London
MCGUIRE, W. (1968) *Steel Structures*, Prentice-Hall, Hemel Hempstead

7

Learning from failures

Objective To report the lessons from past failures that may improve the safety of future design and construction.

Summary After a brief introduction and a discussion of the complexities of contractual relationships this chapter examines three failures: King's Bridge, Melbourne; the Hyatt Regency Hotel, Kansas City; and the Cleddau Bridge, Milford Haven.

7.1 Introduction

Structural failures are rare. Very few are caused by errors in calculation or lack of engineering knowledge. On the few occasions when the design of the complete structure is faulty the errors can usually be attributed to some breakdown in the orderly environment that is necessary for sound design. This might take the form of inadequate communication within the office, a failure to cross-check or excessive haste in an overloaded office. This is not to say that the designer can be absolved from responsibility in many of the failures which are recorded.

Having dismissed as unlikely the failure of completed structures, the obvious implication is that most failures occur during construction. Full details of such failures are rarely made public. In the UK, only if the accident is seen to be of national importance is a Royal Commission appointed, with the power to gather all the evidence and declare the reason for failure, and even then it is necessary for a court of law to apportion blame. During construction at least three parties are concerned with the successful outcome of the job, and each usually measures success in commercial terms, though the basis of that commercial judgement is different in each case. It is difficult for anyone to learn from a failure if the facts are concealed. The lessons to be learned are, however, sufficiently important to make the effort worthwhile.

The most important single reason for structures falling down is lack of communication. As will be seen from the examples briefly detailed below, poor communication will manifest itself in a great number of different ways but the best guard against it is for all the persons involved in a job to know each other, to regard each other as friends as well as colleagues involved in a joint enterprise, and, most of all, to maintain sympathy for one another's views. The difficulty of achieving and maintaining these relationships in a complex contractual situation is discussed in the following section.

Of the many other reasons for failures only a very few recur sufficiently often to warrant specific discussion. Poor detailing may be caused by lack of understanding or omissions in checking. Actual numerical error in calculation rarely leads to failure. The inclination to minimize material use, or maximize stresses, however small the gain in terms of cost and however great the cost in terms of the required accuracy of analysis and/or increased risk of failure, may also be carried too far. One very clear danger lies in using designs which have proved successful at one scale as a basis for larger structures. The main problem here lies in omissions which were unimportant at the smaller scale becoming significant at the larger.

7.2 Contractual matters

Generally a job starts with a client who employs an architect to design the building and control all the other input. The architect will ask a consultant to design the structure, the latter frequently producing an outline of member sizes but no joint details. Tender documents will be sent out for the complete building and each main contractor will ask for subcontract prices for many items of work. Usually, the steel frame will be one of these. In bridgework an engineer will control the work but the main contract–subcontract relationship will still exist.

Once the contracts are let, the steelwork subcontractor will design the connections and begin fabrication, though sometimes even this task is subdivided, with the steelwork subcontractor on the main contract subletting the fabrication work and only doing the erection himself. All the people mentioned, with the addition of 'the engineer', who may or may not be the consultant/designer, are bound in a contractual relationship with one another. The contract is very important, but it is sometimes allowed to disrupt personal relationships between individuals. If a breakdown in the friendship of fellow-professionals can be seen in a job then it may be regarded as a clear indication of danger. It is not possible for everyone to carry out their job effectively if there is animosity at any level. To complicate the problem further, the work is carried out by a labour force which has a corporate identity but which is also a gathering of skilled groups made up of individuals. Safe and economical completion of a job depends upon all the members of the team. Mutual respect of skills and interests is needed, and if it is maintained the chances of failure are reduced to negligible proportions. The courage to question the work of others must be matched by a willingness to accept questions and help from others, but similarly the courage to resist pressure for undesirable change is always necessary.

7.3 King's Bridge, Melbourne

King's Bridge in Melbourne is one of the relatively few examples of failure in service. It was opened in 1961, but only 15 months later, on 10 July 1962 (Melbourne's winter), it failed by brittle fracture when a 45-ton vehicle was passing over it. Collapse was only prevented by walls which had been built to enclose the space under the affected span. Investigation showed that many other spans of the bridge were in danger of similar failure. A Royal Commission was appointed and produced a report of 98 pages plus appendices.

The foundations were in good order. The superstructure consisted of many spans in which each carriageway was supported by four steel plate girders spanning 30 m, topped with a reinforced concrete deck slab; Figure 7.1 shows a typical girder. Each plate girder bottom flange

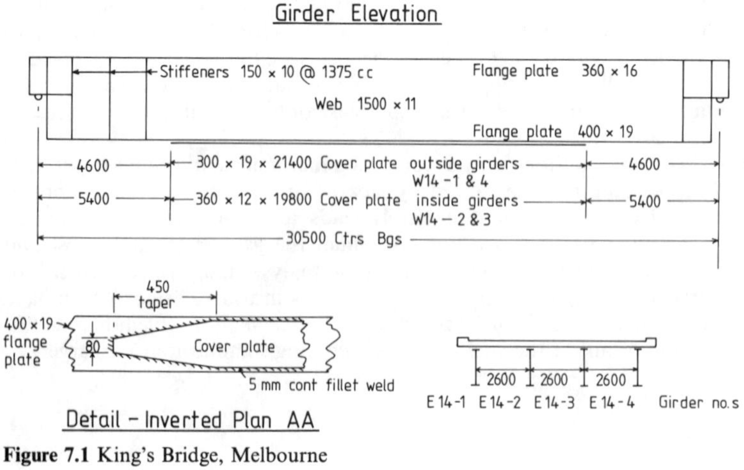

Figure 7.1 King's Bridge, Melbourne

consisted of a 400 × 19 mm plate, supplemented in the region of high bending moment by an additional cover plate which was either 300 × 19 mm or 360 × 12 mm. The cover plate was attached to the flange by a continuous 5 mm fillet weld all round.

The steel specified was to comply with **BS** 968: 1941, an earlier version of **BS** 4360 Grade 50. **BS** 968 at that time contained no requirement for low-temperature notch ductility, but the specification writer for the bridge did add some special requirements of this type. Despite these additional clauses, those who built and inspected the bridge did not understand that high-strength steel needed special care in welding when compared with mild steel (Grade 43, as it is now called). Difficulties were experienced with welding but an expert was not called in time.

The longitudinal welds connecting the cover plates were made before the short 80 mm transverse welds at the ends. Thus when the transverse welds came to be made there was complete restraint against contraction, so that cracking was more likely here than elsewhere. Moreover, a transverse crack in the flange plate would obviously weaken the girder more than a longitudinal one, yet it was never realized that the 80 mm transverse welds required special attention. Inspection was sometimes carried out before the transverse welds had been made.

The transverse welds were made in three passes. In some instances cracks were caused in the main flange plate by the first run and later covered up by a subsequent one. In many other cases a crack was caused by the last run and covered up with priming paint before the girders left the factory. Subsequent investigation showed that priming paint had sometimes spread through the cracks from the unwelded side and passed through the full thickness of the flange even before the girders left the factory. The penetration of later paint coats into the cracks showed that they had often extended further before the bridge was opened to traffic. Thus the most likely and most dangerous cracks were regularly missed by inspectors, who conscientiously had many less harmful longitudinal cracks cut out and repaired.

In the span which failed, cracks existed in the main flange plate under seven of the eight transverse fillet welds. One had extended partly by brittle fracture and partly by fatigue until the tension flange was completely severed, and the crack had extended half-way up the web. All seven cracks developed into complete flange failures on 10 July 1962, under a load that was well within the design one. In some instances the entire girder was severed. Because of the supporting walls underneath, there was no collapse and no loss of life.

Therefore the failure of King's Bridge was due to a poor detail which would not be reproduced now, compounded by poor communication leading to a lack of necessary inspection.

7.4 Cleddau Bridge, Milford Haven

A local failure in the Cleddau Bridge in Milford Haven (one of the sequence of box girder failures that resulted in a radical re-examination of design of thin-plate structures in the 1970s) led to a more general one. The member concerned here was effectively a load-bearing diaphragm.

The bridge (Figure 7.2), which failed during cantilever erection on 2 June 1970, was designed

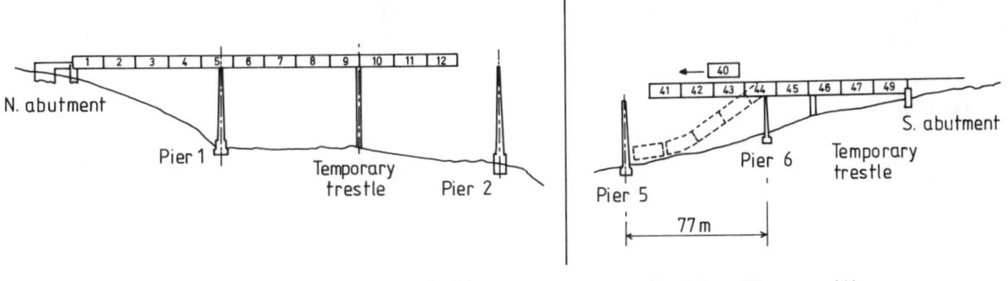

Figure 7.2 Failure of Milford Haven Bridge

as a single continuous box girder of welded steel, although it was rebuilt as a cantilever and suspended span in the main span. The spans measured from the south are 77, 77, 77, 149, 213, 149 and 77 m. The span that collapsed was the second 77 m one on the south side, the first having been erected with the aid of a temporary support. The last section of box for the second span was being moved out along the cantilever, and when the collapse came this section slid forward down the cantilever, killing four men.

No public enquiry was held, so the published facts are less complete than they might otherwise have been. However, from the reports that were made it is clear that failure was initiated by buckling of the support diaphragm at the root of the cantilever being erected (see Figure 7.3). The diaphragm was torn away from the sloping webs near the bottom, allowing buckling of the lower web and bottom flange to take place. The tendency of the bottom flange to buckle was inevitably increased by the reduction of the distance between flanges, as this increased the force needed in each flange to carry the moment with a reduced lever arm.

The support diaphragm was, in effect, a transverse plate girder, which carried heavy loads from the webs at its extreme ends and was supported by the bearings as shown, some distance from its ends. It was therefore subjected to a hogging bending moment and a large vertical shear force. The diaphragm plate near the outer bottom corners was subject to an extremely complex combination of actions. The shear of the transverse girder and diffusion of the point load from the bearings was compounded with the effects of inclination of the webs of the main bridge girder, which produced an additional horizontal compression action and out-of-plane bending effects caused by bearing eccentricity.

The total load transmitted by the diaphragm to the bearings just before collapse was 9700 kN. Allowing for likely values of distortion and residual stress, the calculated design strength, using design rules that were drafted subsequently, would be considerably less, possibly as low as 5000 kN.

Adding material to this diaphragm would carry no secondary penalties of increased load in the main structure. Indeed there is at least one box girder bridge in the UK which has reinforced concrete diaphragms.

Actions for damages were settled out of court in the summer of 1978, when the various parties agreed to share costs.

7.5 Hyatt Regency Hotel, Kansas City

One last example must be cited. On 7 July 1981 a dance was being held in the lobby of the Hyatt Regency Hotel, Kansas City. As spectators gathered on suspended walkways above the dance floor the supports gave way and two levels of bridge fell to the crowded dance floor. One

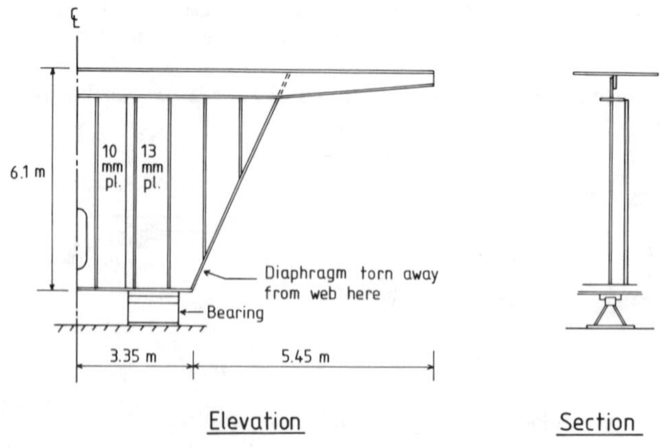

Figure 7.3 Diaphragm over Pier 6 of Milford Haven Bridge

hundred and eleven people died and nearly 200 were seriously injured. Failure occurred at a simple but critical detail.

As shown in Figure 7.4, the walkways that crossed the lobby at the second- and fourth-floor levels were supported above one another by hanger rods from the fifth floor. Floor-to-floor height was 5 m and the walkways hung from three sets of hangers at 9 m centres. In the original design single 15 m long rods supported the two walkways (see Figure 7.5). At each level a cross-beam made from two channels welded toe to toe rested on a nut and washer on the rod. This detail would not have failed under the loading imposed even though its capacity was only one quarter of that required by local design codes.

In the furore which followed the collapse it became obvious that the design had been changed to reduce the cost of the connection. The second-floor walkway was actually suspended from the fourth-floor one (see Figure 7.6). As a result, the connection between the fourth-floor cross-beam and the hanger supported double the load originally intended – or, rather, failed to do so. The alteration seems to have been recommended by an engineer, not party to the original design, who specialized in reducing costs. Unfortunately, in this instance he failed to understand the importance of the details he changed: nor was the effect of the changes noticed by any of the other parties involved.

Once again, a gross underdesign of a detail would not have caused failure had not another factor resulted in a significantly increased load. Here, and in most failures, communication is all-important. Between conception and completion many hands and minds are involved in the production of a working structure.

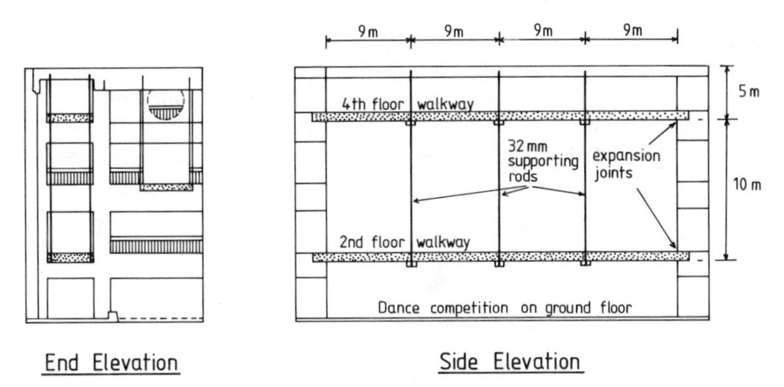

Figure 7.4 Kansas City Hyatt Hotel: arrangement of walkways

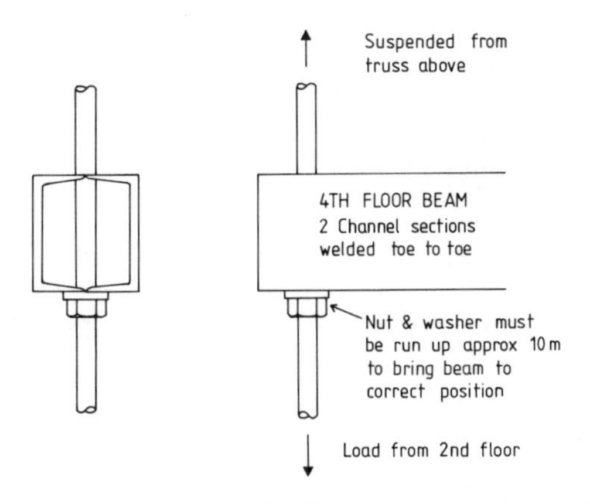

Figure 7.5 Kansas City Hyatt Hotel: original detail

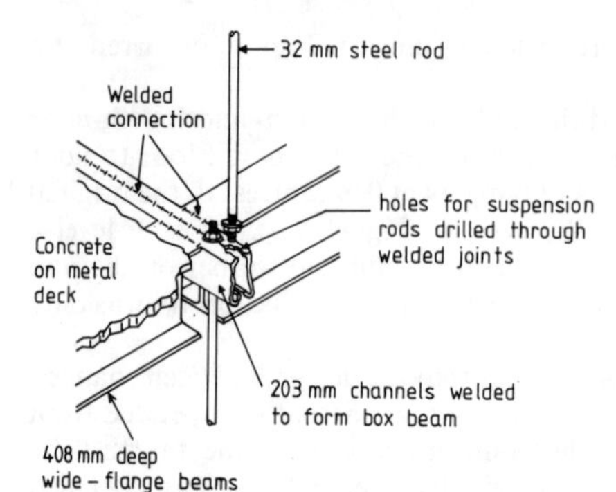

Figure 7.6 Kansas City Hyatt Hotel: failed box beams supported wide-flange beams along edges of suspended walkways

7.6 Concluding summary

1. Society rightly demands a very high standard of safety from civil engineering structures. When a structure fails it may claim many lives, and its reinstatement will require considerable resources.
2. Structures rarely fail from a single cause; there are usually several contributory factors.
3. The structures described failed from one or more of the following causes:
   (a) Poor communications sometimes arising from the failure to perceive the need to communicate;
   (b) Poor details or errors in detailing;
   (c) Material-related problems;
   (d) Inadequate temporary works, or lack of thought about a temporary condition;
   (e) Escalation of scale without pause for thought;
   (f) Design error or lack of understanding of structural behaviour.

8

Characteristics of iron–carbon alloys

Objective To introduce selected, important aspects of the metallurgy of steels.

Summary The chapter commences with a discussion of the crystalline nature of irons and steels and the influence of grain size and composition on properties. It describes iron's ability to have more than one crystalline structure (its allotropy) and the properties of the principal forms of alloys of iron and carbon. The metallurgy and properties of slowly cooled steels are reviewed, including accounts of the influence of grain size, rolling, subsequent heat treatment and inclusion shapes and distribution. Rapidly cooled steels are treated separately; an outline of quenching and tempering is followed by comments on the influence of welding on the local thermal history. Hardenability, weldability and control of cracking are briefly treated. Finally, there is a discussion of the importance of manganese as an alloying element.

8.1 Introduction to the structures of solid steels

If we take a piece of steel bar, cut it to expose a longitudinal section, grind and polish the exposed surfaces and examine them under a microscope, at modest magnifications a few particles will be seen which are extended in the direction of rolling of the bar (Figure 8.1). These

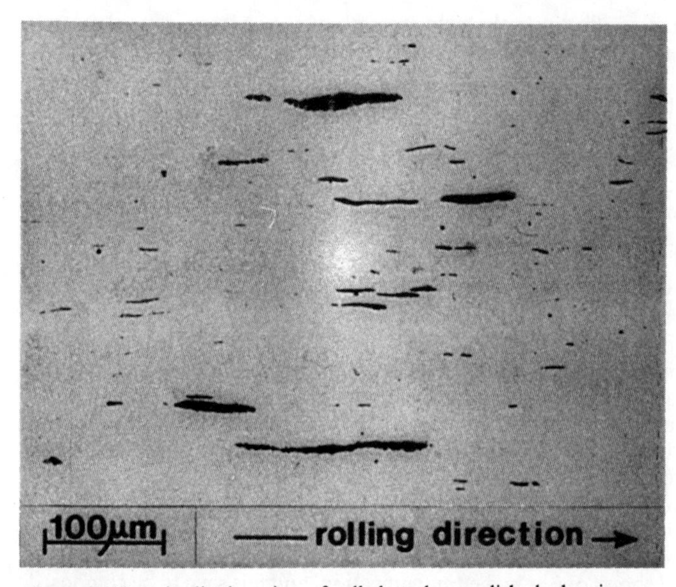

Figure 8.1 Longitudinal section of rolled steel, as polished, showing inclusions

are inclusions, i.e. non-metallic substances which have become entrained within the metal during its manufacture, mostly by accident but sometimes by design. Their presence usually affects toughness adversely, but particular types of inclusion can greatly enhance the machinability of mild steels.

To reveal the true structure of the metal itself the polished surface must be chemically removed by etching. When this is done, a wide diversity of microstructures may be seen, depending on the composition of the steel and on the history of its treatment by heating, deforming, cooling at different rates, etc. These differences in structure cause the properties of steel to range between widely separated limits.

8.2 The components of steel

Steels and cast-irons are alloys of iron with carbon and various other elements, some of them being unavoidable impurities while others are added deliberately. The carbon exerts the most significant effects on the microstructures of the materials and their properties. Steels usually contain less than 1% carbon by weight. Structural steels contain less than 0.25% by weight carbon: the other principal alloying element is manganese, which is added in amounts up to about 1.5%. Cast-irons generally contain about 4% carbon. This very high carbon content differentiates their metallurgy from that of steels.

Each of the microstructures is an assembly of smaller units. For example, most of 0.1%C steel is composed of small, polyhedral grains, in this case $< 20\,\mu m$ in size. Closer examination of one of these grains shows it to be a crystal. However, unlike crystals of quartz or silicon or copper sulphate, it is soft and ductile. We will return to the internal structure of these crystals later.

Steel is an example of a polycrystalline substance. By altering the history of rolling and heating treatments experienced by the steel during its production the grain size can be altered. This is a useful trick. The grain size of mild steel affects its properties – a relationship which it shares with all other metals. In particular, the yield strength is defined by the grain size, according to the so-called Petch equation:

$$\sigma_y = \sigma_0 + kd^{-1/2}$$

where σ_y is the yield strength, σ_0 is effectively the yield strength of a very large isolated crystal: for mild steel this is $\approx 5\,N/mm^2$, d is the grain-size (mm), k is a material constant, which for mild steel is about $38\,N\,mm^{-3/2}$.

Let us now take one of the crystal grains and examine it internally. We find that it is composed of iron atoms arranged according to a regular pattern, illustrated in Figure 8.2. This is the body-centered-cubic (bcc) crystal structure; atoms are found at the corners of the cube and at its centre. Note that the unit cell is only 0.28 nm along its edges. The whole crystal is composed of about 10^{16} repetitions of this unit. This crystal structure of iron at ambient temperature is one of the major factors determining the metallurgy and properties of steels.

Steels contain carbon and some of it (a very small amount) is found within the crystals of iron. The carbon atoms are very small and can fit, with some distortion, into the larger gaps between the iron atoms. This forms what is known as an interstitial solid solution: the carbon is in some of the interstices of the iron crystal.

What, however, of the remaining carbon? Most of it forms a chemical compound with the iron, Fe_3C (iron carbide). This is also crystalline but is hard and brittle. With 0.1%C, there is only a little Fe_3C in steel. The properties of this material are qualitatively similar to those of pure iron. It is ductile, tough but not particularly strong, and it is used for many purposes where its ability to be shaped by bending or folding is its prime attribute.

Suppose we now examine a steel of higher carbon content, say 0.4%. At low magnification we see it to be composed of light and dark regions – about 50:50 in this case. The light regions are iron crystals containing very little dissolved carbon, as in mild steel. The dark regions need closer examination. Figure 8.3 shows one at higher magnification, and it is seen to be composed of alternating laminae of two substances, iron and Fe_3C. The spacing of the laminae is often

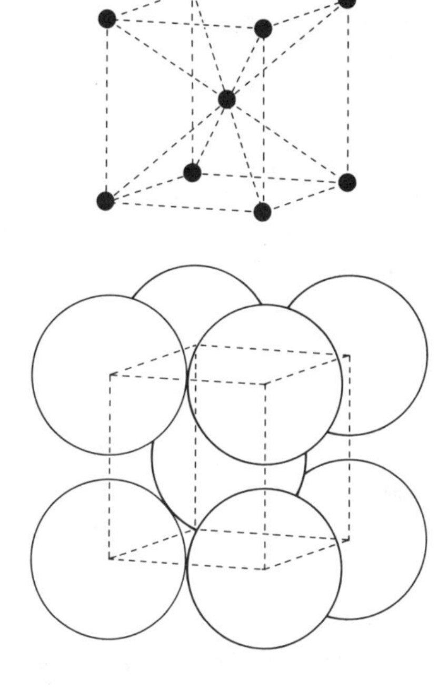

Figure 8.2 Body-centred-cubic crystal: unit cell

close to the wavelength of light. The mixture acts as a diffraction grating, giving optical effects which reminded our Victorian predecessors of mother of pearl. Consequently, this mixture of iron and iron carbide has acquired the name 'pearlite'. How did it get like it? Where did it come from? What are its effects on properties? To answer those questions we need to consider another series of problems.

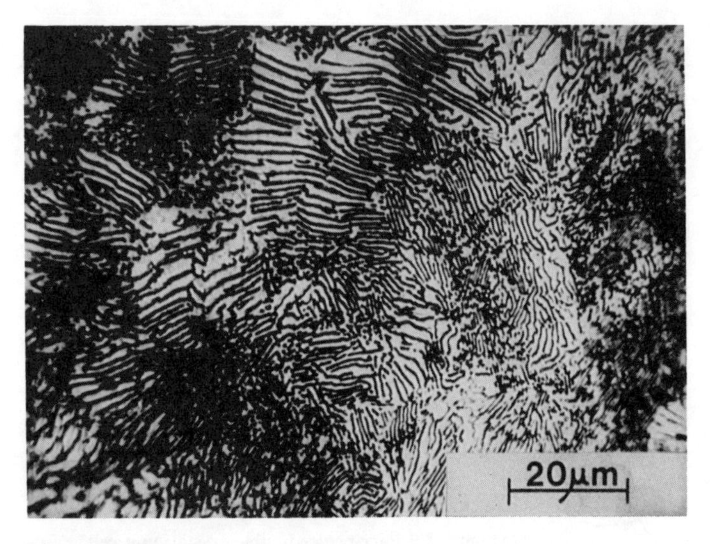

Figure 8.3 Microstructure of pearlite

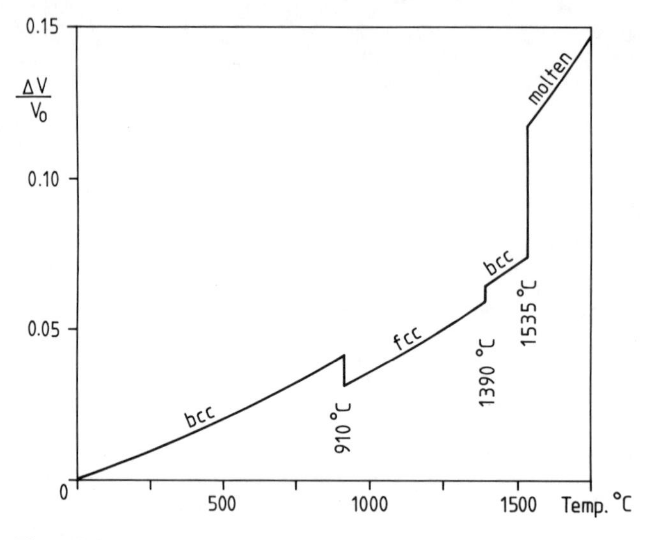

Figure 8.4 Volume changes on heating pure iron

8.3 Allotropy of iron

Let us take some pure iron and observe its volume change on heating. The results are shown in Figure 8.4. It expands in the normal way until a temperature of 910°C is reached. At this temperature there is a step contraction of about $\frac{1}{2}$% in volume. Further heating gives further

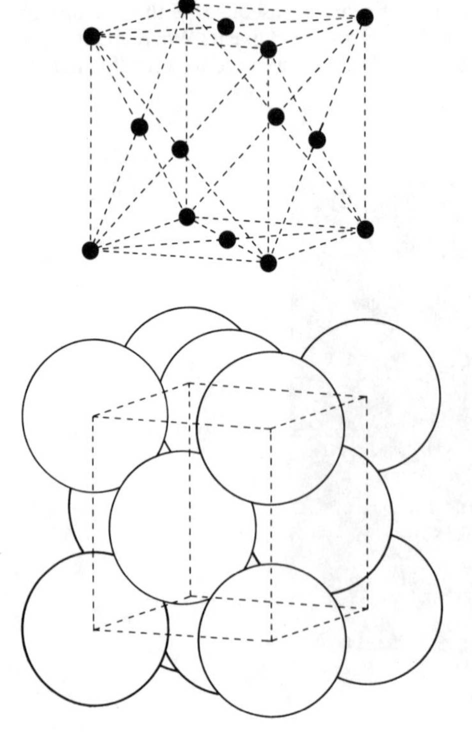

Figure 8.5 Face-centred-cubic crystal: unit cell

thermal expansion until, at about 1400°C, there is a step expansion which restores the volume lost at 910°C. Heating beyond 1400°C gives thermal expansion until melting occurs at 1540°C. The curve is reversible on cooling slowly.

Examination of the structure of the iron during heating shows it to be composed of bcc crystals in the temperature ranges below 910°C and from 1400°C to melting. Between 910°C and 1400°C its crystal structure is different. It is face-centred-cubic (fcc), illustrated in Figure 8.5. Here, the atoms of iron are on cube corners and at the centres of each face of the cube. The body-centred position is empty.

A given number of atoms occupy slightly less volume when arranged as fcc crystals than when arranged as bcc. This is what is responsible for the discrete volume changes at 910°C and 1400°C. Just as carbon atoms can fit between the iron atoms in bcc iron so also can they form interstitial solid solutions in fcc iron. In fact, it is easier to contain carbon in fcc iron. There may be a little less space between the iron atoms, but this smaller space is arranged more conveniently.

The changes of crystal structure described above are influenced by the carbon and other alloying elements in the steel. Their occurrence makes possible a wide range of heat treatments, which give rise to many properties, some desirable, others undesirable. To see what can be done we need to examine the iron–carbon constitutional diagram.

8.3.1 Note on nomenclature

Before looking at the iron–carbon constitutional diagram the following nomenclature of the metallurgist is introduced:

Ferrite or α-Fe	This is bcc iron in which up to 0.08%C by weight may be dissolved.
Cementite	This is Fe_3C (which contains about 6.67%C).
Pearlite	This is the laminar mixture of ferrite and cementite described earlier. The overall carbon content of the mixture is 0.78% by weight.
Austenite or γ-Fe	This is the fcc form of iron which exists at high temperatures and which can contain up to approximately 2%C by weight, although this is dependent on temperature.

Discerning readers may wonder what happened to β-Fe. This disappeared, along with other misconceptions of our predecessors, many years ago. The Curie temperature at which iron loses its ferromagnetism is 770°C. Iron between this temperature and 910°C was called β-iron. We now know that no structural change accompanies the loss of magnetism and so the need for β-iron no longer exists! Incidentally, austenite is not ferromagnetic at any temperature.

8.3.2 The iron–carbon constitutional diagram

This is essentially a map drawn in a space defined by two axes (Figure 8.6). The horizontal axis represents the carbon content of the steel and the vertical axis the temperature. Thus any point in the field of the diagram represents a particular carbon content of alloy at a particular temperature. The diagram is divided into areas showing the structures that are stable at particular compositions and temperatures.

Let us use the diagram to consider what happens when a steel of 0.5%C is cooled from 1000°C. At 1000°C the structure is austenite (i.e. polycrystalline), the crystals being fcc with all the carbon dissolved in them. No change occurs on cooling until the temperature reaches about 800°C. At this temperature a boundary is crossed from the field labelled austenite to that labelled $\alpha + \gamma$, or ferrite + austenite. This means that some crystals of bcc Fe, containing very little carbon, begin to form from the fcc Fe. Because the ferrite contains so little carbon the carbon left must concentrate the austenite. In other words, the ferrite regions contain less than 0.5% but the austenite ones have more than this. Remember that we started with a steel

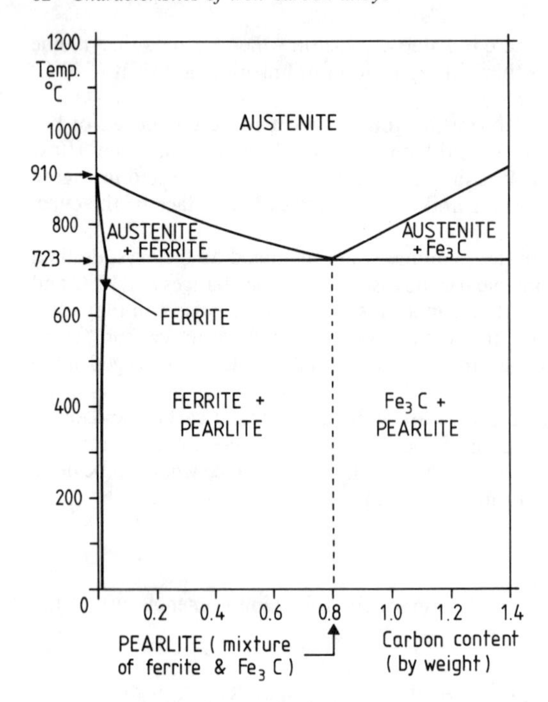

Figure 8.6 Part of the iron–carbon constitutional diagram

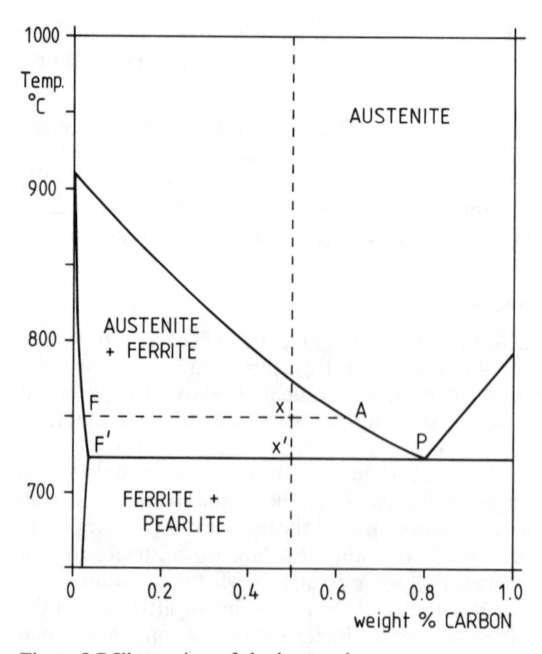

Figure 8.7 Illustration of the lever rule

containing 0.5%C! The carbon contents of the ferrite and austenite and the relative proportions of these two substances in the microstructure are adjusted to maintain the original carbon content overall.

How can we work out these quantities? Consider the expanded part of the diagram shown in Figure 8.7, and imagine that our steel has cooled to 750°C. The combination of overall carbon content and temperature is represented by point X. All the parts of the microstructure are at the same temperature, and so we draw a line of constant temperature through X. This cuts the boundaries of the $\alpha + \gamma$ field at F and A. These intercepts give the carbon contents of ferrite and austenite, respectively.

If now we think of the line FA as a rigid beam which can rotate about a fulcrum at X the weight of austenite hanging at A must balance the weight of ferrite hanging at F. This is the so-called 'lever rule':

wt ferrite × FX = wt austenite × AX

Thus we can see that, as the steel cools, the proportion of ferrite increases as does the carbon content of the remaining austenite until we arrive at 723°C. At this temperature the carbon content of the austenite is 0.78% and it can take no more. Cooling to just below this temperature causes the austenite to decompose into the mixture of ferrite and Fe_3C, which we identified earlier as pearlite. This forms in the regions occupied by the austenite just before it decomposed, i.e. in those not previously occupied by ferrite.

The proportions of ferrite and pearlite in the microstructure, say at 722°C, are virtually the same as those of ferrite and austenite immediately before the decomposition at 723°C. Thus referring to Figure 8.7 and using the lever rule:

wt ferrite × F′X′ = wt pearlite × F′P

In this case there should be about twice as much pearlite as ferrite.

For other steels containing < 78%C the story is identical except for the proportions of pearlite in the microstructure below 723°C. This varies approximately linearly with carbon content between zero at 0.08%C and 100% at 0.78%C.

What of steels containing a greater percentage of carbon, i.e. > 0.78%C? On cooling from high temperatures, where the structure is fully austenitic, the first change to occur is the formation of particles of Fe_3C from the austenite. On further cooling, the carbon content of the austenite follows the line of the boundary between the γ field and $\gamma + Fe_3C$ field. Once again, on reaching 723°C the carbon content of the austenite is 0.78%. On cooling further, it decomposes into pearlite, as before. Therefore the final microstructure consists of a few particles of Fe_3C embedded in a mass of pearlite.

8.4 Slowly cooled steels

As we have seen, the microstructures of slowly cooled steels consist of mixtures of ferrite and pearlite or of cementite and pearlite. Note that, although pearlite is itself a mixture, it is convenient in this context to think of it as a substance in its own right. Ferrite is soft, ductile and not particularly strong. As shown on page 58, its yield strength depends on its grain size. Pearlite is hard, brittle, but strong. Its properties depend on the spacing of the laminae which make it up. The finer the spacing, the stronger it is.

The properties of a steel containing both ferrite and pearlite are roughly the average according to the proportions of these constituents in the microstructure, as seen in Figure 8.8. The effect of carbon content on toughness is shown in Figure 8.9. Increasing pearlite content decreases the toughness at high temperatures and increases the temperature of the ductile–brittle transition to which steels are susceptible. This transformation in properties just happens: there is no structural change to cause it.

Figures 8.8 and 8.9 illustrate one of the eternal difficulties in the selection and design of materials. We see that increasing the carbon content improves yield strength and ultimate tensile strength but reduces ductility and toughness. The choice of a particular steel for a given application is a matter of compromise.

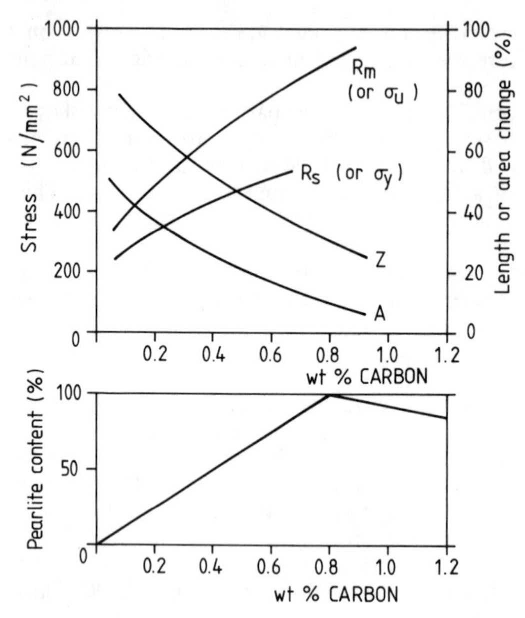

Figure 8.8 The effects of carbon content on the pearlite content and tensile test properties of normalized steels

The lowest carbon content that can be achieved easily on a large scale is about 0.04%. This is characteristic of sheet or strip steels intended to be shaped by extensive cold deformation, as in deep drawing.

Structural steels rarely contain more than about 0.25%C. Figure 8.10 shows the range of carbon contents and strengths specified in BS 4360, 'Weldable Structural Steels'. Note that small changes in carbon content give significant changes in properties. Why not increase the carbon content further? The answer to that question will be dealt with later, but we note that carbon contents of more than 0.25% bring problems when the steel is welded.

Carbon contents of more than 0.25% are used in the wider range of general engineering steels. These are usually put into service in the quenched and tempered state (see later) for a great multiplicity of purposes in mechanical engineering. High-strength bolts for some structural applications would also be steels of this type.

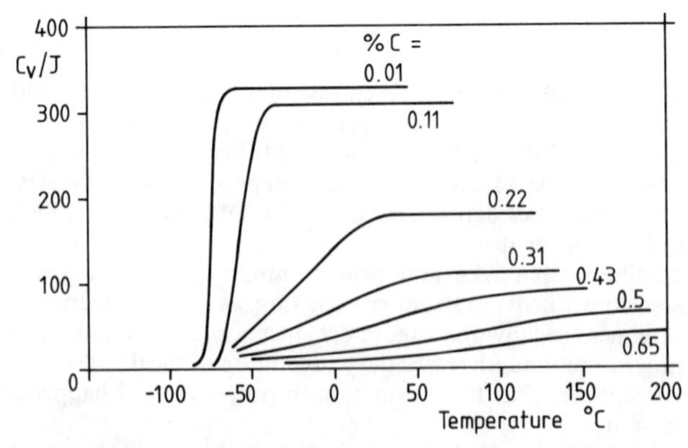

Figure 8.9 Effect of carbon content on the ductile–brittle transition of normalized steels. (After Rinebolt, J.A. and Harris, W.J. (1951) *Transactions American Society for Metals*, **43**, 1175–1214)

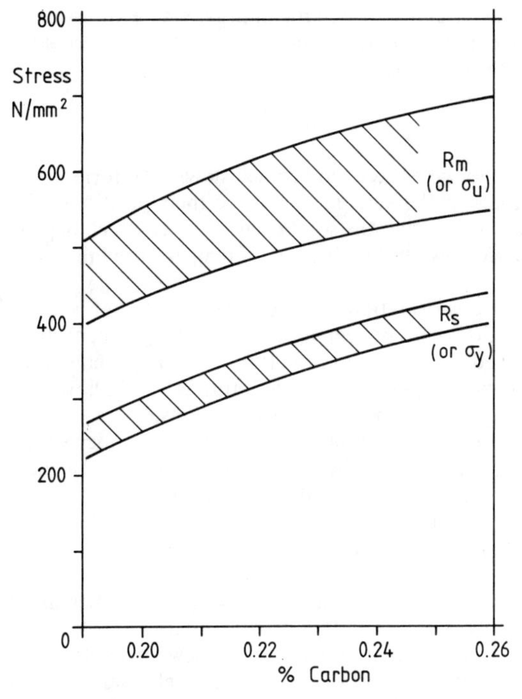

Figure 8.10 Strengths and carbon contents of normalized steels specified in BS 4360

8.4.1 The need for control of grain size

We have seen earlier that the yield strengths of steels are affected by their grain sizes. Other important properties are also improved by a reduction in grain size, i.e. yield strength and ductility. It also has a profound effect on the temperature range of the ductile–brittle transition which is experienced by steels (see Figure 8.11). Thus we have several benefits from the same

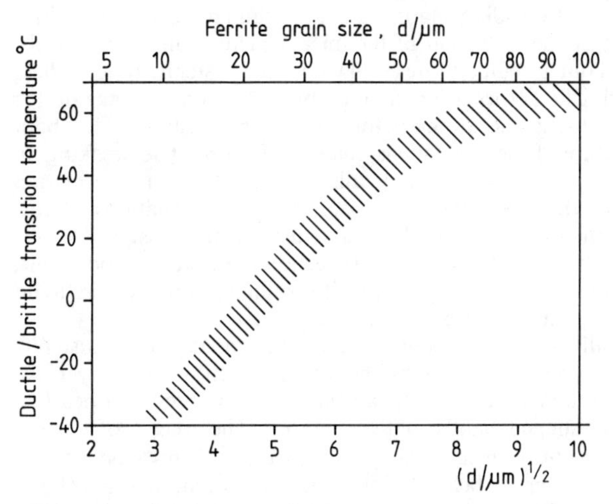

Figure 8.11 Effect of ferrite grain size in the ductile–brittle transition temperature of mild steel. (After Petch, 1959)

adjustment to our material. This is an unusual circumstance in metallurgy, where adjustments to improve one property often mean a worsening of another, and a compromise is necessary. We have already discussed this point in relation to Figure 8.8.

8.4.2 Grain-size control by normalizing

We have seen the transformations that can occur when steels are cooled slowly. To form ferrite and pearlite from austenite the carbon atoms in the steel must change their positions. The diffusion processes which shuffle the atoms about within the solid occur at rates which depend exponentially on temperature. We will now see how altering the rate of temperature fall affects matters.

If we increase the cooling rate a little, the transformations must occur faster. What is more, a hysteresis can arise in which the transformations cannot keep up with the falling temperature. Thus a steel cooled very slowly in a controlled furnace will keep close to the requirements of the constitutional diagram. However, the same steel, removed from a furnace and allowed to cool in air, may undercool before completing its sequence of transformations.

This has two effects. It tends to increase slightly the proportion of ferrite in the microstructure and produces ferrite with a finer grain size and pearlite with finer lamellae. Both these microstructural changes give higher yield strength and better ductility and toughness.

Furnace-cooled steels are known as fully annealed, and air-cooled ones are known as normalized, presumably because cooling in air is the normal thing to do!

Grain size can be affected further by the temperature to which the steel is heated in the austenite range. The grains of austenite coarsen with time, the rate of coarsening increasing exponentially with temperature. The importance of this is that the transformation to ferrite and pearlite on cooling starts at the grain boundaries in the austenite. If the new structures start growing from points which are further apart, the grain size of the result is itself coarser. Thus steels should not be overheated when austenitizing before normalizing.

The temperature to which the steel is heated before cooling in air is usually referred to as the normalizing temperature. The requirements of the last paragraph mean that this should be as low as possible, as long as the structure is single-phase austenite. A glance at the constitutional diagram of Figure 8.6 will show that the normalizing temperature should decrease as the carbon content increases from zero to 0.8%.

8.4.3 Microstructural changes accompanying rolling of steels

Structural steel sections are produced by rolling ingots into the required forms, and these processes have important effects on the development of the microstructures in the materials. The early stages of rolling are carried out at temperatures well within the austenite range, where the steel is soft and easily deformed. The deformation suffered by the material breaks up the structure but, at these high temperatures, the atoms within the material can shuffle about rapidly and the polycrystalline structure of the austenite is quickly reformed. The breaking up and reforming processes keep step with one another, but details of the amounts of deformation and rolling temperatures can affect the austenite grain size. Heavy deformations at low temperatures give finer grains. If the rolling is finished at a temperature just above the ferrite + austenite region of the equilibrium diagram and the section is allowed to cool in air, an ordinary normalized microstructure results. Modern controlled rolling techniques aim to do this, or even to roll at still lower temperatures to give still finer grains.

What if the temperature should fall so that the rolling is finished in the ferrite + austenite range? The ferrite and austenite are each rolled out along the rolling direction and a layer-like structure develops. If, now, the section is air cooled the austenite decomposes into pearlite, which is now present as long, cigar-shaped bands in the material. The weight of current experience is that structural steels are not harmed severely by structures of this kind.

We now let the temperature drop further, to below 723°C. The equilibrium diagram shows us that the structure is now a mixture of ferrite and pearlite. Rolling in this range is usually restricted to low-carbon steels and is mainly practised for mild steels containing less than $< 0.15\%$C. The presence of pearlite makes rolling difficult!

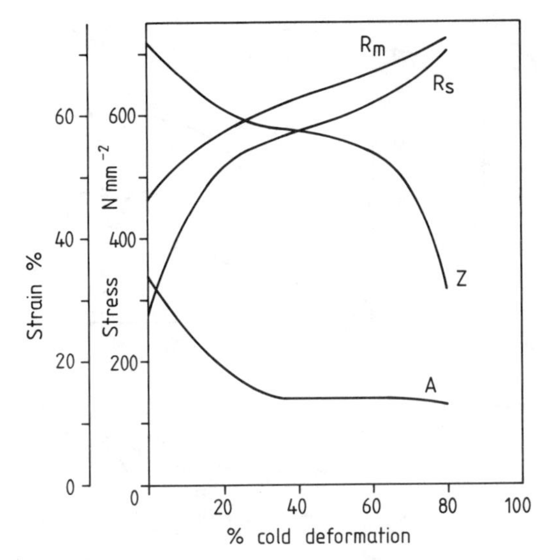

Figure 8.12 Effects of cold work on tensile properties of steel (0.13–0.18%C, 0.6–0.9% Mn). (Data from *Metals Handbook*, Vol. 1, 9th edn (1978) American Society for Metals, p. 225, Figure 5)

If the temperature is above about 650°C the ferrite grains reform as they are deformed, as was the case with austenite. The carbide laths in the pearlite are broken and give rise to strings of small carbide particles extending in the direction of rolling. The ferrite from the pearlite becomes indistinguishable from the rest of the ferrite.

If, say, the rolling was done at atmospheric temperature the pearlite is still broken up in the same way, but the ferrite cannot now reform its grains: it work-hardens. The yield strength and the ultimate tensile strength increase, but the ductility decreases (see Figure 8.12). As the rolling continues, the forces required to continue deformation increase. Furthermore, the steel itself is now less ductile and could begin to split. There is therefore a natural limit to the amount of rolling that can be done.

Of course, the deformation need not be applied by rolling: any method of deforming the material will do. For example, high-strength steel wire is made by cold-drawing, imparting large deformations. In another example, one type of reinforcing bar is made by twisting square section bar into a helical form. The cold-deformation produced by doing this is not large, but it gives some strengthening.

To restore the ductility and, at the same time, reduce the strength of the material it is necessary to reform the isotropic, polycrystalline structure of the ferrite. As-rolled, the original crystals are in a badly mangled state! Reheating to temperatures between about 650°C and 723°C allows the ferrite to recrystallize and the carbide particles are unaffected by this treatment. Here we have yet another chance to control the grain size of our steel. The greater the amount of deformation before the recrystallization treatment and the lower the temperature of the treatment, the finer will be the final grain size.

Because this type of treatment does not involve the formation and decomposition of austenite it is known as subcritical annealing. The resulting microstructures give good ductilities and deep drawing characteristics. Sheet steels of low carbon contents (< 0.1%C) are usually made available in this condition. Objects such as motor car body panels are formed from such steels by pressing. Of course, should we heat the material into the austenite range, subsequent cooling takes us back to normalized microstructures.

8.4.4 Inclusions in steels

We began with a microstructure showing some inclusions in steel, and we now look at these inclusions rather more closely.

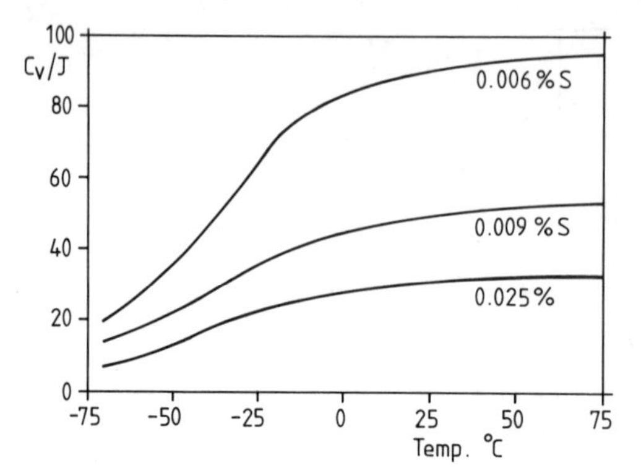

Figure 8.13 Effect of sulphur content on toughness of high-strength low-alloy (HSLA) steels (min R_s = 450 Nmm^{-2}). (Data from *Metals Handbook*, *op. cit.*, p. 694, Figure 9)

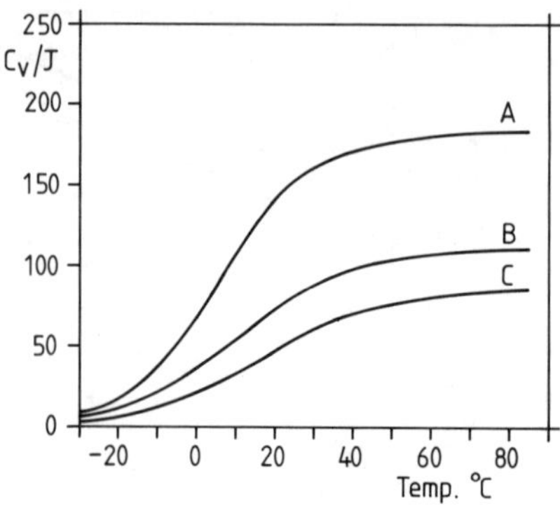

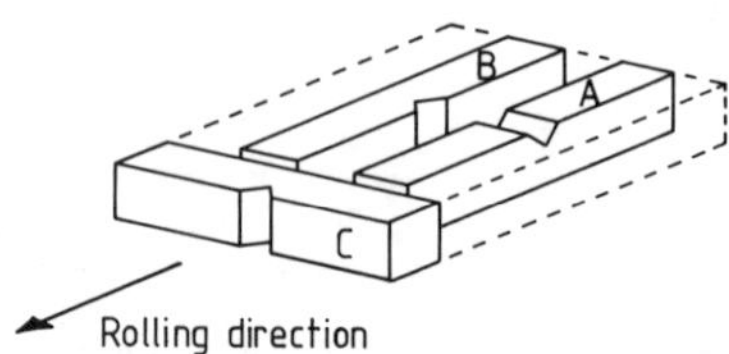

Figure 8.14 Effect of orientation on notch toughness of as-rolled low-carbon steel plate. (Data from *Metals Handbook*, *op. cit.*, p. 700, Figure 21)

One tonne of steel (about half a dustbin-full) contains between 10^{12} and 10^{15} inclusions which occupy up to about 1% of the volume. The total content is largely determined by the origins of the ores, coke and other materials used to extract the metal in the first place and by the details of steelmaking practice.

The principal impurities which worry steelmakers are phosphorus and sulphur. If not at very low concentrations, these form particles of phosphides and sulphides which are harmful to the toughness of the steel. Typically, less than 0.05% of each of these elements is demanded. Low phosphorus contents are relatively easily attained during the refining of the pig iron into steel, but sulphur is more difficult to remove. It is controlled by careful choice of raw materials and, in modern steelmaking, by extra processing steps which remove some of it.

Manganese is always added to steels. This has several functions, but the important one in this context is that it combines with the sulphur to form manganese sulphide (MnS). If the manganese were not present, iron sulphide would form which is much more harmful than MnS.

Some of the inclusions are too small to be seen with optical microscopes and must be detected by more elaborate methods. They are mainly equi-axed in shape. Among this group are oxides and nitrides of aluminium or other highly oxidizable metals which are used to deoxidize the molten steel before casting. Their effects are indirect, since their presence inhibits the processes which lead to coarsening of grain sizes.

Other inclusions, large enough to be seen readily with optical microscopes, include entrained particles of slag and manganese sulphide. At hot-rolling temperatures these inclusions are plastic and are elongated in the rolling direction in concert with the surrounding metal. The result is shown in Figure 8.1. The properties of steels containing such inclusions reflect both the volume of the inclusions in the microstructure and the anisotropy of their shapes (Figures 8.13 and 8.14).

The inclusions, especially the large ones, are unlikely to be uniformly distributed in the steel. In traditional steelmaking the way in which the original ingot solidified tended to drive the inclusions towards the centre line and the top of the ingot. Subsidiary bands of high inclusion content, parallel to the ingot axis, could also be formed in other places. Rolled products derived from that kind of ingot had concentrations of inclusions in bands parallel to the rolling direction and near the centre of the section. They could also be worse in this respect at one end than they were at the other. Bands of high visual inclusion content provided paths for easy crack propagation and could give rise to delamination failures.

In recent years a number of practices have been introduced which aim to reduce the inclusion content in the molten steel before it is cast into ingots. Sulphur contents of 0.01% or less can be produced. These processes produce what have become known as clean steels. The expression is relative: clean steels still contain many inclusions, but are significantly tougher than ordinary ones. Inclusion shape control is also practised in better-quality steels. Additions of calcium or cerium and other rare-earth elements to the refined molten steel cause these to combine with the sulphur in preference to the manganese. Sulphides of these elements appear in the final microstructures as equi-axed particles and are not so deleterious to the through thickness ductility of the material. Steels treated in these ways are used in applications where toughness is of paramount importance and where the extra cost can be justified. Examples include high-integrity pressure vessels, oil and gas pipelines and the main legs of offshore platforms. The introduction of continuous casting has also improved the quality of conventional structural steels.

8.5 Rapidly cooled steels

We have seen already that normalizing causes steels to undercool below the requirements of the constitutional diagram before austenite transforms into fine ferrite and pearlite. Still further increases in cooling rate give further undercooling and still finer microstructures.

If we go to very rapid cooling rates by, for example, quenching into cold water, the formation of ferrite and pearlite is suspended. The internal reshuffling of atoms needed to form those products cannot keep up with the rapid fall of temperature. Instead, we get new products, formed at lower temperatures. Very fast cooling gives martensite. Slightly less rapid cooling can give a product called bainite, although this depends on the composition of the steel.

Large civil engineering structures are not heat-treated by heating to, say, 900°C and quenching into water, but there is one important circumstance which can produce martensite in localized parts of the structure, and that is welding. The weld zone is raised to the melting temperature of the steel and the immediately adjacent solid metal is heated to temperatures well within the austenite range. When the heat source is removed the whole region cools at rates determined mainly by thermal conduction into the cold metal further away. These rates of cooling can sometimes be very large (exceeding $1000°C\,s^{-1}$).

We must therefore consider the properties of rapidly cooled steels. These depend very markedly on the carbon content of the steel and on the nature of the product – ferrite and pearlite, or bainite, or martensite. In this context, hardness measurements are a tolerable index of other properties, irrespective of composition or structure.

Figure 8.15 shows the hardness of martensite as a function of its carbon content. Data for other properties equivalent to hardnesses which bracket the range of structural steels are also given. Reheating martensite at temperatures up to about 600°C softens and toughens it, the extent of these changes increasing as the reheating temperature increases, as shown in Figure 8.16. This reheating is known as tempering. Tempering at 600°C produces an extremely tough material. What is more, its ductile–brittle transition is at a lower temperature than for the same steel in the normalized condition. Bainite has properties similar to those of tempered martensite.

The processes of quenching and tempering, when allied to changes of steel composition, can produce a bewilderingly wide range of properties. Steels heat-treated in this way are used for a great multiplicity of purposes which demand hardness, wear resistance, strength or toughness. Note that, once again, compromises must be struck between these desirable properties. Quenched and tempered plate is used for structural purposes in large storage tanks, hoppers, earth-moving equipment, etc.

Note also that martensite produced as a result of single-pass welding would be in its untempered condition. Furthermore, the formation of martensite from austenite is accompanied by an expansion of approximately 0.3% in volume. This, together with the uneven

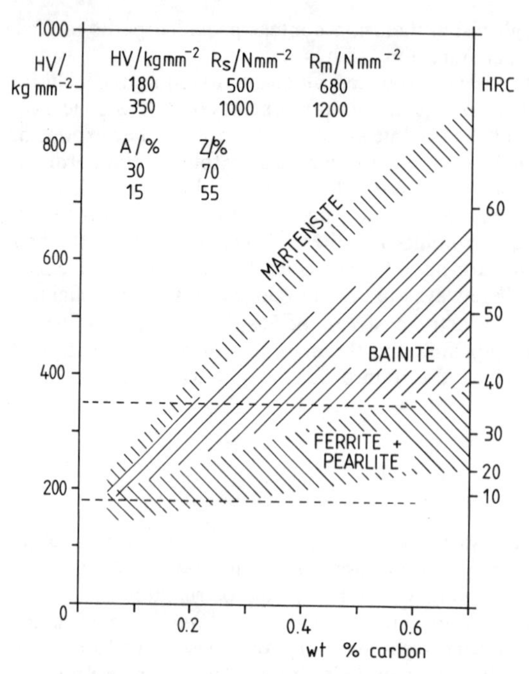

Figure 8.15 Effects of carbon content on the hardnesses of various steel microstructure

thermal contractions taking place as a result of uneven cooling, can produce local stresses of a sufficient magnitude to crack the martensite. Because of the normal processes for producing martensite in other circumstances these are known as quench cracks. The cracking problem can be further aggravated if the steel has picked up hydrogen in some way. Sources of hydrogen during welding might include moisture from the atmosphere or damp welding electrodes. Other sources include long-term exposure to environments which can pump hydrogen into the solid metal. One such source is sour gas emerging from oil- or natural-gas wells, i.e. gas containing hydrogen sulphide. Hydrogen in the metal can move about quite rapidly at atmospheric temperatures and has effects which are best thought of as an internal pressurization. This can

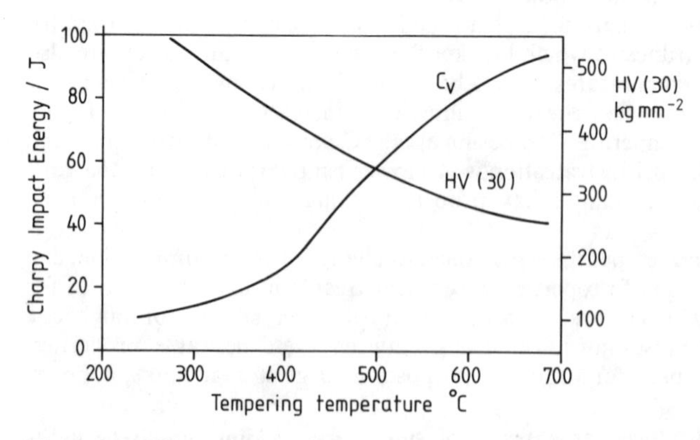

Figure 8.16 Effects of tempering on the hardness and toughness of martensite (0.4%C)

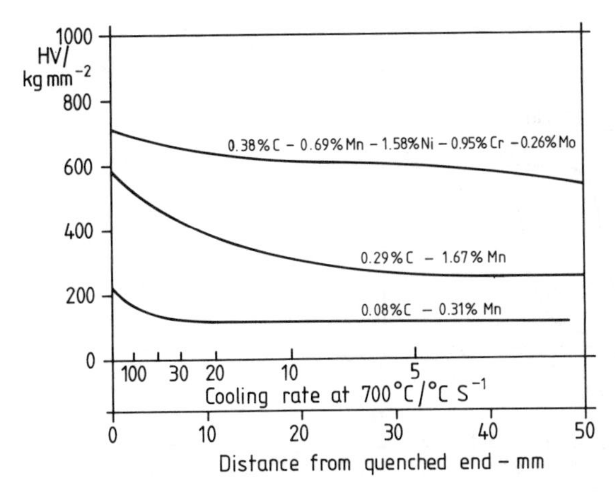

Figure 8.17 Typical Jominy test results

lead to cracking which occurs some time after the hydrogen was picked up. Hard materials of low ductility are less able to cope with this problem than are softer and more ductile ones.

Avoidance of cracking from either source requires that the material should not be over-hardened. As a rule of thumb, 'as-quenched' hardnesses of less than about HV = 350 are considered to be acceptable. This means that either martensite should not be formed or that if it is, it should contain less than about 0.25% carbon.

8.5.1 Conditions for formation of martensite

Martensite forms because ferrite and pearlite did not! It follows that metallurgical factors which promote the formation of ferrite and pearlite will inhibit the production of martensite. The ability of a material to form martensite rather than ferrite and cementite is called the 'harden-ability' of the steel. Note that this term does not refer to the absolute values of hardness obtained but to the ease of formation of martensite.

The most convenient method of assessing hardenability is the so-called Jominy end-quench. A rod-shaped sample is austenitized and then quenched by spraying water onto one end-face. Thus different cooling rates are produced along the length of the bar. Thereafter it is sectioned along its length and the hardness and microstructure assessed as functions of distance from the quenched end.

Some typical results are shown in Figure 8.17 for three different steels. For a plain carbon steel containing 0.08%C and 0.3%Mn cooling rates at 700°C of greater than about 50°C s^{-1} are necessary to form martensite. On the other hand, martensite forms at slower cooling rates in 0.29%C, 1.7%Mn steel. It is the manganese that causes this difference. Very slow cooling rates produce martensite in the alloy steel illustrated.

Responses to these curves would depend very much on what one was aiming to do. If it was producing a tough gearwheel, the alloy steel would be ideal. It could be cooled gently and still produce martensite, the gentle cooling being an advantage because it would reduce stresses arising from differential contraction rates and the possibility of quench cracking. Thereafter it can be tempered to achieve the desired properties. On the other hand, if we were dealing with a welded joint we would prefer to use the plain carbon steel in which it is more difficult to form martensite. What is more, the hardness of the martensite produced is low.

For welding, we want, therefore, a steel of poor hardenability. As we have seen, hardenability is affected by steel composition, including not only carbon content but also other alloying elements. To take all of these into account the concept of the carbon equivalent value is used. Unfortunately, there are a number of ways of calculating carbon equivalents for use in different circumstances. In the context of welding we have:

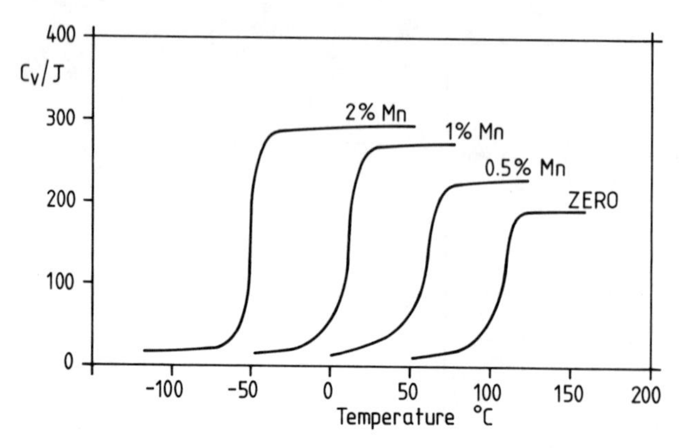

Figure 8.18 Effect of manganese content on the ductile–brittle transition of furnace-cooled 0.05%C steels. (After Rees, W.P., Hopkins, B.E. and Tipler, H.R. (1951) *Journal Iron and Steel Institute*, **169**, 157–168, Figure 14)

$$CE = C + \frac{Mn}{6} + \frac{Cr + Mo + V}{5} + \frac{Ni + Cu}{15}$$

Weldable steels, i.e. steels which can be welded with little or no trouble from martensite, have CEs less than about 0.4%. This is what limits the maximum alloy content of structural steels.

8.5.2 Manganese in structural steels

We have noted earlier that the residual sulphur impurity in steel is less harmful when formed into particles of MnS rather than into iron sulphide. The presence of small amounts of manganese in the steel also confers several other benefits. In normalized steels it tends to increase the amount of undercooling before the start of the formation of ferrite and pearlite. This gives finer grained ferrite and more finely divided pearlite. Both these changes improve strength and reduce the temperature of the ductile–brittle transition (Figure 8.18). What is more, some of the manganese atoms are contained in the ferrite crystals, displacing iron atoms. This also improves the strength of the ferrite.

If the manganese content is increased too much its effects cease to be beneficial and can become harmful. It increases hardenability, i.e. promotes martensite formation, and its presence in the martensite makes it harder and raises its ductile–brittle transition temperature. It is for these reasons that a maximum manganese content is usually specified: in BS 4360 this maximum is 1.5% by weight. A convention has also appeared that distinguishes between plain carbon steels (i.e. steels containing $<1\%$Mn) and carbon manganese steels (i.e. $>1\%$Mn).

8.6 Concluding summary

1. Steels used for structural purposes generally contain up to about 0.25%C, up to 1.5% Mn and have carbon equivalents of up to 0.4%. They are mostly used in the hot-rolled, normalized or controlled-rolled conditions, although low-carbon steels might be used in the cold-rolled one. Production processes aim to produce low inclusion contents and small grain sizes to improve strength, ductility, toughness and the ductile–brittle transition.
2. The elastic modulus of steel is virtually independent of composition and treatment.
3. The upper limits on the proportions of carbon and other alloying elements are determined by the effect of carbon equivalent on weldability and by the effect of carbon on the

ductile–brittle transition. All steels contain manganese, partly to deal with impurities such as sulphur and partly because its presence has a beneficial effect on the ductile–brittle transition. Within the range of allowed carbon and manganese contents compromises must be made between the effects of pearlite on strength, toughness and ductilities.

4. Recent years have seen the development of so-called micro-alloyed or high-strength/low-alloy (HSLA) steels. These are normalized carbon–manganese steels which have been 'adjusted' by micro-alloying to give higher strengths and toughnesses combined with ease of welding. Small additions of aluminium, vanadium, niobium or other elements are used to help control grain size. Sometimes about 0.5% molybdenum is added to refine the lamellar spacing in pearlite and to distribute the pearlite more evenly as smaller colonies. These steels are used where the improved properties justify the extra cost.

Background reading

BS 4360: Weldable Structural Steels

HONEYCOMBE, R.W.K. (1981) *Steels. Microstructure and Properties*, Edward Arnold, London

KNOTT, J.F. (1981) The relationship between microstructure and fracture toughness. In *Steels for Line Pipe and Pipeline Fittings*, The Metals Society, London

ROLLASON, E.C. (1973) *Metallurgy for Engineers*, 4th edn, Edward Arnold, London

9

Engineering properties of metals

Objective To introduce the basic characteristics of metals and describe the important mechanical and physical properties relevant to their performance in structural engineering applications.

Summary The chapter begins with an explanation of structure sensitivity. Tensile and compressive properties are then reviewed with discussion of real and nominal stresses and strains, ductile and brittle behaviour, Young's modulus, Poisson's ratio, upper and lower yield stresses, proof stress and ultimate tensile strength. Temperature dependence, impact strength and fracture toughness are also presented. Fatigue properties are reviewed with a discussion of fatigue lives, endurance limits, *S–N* curves and Goodman/Gerber diagrams. Thermal expansion and corrosion resistance are also summarized.

9.1 Structure-sensitive and structure-insensitive properties

Before embarking on an examination of the properties of interest the meaning of structure-sensitivity and structure-insensitivity, in the context of material properties, must be clarified. Structure-insensitive properties are those which are not influenced significantly by changes in microstructure or macrostructure. It is recognized that many of the physical properties of a material (for example, elastic modulus, bulk density, specific heat, coefficient of thermal expansion) do not vary other than by small amounts from one specimen to another of a given material, even if the different specimens have been subjected to very different working and/or heat-treatment processes. This is despite the fact that these processes may have produced quite substantial micro- and macrostructural modifications. On the other hand, most of the mechanical properties are very dependent on these modifications. Thus, for instance, the yield strength, ductility and fracture strength are seen to be structure-sensitive.

9.2 The mechanical properties of metals

9.2.1 Introduction

There are several mechanical properties which determine the suitability of a given material for a given application. These can be either monotonic, such as those describing the behaviour under simple tensile or compressive loading (including creep behaviour during deformation at elevated temperature), or cyclic properties associated with fluctuating loading under fatigue conditions. In all cases mechanical properties are most properly determined using standardized test pieces and testing procedures. In considering the different properties it is conventional to distinguish between the time-independent and the time-dependent behaviours characteristic of creep deformation under load.

In identifying the mechanical properties of interest attention must be given to the stress state to which the material is to be subjected. There are two simple stress states:

1. Tension or compression produced by uniaxial loading; and

2. Shear produced by torsional loading.

Other common states of stress are biaxial stress, such as that occurring in a vessel subjected to internal pressure, and hydrostatic stress resulting from immersion in a fluid. Although service conditions usually involve complex combinations of these various stress states it is quite usual for one of the stress conditions to be dominant, so that one particular mechanical property is of overriding importance.

When under stress, a material which is not stressed to fracture will deform either elastically and/or plastically. Elastic deformation is reversible on unloading while plastic deformation is permanent and is not removed on unloading. Frequently the elastic behaviour is linear and described as Hookean in that it obeys Hooke's law, i.e. that the deflection or displacement produced under load is directly proportional to that load. Non-linear elastic behaviour can also occur. Further, it is also possible for materials to behave anelastically in that the load-deflection path followed on loading is different from that followed on unloading, and hysteresis occurs. In practice, all metals are anelastic to some degree, even in the apparently truly elastic regimen.

9.2.2 Tensile properties

Tensile properties are usually measured using shaped test pieces of circular or rectangular cross-section. The load is applied through threaded or shouldered ends. The dimensions are commonly specified in detail and typical test-piece geometries are shown in Figure 9.1. The important dimensions are:

1. The initial gauge length, L_0, which is the prescribed part of the cylindrical or prismatic section of the test piece on which the elongation (the increase in length) is measured at any moment during the test; and
2. The initial cross-sectional area, S_0.

These dimensions are not usually treated independently. Rather, the initial gauge length is commonly taken to be $5.65\sqrt{S_0}$, which, for a circular cross-section test piece, is a gauge length of 5 diameters. In the most general case the elongation of the test piece consists of two principal

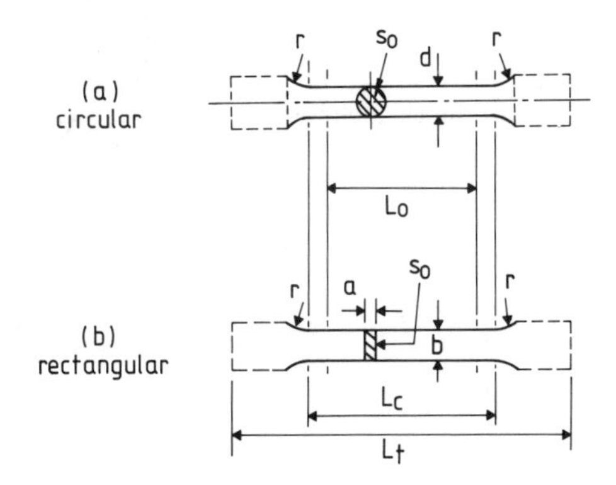

Figure 9.1 Standard test piece geometries. (After BS 18: Part 1, 1970)

contributions, an elastic and a plastic one. The elastic contribution contains both time-independent and time-dependent (anelastic) parts, although the latter are normally neglected. The plastic contribution involves both a region of uniform deformation in which all parts of the gauge length elongate to the same amount and a non-uniform one in which localized deformation or necking occurs. These three main regions are indicated on the load–elongation curve shown schematically in Figure 9.2, while the form of a necked but unfractured test piece is shown in Figure 9.3.

The deformation process is terminated by fracture in the necked portion. Occasionally, fracture will occur in the elastic region, and in such a case the metal under test would be described as being 'brittle'. This is also the term normally applied to specimens which fracture after only a very limited amount of plastic deformation. When considerable plastic deformation occurs the metal is described as 'ductile'. Even then, the final fracture can occur when little reduction has occurred in the necked portion.

Load–elongation curves are normally converted to stress–strain curves, taking into account the test piece dimensions. The various parameters of interest are defined as follows:

1. The nominal or engineering stress (σ_n) is the load at any instant divided by the original cross-sectional area;
2. The true stress (σ_t) is the load divided by the instantaneous cross-sectional area;
3. The nominal or engineering strain (ε_n) is the ratio of the change in length to the original length;
4. The true strain (ε_t) is the incremental instantaneous strain integrated over the whole of the elongation.

Thus for a specimen of initial cross-sectional area (S_0) and initial gauge length (L_0) elongated uniformly under a load (P) to a gauge length (L) with a corresponding cross-sectional area (S) the appropriate relationships are:

$$\sigma_n = \frac{P}{S_0} \qquad \sigma_t = \frac{P}{S}$$

$$\varepsilon_n = \frac{L - L_0}{L_0} \qquad \varepsilon_t = \int_{L_0}^{L} \frac{dl}{l} = \ln\left(\frac{L}{L_0}\right) \tag{9.1}$$

In the elastic region where stress is proportional to strain the elastic modulus of Young's (E) is defined as:

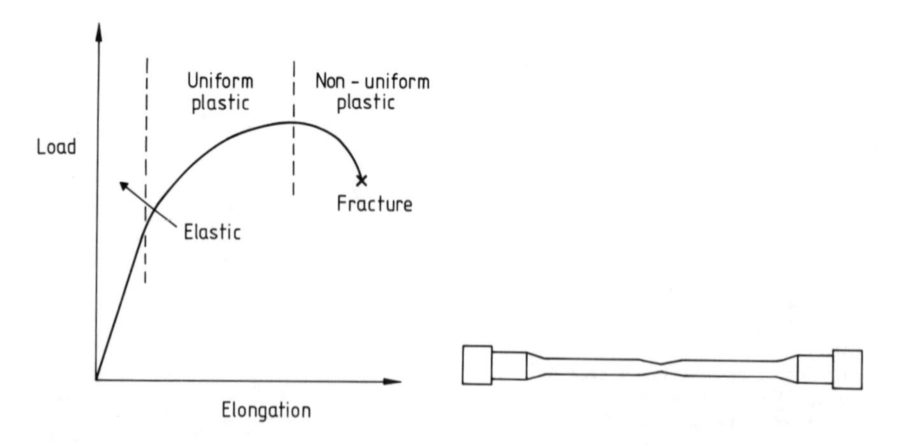

Figure 9.2 Typical load–elongation curve (schematic) for tensile deformation

Figure 9.3 Tension coupon at point of failure

$$E = \frac{\text{stress}}{\text{strain}} = \frac{\sigma_n}{\varepsilon_n} \tag{9.2}$$

During elastic deformation there are small changes in width as well as in the length. These are related by Poisson's ratio (ν), given by

$$\nu = \frac{\text{width strain}}{\text{length strain}}$$

During uniform plastic elongation the volume remains constant, i.e. $S_0L_0 = SL$. Using this relation, it can easily be shown that for tensile deformation:

$$\sigma_t = \sigma_n(1 + \varepsilon_n) \tag{9.3}$$

$$\varepsilon_t = \ln(1 + \varepsilon_n) \text{ (see Table 9.1)}$$

For many metals the transition from elastic to plastic deformation is not clearly evident but occurs progressively. Then the stress at which the metal is said to yield, or flow plastically, is defined in terms of a proof stress. This is the stress at which a permanent elongation, or permanent set, of a specified percentage of the initial gauge length (usually 0.2%) occurs. The method of determining the proof stress is given in Figure 9.4.

Table 9.1 Relationships between true stress and true strain and nominal stress and nominal strain

Nominal stress	$\sigma_n = \dfrac{P}{S_0}$	True stress $\sigma_t = \dfrac{P}{S}$
Nominal strain	$\varepsilon_n = \dfrac{L - L_0}{L_0}$	$= \dfrac{L}{L_0} - 1$
Then	$\sigma_t = \dfrac{P}{S} = \dfrac{P}{S_0} \cdot \dfrac{S_0}{S}$	$= \dfrac{P}{S_0} \cdot \dfrac{L}{L_0}$
since	$S_0L_0 = SL$ (ignoring elastic changes in volume)	
Therefore	$\sigma_t = \dfrac{P}{S_0} \cdot \dfrac{L}{L_0}$	$= \sigma_n(1 + \varepsilon_n)$
True strain	$\varepsilon_t = \ln \dfrac{L}{L_0}$	
but, as above,	$\dfrac{L}{L_0} = 1 + \varepsilon_n$	
Thus	$\varepsilon_t = \ln(1 + \varepsilon_n)$	

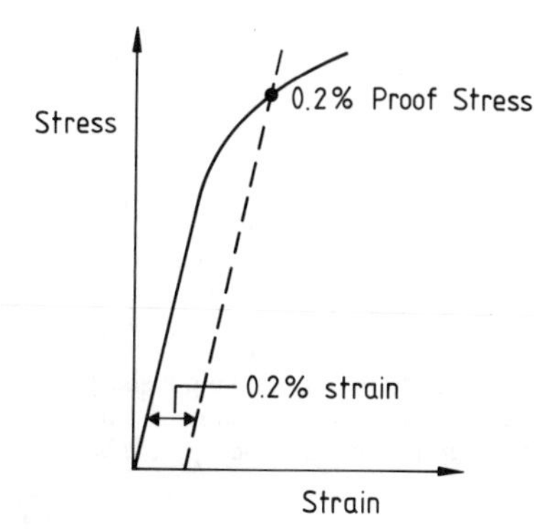

Figure 9.4 Determination of the proof stress from a tensile stress–strain curve

Some metals, especially mild steels, show an abrupt yield point followed by a short period of non-uniform plastic strain. The stress–strain curve then appears as shown in Figure 9.5. During the extension immediately after yielding a series of markings known as Lüder's lines or Lüder's bands are observed on the surface of the test piece. These indicate the regions which are deforming plastically, and they broaden until they occupy the whole of the gauge length. This occurs over the Lüder's strain (normally about 1–2% for structural mild steel) (see Figure 9.5). Once the whole of the gauge length has yielded, the subsequent behaviour is similar to that of metals which do not exhibit a yield point phenomenon.

In a steel which shows a yield point phenomenon two parameters are defined to describe yielding:

1. The upper yield stress, which is the stress at the initiation of yielding; and
2. The lower yield stress, which is the lowest value of stress during propagation of the Lüder's bands, ignoring any initial transient effects which might occur.

When the transition from uniform to non-uniform plastic deformation occurs and necking begins the load–elongation curve goes through a maximum. This leads to the definition of the ultimate tensile stress, (σ_u) given by:

$$\sigma_u = \frac{\text{maximum load}}{\text{initial cross-sectional area}} = \frac{P_{max}}{S_0} \qquad (9.4)$$

Two other important parameters are determined from the tensile test:

1. The percentage elongation to fracture, which is the permanent elongation of the gauge length after fracture expressed as a percentage of the original gauge length; and
2. The percentage reduction in area, which is the ratio of the maximum change in cross-sectional area that has occurred during the test to the original cross-sectional area.

Then if (L_u) is the gauge length after fracture and S_u the minimum cross-sectional area in the necked region:

$$\text{Percentage elongation} = \frac{100(L_u - L_0)}{L_0}$$

$$\text{Percentage reduction in area} = \frac{100(S_0 - S_u)}{S_0} \qquad (9.5)$$

Data for mechanical properties are available from a variety of sources, including published standards and reference books. Some typical data for various common metals and alloys (including steels) are given in Tables 9.2 and 9.3.

The mechanical properties are normally structure-sensitive, being altered by treatments such as work-hardening or controlled heat-treatments.

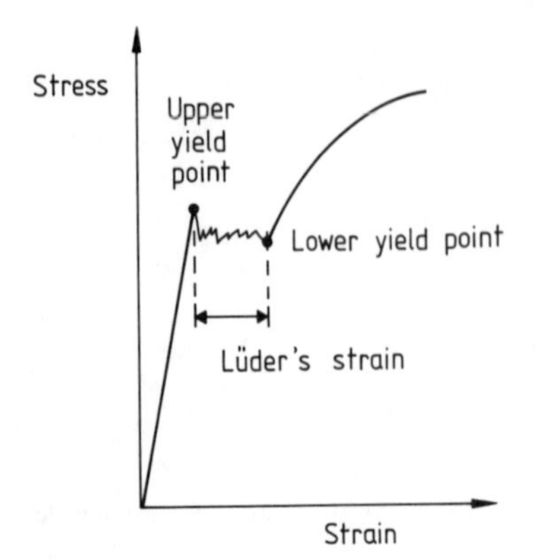

Figure 9.5 Schematic stress–strain curve showing a sharp, upper yield point and a lower yield point. (Lüder's strain, during which the plastic regions spread to occupy the whole specimen, is shown)

Table 9.2 Typical mechanical property data for weldable structural steels

Steel (grade according to BS 4360: 1979)	*Condition*	*Yield stress* (N/mm^2)	*Ultimate tensile strength* (N/mm^2)	*Elongation to fracture (%)* ($L_0 = 5.65\sqrt{S_0}$)
Grade 43A (0.25% carbon)	Hot-rolled	240	470	22
Grade 43C (0.19% carbon)	Normalized	240	470	22
Grade 50A (0.23% carbon)	Hot-rolled	345	550	20
Grade 50D (0.18% carbon, 0.10% vanadium)	Normalized	345	550	20

Table 9.3 Typical mechanical property data for some common non-ferrous metals and alloys

Metal or alloy	*0.2% Proof stress* (N/mm^2)	*Ultimate tensile strength* (N/mm^2)	*Elongation to fracture (%)* ($L_0 = 5.65\sqrt{S_0}$)
Aluminium (annealed)	34.0	77.2	47
Aluminium (cold worked)	94.2	115.8	13
Duralumin (annealed)	123.5	231.6	15
Duralumin (age-hardened)	278.0	432.4	15
Copper (annealed)	54.1	223.9	56
Copper (cold worked)	285.7	316.6	13
70–30 brass (annealed)	84.9	319.7	65
70–30 brass (cold worked)	378.4	463.3	20

9.2.3 Compressive properties

Compressive mechanical property data are more difficult to determine than tensile data. This is because compressive testing is normally carried out on cylindrical specimens that are short enough to avoid buckling when axial load is applied. Then, however, there are frictional end-effects which influence the behaviour. There are various ways to minimize these but none is completely satisfactory. For compressive loading, nomal stress and strain and true stress and strain are defined in similar ways to the corresponding definitions for tensile loading, with the exception that the strains must be arranged so as to produce positive values. Then:

$$\sigma_n^c = \frac{P}{S_0} \qquad \sigma_t^c = \frac{P}{S}$$

$$\varepsilon_n^c = \frac{h_0 - h}{h_0} \qquad \varepsilon_t^c = \int_h^{h_0} \frac{dl}{l} = \ln\left[\frac{h_0}{h}\right] \tag{9.6}$$

for compressive deformation of a cylinder with initial cross-sectional area (S_0) and initial height (h_0) deformed to a height (h) with a corresponding area (S). There is no region of non-uniform plastic deformation in compression to correspond to that of necking in tensile deformation, although the frictional end-effects can give rise to 'barrelling' of the specimen at large strains. For that part of the deformation where uniform compression occurs, constant volume can again be assumed which leads to the relations (see Table 9.1):

$$\sigma_t^c = \sigma_n^c(1 - \varepsilon_n^c)$$

$$\sigma_t^c = - \ln(1 - \varepsilon_n^c) \tag{9.7}$$

For a given material the nominal or engineering stress–strain curves in tension and compression and the true stress–strain curve, which is a material property, are related in the way shown in Figure 9.6.

9.2.4 Hardness

Hardness is a measure of resistance to deformation under conditions in which a loaded indenter is forced to penetrate the surface of the metal under test. The deformation during indentation is again a combination of elastic and plastic behaviour. However, as measured, the hardness is largely related to plastic properties and only to a secondary extent to elastic ones. Other measures of hardness are sometimes used such as scratch hardness, which depends upon the ability of one solid to scratch or be scratched by another, or rebound (dynamic) hardness, which involves the dynamic deformation of a specimen expressed in terms of the part of the energy of impact absorbed when an indenter drops upon a specimen. However, indentation hardness is by far the most important of the measures in current use.

The different methods of indentation hardness testing differ in the form of indenter which is forced into the surface:

1. The Brinell hardness test uses a hardened steel ball as the indenter.
2. The Vickers hardness test is based upon the use of a square-based diamond pyramid of 136° included angle.
3. The Rockwell hardness test involves a diamond cone indenter with 120° included angle and a slightly rounded point.

In all cases the hardness number recorded is related to the ratio of the applied load to the surface area of the indentation formed. The testing procedure involves bringing the indenter into contact with the surface of the material to be indented followed by a cycle of loading–holding under load–unloading. On removal of the load the size of the indentation is measured using a microscope. Expressions for the two most commonly used hardness numbers are given in Table 9.4.

For hardness measurements to be valid they should not be sensitive to the conditions of

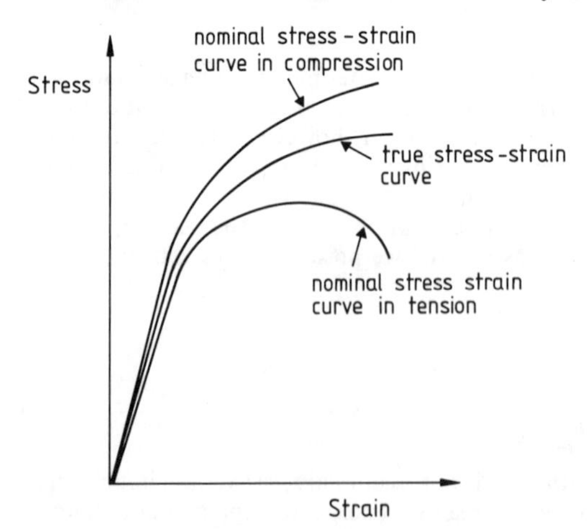

Figure 9.6 Schematic nominal stress–strain curves in tension and compression with a common true stress–true strain curve derived from them

Table 9.4 Hardness

$$\text{Brinell Hardness Number (BHN)} = \frac{P}{(\pi D/2)\,[D - \sqrt{(D^2 - d^2)}]}$$

where P is the load (kg), D is the ball indenter diameter (mm) and d is the indent diameter (mm), and

$$\text{Vickers Hardness Number (VHN)} = \frac{1.854P}{L^2}$$

where, again, P is the load (kg) and L is the average length of the diagonals of the indentation.

Typical values of hardness for some metals and alloys

Metal	*BHN*	*VHN*
Copper (annealed)	49	53
Brass (annealed)	65	70
Grade 43A structural steel	152	157
Grade 50A structural steel	190	195

testing. Normally there is a range of loads for which this requirement is met for a given range of hardness. For this reason, it is common to impose restrictions on the permissible load ranges. The different hardness measurements correlate quite closely especially at lower values. Some typical values are also given in Table 9.4.

The correlation of hardness values with other measures of resistance to deformation, such as tensile properties, is more difficult. For example, the deformation in a Vickers indentation is equivalent to a tensile strain of approximately 8%. At best, the correlations are empirical, but even then they must be treated with caution, since the relationships derived assume that the materials are of uniform composition and have been subjected to uniform heat and/or mechanical treatments.

9.2.5 The temperature dependence of tensile properties and hardness

As the temperature is increased above ambient the tensile properties such as yield stress and ultimate tensile stress decrease. So also does the hardness.

9.2.6 Impact strength and fracture toughness

Materials that show quite acceptable properties when tested in tension at slow loading rates often fail in a brittle way when subjected to rapid loading. Susceptibility to failure under conditions of fast fracture is enhanced if notches or other defects are present on the surface of the sample. Ductile face-centred-cubic materials such as copper and aluminium normally resist fast fracture under all loading conditions and at all temperatures. This is not the case, however, for many ferrous alloys, particularly plain carbon and low-alloy steels. Of particular note is the occurrence of a ductile-to-brittle transition in these alloys as the temperature is lowered. Resistance to fast fracture is commonly referred to as toughness, and loss of toughness in service can have catastrophic effects.

It is important to be able to quantify toughness, and conventionally this has been done by carrying out high strain rate 'impact' tests on standardized notched specimens. The most common is the Charpy test, in which a 10 mm square bar with a machined notch is struck by a calibrated pendulum. The energy absorbed from the swinging pendulum during the deformation of fracture of the test specimen is used as a measure of the impact strength.

The nature of the resulting fracture surface can also provide information of value in assessing impact resistance. For instance, high-impact strength is normally associated with a ductile, fibrous fracture, while low-impact strength corresponds to a brittle, cleavage fracture.

The ductile-to-brittle transition in steels can best be demonstrated by the change in the energy absorbed in fracture as a function of temperature. This is shown schematically in Figure 9.7.

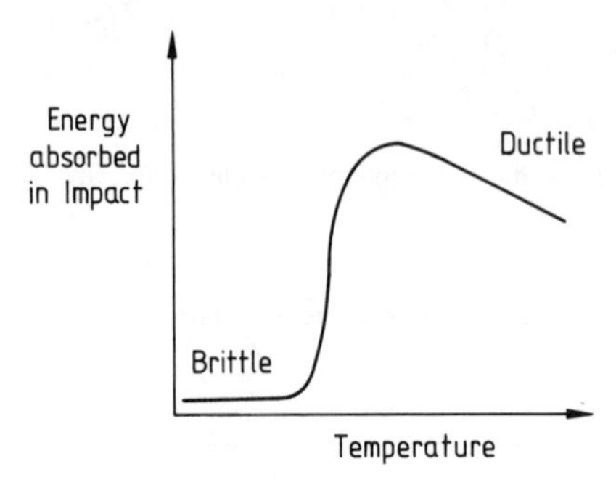

Figure 9.7 Ductile-to-brittle impact transition curve for steel (schematic)

The transition temperature is normally lower than room temperature. When it is near to ambient this can have very important consequences in so far as service behaviour is concerned. Thus a structure or an engineering component can suffer sudden failure under service loading if temperature changes occur that move the operating conditions to below the transition temperature. This temperature is affected by both metallurgical variables such as grain size (increasing the grain size raises the transition temperature) and by service variables (for example, the presence or absence of notched and/or residual stresses). The use of the notched impact test was adopted because this involved a combination of conditions most favourable to impact failure with low energy absorption. However, the impact energy is only a relative measure of impact strength. For many years, efforts were made to identify a quantitative measure of toughness which could be used by design engineers, and this was eventually provided through the advent of fracture mechanics.

The fracture of ductile or semi-brittle materials involves the dissipation of energy through plastic work additional to that needed to create the new surfaces produced by the fracture process. The important material property parameter is the toughness (G_c) which is a measure of the total energy absorbed in making unit area or crack. The relevant equation is:

$$\sigma\sqrt{(\pi_a)} = \sqrt{(EG_c)} \tag{9.8}$$

where E is Young's modulus and σ is the applied stress. The assumed crack length is:

a for a surface crack,

$2a$ for an internal one.

This equation states that fast fracture will occur in a given material subjected to a stress (σ) when a crack reaches a critical size (a) or, alternatively, it will occur in the presence of a crack of size (a) when a critical stress (σ) is reached.

The materials parameter $\sqrt{(EG_c)}$ which clearly plays a critical role in determining the likelihood of fast fracture is usually given the symbol K_c and is called the fracture toughness. The determination of the fracture toughness is also carried out using standardized test specimens and procedures. Values vary substantially from one material to another; some representative values are given in Table 9.5.

Table 9.5 Fracture toughness values of different materials

Material	$K_c\,(MNm^{-3/2})$	*Material*	$K_c\,(MNm^{-3/2})$
Ductile metals (e.g. Cu)	200	Cast iron	12
Grade 43A structural steel (room temperature)	140	Glass-reinforced plastic	40
		Reinforced cement	12
Grade 43A structural steel (– 100°C)	10		
Pressure-vessel steels	170		

9.3 Fatigue properties

When discussing the tensile and compressive properties of metals and alloys it was assumed that the loading was monotonic, i.e. it increased steadily until failure occurred. In most practical situations this type of loading does not occur. Instead, the applied stresses fluctuate, often in a random way. In some cases, such as a rotating shaft subjected to bending stresses, the stress at a given point on the surface varies cyclically from tension to compression and back again. Various forms of cyclic stress loading are shown schematically in Figure 9.8. In the presence of cyclic stress it is frequently found that premature failure occurs at stress levels much lower than those required for failure under a steady applied stress. This phenomenon of premature failure under fluctuating stress is known as fatigue, and involves processes of slow crack growth.

Fatigue strength is a very variable quantity, and, as such, its measurement requires extensive testing. Usually, and most simply, this is achieved by preparing a substantial number of identical test specimens, subjecting them to defined cyclic conditions and measuring the number of cycles required for failure. The cyclic loading conditions in most common use involve either fluctuating tension (tension–zero–tension) or push–pull (compression–tension–compression) stresses. The mean stress (σ_m) and the fluctuating stress amplitude (σ_a) are then the important test parameters. The test data are plotted as an *S–N* curve as shown in Figure 9.9, usually for zero mean stress loading. With ferrous alloys a limiting fluctuating stress amplitude is recorded, below which failure does not occur. This is known as the endurance limit, and is a function of mean stress. However, this concept must be treated with caution, since the fatigue behaviour is very sensitive to operating conditions. Thus:

1. Random stress fluctuations;
2. Stress concentrations, such as changes in section;
3. Surface roughness;
4. Residual stress; and
5. The presence of a corrosive environment

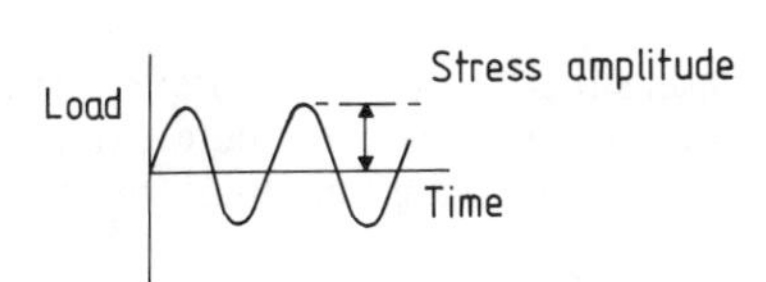

(a) Fluctuating tension – compression
(mean stress = 0)

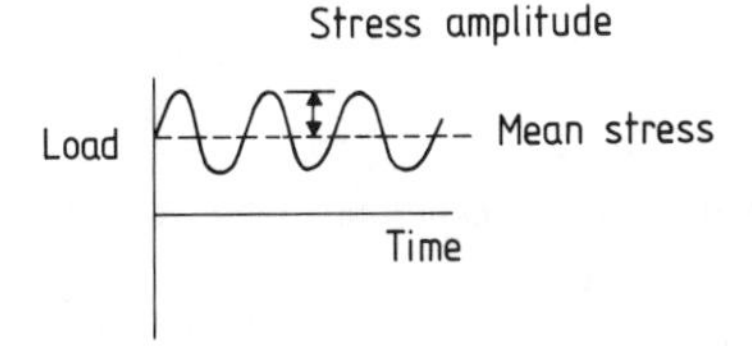

(b) Fluctuating tension (mean stress >0)

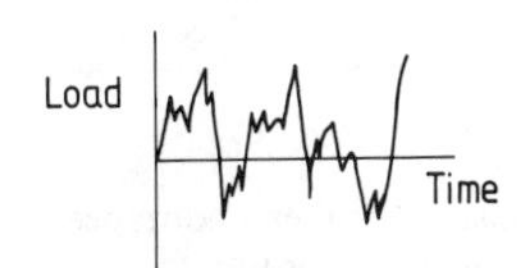

(c) Random tension – compression loading

Figure 9.8 Forms of cyclic loading

can all have a deleterious effect on the endurance limit and fatigue life. With non-ferrous metals and alloys no endurance limit is recorded, there being only a steady increase in life as the cyclic stress amplitude falls. This is also shown in Figure 9.9.

Fractures caused by fatigue usually have a characteristic appearance. The point of crack initiation is surrounded by a flat region where slow crack growth has occurred. Microscopically, this region usually exhibits a series of well-defined markings known as 'beach marks'. On a microscopic scale slow crack growth is accompanied by the formation of fatigue striations. As the fatigue crack propagates, the effective stress on the remaining part of the cross-section increases until there is sudden fast fracture, giving rise to a typically fibrous ductile fracture.

Apart from the factors described above, the mean stress has a significant effect on the fatigue life. As an example, its influence on the endurance limit is shown by the Goodman relationship or some equivalent relationship such as that defined by Gerber. These are shown in Figure 9.10.

None of these considerations makes allowance for the influence of metallurgical factors. It is well established that microstructural variables such as changes in grain size, alloying and, most particularly, the presence or absence of non-metallic inclusions also have significant effects on fatigue life.

9.4 Physical properties

Although in the majority of structural engineering applications mechanical properties are of great importance it is necessary also to have a knowledge of various physical properties which can affect in-service performance, such as the elastic modulus or the coefficient of thermal expansion.

The constant of proportionality between stress and strain in the linear elastic region during the initial stages of loading a specimen either in tension or compression is known as Young's modulus. This is usually given the symbol E, and is defined as:

$$E = \frac{\sigma}{\varepsilon} \tag{9.9}$$

The elastic modulus is normally taken to be structure-insensitive for a given material. The variations caused by alloying are small and thus, on both counts, it is usual to attribute a unique value of the elastic modulus to a given material. Reported values of Young's modulus for some common metals are given in Table 9.6.

When the temperature of a metal or alloy is raised the atoms undergo increased thermal vibration, which leads to an expansion of the lattice. The resulting strain increment is given by:

$$\Delta\varepsilon = \alpha \Delta T \tag{9.10}$$

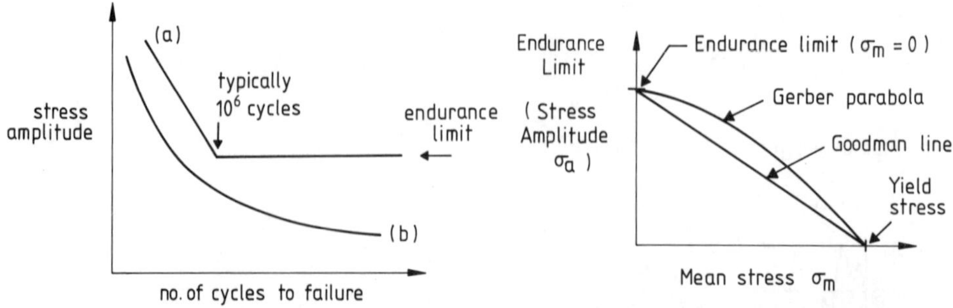

Figure 9.9 *S–N* curve for zero mean stress loading for (a) steel and (b) aluminium

Figure 9.10 Relationship between endurance limit and the mean stress (according to Goodman and Gerber)

Table 9.6 Typical values of Young's modulus for metals

Metal	Al	Au	Ti	Cu	Fe	Ni
kN/mm^2	71	79	120	130	211	200

where α is the coefficient of thermal expansion and ΔT the temperature change. For most metals and alloys α lies in the range 10×10^{-6} to $20 \times 10^{-6} K^{-1}$. Changes in temperature can give rise to very substantial thermal stresses through restraint of the consequent dimensional changes, and these must normally be taken into account in design. Non-uniform temperatures often produce a self-equilibrating set of stresses or stress resultants in a structure. Where local yielding occurs (for example, during welding) a strain incompatibility is established which leaves a self-equilibrating set of stresses after the temperature variation has dissipated: these are called residual stresses. Either set of stresses may influence structural stiffness and strength.

9.5 Corrosion resistance

In the presence of aggressive gaseous environments (for example, hot exhaust gases) or aggressive aqueous environments (for example, salt water) considerable and irreversible surface damage can occur, and the responses are known as oxidation and corrosion, respectively. In order to preserve component integrity it is therefore essential to ensure that the surface exhibits resistance to the environment.

Commonly, the component surface is provided with some form of protection either through a surface treatment or by surface coating. With aqueous corrosion, special techniques, such as those involving cathodic protection or the use of corrosion inhibitors, can be used to minimize or eliminate the interaction between component and environment.

In aqueous environments the most effective methods of gaining corrosion resistance involve modification of the properties of the parent metal or alloy or the use of protective coatings. The principles underlying modification of properties are similar to those involved in producing oxidation resistance. A primary aim is to produce some form of surface film which is stable in the corrosive environment (be it acidic, neutral or alkaline) and which affords protection. Thus with the stainless steels this is achieved by producing a passive chromic oxide surface film. As knowledge of corrosion mechanisms has been gained it has become possible to design alloys to obtain acceptable behaviour under different conditions. For example, it is necessary to add a small amount of molybdenum to stainless steels to produce the required performance in acidic environments.

Protective surface coatings can be either metallic or non-metallic. Here the aim is to produce changed surface properties. One of the simplest procedures is to apply a corrosion-resistant surface coating to the material prone to corrosion. This can be done in a variety of ways – for example, by cladding, dipping, spraying, electrodeposition or cementation. It is common practice to deposit surface coatings of aluminium, zinc, tin and chromium using these processes. Alternatively, inorganic coatings (for example, phosphates), anodized oxides and glassy enamels, or organic coatings (for example, polymeric resins, oils and bitumen) can be applied to the surface. In some cases, as with paints, a combination of metallic and non-metallic components is used. Surface coatings are usually effective without having any effect on the intrinsic properties of the coated artefact. This is undoubtedly one of the reason why they are so widely used. However, it is important to recognize that they provide effective protection only while they are undamaged. In many cases some short-term resistance to minor damage can be achieved, but in the longer term repair and/or replacement of the surface coating is necessary.

In general, it needs to be recognized that it is not necessarily the amount of corrosion that is of greatest importance but its distribution. Thus localized attack can have very damaging effects, even though the total amount of damaged material is small.

9.6 Concluding summary

1. Most engineering properties of metals are structure-sensitive.
2. The principal quantities defining the mechanical properties of metals under non-repeating loading are:
   (a) Young's modulus;
   (b) Poisson's ratio:
   (c) Nominal yield stress, or proof stress for metals without a defined yield point;
   (d) Ultimate strength;
   (e) Percentage elongation to fracture or percentage reduction in area;
   (f) Hardness;
   (g) Impact strength or fracture toughness.
3. Fatigue life is a function of:
   (a) Fluctuating stress;
   (b) Mean stress:
   (c) Conditions at the point of crack initiation.
4. Other engineering properties that may be of significance are:
   (a) Thermal explansion;
   (b) Wear resistance;
   (c) Corrosion resistance.

10

The manufacture and forming process

Objective To give an appreciation of the casting and forming process in the manufacture of steel products and to define the products available to the designer of structural steelwork.

Summary The chapter starts with a brief review of ingot and continuous casting. It describes hot rolling and the variety of hot-rolled products; plates, structural sections, merchant bar and strip. It also summarizes forging, extrusion, tube forming, cold rolling and cold forming. The principal hot-rolled products are presented: universal beams, universal columns, joists, channels, trees, angles and structural hollow sections. Aditional finishing processes that can be specified are reviewed. Tolerances, steel qualities and material selection are also discussed.

10.1 Ingot and continuous casting

At the end of the steel-refining process, whether by basic oxygen or electric arc steelmaking, the liquid steel is 'tapped' into a large refractory lined ladle. To enable this molten steel to be rolled or formed into finished products suitable for the construction industry it must be solidified to form basic shapes, either ingots or continuously cast slabs, blooms and billets, as shown in Figure 10.1.

In the continuous casting process the molten metal is poured directly into a casting machine to produce the slabs, blooms or billets, as shown in Figure 10.2.

Many technical problems associated with casting have been solved to make the process reliable. Defects associated with the ingot route, which have to be removed expensively further down the line, are eliminated. Ingot equipment and primary rolling are eliminated and product yields improved. These are important economies, and modern works have continuous casting operations for up to 100% of their steel output.

10.2 Hot-rolling processes

10.2.1 Rolling mills

There are various methods for forming steel into finished products, with hot rolling being the one most extensively used for constructional material. Others will be described later. Hot rolling involves initial reheating of ingots, slabs, blooms or billets to the region of 1200–1300°C and passing the material between two rolls, where the gap between the rolls is lower in height than the material entering: the piece of steel may be passed repeatedly back and forth through the same rolls. The number of passes between rolls depends upon the input material and the size of finished product; it can be as high as 70 in certain cases before the material becomes too cold to roll down further. Plain barrel rolls, as shown in Figure 10.3, are used for flat products such as plate, strip and sheet, while grooved rolls are used for structural sections, rails, rounds, squares and special shapes.

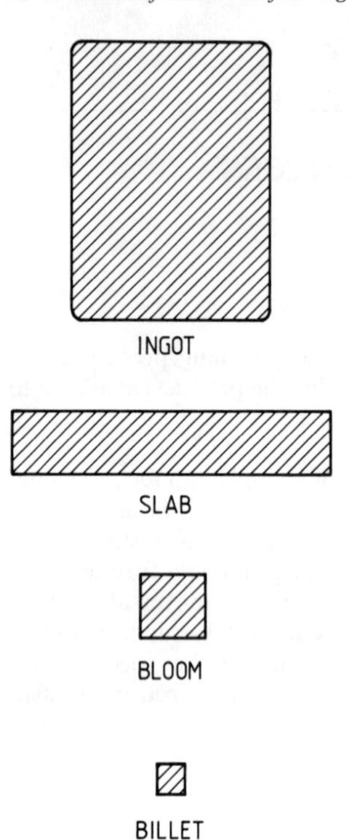

Figure 10.1 Basic shapes in relative proportion

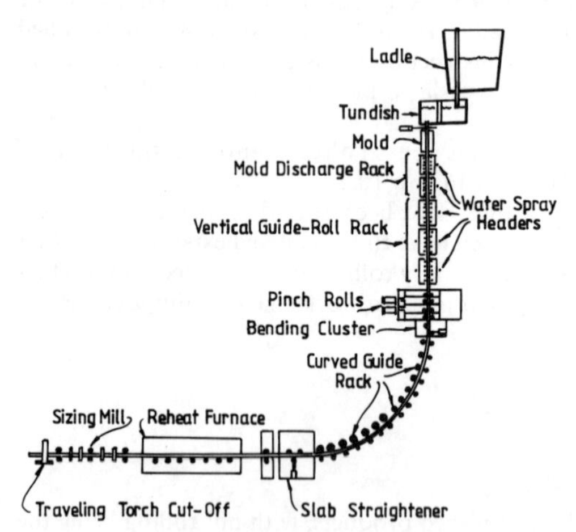

Figure 10.2 Molten metal poured direct into casting machine

The reduction in thickness of the hot material results in length increase and sideways spread. The spreading, which depends mainly upon the amount of reduction, temperature, roll material and roll diameter, must be controlled to give the correct dimensions and cross-section. This is achieved by the use of vertical rolls in some cases and by correct design of successive roll grooves in others.

In addition to its function of shaping the steel into the required size, hot rolling also improves mechanical properties by refining the grain size of the material. Correct control of the cast steel chemical composition, finish rolling temperature and amount of material reduction is necessary

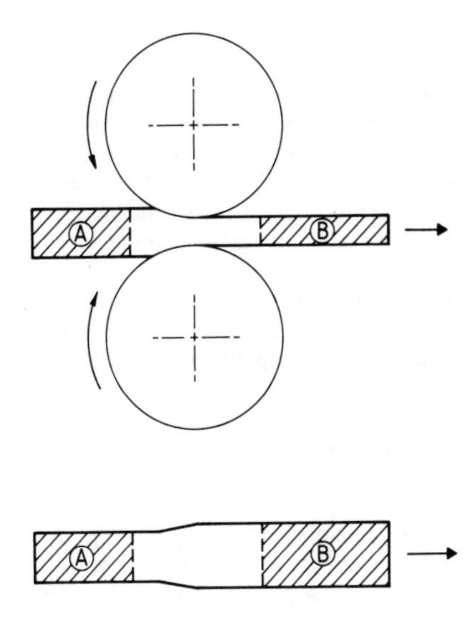

Figure 10.3 Basic rolling process

to give products with the required physical properties. For certain steel qualities (for example, high strength with good impact properties at low temperature) 'controlled rolling' is employed. This involves delaying material prior to finish rolling until a specified lower temperature is reached for the final passes through the mill. Figure 10.4 summarizes the main product routes for structural steelwork.

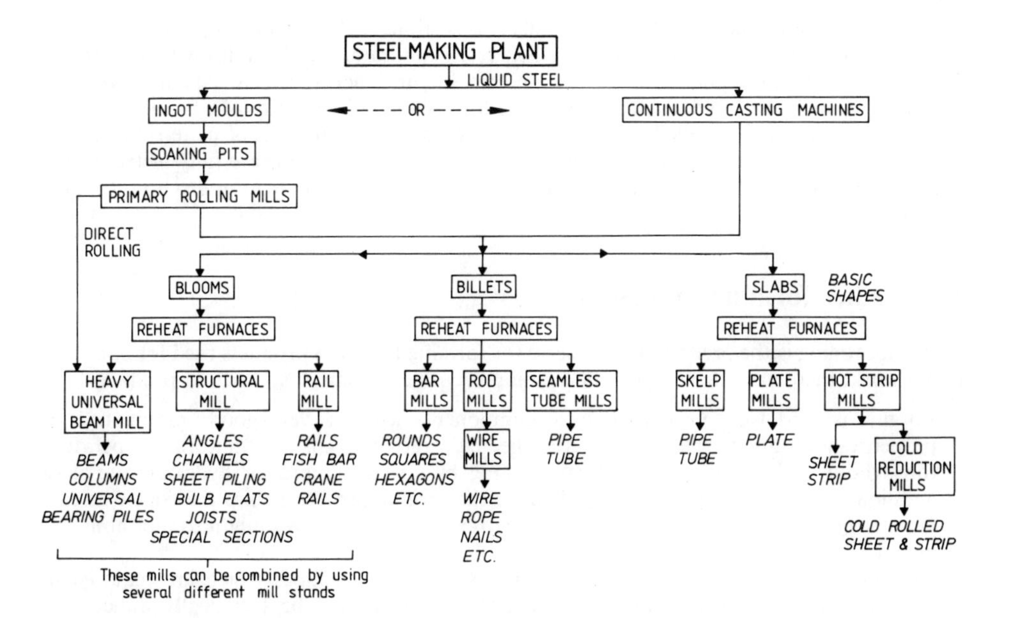

Figure 10.4 Principal product routes

10.2.2 Application of hot-rolling to various products

The first hot-rolling operation is to convert ingots into the basic shapes shown in Figure 10.1. This is generally carried out on a large single-stand, two-high reversing mill, known as a primary mill. In between the steelmaking plant and the primary mill there is a bay for stripping moulds from the ingots and a battery of soaking pits. Each pit may hold up to 150 tonnes of ingots and serves to bring the ingots up to a uniform temperature for rolling and act as a reservoir to ease out fluctuations in the flow of ingots. Soaking pit temperatures are generally controlled to 1300°C.

Primary mills are equipped with manipulators for positioning and turning the ingots to enable work to be done on each face as rolling proceeds. Roll grooves are arranged to enable a variety of basic shapes to be made (Figure 10.5), in some cases for rolling directly into another mill to produce billets or finished sections such as rails or 'structurals', or, more usually, for rolling into stock at other mills to be inspected and conditioned prior to reheating and rolling into finished products. It is at this reheating stage that the economic and technical advantages of continuous cast basic shapes become apparent. Subject to size and type, primary mill outputs are typically from 500 000 to 5 million tonnes per year.

The finish rolling of products for construction work divides broadly into four groups: plates, structural sections, merchant bar and strip. Structural sections comprise standard shapes (for example, beams, channels, angles, bulb flats, etc.) and special sections (i.e. custom-designed products for individual applications). They are rolled on a variety of mill types, depending mainly upon the product size range. As a general rule, large sections are rolled directly from ingot, intermediate-size ones from reheated blooms and small sections from reheated billets. In all cases the process begins with roughing down, in which the initial square or rectangular cross-section is gradually shaped in successive roll passes into an outline of the required product; this is followed by finish rolling in successive passes to give the final standard shape and dimensions after cooling. Finishing mill rolling temperatures are usually in the region of 900–1000°C. An example of the pass sequence for angle rolling is given in Figure 10.6 and the rather more specialized method for universal beams and columns is shown in Figure 10.7. Subject to mill size and type, section mill outputs are typically from 200000 to 1 million tonnes per year.

Merchant bar is a traditional term for small cross-sections such as rounds, squares, hexagons, flats, etc. which are rolled from reheated billets from continuous in-line mills with as many as 23 rolling stands. Feedstock is generally 100 mm square billet and pass sequences are of a square, diamond, oval type, culminating at the last mill stand with a finished round.

The production of hot-rolled strip is, in many respects, an extension of plate rolling, with thicknesses in the range 2–16 mm and widths up to 2 m. Modern mills are fully instrumented and computer-controlled to give a high standard of dimensional accuracy and finished properties.

10.3 Other forming processes

While hot rolling is the predominant process for forming the main products used for constructional work, there are other important forming processes for more specialized products:

1. Forging is used for heavy mechanical components where a very good combination of properties is required. The as-cast ingot is heated carefully to ensure a uniform temperature throughout, following by mechanical working by steam hammer or hydraulic press. In addition to roughing out the required shape the forging process refines the coarse as-cast grain structure, improves the material properties and ensures the soundness and homogeneity of the component.
2. Extrusion is generally used for the production of special shapes which are required in relatively small quantities. In some cases the product may be technically difficult or impossible to form by any other process. The input stock is heated to forging temperature

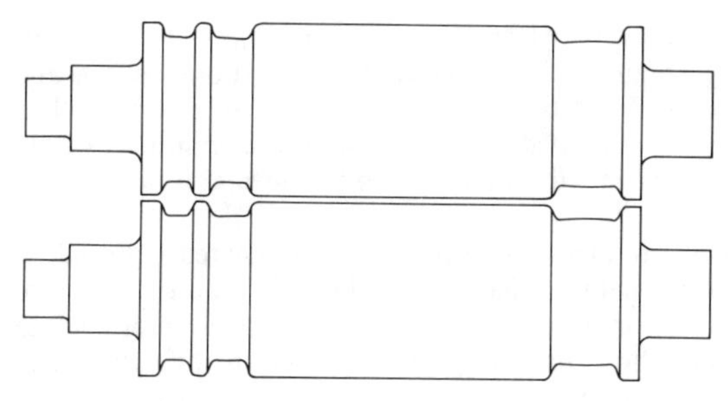

Figure 10.5 Primary mill rolls for slabs and blooms

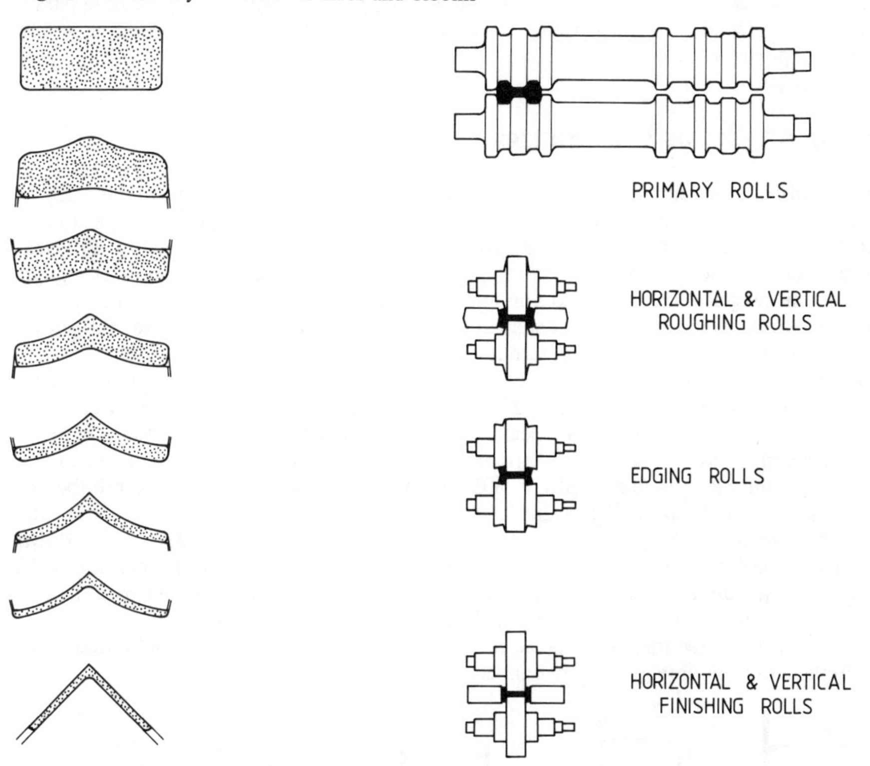

Figure 10.6 Pass sequence for angles

Figure 10.7 Sequence of operations for universal beams

and enclosed in a container which has a die at one end shaped to the required profile and a compression head at the other which applies high pressure. The metal is forced through the die and flows plastically to form the shape required. By using a mandrel the process can be used for seamless tube production.

3. Tube forming is carried out by several different processes involving hot working, cold working or welding, depending upon the tube size and its application. Most pipes are produced by the electric resistance welding process, in which coils of strip or sheet are decoiled directly into a cold-forming mill, followed by edge heating and welding by low-frequency power through wheel electrodes. On-line stretch reduction can, if necessary, be carried out to produce the required diameter and gauge of pipe and also square or rectangular cross-sections. Seamless tubes are produced by hot-working methods which involve

piercing a billet with a mandrel and rolling it externally in a helical way. Heavy-duty, thick-wall pipes are produced from single plates which are cold formed to tubular shape and seam welded using submerged-arc welding. In addition to circular tubes, structural hollow tubing is formed in square, rectangular and rounded shapes by either the continuous weld or the seamless tube process. The forming is an additional step to the normal tube-working process.

4. Cold-rolling, as the term implies, involves reducing the thickness of unheated material by rolling pressure. The main products in this field are flat bars and sheets. The reduction to give the required thickness is small and the main purpose is to obtain a smoother surface and improve mechanical properties. On the other hand, cold reduction involves much greater reduction to deliver sheet as low as 1 mm in thickness.
5. Cold forming is used to form thin strip into light section profiles by a series of forming passes in a continuous train of roll sets. Such products are too thin to be produced by hot rolling and may be particularly complicated in shape.

10.4 Section types and size ranges

Structural sections, also known as profiles or shapes, differ from solid steel products such as plate, strip, rounds, squares and hexagons by their function and their strength-to-weight ratios. They can be broadly divided into standard items comprising joists, universal beams, universal columns, channels, angles and hollow sections, which are rolled by the various mills at regular intervals, and more specialized items comprising tees, bulb flats, sheet piling, rails and special application sections, which are rolled to suit individual customer demand. Figure 10.8 shows the various section shapes with their designations in common usage.

The most commonly used sections are universal beams and columns, known internationally by the American terminology of wide-flange beams and columns. They are manufactured in the UK to BS4: Part 1: 1980, which is a selection of the most useful sizes from the American imperial range expressed in equivalent metric terms, and are rolled with parallel flanges. Each serial size (for example, 914 mm depth × 305 mm flange) is subdivided into a number of standard masses per unit length (for example 289, 253, 224 and 201 kg/m for 914 × 305). The total range available is from 914 × 419 down to 203 × 133 for beams and 356 × 406 down to 152 × 152 for columns. Included in the same group of sections are universal bearing piles, which are obtained from the same roll sets as columns. The essential difference between the two sections is that the web and flanges have equal thicknesses for bearing piles.

Joists, which are an I-section with tapered flanges, continue to be produced in a limited range which complements the universal beam range.

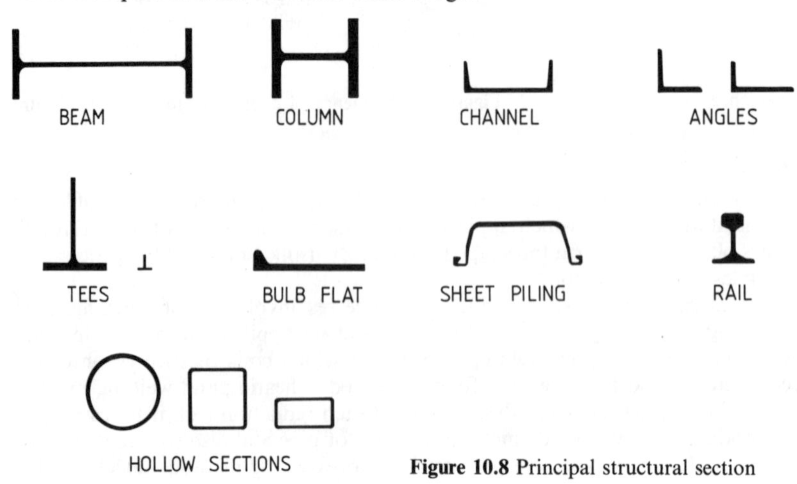

Figure 10.8 Principal structural section

Channels are manufactured with a 5° taper flange in eleven widths ranging from 76 to 432 mm, of which six sizes are available with two different leg lengths.

Structural tees are mainly produced by slitting down the centre of the web of universal beams or columns. With the exception of 914 × 419, all standard beam and column sizes designated in BS4: Part 1: 1980 can be slit in this way. To complement the lower end of this tee range, two smaller tees are rolled (51 × 51 mm and 44 × 44 mm). Both the stalk and the table have a $\frac{1}{2}$° taper on faces tangential to the radius between stalk and table.

Equal and unequal angle sections are manufactured in the UK to a coordinated, metric size range based on a selection of sizes from the internationally agreed range (ISO/R657). Angle legs have no taper and each designated size is produced in a number of standard thicknesses.

Structural hollow sections are produced in accordance with BS 4848, Part 2, with the addition of a small number of sizes and thicknesses not included in the standard. The circular hollow sections range from 21.3 to 508 mm O/D, squares from 20 × 20 to 400 × 400 mm and rectangulars from 50 × 25 to 450 × 250 mm. The square and rectangular size range is in modular metric steps.

Bulb flats, rarely used in construction work, are flats with a bulb at one side and are produced to BS 4848: Part 5. The size range is 120 × 6 mm to 430 × 20 mm.

10.5 Additional finishing processes and tolerances

In addition to the normal finishing processes which are necessary to bring hot-rolled steel to the required standard for despatch, a number of other services are available. The main examples are:

1. Exact cutting which improves on normal hot-sawing tolerances by cold-sawing lengths to within ±3 mm.
2. Line straightening of universal beams and columns which, by definition, improves on BS 4: Part 1: 1980, Straightness Tolerances.
3. Cambering of universal beams, which approximates to a simple regular curve nearly the full length of the beam. There is a limitation of camber which can be applied to various sizes and weights of beams.
4. Sections can be supplied shot-blasted and primed. Protective paints and mastics are also commonly applied for sheet piling and other uses.
5. Plates can be given various types of heat treatment such as annealing, normalizing, stress relieving and quenching and tempering to improve upon the as-rolled material properties.
6. Rolled sheet which is to be cold reduced must have the surface oxides, which are formed during hot rolling, removed. This is done by decoiling the sheet and passing it through a pickling line.

Rolling tolerances relate to such items as mass per unit length, geometry of cross-section, dimensions, straightness and length supplied. They are to be found in the standard specification for the section being used. While these tolerances generally have very little effect when the products are used in normal fabrication and in structural applications, they should be considered when special applications are involved. These may include guides for tracked vehicles, frames fitted tightly around equipment, sections butted alongside one another, etc.

Engineers and designers should be aware of these rolling tolerances, and where a special application could encounter difficulties arising from rolling tolerances it is worthwhile to arrange discussions with the steel supplier at the design stage. Very often arrangements can be made to overcome potential problems at an early stage of development.

10.6 Steel qualities and material selection

The plates and structural sections described earlier, which are used for normal constructional steelwork, can be supplied in a number of different grades. In the UK these steel quality grades are supplied in accordance with BS 4360: 1979, Weldable Structural Steels. This standard includes four tensile strength ranges, 40, 43, 50 and 55, indicating minimum ultimate tensile strengths of 400, 430, 490 and 550 N/mm^2.

Each strength grade has several subgrades indicated by a letter between A and E. As the subgrades proceed from A to E the specification becomes more stringent, the chemical composition changes and the notch ductility of the steel improves. The improvement in notch ductility, particularly at low temperatures, assists in the design of welded joints and reduces the risk of brittle and fatigue failure.

In practice, around 80% of the steel used in the UK for general constructional purposes is required in Grade 43A, 15% is required in Grades 50B and C and the remaining 5% comprises the higher-strength grades and a fifth category known as weathering steels (WR50). Compared with 43A, the other steel grades and subgrades are supplied at an additional cost per tonne which must be evaluated against benefits and other considerations at the design stage.

Weathering steel is usually supplied to Grade WR50B and under the brand-names of Cor-Ten and Stalcrest. Under the action of weather this steel forms a patina of fine and adhering rust which then inhibits further corrosion. To obtain the visual attractiveness of a uniform patina care is needed in detailing and erecting structures in this material.

10.7 Concluding summary

1. Following the production of liquid steel there are several methods for forming finished products, with the main route being continuous casting and hot rolling.
2. Modern rolling mills are moving increasingly to continuous rolling lines equipped with computer tracking and process control.
3. A wide range of section sizes in several steel quality grades is available with optional additional finishing processes.
4. Familiarity with standard specifications, including rolling tolerances, is of considerable advantage to the designer of constructional steelwork.

Background reading

UNITED STATES STEEL CORPORATION (1971) *The Making, Shaping and Treatment of Steel,* 9th edn.

11

Principles of welding

Objective To present an overall view of the implications of jointing by welding.

Prior reading Chapter 8 (Characteristics of iron–carbon alloys).

Summary The chapter describes the basic principles of making a welded joint and discusses the structure and properties of both weld metal and the heat-affected zone, It explains the necessity for, and types of, edge preparations for butt welds and gives the rationale for choosing welding procedures for particular joints.

11.1 Types of joint

Welding offers a means of making continuous, load-bearing metallic joints between the components of a structure. In structural work a variety of joints is found, but all joints can be typified by or made up from the four basic configurations shown in Figure 11.1. These are:

1. Butt;
2. Tee;
3. Lap;
4. Corner.

11.2 Method of making a welded joint

A welded joint is made by fusing (melting) the steel plates or sections (the parent metal) along the line of the joint. The metal melted from each member of the joint unites in a pool of molten metal which bridges the interface. As the pool cools, molten metal at the fusion boundary solidifies, forming a solid bond with the parent metal (Figure 11.2). When the solidification is complete there is a continuity of metal through the joint.

Two types of weld are in common use, butt and fillet. In the former the weld metal is generally contained within the profiles of the welded elements. In the latter deposited weld metal is external to the profile of the welded elements. Obviously, the complete length of joint cannot be melted simultaneously. In practice, a heat source is used to melt a small area and is then moved along the joint line, progressively fusing the parent metal at the leading edge of the weld pool, as in Figure 11.3. At the same time, the metal at the trailing edge of the pool solidifies. Most people are familiar with the use of an oxyacetylene flame to weld sheet metal (for example, a car repair), but the most commonly used heat source in structural work is a low-voltage (15–35 V), high-current (50–1000 A) arc. As shown in Figure 11.4, the arc operates between the end of a steel rod (electrode) and the workpiece and melts both the parent metal and the electrode. Molten metal from the latter is added to the weld pool.

The molten steel in the pool will readily absorb oxygen and nitrogen from the air. This would lead to porosity in the solidified weld and possibly to metallurgical problems. Figure 11.5 shows

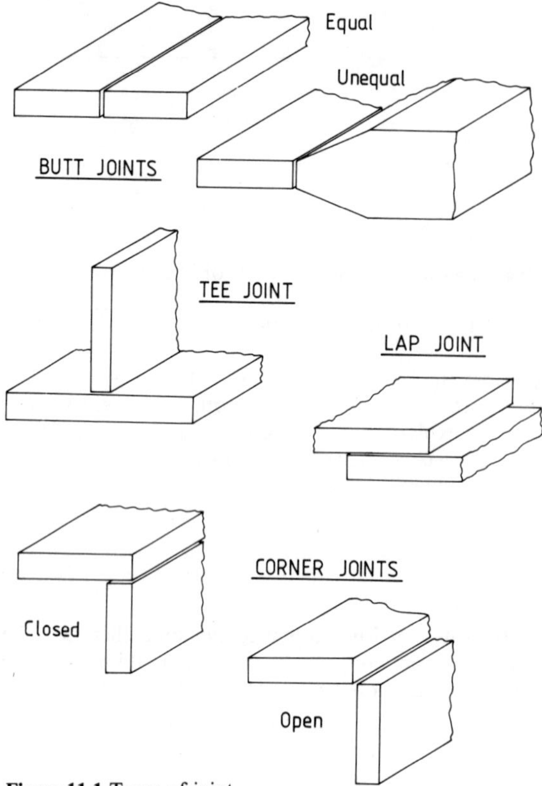

Figure 11.1 Types of joint

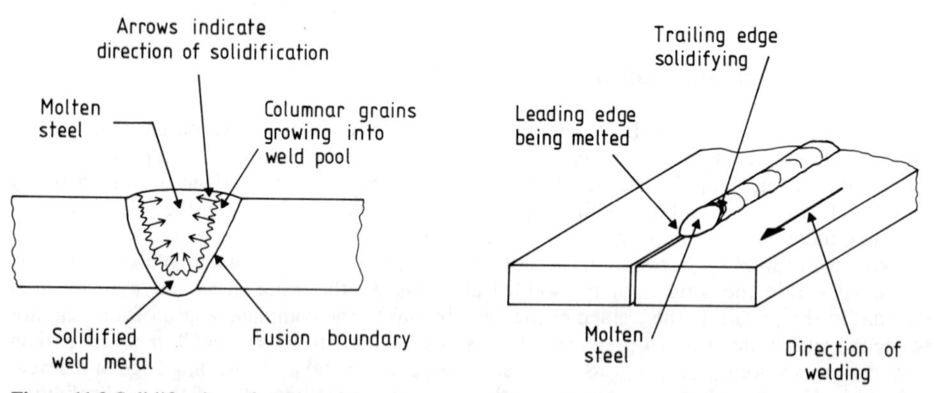

Figure 11.2 Solidification of weld metal

Figure 11.3 Progressive fusion and solidification

how this is avoided by covering the pool with a molten flux (MMA and SAW) or by replacing the air around the arc by a non-reactive gas (MAG welding).

11.3 Structure and properties

The solidified weld metal has a cast structure and properties characteristic of cast steel, i.e. higher yield to ultimate ratio and lower ductility compared with structural steel. The weld metal

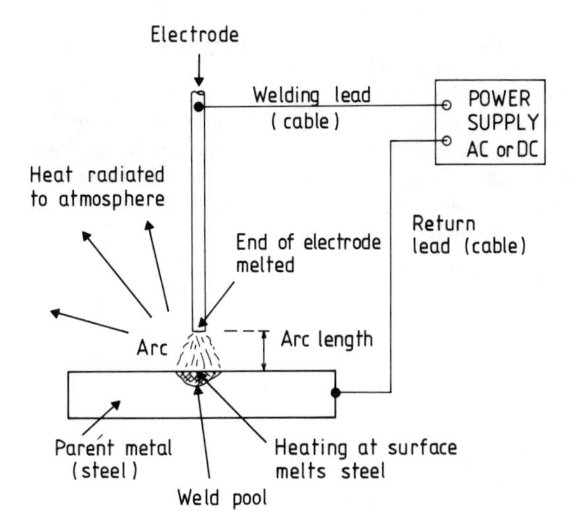

Figure 11.4 Welding arc

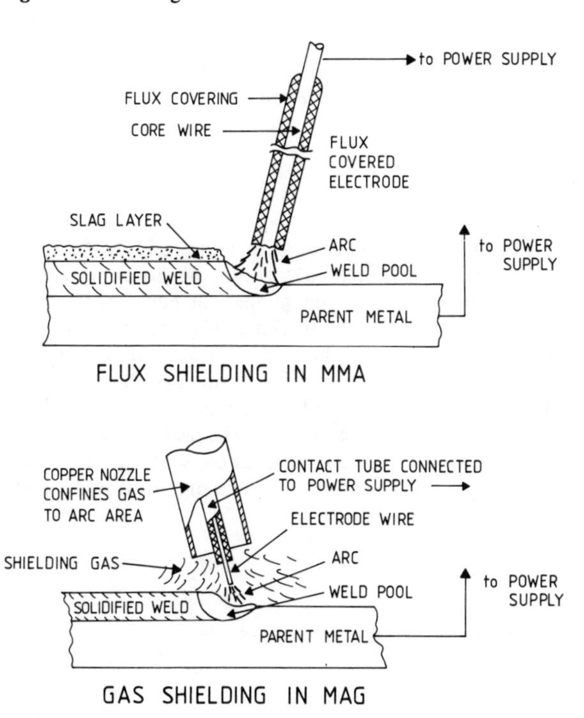

Figure 11.5 Shielding methods

is a mixture of parent metal and steel melted from the electrode. In structural work the composition of the electrode is usually chosen so that the resultant weld metal is stronger than the connected elements. For example, an E51 electrode (BS 639), which would be used to weld a BS 4360 Grade 50 steel, gives a weld deposit which has a minimum yield stress of 360 N/mm^2, with a tensile strength within the range 510–650 N/mm^2. Occasionally specific conditions may override this. For example, when joining stainless steel to carbon–manganese steel a highly alloyed electrode must be used to avoid cracking in the weld metal.

When the weld pool is cooling and solidifying, the majority of the heat flows through the parent metal alongside the joint. The steel is thus subjected to heating and cooling cycles similar to those experienced in heat-treatment practice. As shown in Figure 11.6, the structure of the steel will be changed in this region (called the heat-affected zone, HAZ) and this must be taken into account in the design stages in terms of notch toughness (Charpy value), corrosion resistance, etc. The structure of the HAZ will be dictated by:

1. The composition of the steel (carbon equivalent); and
2. The cooling rate in the HAZ.

In turn, the cooling rate is determined by:

1. Arc energy (i.e. heat input to the joint);
2. Type of joint;
3. Thickness of steel;
4. Temperature of steel plate or section during welding (i.e. preheat).

In addition to its effect on the cooling rate, preheat is used to:

1. Disperse hydrogen from the weld pool and HAZ. (Hydrogen in the HAZ increases the risk of cracking if hardening has occurred. The hydrogen comes principally from the flux. An appropriate electrode, correctly stored, especially on site, will reduce the risk of hydrogen pick-up);
2. Remove surface moisture in highly humid conditions or on site;
3. Bring the steel up to 'normal' ambient conditions (20°C) on site.

11.4 Edge preparation for butt welds

The depth the arc melts into the plate is called the depth of penetration (Figure 11.7). As a very rough guide, the penetration is about 1 mm per 100 A. In manual welding the current is usually not more than 350 A; more commonly it is 150–200 A. This means that the edges of the plate must be cut back along the joint line if continuity through the thickness is to be achieved. The groove so formed is then filled with metal melted from the electrode. Various edge forms are used and are illustrated in Figure. 11.8; the edges may be sawn, guillotined or flame cut.

The first run to be deposited in the bottom of the groove is called the root run. The root faces must be melted to ensure good penetration, but at the same time the weld pool must be controlled to avoid collapse, as seen in Figure 11.9. This task requires considerable skill, and the difficulties can be eased by using a backing strip. The choice of edge preparation depends on:

1. Type of process;
2. Position of welding (Figure 11.10);
3. Access for arc and electrode;
4. Volume of deposited weld metal, which should be kept to a minimum;
5. Cost of preparing edges;
6. Shrinkage and distortion (see below).

11.5 Welding procedure

11.5.1 Introduction

The term 'welding procedure' is used to describe the complete operation of making a weld, and it covers choice of electrode, edge preparation, preheat, welding parameters (voltage, current

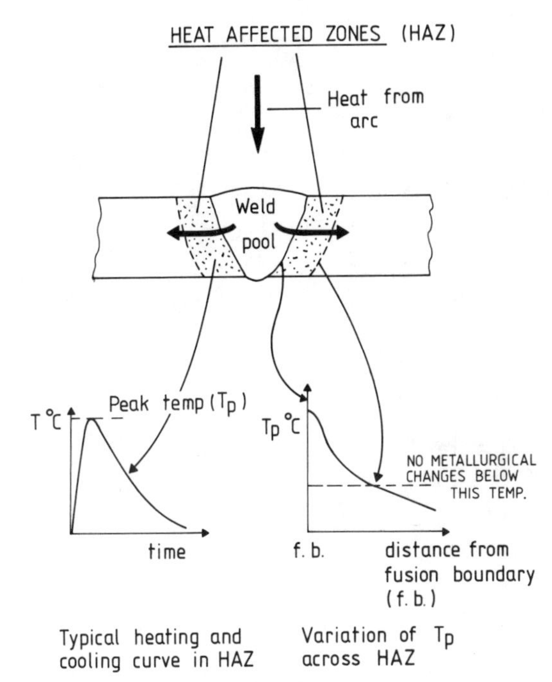

Figure 11.6 Formation of HAZ

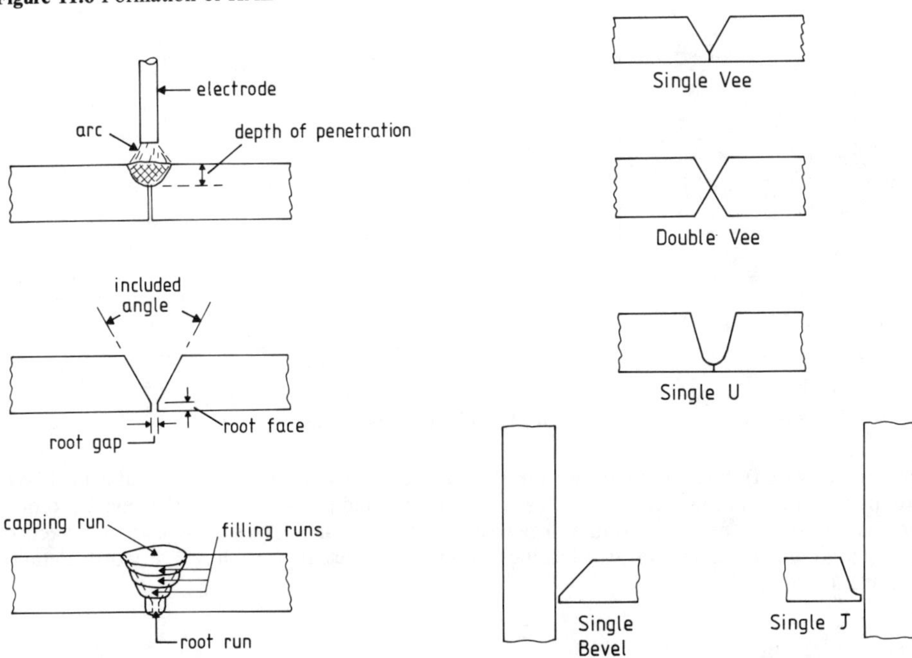

Figure 11.7 Penetration in arc welding

Figure 11.8 Edge preparations

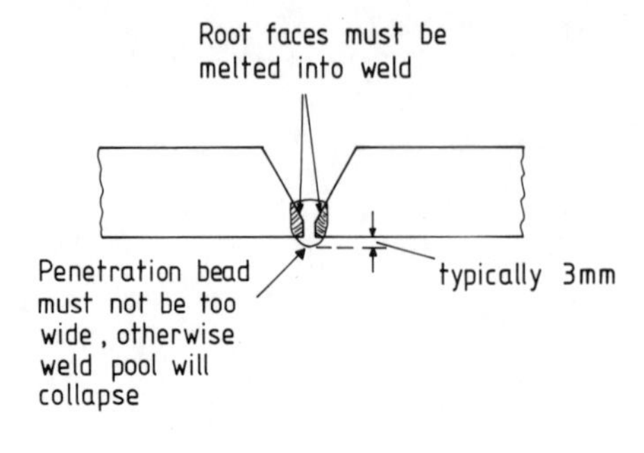

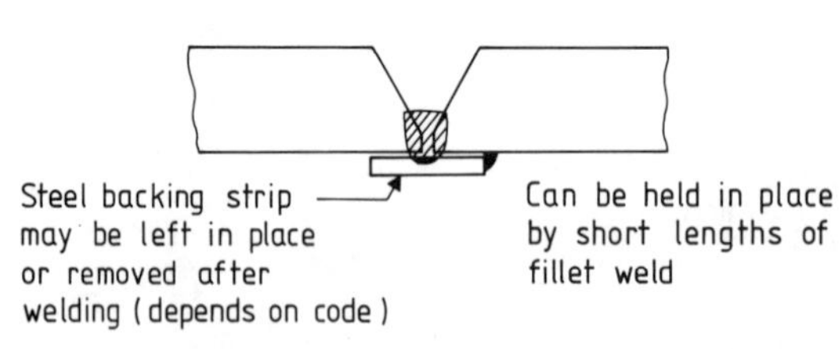

Figure 11.9 Root-run techniques

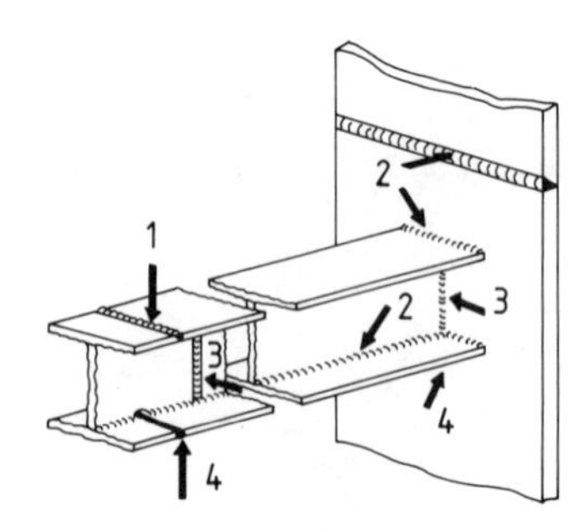

Figure 11.10 Welding positions

and travel speed), welding position, number of weld runs to fill the groove and post-weld treatments (for example, grinding or heat treatment). Weld procedures may be devised to meet various needs – for example, to minimize costs, control distortion (see below), avoid defects or achieve good impact properties. Specific aspects of the weld procedure are worth detailed comments.

11.5.2 Current

The current controls heat input. The minimum value is fixed by the need to fuse the plate and to keep the arc stable, but the specified minimum may be higher to avoid HAZ cracks. The maximum current depends on operating conditions. Usually, as high a current as possible is used to achieve faster welding and hence lower costs. The use of maximum current may be restricted by position; in the overhead position, currents above 160 A cannot be used. High currents usually give low impact properties. Note that the electrode diameter is chosen to match the current being used.

11.5.3 Welding position

The effect of position on current is noted above. Welding in the overhead position requires greater skill to avoid defects such as poor profile and should only be allowed when absolutely necessary. Vertical welding is slower than welding in the flat position.

11.5.4 Environment

Work on site creates constraints which must be accommodated by the weld procedure. In cold weather steel may need to be heated to bring it up to 20°C. Overnight condensation and high humidity can lead to porosity; note the problems of keeping electrodes dry in store. It is often difficult to achieve accurate fitting of the joint, and variable and/or large root gaps result in defective welds.

11.6 Shrinkage

During cooling, the hot metal in the weld zone contracts, causing the joint to shrink. The contraction is constrained by the cold metal surrounding the joint; stresses are set up which, being in excess of the yield stress, produce plastic deformation, and this can lead to the distortion or buckling shown in Figure. 11.11. Distortion can be controlled by choice of edge preparation and weld procedure.

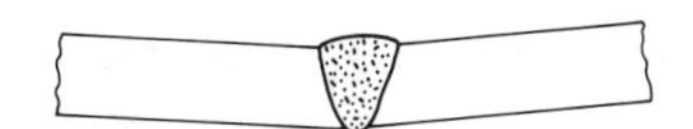

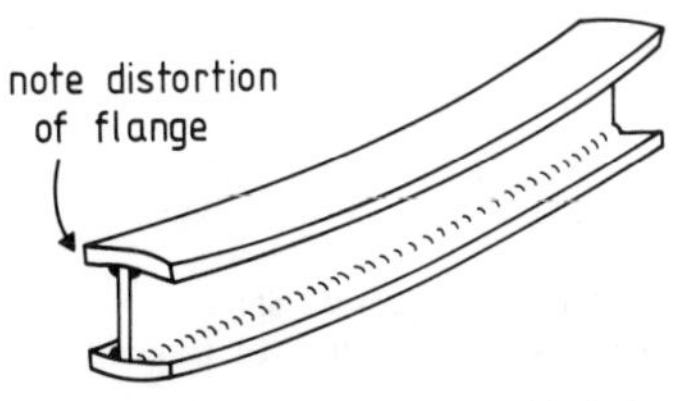

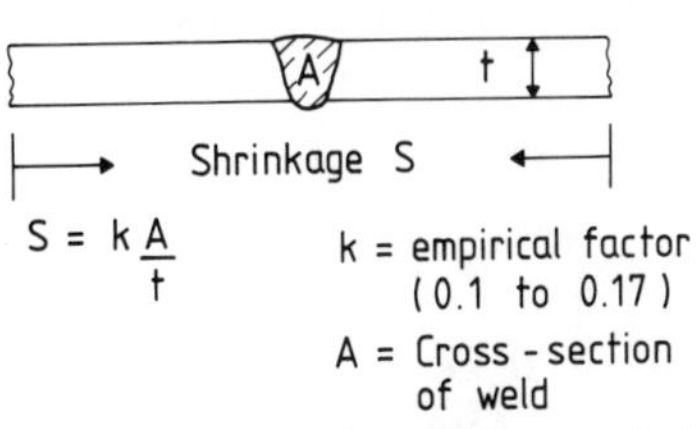

Figure 11.11 Distortion in welding

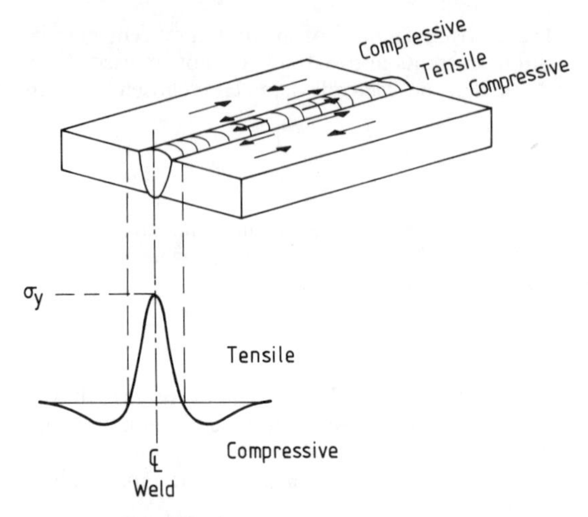

Figure 11.12 Residual stress

When the plastic deformation has ceased the joint is left with the residual stress pattern of Figure 11.12 with tension in the weld metal and HAZ and compression in the surrounding steel. The significance of these residual stresses is discussed in other chapters.

11.7 Concluding summary

1. A welded joint is made by fusing parent metal from both components being joined, possibly with added weld metal.
2. The properties of both the weld metal, which has melted and solidified, and the surrounding heat-affected zone may differ from those of the parent metal.
3. Welding procedures should be properly specified to give a satisfactory welded joint. The major parameters are: welding position, electrode type, edge preparation, preheat, voltage, current, travel speed, number of runs and post-weld treatments.
4. Hot metal in the weld zone contracts during cooling, causing residual stresses. Distortion will occur if appropriate remedial measures are not taken.

Background reading

BRITISH STANDARD BS 5135 (1974) Metal-arc Welding of Carbon and Carbon Manganese Steels
BRITISH STANDARD BS 639 (1976) Covered Electrodes for the Manual Metal-arc Welding of Carbon and Carbon Manganese Steels
GOURD, L.M. (1980) *Principles of Welding Technology*, Edward Arnold, London
HICKS, J. (1979) *Welding Design* Granada, St Albans
MILNER, D.R. and APPS, R.L. (1968) *Introduction to Welding and Brazing*, Pergamon, Oxford
PRATT, J.L. (1979) *Introduction to the Welding of Structural Steelwork*, Constrado, London

12

Welding processes

Objective To review the welding processes commonly used in construction and to highlight some practical considerations.

Prior reading Chapter 11 (Principles of welding).

Summary The chapter describes the welding processes commonly used in construction: manual metal-arc welding, dip and spray transfer metal-active gas welding, submerged-arc welding and stud welding. Each process is briefly described and its advantages, disadvantages and limitations of use are summarized. The choice of welding process for a particular situation is reviewed.

12.1 Manual metal-arc (MMA) welding

This is the most widely used arc welding process. It is manually operated and requires considerable skill to produce good-quality welds. The electrode consists of a steel core wire (3.2–6.0 mm diameter) and the covering flux contains alloying elements (for example, manganese and silicon). The arc melts the parent metal and electrode. As metal is transferred from the end of the core wire to the weld pool the welder moves the electrode to keep the arc length constant. This is essential, as the width of the weld run is largely governed by the arc length. The flux melts with the core wire and flows over the surface of the pool to form a slag, which must be removed after solidification (Figure 12.1).

MMA has many attractions: it has low capital cost and freedom of movement (up to 20 m from power supply – useful on site), it can be used in all positions and it is suitable for structural and stainless steels (but not aluminium). Its main drawback is a low duty cycle, i.e. only a small volume of metal is deposited before the welder has to stop and insert another electrode. This is not a problem on short welds but becomes a consideration on long ones, especially when labour costs are high.

The operating characteristics of the electrode are controlled by the composition of the flux covering. A variety of electrodes is available to suit different applications and MMA electrodes for structural welds are specified in BS 639. The diameter of the wire is chosen to match the current being used. When low hydrogen contents in the weld pool are necessary to avoid cracks in the heat-affected zone (HAZ) on cooling, MMA electrodes must be stored at 150°C to ensure that they are free from moisture.

12.2 Metal-active gas (MAG) welding

This process is sometimes referred to as metal-inert gas (MIG) welding, although strictly the term MIG should be limited to the use of pure argon as a shielding gas, which is not used for steel.

MAG is also manually operated. The arc and weld pool are shielded by a gas which does not react with molten steel; in current practice the shielding gas is carbon dioxide or a (80%:20%)

Circuit:

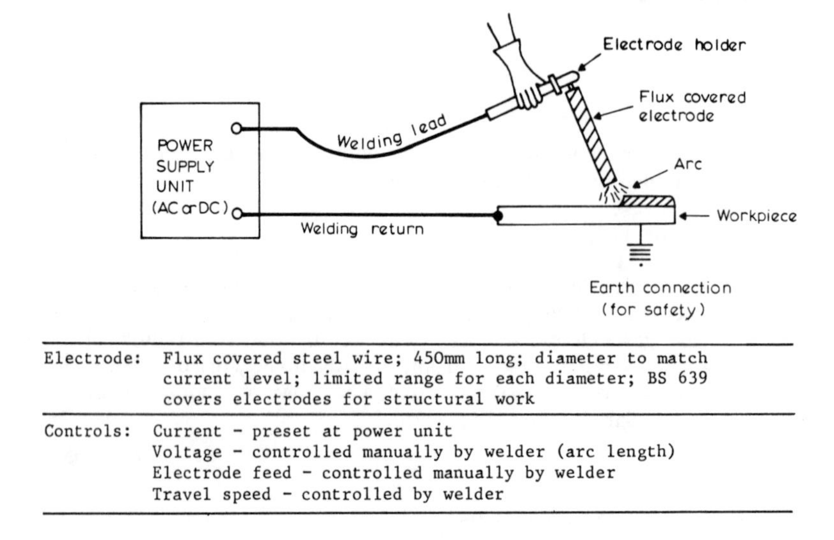

Electrode:	Flux covered steel wire; 450mm long; diameter to match current level; limited range for each diameter; BS 639 covers electrodes for structural work
Controls:	Current - preset at power unit Voltage - controlled manually by welder (arc length) Electrode feed - controlled manually by welder Travel speed - controlled by welder

Figure 12.1 Manual metal-arc (MMA) welding

mixture of argon and carbon dioxide. No flux is necessary to shield the pool (the alloying elements are in the electrode wire), but sometimes a flux-cored electrode is used to produce a slag which controls the weld profile (for example, in large fillet welds in the horizontal–vertical position). The arc length is controlled by the power supply unit. Although MAG welding is somewhat easier to use than MMA, considerably more skill is required to set up the correct welding conditions (Figure 12.2).

For 'positional' welding, i.e. vertical and overhead, the current must be kept below 180 A (so that welding takes place in the 'dip transfer' mode) and welding speeds are comparable with MMA. Overall times for a joint, and hence productivity, are better since there is no need to deslag or change electrode. In the flat position currents up to 400 A ('spray transfer') can be used to give high welding speeds. MAG welding is especially suitable for fillet-welded joints (for example, beam-to-beam and stiffener-to-panel connections). It is not easy to use on site because of problems of equipment movement and the need to provide screens to avoid loss of the gas shield in windy conditions.

12.3 Submerged-arc welding (SAW)

This is a fully mechanized process in which the welding head is traversed along the joint by a tractor or lead-screw. The electrode is a bare wire which is advanced by a governed motor. The voltage and current are selected at the beginning of the weld and are maintained at the preselected values by feedback systems which, in practice, vary in sophistication. The flux is in the form of granules and is placed on the surface of the joint. The arc operates below the surface of the flux, melting a proportion of it to form a slag. Unfused flux is collected and re-used for the next weld (Figure 12.3).

Submerged-arc welding operates best at currents between 400 and 1000 A. This means that

Circuit:

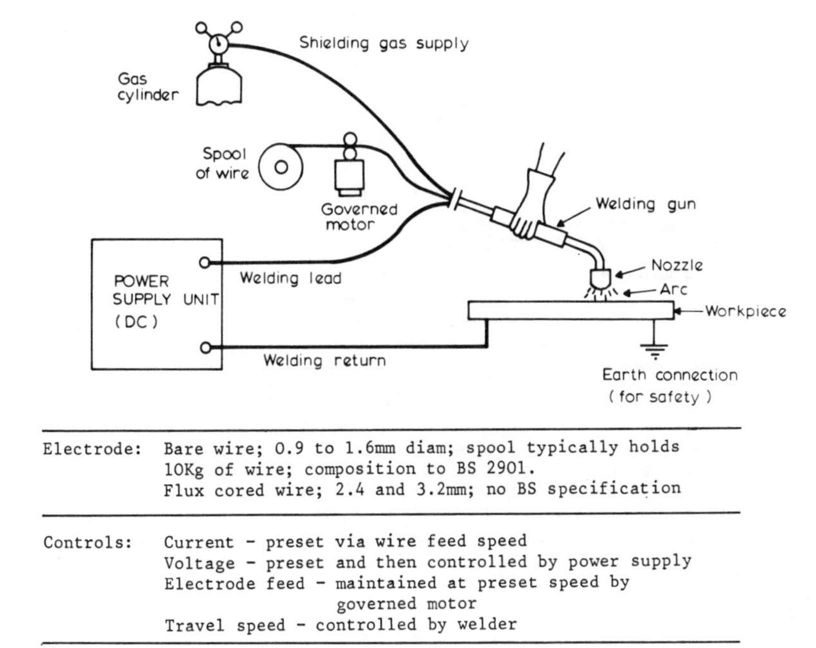

Electrode:	Bare wire; 0.9 to 1.6mm diam; spool typically holds 10Kg of wire; composition to BS 2901. Flux cored wire; 2.4 and 3.2mm; no BS specification
Controls:	Current - preset via wire feed speed Voltage - preset and then controlled by power supply Electrode feed - maintained at preset speed by governed motor Travel speed - controlled by welder

Figure 12.2 Metal-active gas (MAG) welding

weld pools are large and can only be controlled in the flat position, although fillets can be deposited in the horizontal–vertical position up to 10 mm leg length in one run. It is also difficult to control penetration in the root run, and a backing strip is usually used for butt welds. Alternatively, the root run can be made by MMA or MAG and the groove filled with SAW. SAW offers considerably advantages when welding long joints (i.e. those in excess of 1 m). The high welding speeds and continuous operation lead to high productivity. An accurate joint fit-up is, however, a prime requirement.

Other mechanized welding systems, similar to SAW, use a flux-cored electrode or, more rarely, a flux-covered electrode. The latter is usually called Fusare welding (trade name).

12.4 Stud welding

This is a variant of arc welding which mechanizes the welding of studs to plane surfaces. The stud, which may be a plain bar with an upset head or threaded, is the electrode, and it is held in a chuck which is connected to the power supply. The stud is first touched onto the surface of the steel plate or section. As soon as the current is switched on, the stud is moved away automatically to establish an arc. When a weld pool has been formed and the end of the stud is molten the latter is automatically forced into the steel plate and the current is switched off. The molten metal which is expelled from the interface is formed into a fillet by a ceramic collar or ferrule which is placed around the stud at the beginning of the operation. The ferrule also provides sufficient protection against atmospheric contamination (Figure 12.4).

Stud welding offers an accurate and fast method of attaching shear connectors, etc. with the minimum of distortion. While it requires some skill to set up the weld parameters (voltage, current, arc time and force), the operation of the equipment is relatively straightforward.

Circuit:

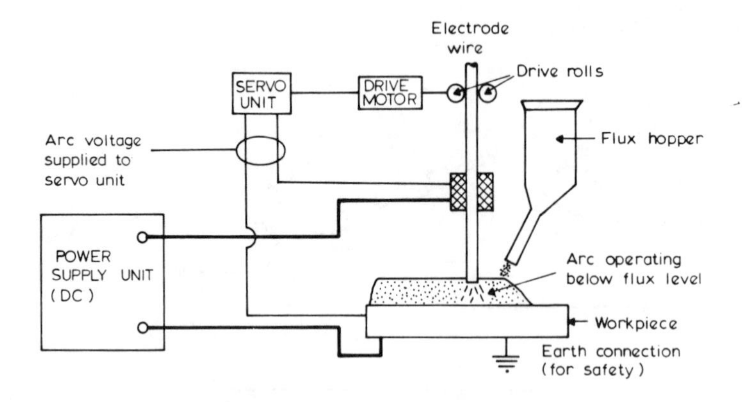

Electrode:	Bare wire; 2.4 to 6mm diam; composition to BS 4165
Flux:	Fused (silicate) or agglomerated (basic); BS 4165
Controls:	Current - maintained at preset values Voltage - by servo-unit Electrode feed Travel speed - preset on traverse unit. Can be altered during welding

Figure 12.3 Submerged-arc welding (SAW)

12.5 Choice of process

A number of factors must be taken into account:

1. Where the weld is to be carried out. SAW and MAG are best done in the protected environment of the fabrication shop. MMA may more readily be used on site.
2. Accuracy of fit-up and possibility of misalignment. SAW and spray transfer MAG require good fit-up; they are particularly sensitive to variations in root gap and/or root face dimensions.
3. Access to joint. It is necessary to ensure that both the welding plant and the welding torch or head can be properly positioned.
4. Position of welding. SAW and spray transfer MAG are not suitable for vertical or overhead positions. Dip transfer MAG is acceptable for vertical and overhead welding, but MMA is probably best for overhead work, especially on site.
5. Steel composition. MAG and SAW are less likely to lead to HAZ cracking. The consequent reduction in preheating requirements could offset the other disadvantages of MAG for site work.
6. Comparative cost. The cost per unit length of weld can be readily calculated but must allow for differences in duty cycle (idle time between electrodes for MMA, etc.).

Circuit

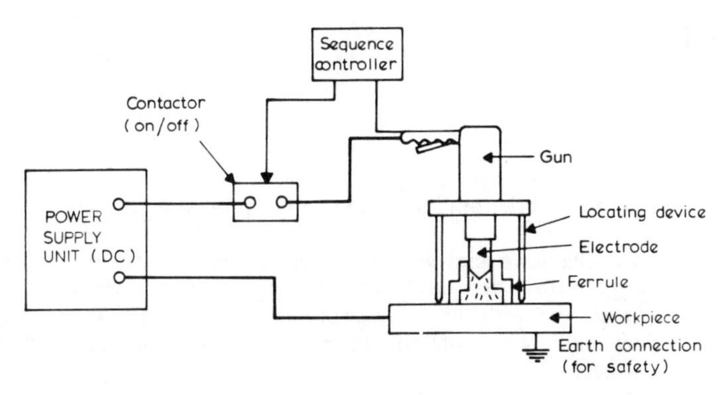

Electrode:	Studs with ends prepared for welding by manufacturer
Ceramic ferrules:	Shaped to suit stud profile
Controls:	Current - Preset at power supply unit Voltage - Vary during welding in a sequence Electrode feed - determined by the control unit Location - controlled by operator

Figure 12.4 Stud arc welding

12.6 Concluding summary

1. The welding processes commonly used in constructional steelwork are: manual metal-arc welding, dip and spray transfer metal-active gas welding, submerged-arc welding and stud welding.
2. Stud welding is used for attaching shear connectors and other studs to structural steelwork.
3. The correct choice of process depends on: situation, fit-up, access, position, steel composition and economics.

Background reading

BSC, *Structural Hollow Sections*
GOURD, L.M. (1980) *Principles of Welding Technology,* Edward Arnold, London
HOULDCROFT, P.T. (1977) *Welding Process Technology,* Cambridge University Press, Cambridge
Welding Handbook, Vol. 1, *Welding Technology*, 8th edn (1987), American Welding Society, Miami

13

Fabrication

Objective To give a brief outline of fabrication processes.

Summary The chapter begins with a brief summary of forms of contract and organization for steel construction. It then reviews fabrication processes, with brief descriptions of the main operations and concludes with notes on fabrication tolerances, trial erection and inspection.

13.1 Forms of contract and organization

13.1.1 Contract scope

It is common practice for the fabrication company to receive a contract for both fabrication and erection which includes the preparation of detail drawings. Package-deal contracts, where the design is undertaken by the fabricator, are also common in the UK, particularly for single-storey industrial buildings. Major high-rise structures are generally designed by a consulting engineering practice. The detail drawings used in fabrication are, however, traditionally made by the steelwork fabricator.

13.1.2 Contract procedures

After the receipt of the order and agreement of terms, copies of the principal documents are given to the drawing office. These are usually:

1. The engineer's drawings;
2. The conditions of contract;
3. The technical specification;
4. The contract programme.

The programme is the essential factor on which the success of the contract depends and it has the greatest effect on costs. The consequential effects of delays can be traumatic to following trades, and subsequently the client. Programmes are normally made out in bar-line form or by using computer-based techniques of critical path analysis. Essential elements are:

1. Erection sequence, for programme planning;
2. Ordering of material from the mills;
3. Preparation of fabrication drawings;
4. Material preparation;
5. Fabrication;
6. Assembly;
7. Protective treatment;
8. Delivery to site.

Each element is planned to a set time scale which must take into account other contracts being handled in the same period. If the erection programme imposes shop fabrication demands in excess of the fabricator's capacity, then subletting of work (coupled with the requisite quality assurance and quality control assessment) will be necessary.

13.1.3 Drawing office control

It is the responsibility of the drawing office to interpret the engineer's design drawings and calculations and provide the material ordering department with complete lists for all steel, fittings and bolts. The drawing office is also responsible for connection design, although the competent engineer will illustrate on his design drawings the type of connection he had in mind when designing the structure. If the engineer has not considered the connection details heavy costs may be incurred by the fabricator in providing stiffening to the basic sections.

13.2 Fabrication procedures

13.2.1 Fabrication shop layout

All fabrication should ideally be planned for material to pass through a one-way system from receipt to final despatch. A flow-line as indicated in Figure 13.1 shows the main areas of activity in a modern fabrication shop. All fabrication shops are equipped with overhead travelling cranes and some of these are radio-controlled from floor level. Mechanized conveyor systems, which are becoming more common in larger shops, can greatly reduce handling costs. Some operations, such as mechanical chipping and blast-cleaning, are excessively noisy, and are best kept clear of the main buildings.

Material is placed in a stockyard so that it can be easily identified and moved. A high degree of automation is now utilized in material handling based on magnetic cranes and conveyors. The key factor in the stockyard illustrated in Figure 13.1 is a travelling Goliath Magnet Crane, which can lift both plates and sections. Computer records hold details of member sizes, lengths, weights and steel quality, all related to an identification mark. When required, the steel is shot-blasted in an adjacent plant which automatically senses the member size. Automatic paint spraying and drying can follow blast-cleaning, although this is not always done if welding is required. In the stockyard illustrated, the steel is then sorted onto transporter trestles by a fixed portal crane and moved into the preparation bays.

13.2.2 Templates and marking

Steel may be marked directly with scribe lines and hole centres, but this is not necessary with automatic plant. Some templates are still made today, particularly for small plate fittings and gussets, and it is now possible for them to be plotted in the drawing office using computer

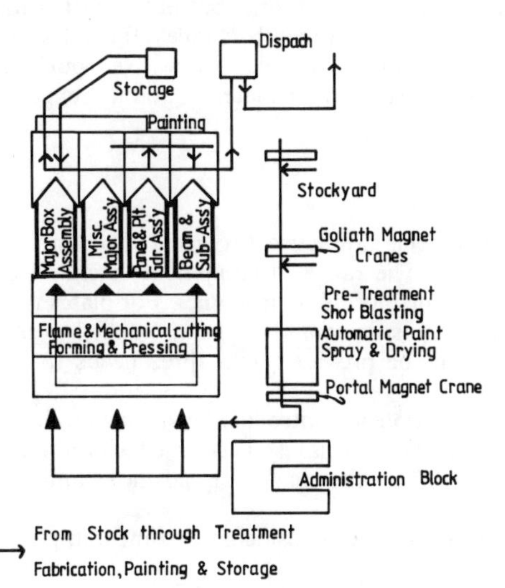

Figure 13.1 Production flow through fabrication works. (Reproduced with permission of Cleveland Redpath Ltd)

graphics workstations. Automation in the fabrication shop is now, however, greatly reducing the work of the template maker.

13.2.3 Sawing line and rolled sections

Most rolled sections are sawn to length, and three types of saw are available to the fabricator:

1. The circular cold saw;
2. The bandsaw;
3. The motor-operated hacksaw.

By far the most popular is the circular cold saw, which is more productive than the bandsaw. A saw can perform within an accuracy of a fraction of a millimetre on length and within a squareness of 0.2% of the depth of the cut. The most accurate cold saw is a swivelling-arm type, where the blade descends onto the bar. The blade speed of cutting is automatically adjusted as it passes through the workpiece.

Automatic sawing lines are equipped with mechanized longitudinal and transverse conveyors and have a measuring device working in conjunction with the saw. A typical arrangement is shown in Figure 13.2. A fully automated system is operated through a computer-produced tape.

13.2.4 Drilling and the beam line system

Larger fabrication shops now have automatic beam-line systems which are generally linked to the sawing line with conveyors. The beam moves on longitudinal conveyors along the x-axis, the flange drilling heads being adjustable along the y-axis, denoted Y and V for each flange, and the web drilling heads along the z-axis (see Figure 13.3).

Like the sawing line, the system is operated by a computer tape. Some machines have multi-heads which can drill three holes simultaneously in each axis.

13.2.5 Cropping, guillotines and punching

Cropping shears are used for cutting small sections. Guillotines are used for shearing plates up to 25 mm thick when specifications allow, but the plate in contact with the bottom blade is usually distorted.

Punching holes in steelwork is, of course, much faster than drilling, but in the past British Standards have limited its use to Grade 43 steel and secondary members unless the holes were being reamed out to a larger size; these restrictions are currently under review. New punching machines which operate at high speeds induce less distortion in the material, and it is expected that more punching will be allowed in the future.

13.2.6 Flame burning of plates

Splitting and shaping plates by flame cutting is now general practice. Oxygen and propane are stored in bulk with pipelines laid to each machine. The range of equipment extends from hand-held torches to multi-torch, numerically controlled, profiling machines. For plates up to 80 mm thick as many as nine heads are arranged to ensure equal heat being applied to both edges to avoid distortion. The cutting carriage can be provided with three heads to give double-bevel edge preparations.

Other single-head machines can be operated by an optical controlling head which follows a one-in-ten, or full-size, outline drawn on paper. Profile cutting is also performed with numerically controlled (NC) tape-operated machines, which can also mark hole positions and hard stamp identification marks.

For accelerated cutting speeds where edge hardness is not considered detrimental, plasma

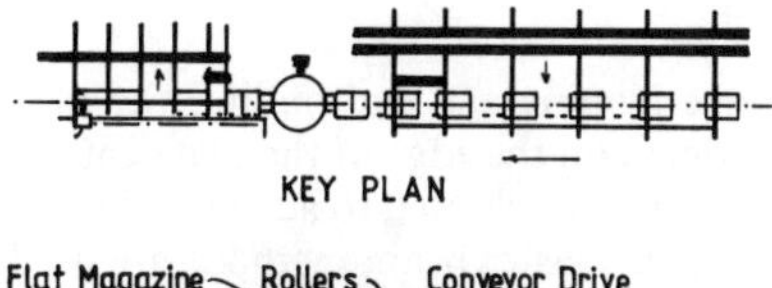

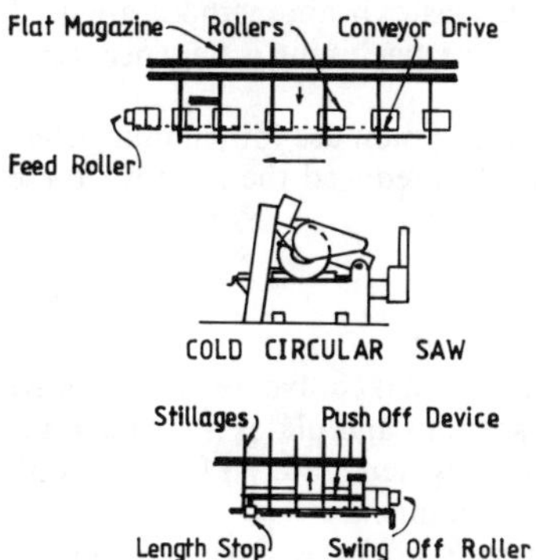

Figure 13.2 Automatic sawing line. (Reproduced with permission of Kaltenbach (GB) Ltd)

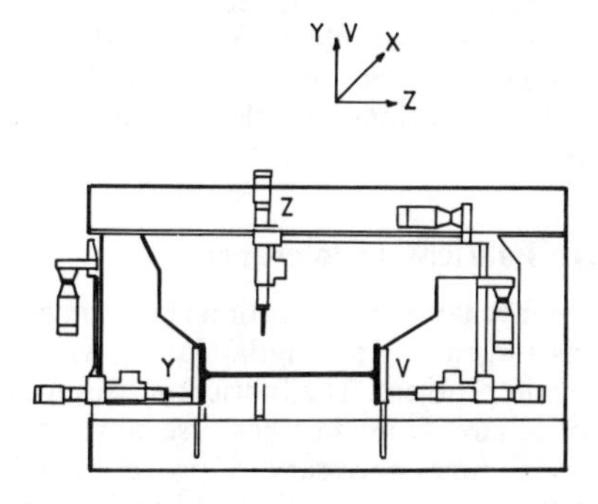

Figure 13.3 Multiple-axis drilling machine. (Reproduced with permission of Kaltenbach (GB) Ltd)

cutting under water or an inert powder is now available in some fabrication shops. Laser cutting is beginning to come into use but is at present restricted to thin plates, while the resulting edge hardness makes it unsuitable for many purposes.

13.2.7 Pressing and forming

The most important plate pressing and forming operations for the modern fabricator are those used in creating shapes to supplement the range of rolled sections. The trapezoidal shaped trough, used to stiffen bridge decks, is a good example. Circular sections of greater than standard dimensions and cold-formed sections made from thin-gauge plates are also prepared.

Bending sections and plates is also achieved with rolling machines, generally with three rolls in a pyramid configuration. Modern rolls can accommodate plates up to 175 mm thick and have, for example, an accuracy on tubular rolling of around 6 mm 'ovality' on a diameter of 1.5 m.

13.2.8 Welding and assembly

Three processes, described in Chapter 12, are regularly employed:

1. Manual metal-arc welding for fittings, and for some profile and positional welding;
2. Metal-active gas (MAG) welding which can carry out many of the tasks formerly done by manual welding.
3. Submerged-arc welding for fully automatic processes. This is particularly useful for welding in the flat or horizontal–vertical position and for long-run weld lines for fabricating plate and box girders.

Other processes used for particular purposes are electroslag welding for vertical joints, electric-arc welding of studs and thermit welding of thick steel sections such as rails.

Welding procedures are decided by the welding engineer, who will prepare a procedure sheet. Ensuring that the welder is qualified to the required standard is the responsibility of the welding engineer. He will also decide the method of non-destructive testing (NDT), whether by radiographic, ultrasonic, magnetic particle or dye-penetrant means.

13.2.9 Machine operations

Most fabrication shops are equipped with facilities for edge-planing and end-milling and also for surface machining of slabs. Unacceptable levels of hardness on the edge of the plate caused by burning may be removed by planing. End-planing of members is adopted to achieve a higher standard of squareness than is possible with a sawn end. Optical laster beam methods are used to align the axis of the member normal to the cutting head. Surface machining is only necessary on column slab base plates and special bearing surfaces.

Chipping with a pneumatic chisel and hand-grinding were in common use for trimming steel for weld preparations, but the advent of machine gas cutting has reduced the need for these procedures.

13.2.10 Fabrication tolerances

Modern mechanized fabrication shops have very accurate dimensional control over fabricated sections and cutting to length. Coping with the tolerance on sections and plates from the mills is the main problem. The fabricator uses his bending rolls to straighten material from the mill and to 'square' flanges of beam sections at critical connection points.

It is sometimes necessary to 'prove' assembly in the fabricator's works by a trial erection of one section of the work. Parts of bridge structures, particularly those for overseas locations, and structures for supporting intricate industrial plant are examples. Trial erection is expensive and should be avoided where possible by building methods of site adjustment into the design.

13.3 Concluding summary

1. A design engineer should be mindful of the processes to be used in fabrication and in erection to achieve his design, and should ensure that unnecessary cost is avoided.
2. He must remember:
   (a) That material is cheaper when bulk ordered, and so avoid small quantities of different sizes;
   (b) To reduce to a minimum the number of pieces to be handled, and avoid the need for excessive stiffening;
   (c) To make allowance for weld distortion and fabrication tolerances;
   (d) That the cost will be reduced where automatic fabrication techniques can be employed;
   (e) That the cost of delivery, particularly overseas, can be reduced by careful design;
   (f) That good quality control is essential but specifications should not be unnecessarily stringent since this will increase costs.

14

Erection

Objective To give an understanding of the extent to which planning, basic design decisions and the design of details must be influenced by the erection process.

Summary The chapter begins with a review of the requirements for economic and safe erection. Then the significance of the erection process to the designer is emphasized, both for overall design concepts and detailed design.

14.1 Erection costs

The cost of erection can be a significant proportion of the total project budget and its estimation requires careful consideration of all aspects of the erection scheme. For example, questions concerning the ability of the structure to support the erection cranes as they climb ever higher must be addressed early in the design process. A structure may have to sustain dynamic crane loads during its erection which it will never again be asked to carry, and that with some parts left out.

14.2 The safety of the workforce

Steelwork erection is potentially a very hazardous task, and it is most important to make proper provision for the safety of the steel erector. The Health and Safety Executive and Statutory Regulations lay down the requirements of the law regarding the provision of equipment and arrangements, which will minimize the risk of an accident.

The regulations stipulate requirements for people working in places where they can fall more than 2 m. Provision must be made for safe access to and from the working place, as well as the security of the working place itself. The width of the walkway, the height of the handrail and means of preventing small tools being kicked off the platforms are all defined. Cranes, slings and lifting devices in general are all subject to inspection and to testing at regular intervals.

14.3 Design for erection

14.3.1 Introduction

The main aim during erection is the preservation of the stability of the structure at all times. Most structures which collapse do so during erection, and these failures are very often due to a lack of understanding on someone's part of what another has assumed about the erection procedure (i.e. a communication problem).

14.3.2 Temporary stress reversals and stability

Stresses can be reversed during erection, and every reversal, no matter how transient, must be

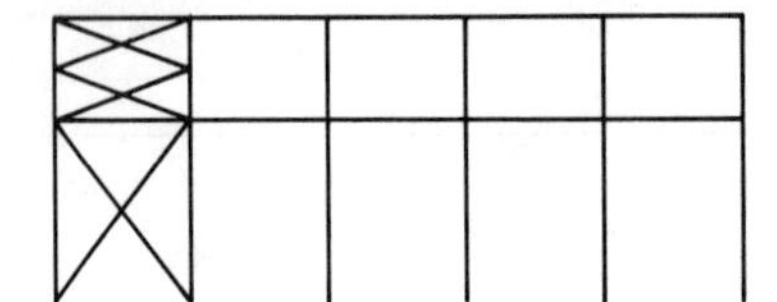

Figure 14.1 Stability during erection

considered at the design stage in order that provision can be made for coping with that temporary condition. Questions concerning which part of a structure must be built first to give initial stability or whether parts that have to be temporarily left out will impair that stability must be resolved at an early stage.

14.3.3 Temporary loads

If a crane or other heavy load is to be carried on the structure, or if the structure itself requires temporary supports to maintain its stability during erection, there will be a need for the design of temporary works. In some cases these temporary works will require the design of foundations and a minor structure in its own right. The same care must be devoted to the design of temporary structures as to that of the main project. A temporary structure has to cope with live as well as dead loads. Often the structure being supported is subject to temperature changes or vibrations resulting in movements. Failure to design means of accommodating these movements and loads can lead to collapse.

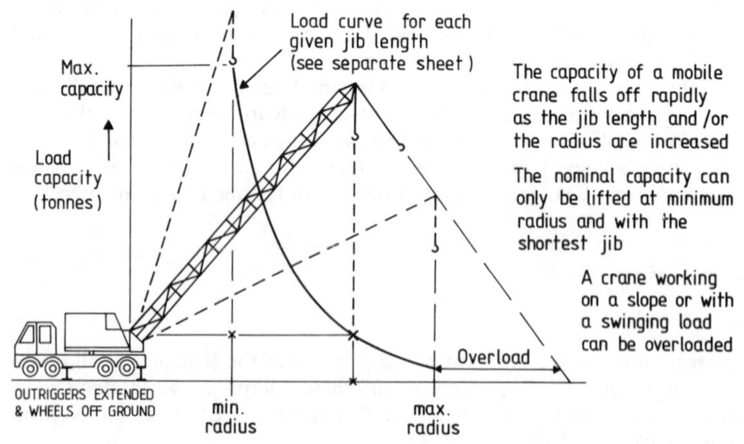

Figure 14.2 Mobile crane

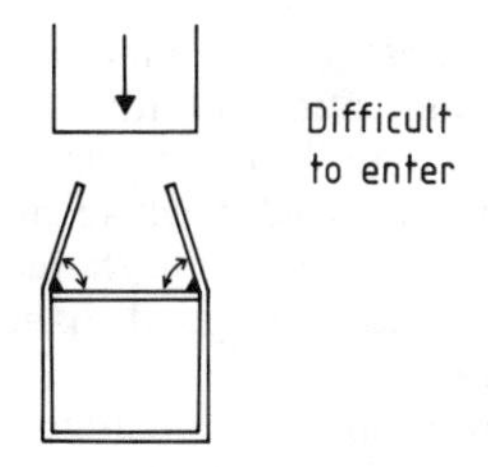

Weld distortion at a splice detail can prevent the entry of the joining member or may prevent the development of friction across the faying surfaces

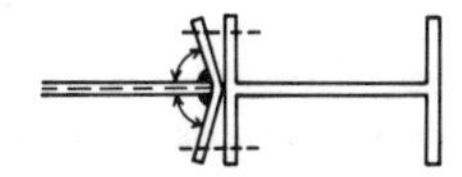

It can also lead to an accident as the erector struggles to make the joint and bolt it up

Figure 14.3 Problems caused by weld distortion

14.3.4 Permanent and temporary bracings

The designer of a structure should ensure that he positions braced bays in such a way as to ensure that they are the first to be erected. It is essential that the structure is braced and true before erection proceeds further. The safety and stability of the frame is of paramount importance, as seen in Figure 14.1.

The general arrangement drawing should be annotated by the designer in order to transmit his intention to the erector. A note stating 'Erection starts from this end' or 'Steelwork between lines 1 and 2 must be erected and braced first' will draw attention to the requirement. Only where stability is to be achieved by horizontal or vertical diaphragms that are to be added later can the use of temporary bracing be justified. In these cases the designer should have been aware of the need for stability and should have incorporated these temporary bracings into his design.

14.4 Practicalities of erection

14.4.1 The choice of craneage

Mobile cranes are most commonly used for steel erection, mounted on either wheels or on tracks. Tower cranes are also frequently used; these are often mounted on rail tracks to achieve a degree of mobility. Cranes up to 50 tons capacity are commonly found as part of the contractor's own plant fleet. Larger cranes (up to the 300-ton giants) are more often hired from specialist companies. Figure 14.2 shows the relationship between boom length and capacity for a typical mobile crane.

14.4.2 Maintaining tolerances

The finished work must be within a designated range of dimensional tolerances, and the best way of complying with these is to make sure that accuracy is achieved from the commencement of the job. Setting out the foundations and the holding-down bolts is often done at a stage when close tolerances and the thick mud on site seem incompatible. However, time spent on marking the centre lines and the levels of the concrete bases accurately is time well spent.

During fabrication, the heat input from welding, and the subsequent cooling, will inevitably cause distortion. Only with proper procedures will the final fabrication be within tolerance and be able to be fixed into the framework in its correct relationship to the mating half of the joint.

Details which are sensitive to welding distortion should be avoided where possible. For example, splice plates and gussets which are to be fixed and bolted on site should be wholly bolted, since if they are welded to one of the components there is a very good chance that they will not be able to be entered onto the other (see Figure 14.3).

End-plates are particularly prone to distortion and require careful fabrication. If the end-plates of the components of a portal frame are distorted the whole geometry of the frame will be affected. If the two mating parts of a built-up column section, or of a heavy lattice girder boom, have distorted unequally it may not be possible to splice them within permitted tolerances.

14.5 Concluding summary

1. The erection of a structure is an operation requiring meticulous planning at all stages, from conception to completion.
2. It is an operation involving many disciplines and requiring cooperation and communication between all those involved.
3. It is, furthermore, an operation that highlights the personal responsibilities of all those involved in ensuring that it is completed without accident, to time and within the contract budget.

15

Corrosion protection

Objective To develop an understanding of the corrosion process and the means of achieving satisfactory performance of structural steelwork.

Prior reading Chapter 9 (Engineering properties of metals).

Summary The chapter presents the basic theory of corrosion and indicates the conditions under which it may occur. It emphasizes that the design of a structure can have a significant influence on its corrosion resistance. It lists the components of paint coatings and describes the most important paint systems, metal coatings, concrete encasement and weathering steel.

15.1 Basic theory of corrosion

Corrosion of steel is an electrochemical process. The mechanism is basically the same as in a simple battery – two metals, called the anode and the cathode, when immersed in a conductive solution (the electrolyte) will react so that the anode is dissolved and an electric current is generated.

Atmospheric corrosion of steel occurs when moisture acts as the electrolyte. Pure water is a poor conductor of electricity, but when its conductivity is increased by salt or acid contamination the reaction is accelerated. Thus corrosion rates are relatively low in clean rural environments but are much higher in industrial atmospheres polluted with acidic sulphur dioxide, or in marine environments where chloride ions are present.

Anodes and cathodes are formed on a steel surface by slight changes in steel composition or variations in temperature or in the environment. For example, areas covered by dirt particles or rust would be anodic while areas more freely exposed to oxygen would be cathodic. These positive and negative areas shift and change as the corrosion reactions proceed.

The reactions that occur are:

at the anode: $\underset{\text{(steel)}}{Fe} = \underset{\text{(ferrous ions)}}{Fe^{++}} + \underset{\text{(electrons)}}{2e}$

at the cathode: $\underset{\text{(oxygen)}}{O_2} + \underset{\text{(moisture)}}{2H_2O} + \underset{\text{(electrons)}}{4e} = \underset{\text{(hydroxyl ions)}}{4(OH)^-}$

The ferrous ions react with the hydroxyl ions:

$$\underset{\text{(ferrous ions)}}{2Fe^{++}} + \underset{\text{(hydroxyl ions)}}{4(OH)^-} = \underset{\text{(ferrous hydroxide)}}{2Fe\,(OH)_2}$$

The ferrous hydroxide then oxidizes to hydrated ferrous oxide ($Fe_2O_3.H_2O$) – otherwise known as rust (Figure 15.1).

In order for the corrosion reactions to take place it is essential that both oxygen and water are present simultaneously. In the absence of either, corrosion will not occur. It follows that the rate at which corrosion occurs will depend upon the availability of oxygen and water. In

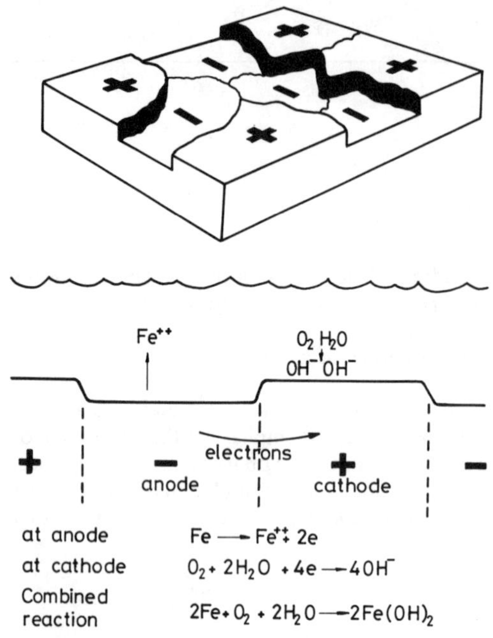

Figure 15.1 Basic chemistry of corrosion

applications divorced from the atmosphere (for example, underwater and underground) the supply of oxygen is the determining factor (corrosion is normally insignificant on steel piling driven into undisturbed ground because of the low oxygen content).

In the atmosphere, where oxygen is freely available, moisture is the determining factor and corrosion rate is governed by the period of wetness, i.e. a steel surface that is wet for 2 h per day will corrode twice as fast as one that is wet for 1 h per day (other parameters, such as pollution, being constant).

In some circumstances, however, the electrical part of the reaction is an important consideration. The electric currents that are generated between the positive and negative areas on the steel surface are extremely small. However, when two different metals are in contact, in the presence of water, the current is magnified and may be sufficiently high to change the whole surface of one metal into the anode and that of the other metal into the cathode of the corrosion cell. In other words, corrosion of one metal (the anode) will be accelerated while corrosion of the other (the cathode) will be suppressed.

This galvanic action clearly offers a method of protecting structural steel from corrosion by coating with anodic metals such as zinc or aluminium. However, designers must also take account of the reverse reaction, i.e. contact between structural steel and cathodic metals such as copper or stainless steel. In these cases corrosion of ordinary steel will be accelerated and electrical insulation must be provided between the two metals to avoid excessive attack.

The structural engineer or architect has little opportunity to modify the environment by eliminating pollutants or oxygen from the atmosphere. The practical attack on corrosion is confined to reducing the time of wetness or stopping water from reaching the steel surface.

15.2 Building interiors

For the great majority of buildings the time of wetness of interior steelwork is so short that corrosion is insignificant. In dry heated areas of buildings such as offices, hospitals and schools steelwork will not corrode. Even in unheated areas, the time of wetness from condensation is likely to be so short that uncoated steel will remain structurally sound for the lifetime of the structure. Where steelwork inside a dry building is hidden, perhaps by fire-protection materials or suspended ceilings, it can be left bare.

Where steelwork is visible it may be necessary to paint it for decorative reasons; the preparation and paint systems used in such cases can be simple and cheap, as they have only to cope with superficial aesthetic deterioration and not aggressive corrosion.

Only a small minority of interior environments require corrosion protection treatments, – for example, those containing wet or chemical processes such as kitchen areas, swimming pools, electroplating tanks or dyeing vats. In these cases (and, of course, any exterior steelwork that will become wet by exposure to the weather) some form of coating will be necessary to keep water from the steel surface.

15.3 Design

Before describing the various coatings that are available it is worth pointing out that the design of a structure can have a significant influence on corrosion resistance. Since the longer the period of wetness, the greater the corrosion it is important that structures should be designed as far as possible to shed water rather than retain it. Any situation that is likely to lead to the entrapment of moisture and dirt should be avoided. Where necessary, drainage holes should be provided of a size and location sufficient to ensure that all water drains from the member. Where possible, corners should be designed to allow a free flow of air to promote rapid drying (Figure 15.2).

Design for maintenance is also important. If access is difficult or space is cramped it is unlikely that maintenance work will be properly carried out.

15.4 Paint coatings

Paint coatings have three main components:

1. A pigment – fine solid particles which give the paint film colour and hardness, and which may inhibit corrosion by impeding the chemical or electrical processes;
2. A binder (or medium) – a resin or oil which binds the pigment particles together and provides a tough, flexible adhesive film;
3. A solvent (or thinner), which reduces the viscosity of the binder to allow even application. Modern developments include solventless formulations or the use of water as the solvent.

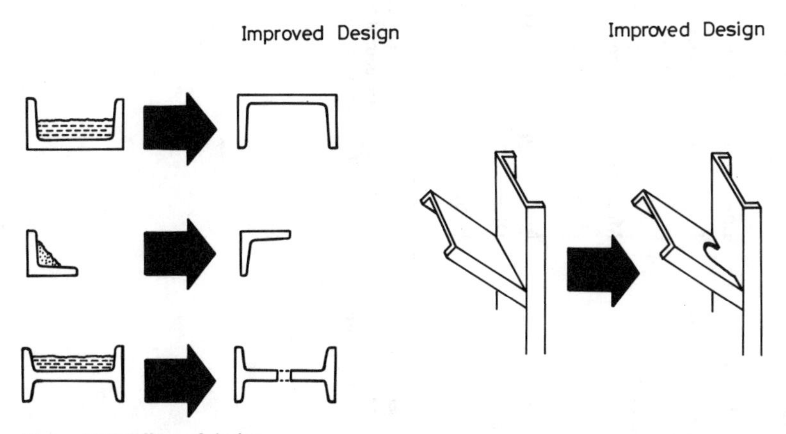

Figure 15.2 Effect of design

Table 15.1 Main generic types of paint and their properties

Paint type	*Cost*	*Tolerance of poor surface preparation*[a]	*Chemical resistance*	*Solvent resistance*[b]	*'Overcoatability' after ageing*[c]	*Comments*
Bituminous	Low	Good	Moderate	Poor	Good – with coating of same type	Limited to black and dark colours. Thermoplastic
Oil based	Low	Good	Poor	Poor	Good	
Alkyd, epoxyester, etc.	Low/medium	Moderate	Poor	Poor/moderate	Good	Good decorative properties
Chlorinated rubber	Medium	Poor	Good	Poor	Good	High build films remain soft and are susceptible to 'sticking' during transport
Vinyl	High	Poor	Good	Poor	Good	
Epoxy	Medium/high	Very poor	Very good	Good	Poor	Susceptible to 'chalking' in UV light
Urethane	High	Very poor	Very good	Good	Poor	Better decorative properties than epoxies
Inorganic silicate	High	Very poor	Moderate	Good	Moderate	May require special surface preparation

[a] Tolerance of poor surface: Types rated poor or very poor should only be used on blast-cleaned surfaces.
[b] Solvent resistance: Types rated poor or very poor should not generally be overcoated with any other type.
[c] Overcoating after ageing: Types rated poor or very poor require suitable preparation of the aged surface if they are to be overcoated after an extended period.

The different types of paint are often classified by the binder used in the formulation. A rough general guide to the characteristics of the more common types of paint is given in Table 15.1.

All paint films are permeable to moisture and oxygen to some extent, and may be subject to degradation by chemicals and by the UV radiation in sunlight. It is these factors which set the ultimate limit on the life of a paint system. Generally speaking, increasing the thickness of a paint coating will lengthen its life.

15.5 Paint systems

Paint systems are made up of a number of layers, each of which has a specific function (Figure 15.3).

1. *The primer*. This is the first coat, applied directly onto the steel surface. Its main function is to wet the surface and provide adhesion. In addition, it commonly contains inhibitive pigments which impede the corrosion process. Some of these pigments (for example, red lead, zinc chromate, zinc phosphate) partly dissolve in the moisture which diffuses through the upper layers of paint and supply traces of chemicals, which make the water less corrosive. This type of inhibitive pigment (in particular, red lead in oil binder, BS 2523: Type B) is preferred when the steel surface is prepared by wire brushing.

   Metallic zinc pigment in high concentration (90% or more) reduces corrosion at the surface of galvanic action. This requires good electrical contact and therefore a high standard of surface preparation by shot-blasting.
2. *The intermediate layer*. This may be applied in one or more coats, and its function is to build the thickness of the paint system. Two developments have taken place which are useful in this respect. 'High-build' formulations are now in common use; these allow thick layers to be applied in one coat, where previously two coats would have been required, and this gives a saving in labour cost. Second, lamellar pigments such as micaceous iron oxide (MIO) have the effect of lengthening the moisture diffusion path without increasing the actual film thickness. They also reinforce the film, making it less susceptible to mechanical damage.
3. *The finishing layer*. This is normally an inert barrier to the environment which also provides the cosmetic appearance. The number of coats and thicknesses of the system depend on the severity of the environment. For dry interior conditions, which are to be painted for cosmetic reasons, the intermediate and finishing layer may be interchangeable and the thickness is unimportant. For aggressive exterior conditions at least 200 μm would be required. Whatever paint system is adopted, all materials should be obtained from the same manufacturer to ensure compatibility.

   The largest element in the cost of painting is labour. Costs are lowest and quality is highest when painting is carried out in the fabricator's works, where automatic machinery and a weatherproof environment is available. It is common for all but the final one or two coats to be applied in the fabricator's works, and for the top coat(s) to be applied on site after erection. While there is a risk of mechanical damage during transport and erection, the cost of repair is likely to be less than full application on site.

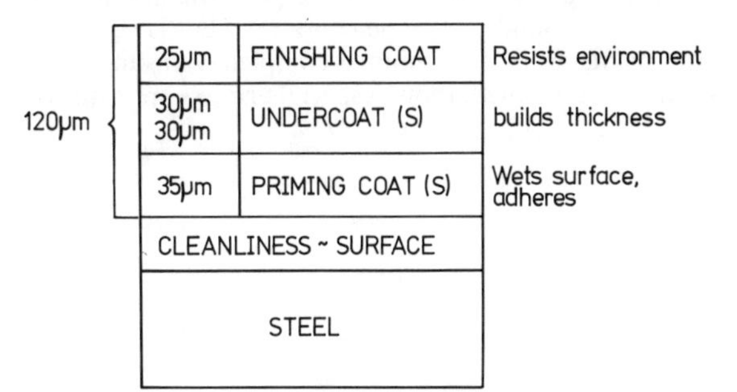

Figure 15.3 Paint systems

15.5.1 Surface preparation

The performance of all paint systems is improved by good surface preparation (Figure 15.4). Steel delivered from the rolling mill to the fabricator is covered with iron oxides in the form of a grey 'mill scale', and rusting occurs in the fabricator's stockyard. For mild conditions (for example, inside buildings when the steelwork is to be regularly overpainted for decorative reasons) manual preparation by scraping and wire brushing will usually be satisfactory, if primers that are tolerant of poor surface conditions, such as red lead in oil-based medium, are used. However, for external exposure and aggressive conditions mill scale and rust should be removed by blasting.

The blast-cleaning process, in which abrasive particles are projected at high speed onto the steel, produces a clean but rough surface of minute peaks and troughs. The depth of this profile depends largely on the size and shape of the abrasive used. Grit-blasting, using angular particles, tends to produce a sharp deep profile while shot-blasting, using spherical particles, tends to produce a smoother, rounded one. Shot-blasting is to be preferred for most paint coatings, but grit-blasting must be used for sprayed metal coatings since they rely largely on mechanical grip for adhesion.

15.6 Metal coatings

The water barrier can be provided by coating the steel surface with another metal. Zinc and aluminium are the most common metals used on structural steel, and two methods of application are used – hot dipping and metal spraying.

Galvanizing, i.e. dipping steelwork into a bath of molten zinc, is the cheapest and most common method of metal coating. Hot dip aluminizing is only used for light sheet materials for high-temperature use (for example, motor car exhausts). In the galvanizing process steel members are chemically cleaned by 'pickling' (immersion in dilute hydrochloric acid) to remove all traces of rust and mill scale before being dipped into molten zinc which is held at a temperature of about 450°C.

Because of the high temperatures used it is important to ensure that any enclosed members (for example, tubes) are vented to avoid the risk of explosion from vaporization of trapped moisture. A zinc/iron alloy layer is formed at the steel surface which gives perfect adhesion and total impermeability to moisture. In addition, this alloy, which is harder than the underlying steel, is resistant to impact damage. Galvanizing is particularly cost-effective for light steelwork (for example, lattice structures).

Metal spraying, using a special spray gun to throw molten metal particles onto a clean, grit-blasted, steel surface, is used for both zinc and aluminium coatings. The metal droplets freeze onto the surface and form a slightly porous layer of overlapping plates, which should be impregnated with a sealing treatment to give maximum corrosion resistance. No alloying takes place; adhesion is achieved by mechanical bonding to the roughened surface. The cost of sprayed coatings is high because of the extreme demands of surface preparation and cleanliness.

If a metal coating is damaged both zinc and aluminium will continue to protect the underlying steel by galvanic action, and both metals will give very good performance in clean, neutral conditions. Zinc will also perform well in mildly alkaline conditions but is subject to chemical attack in acid (industrial) environments, while aluminium is completely unsuitable for use in alkaline conditions. Both metals can be overpainted, though special primers are required.

METHOD	YEARS TO FAILURE	
	UNPOLLUTED	POLLUTED
Wirebrush	6¼	2¼
Blast Clean	16½	11

Figure 15.4 Surface preparation

15.7 Concrete encasement

Steel in an alkaline environment becomes 'passive' and will not corrode. Concrete provides such an alkaline environment and no deterioration of steel coated with concrete will take place as long as the pH of the concrete remains high (Figure 15.5). However, water will diffuse through concrete, carrying with it SO_2 and CO_2 gases from the air in the form of weak acids, particularly if poorly graded aggregate or a high water/cement ratio has been used. The effect, known as 'carbonation', is to reduce the alkalinity of concrete and promote corrosion of the underlying steel.

15.8 Weathering steel

Structural steel can be chemically modified so that it forms a protective surface layer which not only eliminates the need for and cost of applied coatings but which is also self-sealing if it is damaged. Normal rust layers are porous and allow water to pass through to continue to attack

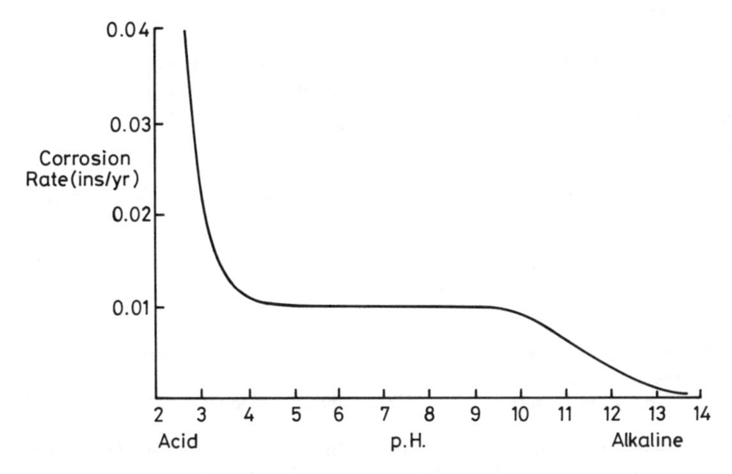

Figure 15.5 Effect of environment on corrosion rate

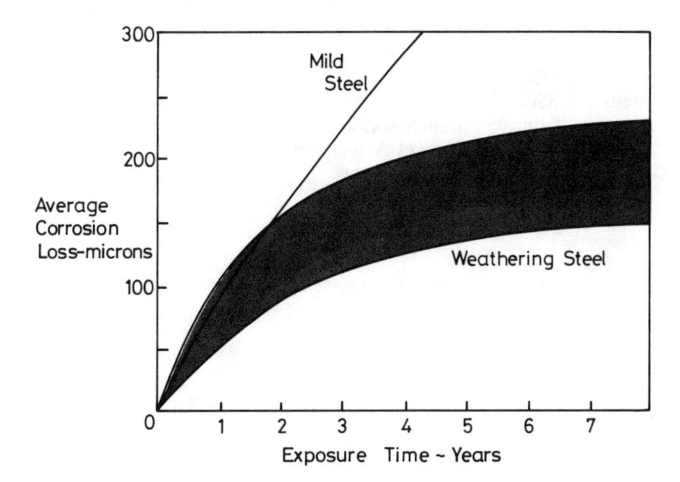

Figure 15.6 Weathering steel: UK performance

the underlying steel surface. The addition of small quantities (up to 3%) of certain elements (commonly Cu, Ni and Cr) to the steel modifies the structure of the rust layer that is formed. The alloying elements encourage a denser, more finely grained, rust film and also react chemically with sulphur in the atmosphere to form insoluble, basic, sulphate salts which block the pores in the rust films formed on these low-alloy steels. These two mechanisms produce thin, tightly adherent and self-sealing protective rust films of such low permeability that the corrosion rate is reduced almost to zero (Figure 15.6).

There are, however, circumstances in which the coating does not achieve its full protective potential. First, the low-permeability layer is only formed under wet/dry cyclic conditions. In conditions approaching constant wetness (for example, immersion, or sheltered north-facing exposures) the porosity of the film remains relatively high. Second, since most chloride compounds are soluble the 'pore-blocking' mechanism, which occurs in sulphurous environments, does not fully develop when the steel is exposed to severe marine or salt spray conditions. The optimum condition for the use of weathering steel is in boldly exposed 'skeletal' structures inland.

15.9 Concluding summary

1. Corrosion will not occur unless water and oxygen are both present on the steel surface.
2. The potential corrosion rate depends on the degree of atmospheric pollution.
3. The actual corrosion rate depends on the time of wetness of the steel surface.
4. Local corrosion rates are influenced by contact with other metals.
5. Corrosion rates inside most buildings are insignificant.
6. The most important factor governing the performance of coatings is correct surface preparation.

Background reading

BS 5493: Code of Practice for Protective Coating of Iron and Steel Structures against Corrosion (1977). Published by British Standards Institution

Controlling Corrosion, series of booklets published by the Department of Industry – Committee on Corrosion

Durability of Steel Structures (1983) Published by Commission of European Communities, Directorate General Information Market and Innovation, Bâtiment Jean Monnet, Luxembourg

Galvanizing, literature available from the Galvanizers Association

Protection against Corrosion of Reinforcing Steel in Concrete (1965) BRE Digest 59, Building Research Establishment

Recommended systems:		Joint publications by:
(a) *Building Interiors* (1986)		BCSA
(b) *Wall Cavities*	in prep-	BSC
(c) *Exterior Steelwork*	aration	Paint Research Association
		Zinc Development Association

UHLIG H.H. (1985) *Corrosion and Corrosion Control*, Wiley, Chichester

Fire protection

Objective To illustrate the means of achieving satisfactory performance of structural steel in fire.

Summary The chapter reviews the regulatory requirements for fire resistance in the UK and explains how the fire resistance of steel elements is measured. It summarizes the variations in the properties of steel elements with temperature and describes how individual members are protected, with a brief reference to tubular elements. The development of fire engineering concepts is briefly introduced.

16.1 Regulation requirements

All buildings in the UK are required to comply with the Building Regulations, which are concerned with personal safety. The provisions of the Regulations which deal with structural fire precautions are aimed at reducing the danger to people who are in or around a building when a fire occurs. Their objectives are to contain the fire and ensure stability of the structure for sufficient time to allow everyone to reach safety.

The degree of fire resistance required of a structural member is governed by the building function (offices, shop, factory, etc.), the building height and the size of the compartment in which the member is located.

16.2 Fire resistance

Fire-resistance requirements given in the Building Regulations are defined in units of time: $\frac{1}{2}$, 1, 2 or 4 h. These periods refer to the time in a standard test defined by BS 476: Part 8, for which a loaded member can demonstrate stability (i.e. it will support its load), integrity (i.e. it will not allow the passage of flame) and insulation (i.e. it will not allow passage of heat by conduction). The values are based largely on the findings of the Fire Grading of Buildings report of 1946. Traditionally, columns in the standard test are loaded vertically and exposed to fire on all four sides, while beams are loaded horizontally and exposed on three, the upper flange being in contact with the loaded furnace roof.

Elements under test are heated at a standard rate, following the curve (Figure 16.1):

$$\theta - \theta_0 = 345 \log_{10}(8t + 1)$$

where

θ = furnace gas temperature

θ_0 = ambient temperature

t = heating time in minutes

The stability failure criteria are a deflection of $L/30$ for beams and collapse for columns. The

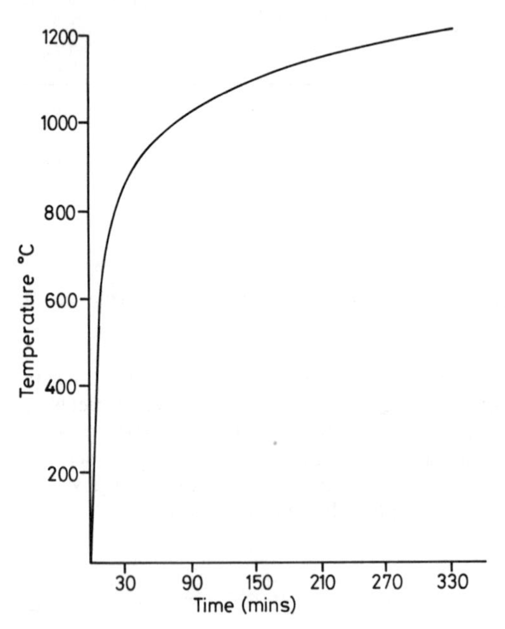

Figure 16.1 Standard time/temperature curve in fire test

measured strains in beams at the point of failure have been measured at between 2% and 4%.

This standard test has been used for many years to evaluating insulating or fire-protection materials for use on structural steelwork and to provide a 'rank order', or comparison of their performance. However, it is only a coarse indication of actual behaviour in service conditions. The achievement of one-hour fire resistance in the standard test does not guarantee structural stability for one hour in a real fire, which may be of greater or lesser severity.

16.3 Properties of steel

Tensile tests carried out at elevated temperature on samples of structural steel using test pieces of 10 mm diameter show that yield (plastic strain) occurs at progressively lower levels of stress as the test temperature increases (Figure 16.2). This phenomenon is demonstrated in both standard, isothermal, tensile tests (in which the stress is raised while the temperature remains constant) and in anisothermal creep tests (in which the stress remains constant while the temperature is raised) which relate more closely to structural practice (Figure 16.3). In both cases the test data indicate that steel, under small-scale test conditions and subject to a stress of approx. 0.6 × ambient temperature yield stress, exhibits excessive plastic strain (say, more than 3%) at about 550°C. This temperature is often referred to as the 'critical temperature' for steel structures – terminology that implies dire consequences if it is exceeded.

Clearly, this is an oversimplification. The temperature at which any given strain occurs depends on the applied load and the conditions under which the load is applied. In the laboratory tests are carried out under conditions of axial stress (uniform stress distribution) and uniform temperature distribution. Real structural members, which may be subject to bending or to temperature gradients, cannot realistically be judged by this single criterion.

16.4 Protection of members

The conventional approach to maintaining structural stability in fire conditions is to apply

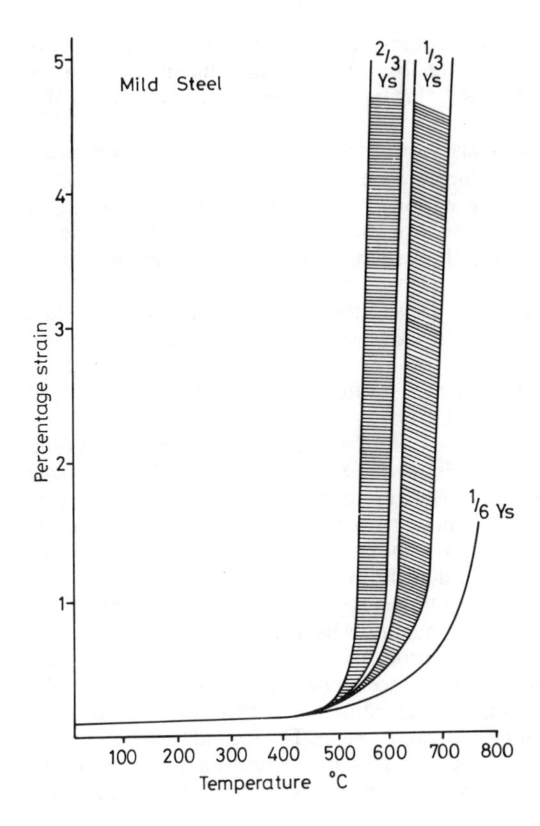

Figure 16.2 Variation in proof stress with temperature

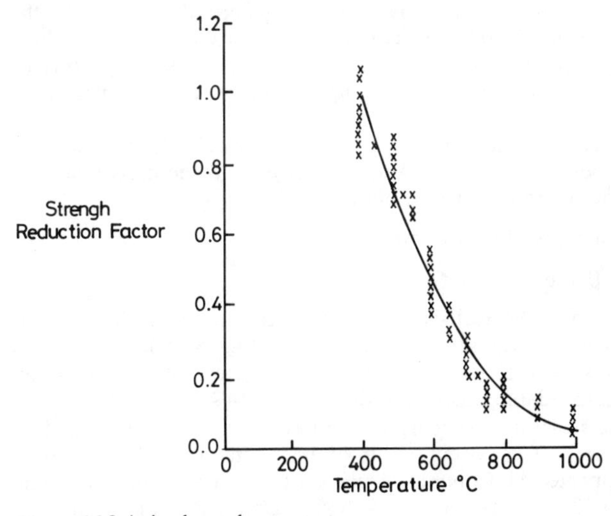

Figure 16.3 Anisothermal creep tests

fire-protection materials which have been deemed to meet Building Regulations requirements as a result of BS 476, Standard Tests on Individual Members. The fire protection is applied after erection of the frame and takes the form of an insulating barrier between the steel and the fire to slow down the transfer of heat.

Passive protection can be by traditional materials, such as concrete or bricks. However, there are newer, lightweight, asbestos-free materials which are almost invariably more cost-effective and impose less load on the structure. Asbestos is no longer used in fire protection because of health hazards; most modern materials gain their insulating properties from rock-fibre or exfoliated vermiculite. They are supplied either in the form of a spray for direct application or boards for mechanical fixing.

Sprays are the cheapest method, typically adding 15% to the frame cost. Application is fast and it is easy to coat complex shapes or connections. However, sprays are applied wet, which can create problems in freezing winter conditions, they can be messy and the appearance is often poor. For the latter reason, spray coatings are most often used in hidden areas, such as on beams above suspended ceilings or plant rooms.

Boards tend to be more expensive, up to twice the cost of a sprayed system, because of the higher labour content in fixing. The price depends on the rating required and the surface chosen, but tends to be less sensitive to job size. They are dry fixed by glueing, stapling or screwing, so there is less interference with other trades on site; the hollow box appearance is often more suitable for frame elements, such as free-standing columns, which will be in view.

Intumescent coatings are somewhat different in that the insulating layer is only formed by the action of heat when the fire breaks out. The coating is applied as a thin layer, perhaps as thin as 1 mm, but it contains a compound in its formulation which releases a gas when heat is applied. This gas inflates the coating into a thick carbonaceous foam which provides heat insulation to the steel underneath. Two types of intumescent coating are currently available. The first, based on epoxy or vinyl resins, has a maximum rating of 2 h. It is water-resistant but expensive, approximately four times the cost of a sprayed system. The second type is based on other resins and has a maximum rating of 1 h. These resins are not so resistant to moisture and are not recommended for wet applications such as swimming pools but are satisfactory in dry buildings.

In comparison with these lightweight materials, protection by *in situ* concrete would cost a similar amount to board protection, but would add considerably to the timescale.

The thickness of insulation required to provide a given fire rating depends not only on the thermal conductivity of the insulating material but also on the dimensions of the steel to which it is applied. The rate of temperature rise of a steel member depends on its mass and its surface area. Light members such as purlins or lattice girders heat up much more quickly than heavy columns.

The rate of heating of a given section is described by its 'section factor', which is the ratio between the surface perimeter exposed to radiation and convection and the mass (which is directly related to cross-section area), Hp/A, where (Figure 16.4):

Hp = perimeter of the section exposed to fire (m)

A = cross-sectional area of the member (m^2)

Thus a member with a low Hp/A ratio will be heated at a slower rate than one with a high Hp/A, and will require less insulation to achieve the same fire-resistance rating. Standard tables are available listing Hp/A ratios for structural sections, and manufacturers of fire-protection products now give recommendations relating insulation thickness to this section factor.

Sections at the heavy end of the structural range have such low Hp/A ratios, and therefore such slow heating rates, that failure does not occur within $\frac{1}{2}$ h under standard BS 476 heating conditions, even when they are unprotected. Columns, which are heated on all four sides, and which have Hp/A ratios less than $50\,m^{-1}$, can be used without added insulation in $\frac{1}{2}$ h rated conditions (Figure 16.5). Beams, which are heated on three sides, may be used unprotected in $\frac{1}{2}$ h rated situations when their Hp/A ratio is less than $110\,m^{-1}$ (Figure 16.6).

These conventional methods of achieving fire resistance are based largely on standard tests,

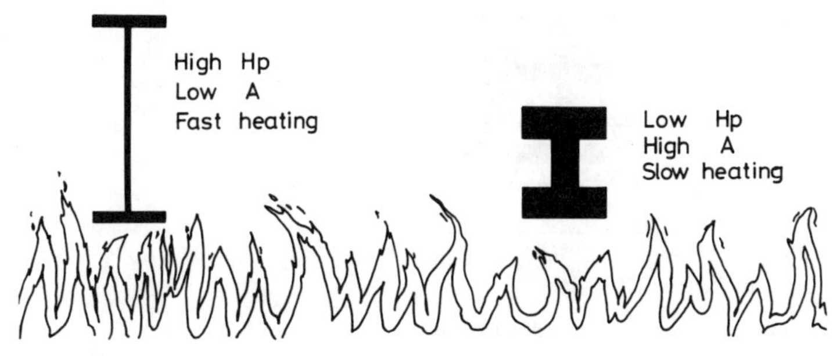

Figure 16.4 *Hp/A* concept

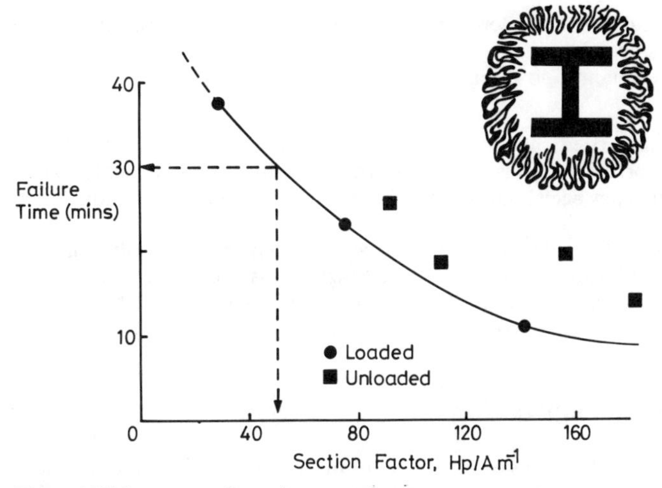

Figure 16.5 Summary of test data on columns

carried out by manufacturers of fire-protection material, in which insulated individual members have been subjected to the maximum permissible design stress with the maximum surface area exposed to heat. In real structures these conditions do not necessarily apply.

It has already been implied that the failure temperature of an element depends on the applied load, the stress distribution and the temperature distribution through the cross-section. We know that the yield strength of steel decreases with increasing temperature. Clearly, the lower the applied stress in a member, the higher will be the temperature at which plasticity occurs. Consequently, the lower the stress in a member, the longer it will survive in a fire.

16.5 Tubular structures

The fire resistance of tubular members can be improved by making use of the hollow interior – either for cooling by water filling or for load transference by filling with concrete. Water filling

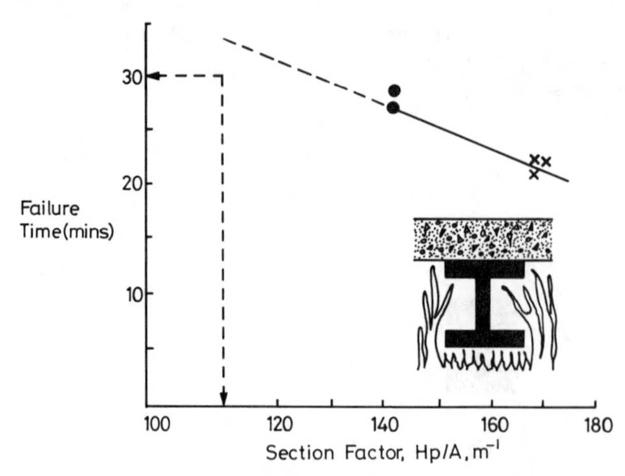

Figure 16.6 Summary of test data on beams – three-sided exposure

gives unlimited fire resistance when circulation is maintained. This can be done by:

1. Natural convection using a number of interconnecting members, not all of them exposed to the fire, and a high-level storage tank;
2. Direct connection to water mains and drainage;
3. Pumps.

Research is currently being carried out into static unreplenished systems. Chemicals are added to the water to inhibit corrosion (potassium nitrate) and freezing (potassium carbonate). One water-filled structure which has been built in the UK is Bush Lane House in Cannon Street, London.

The performance of concrete-filled tubular members depends largely on the member size and the tensile and flexural properties of the concrete. Fire resistance is improved markedly by inclusion of reinforcing bars or steel fibre reinforcement, and ratings approaching 2 h have been achieved on fully loaded 305 mm square columns.

16.6 Actual fires and fire engineering

The standard BS 476 fire test, and the tables of requirements in the Building Regulations which derive from it, have been a simple and convenient means of ensuring satisfactory performance of structures in fire. This classification method will continue to be used for small structures where the expense of more detailed analysis is unjustified.

However, it is clear that the conditions during a real fire differ greatly from those during a standard fire test. Methods have been developed since the 1960s which enable the behaviour of real structures in real fires to be predicted with greater accuracy, and they will almost certainly become a feature of UK regulations in the future.

Essentially, the fire engineering design method can be divided into four main steps:

1. Determination of the fire load;
2. Prediction of maximum gas temperature;
3. Prediction of maximum steel temperature;
4. Prediction of structural stability.

Much of the work on fire engineering has been carried out in Sweden, but the concept is being accepted and adopted in other countries. It is not one that can be recommended for buildings that are subject to change of use, such as advance factory units, but many buildings are 'fixed' in terms of their occupancy (for example, car parks, hospitals, swimming pools), and in such cases fire engineering is a valid approach.

In the UK a number of buildings have been built using unprotected steel on fire-engineering principles. One example is the North Stand at Ibrox football ground in Glasgow. Although a $\frac{1}{2}$ h regulation requirement applied to the building, fire-engineering studies showed that a total burnout would not raise steelwork temperatures to the point of collapse. Some £40000 was saved in 1979.

16.7 Concluding summary

1. Steel strength decreases with increasing temperature.
2. Elements of buildings in the UK are generally required to have a specified fire resistance, i.e. length of time that must elapse prior to collapse.
3. The fire resistance that is required varies with building function, length and compartment size. Values are $\frac{1}{2}$, 1, 2 or 4 h.
4. Steel elements must generally be protected by a cover that insulates the steel from the fire. However, tubular elements may be water- or concrete-filled.
5. Fire-engineering techniques, which consider behaviour from first principles, are currently under development.

Background reading

BOND, E.V.L. (1975) *Water Cooled Hollow Columns,* The Steel Construction Institute
Building with Steel (1982) Vol. 9, No. 2, December, TheSteel Construction Institute
LAW, M. and O'BRIEN, T. (1981) *Fire Safety of Bare External Structural Steel,* The Steel Construction Institute
The Behaviour of Steel Portal Frames in Boundary Conditions (1980) The Steel Construction Institute
The Fire Protection of Structural Steel in building (1983) ASFPCM and Constrado, London

17

Tension members

Objective To describe the types and uses of tension members and set out their behaviour and design.

Prior reading Basic strength of materials (i.e. elasticity and plasticity is essential).

Summary The chapter describes the principal uses for and types of tension members in steel structures. It discusses the particular problems of connections and splices for such members and summarizes behaviour and analysis both under axial tension and under combined tension and bending. Practical considerations are reviewed and design procedures for current codes of practice are presented, with examples illustrating their application. For worked examples see Appendix, p. 366.

17.1 Introduction

The tension member transmitting a direct pull between two points in a structure is theoretically the simplest and most efficient structural element. In theory, tension members may be thin and of small cross-sectional area, in contrast to the more stocky compression members discussed in a later chapter, where the cross-section must be larger in order to resist buckling.

In many cases, in practice the simple theoretical efficiency is seriously impaired by the end-connections required to join tension members to other members in the frame. In other situations the load in the member can reverse, usually due to the action of wind, and again the efficiency of the simple tie is lost when the member must also act as a strut. Where the load can reverse, the designer often permits the member to buckle, with the load then being taken up by another member. This action occurs in cross-braced panels.

Some of the main uses of tension members are shown in Figure 17.1. These are:

1. Tension chords and internal ties in trusses and lattice girders in buildings and bridges;
2. Tension bracing members in buildings;
3. Hangers in suspended buildings;
4. Main cables and deck-suspension cables in suspension bridges.

There are many other uses for tension members that have not been listed.

17.2 Types of tension members

The main types of tension members and some comments on their use and behaviour are given below, and are illustrated in Figure 17.2:

1. Open and closed single rolled sections such as angles, tees, channels and the structural hollow sections. These are the main sections used for tension members in light trusses and lattice girders and for bracing in buildings.
2. Compound sections consisting of double angles and channels. These are used in the same

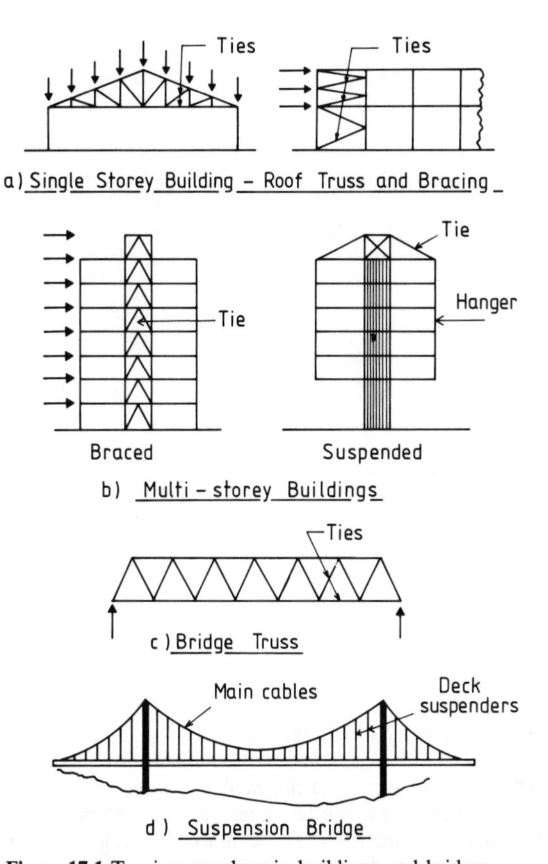

Figure 17.1 Tension members in buildings and bridges

situations as the single open sections mentioned above but are specified when the loads are heavier than can be carried by a single section.

3. Heavy rolled sections, heavy compound sections and built-up H- and box sections. These are used for tension members in bridges and heavy lattice girders in industrial buildings.
4. Bars and flats. A range of high-strength, precision, round tie bars is available commercially. Flats are bars of rectangular section and are often used for bracing buildings. Neither of these member types can carry reversal of load.
5. Ropes and cables. These are constructed in a variety of ways from hard-drawn, high-strength wires. The cross-sections for a round strand and a locked coil rope are shown in Figure 17.2. Cables formed from numbers of individual wires are the only practical way of constructing long and very strong continuous tension members. The main cables in suspension bridges are good examples of such members. For instance, each main cable in the Humber Bridge contains 14 948 wires of 5 mm diameter and has a strength of 19 000 tonnes.

17.3 End-connections and splices

End-connections for tension members create problems which affect the design and often result in a loss of efficiency. This is illustrated by the end-connections for angles shown in Figure 17.3(a). In bolted joints the member strength is reduced, first, by the bolt hole, and second,

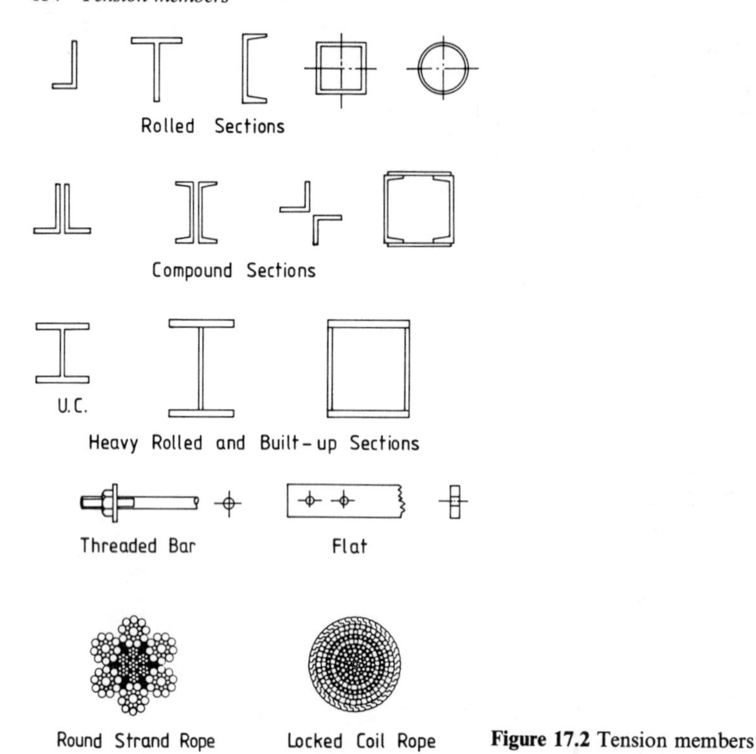

Figure 17.2 Tension members

because the outstanding leg is not fully effective and there is a moment due to eccentricity in the connection. The welded joint for a single angle has a similar eccentricity problem.

Full-strength joints where the member is fully effective may be made by welding. For practical and economic reasons, welding is not used in many cases; for example, the bracing member shown in Figure 17.3(b) would not be welded to the column but would be connected by bolting. However, internal joints in lattice girders constructed in structural hollow sections can be made by welding to develop the full strength of the member, as shown in Figure 17.3(c).

Bars have threaded ends and may be connected directly through a nut. The strength is determined by the net tensile area at the thread. Ropes are connected through end-sockets or terminals which will develop the full static strength. Examples of end-connections for bars and ropes are shown in Figure 17.3(d).

Splices are needed to re-connect long tension members that have been broken into sections for ease of fabrication and transport (for example, tension chords in trusses). Splices for open or closed sections may be bolted or welded, as shown in Figure 17.3(e).

17.4 Behaviour and analysis of tension members

17.4.1 Direct tension

The behaviour of rolled sections in direct tension is similar to that of a tensile test specimen. The member behaves elastically up to the yield point, when it enters the plastic range where, if the member is made of mild steel, the stress remains the same as the elongation increases. Following this stage, the strength again increases due to strain-hardening until failure occurs. Typical stress–strain diagrams for the structural steels and a high-strength roping wire are shown in Figure 17.4(a).

For the straight member subjected to direct tension F, as shown in Figure 17.4(b):

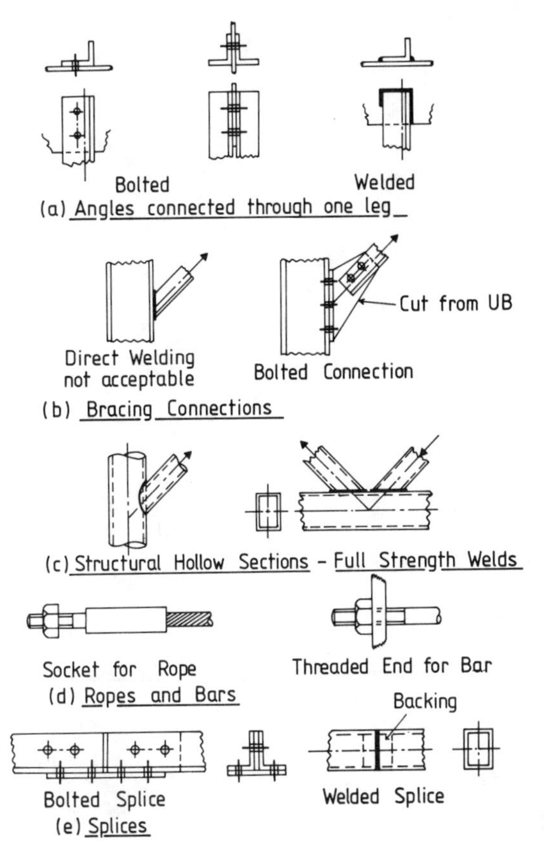

Figure 17.3 End-connections and splices

Tensile stress	$f_t = F/A$		(17.1)
Elongation	$\delta_L = FL/AE$	(prior to yielding)	(17.2)
Load at yield	$P_y = Y_s A$		(17.3)

17.4.2 Stress concentrations

Stress concentrations occur at holes and changes in cross-sections in members. They are not usually important in ductile materials but can be the cause of failure due to fatigue or brittle fracture in certain conditions. A hole in a flat member will increase the stress locally on the net section by a factor of 2 or 3, depending on the ratio of hole diameter to net plate width. Stress concentration factors are given in Figure 17.5(a) while the stress distributions adjacent to a hole in a flat plate are shown in Figure 17.5(b).

17.4.3 Direct tension and moment:elastic analysis

Moments in tension members are caused by eccentric connections and/or rigid connections and lateral loading. In a rigorous elastic analysis the moment along the member due to lateral load or applied end-moments is reduced by the direct tension. Figure 17.6 shows a pin-ended tie subjected to a direct tension (F) and uniform lateral load (W). The net moment at any point at distance Z from the end is given by:

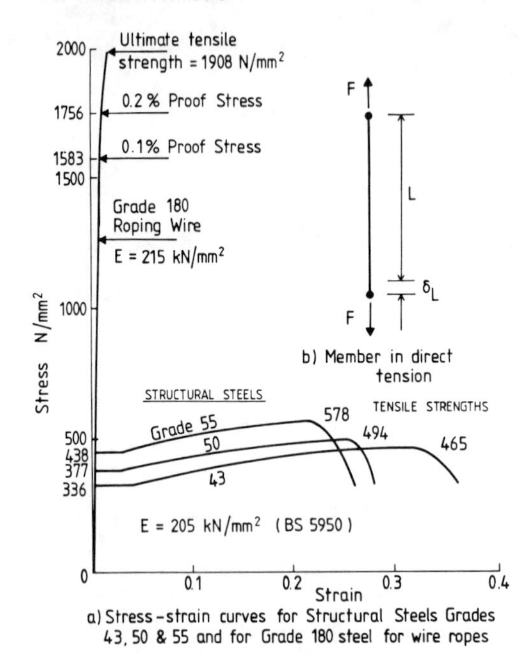

Figure 17.4 Behaviour in direct tension

$$M = WZ/2 - WZ^2/2L - Fv \qquad (17.4)$$

The maximum moment at the centre is:

$$M_c = \frac{WL}{4}\left[\frac{\cosh U - 1}{U^2 \cosh U}\right] \qquad (17.5)$$

$$\text{where } U = \sqrt{(FL^2/4EI)} \qquad (17.6)$$

The solutions for other load cases are available. It is safe practice to neglect this interaction effect and combine stresses from independent analyses for direct tension and bending.

Combined stresses are discussed first for a section with two axes of symmetry:
Axial tension and bending about the XX axis, as shown in Figure 17.7:

$$\text{Tensile stress} \quad f_t = F/A \qquad (17.7)$$

$$\text{Bending stress} \quad f_{bx} = M_x/Z_x \qquad (17.8)$$

$$\text{Maximum stress} \quad f_{max} = f_t + f_{bx} \qquad (17.9)$$

The stress diagrams are shown in the figure. The variation of axial load and moment for the case where the maximum combined stress is equal to the yield stress is shown in the interaction diagram. The values of the separate actions causing yield are:

$$\text{Load at yield } P_y = Y_s A \qquad (17.10)$$

$$\text{Moment at yield about the XX axis } M_{xy} = Y_s Z_x \qquad (17.11)$$

Then any point A on the line joining P_y and M_{xx} in the figure gives values of F and M_x, which will give a maximum stress equal to the yield stress. The diagram could also be constructed to make the maximum stress equal to the allowable stress.

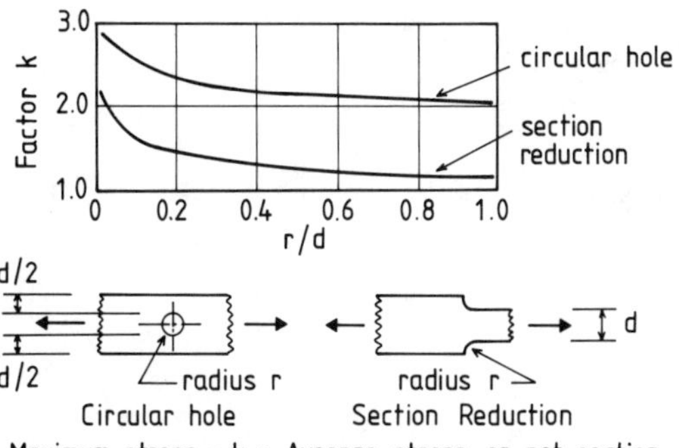

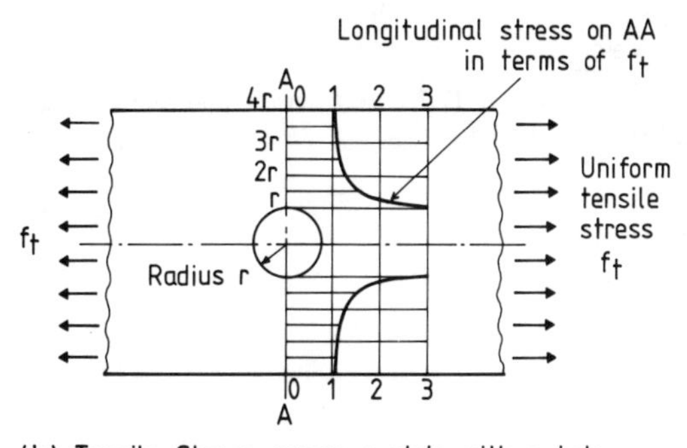

Figure 17.5 Stress concentration factors and stress across a plate with a hole.

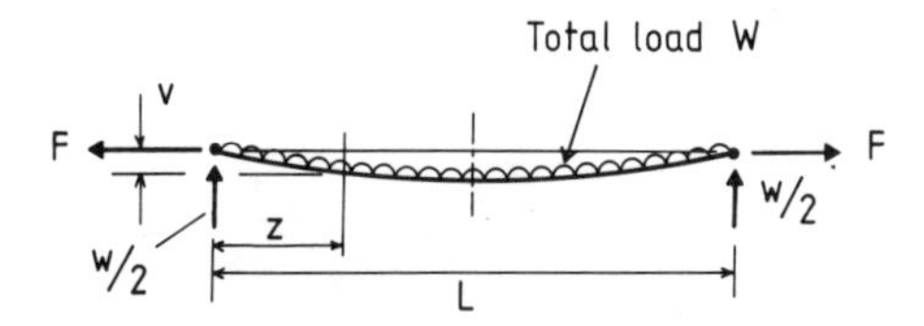

Figure 17.6 Combined bending and tension

Axial tension and bending about both axes XX and YY (see Figure 17.8):

Tensile stress	$f_t = F/A$	(17.12)
Bending stress XX axis	$f_{bx} = M_x/Z_x$	(17.13)
Bending stress YY axis	$f_{by} = M_y/Z_y$	(17.14)
Maximum stress	$f_{max} = f_t + f_{bx} + f_{by}$	(17.15)

The separate stress diagrams are shown on the figure.

The values of the separate actions causing yield are:

P_y and M_{xy}, as defined above, and, in addition,

Moment at yield about the YY axis $M_{yy} = Y_s Z_y$

Then the plane defined by P_y, M_{xy} and M_{yy} forms the interaction surface shown on the figure. Any point A on the plane gives co-existent values of F, M_x and M_y for which the maximum stress equals the yield stress.

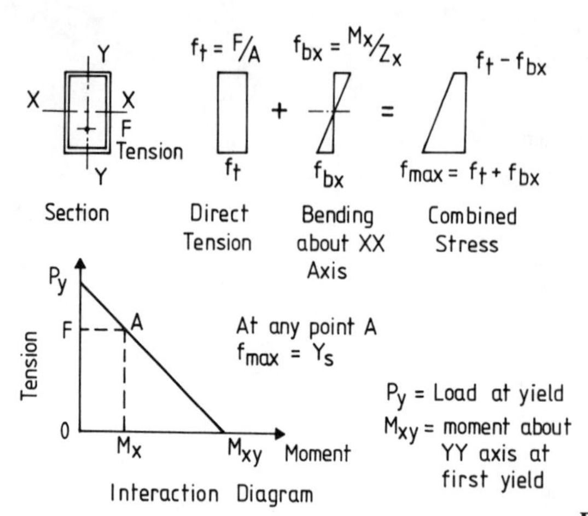

Figure 17.7 Direct tension and moment–elastic analysis

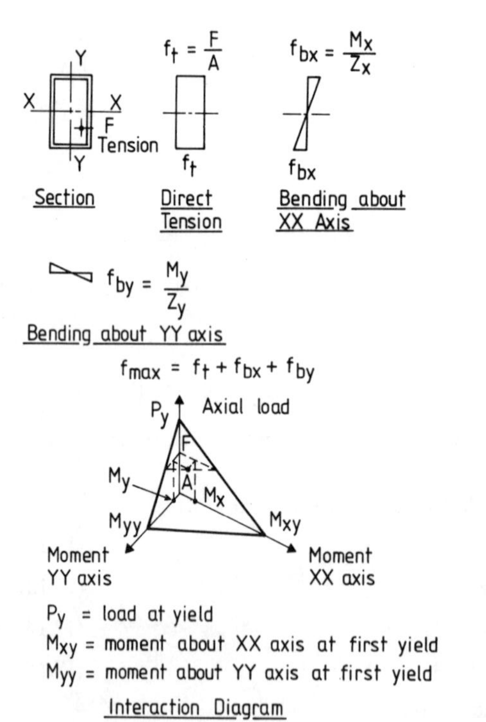

Figure 17.8 Direct tension and moment–elastic analysis

Sections with one axis of symmetry, or unsymmetrical sections where the unrestrained section will bend about the principal axes, can be treated in a similar way.

17.4.4 Direct tension and moment: plastic analysis

An I-section with two axes of symmetry subjected to axial tension and bending about the XX axis is considered. Referring to Figure 17.9(a), two equal areas extending inwards from the

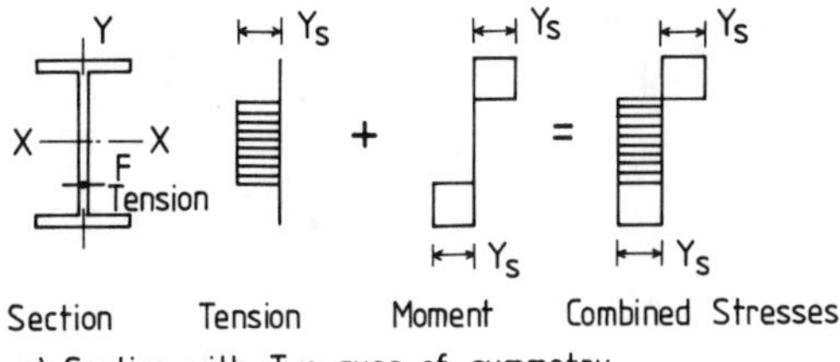

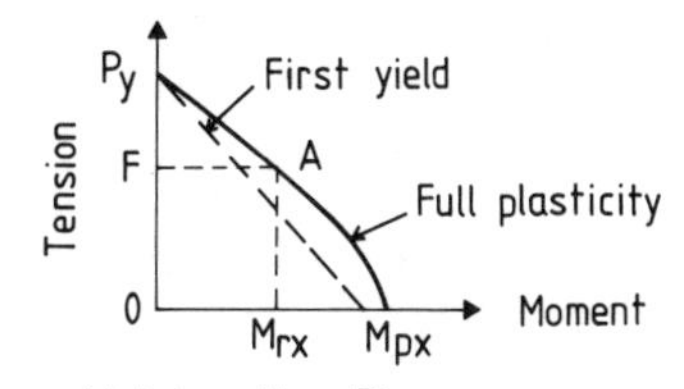

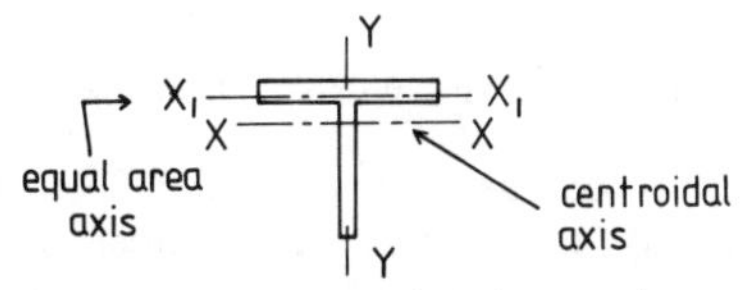

Axial tension and bending about one axis

Figure 17.9 Direct tension and moment–plastic analysis

flanges, one in tension and one in compression at the yield stress, resist moment while a symmetrically placed central area resists tension. In the case shown the plastic neutral axis lies in the web. Increasing the axial tension will cause the plastic neutral axis eventually to enter the compression flange. The moment reduces the tension capacity of the section.

The separate maximum values for axial tension and plastic moment are:

Axial tension $P_y = Y_s A$ (17.16)

Plastic moment $M_p = Y_s S$ (17.17)

The value of M_{rx}, the plastic moment of resistance in the presence of axial load, may be calculated for any value of F less than P_y. The interaction curve is plotted as shown in Figure 17.9(b). The elastic interaction curve is shown dotted on the same figure. The difference between the two curves represents the additional design strength available if plasticity is taken into account. Other shapes may be treated in a similar way. The analysis for a section with one axis of symmetry such as the T-section shown in Figure 17.9(c) is more difficult.

Sections subjected to biaxial bending at ultimate load give convex failure surfaces, as shown in Figure 17.10. The degree of convexity depends on the section considered. At point A the section can sustain an ultimate axial load F and simultaneous plastic moments of M_x and M_y about the XX and YY axes, respectively.

17.5 Other considerations

17.5.1 Serviceability, fatigue and corrosion

Ropes and bars are not normally used in building construction because they lack stiffness, but they have been used in some cases as hangers in suspended buildings. Very light, thin tension

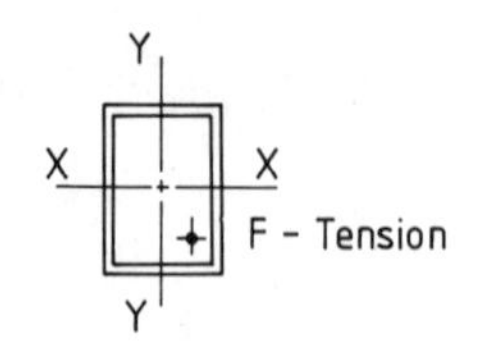

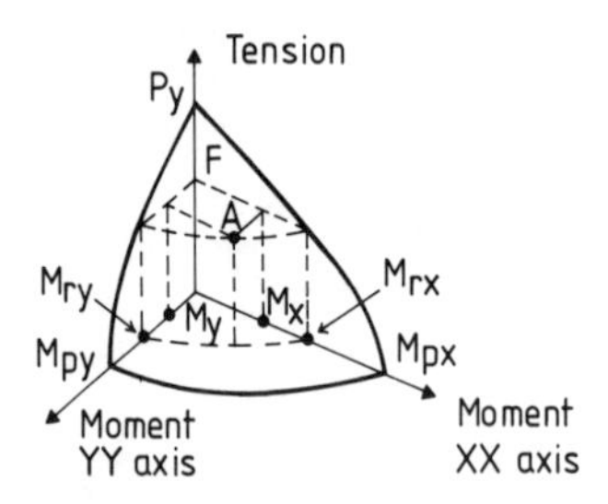

Axial tension and bending about two axes **Figure 17.10** Direct tension and moment–plastic analysis

members are susceptible to excessive elongation under direct load as well as lateral deflection under self-weight and lateral loads. Special problems may arise where the members are subjected to vibration or conditions leading to failure by fatigue, such as can occur in bridge deck hangers. Damage through corrosion is also important and adequate protective measures must be adopted. All these factors can make the design of tension members a complicated process in some cases.

The light-rolled sections used for tension members in trusses and for bracing are easily damaged during transport. It is customary to specify a minimum size for such members to prevent this happening. For angle ties, a general rule is to make the leg length not less than one-sixtieth of the member length.

17.5.2 Fabrication and erection

The behaviour of tension members in service depends on the fabrication tolerances and the erection procedure. Provision must be made for tensioning bars and cables through screwed ends, as shown in Figure 17.2, to ensure that these members are immediately active in resisting applied loads.

Where possible, bracing members fabricated from rolled sections should be installed before other connections and column base plates in the building are bolted up. Bracing members also serve to pull the building into line and square. Sometimes tension members are made slightly shorter than the exact length so that they can be bolted tightly into position.

Trial shop assembly is often specified for the members in bridge and heavy industrial trusses to ensure that the fabrication is accurate and that erection is free from problems.

17.6 Design of tension members to BS 5950: Part 1

The design is made for factored loads and resistance capacities are determined using design strengths of materials.

17.6.1 Axial tension

The design of axially loaded tension members is given in **Clause 4.6**. The tension capacity is given by

$$P_t = A_e p_y \tag{17.22}$$

where A_e is the effective area defined in **Clause 3.3.3**. This clause states that the effective area of a member with holes may be taken as K_e times its net area but not more than its gross area. The net area is defined as the gross area less deductions for fastener holes. Reference should be made to the code for members with staggered holes.

The factor K_e, values of which are given in **Clause 3.3.3**, has been introduced as a result of tests. These show that the presence of holes does not reduce the effective capacity of a member in tension provided that the ratio of the net area to the gross area is suitably greater than the ratio of the yield strength to the ultimate strength.

The moment due to eccentric connections must be taken into account as set out below, except in the case of angles, channels and tees, which may be designed as axially loaded members. Design rules for two common tension members are given below and reference should be made to the code for other cases:

1. Single angles connected through one leg as above are shown in Figure 17.9(a). The effective area is to be taken as:

$$A_e = a_1 + \left(\frac{3_{a_1}}{3_{a_1} + a_2}\right) a_2 \tag{17.23}$$

2. Double angles connected to both sides of a gusset or section may be designed using the effective area for axially loaded members as defined above (see Figure 17.11).

17.6.2 Combined bending and tension

The design of tension members with moments is given in **Clause 4.8.2**. This states that tension members should be checked for capacity at the points of greatest bending moments and axial loads, usually at the ends. The following relationship should be satisfied:

$$\frac{F}{A_e p_y} + \frac{M_x}{M_{cx}} + \frac{M_y}{M_{cy}} \leqslant 1 \tag{17.24}$$

where F = the applied axial load in the member,
A_e = the effective area,
p_y = the design strength,
M_x = the applied moment about the major axis at the critical region,

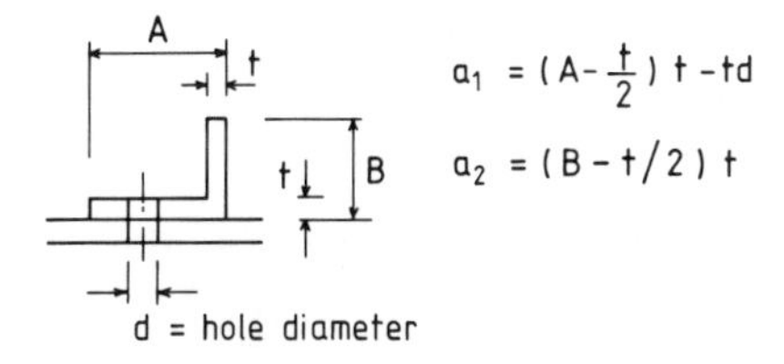

a) Single Angle, connected through one Leg

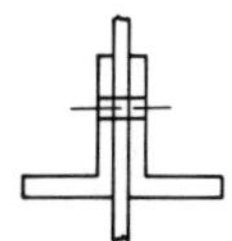

b) Double Angles, connected to each side of a Gusset

Figure 17.11 Angle connections

M_{cx} = the moment capacity about the major axis in the absence of axial load,
M_y = the applied moment about the minor axis,
M_{cy} = the moment capacity about the minor axis in the absence of axial load.

For plastic or compact sections with low shear load:

$$M_c = P_y S \leqslant 1.2\, p_y Z \tag{17.25}$$

where S = plastic modulus about the relevant axis,
Z = elastic modulus about the relevant axis.

The interaction diagrams for tension and moment for bending about one and two axes are shown in Figure 17.12. These are derived from the diagrams for plastic analysis shown in Figures 17.9 and 17.10, respectively. Equation 17.24 and the diagrams in Figure 17.12 give a linear variation between the separate capacity values for axial load and moments.

An alternative relationship to Equation 17.24 is given in the code: its use gives greater economy in design of plastic or compact sections. Reference should be made to the code for this expression.

17.7 Design of tension members to BS 5400: Part 3

The design of tension members in steel bridges is given in **Clause 11.** This clause states that:

> Tension members should be designed for the ultimate limit state. The fatigue endurance should be in accordance with recommendations in **Part 10**. The serviceability limit state need not be considered.

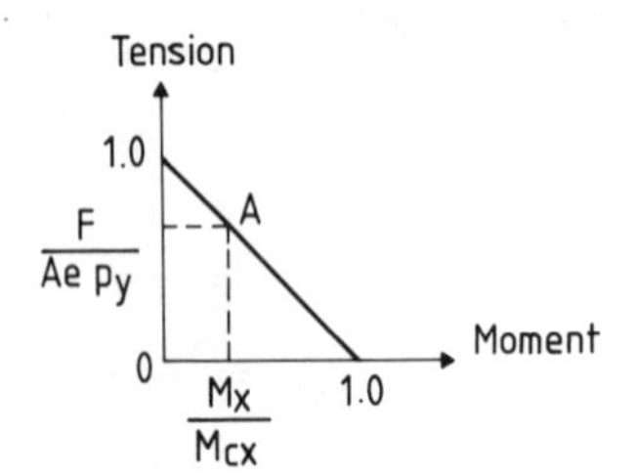

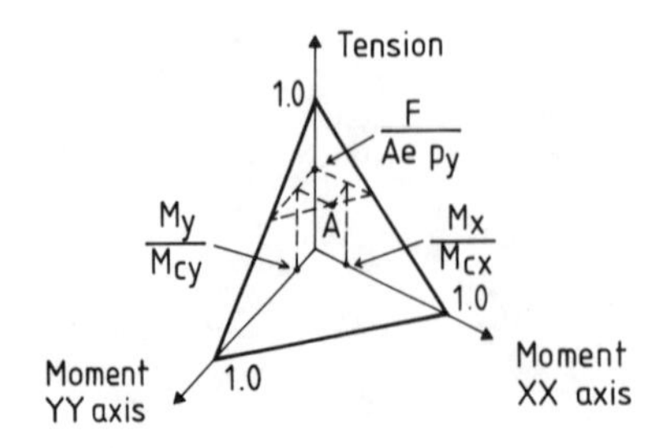

BS 5950 - Simplified Expression from Plastic Theory

Figure 17.12 Interaction diagrams, tension and moment

17.7.1 Axial tension

The member should be such that the design ultimate axial load does not exceed the tensile resistance (P_D), given by:

$$P_D = \frac{\sigma_y A_e}{\gamma_m \gamma_{f3}}$$

where A_e = effective cross-sectional area
= $k_1 k_2 A_t \leqslant A$,
A_t = net cross-sectional area,
A = gross area,
σ_y = nominal yield stress for steel given in BS 4360,
*γ_m = partial safety factor. Values are given in **Table 2**. This takes account of material variation.
*γ_{f3} = partial safety factor. Values are given in **Clause 4.3.3**. This is often called the 'gap factor'.
k_1, k_2 = factors. Values are given in **Clause 11**.

*Note that the design ultimate load incorporates other safety factors (γ_{fL}).

17.7.2 Combined tension and bending

A member subjected to co-existent tension and bending should be such that at all cross-sections the following relationship is satisfied:

$$\frac{P}{P_D} + \frac{M_x}{M_{Dxt}} + \frac{M_y}{M_{Dyt}} \leqslant 1.0$$

where P = axial tensile force,
P_D = derived as set out above,
M_x, M_y = bending moments at the section about XX and YY axes, respectively,
M_{Dxt}, M_{Dyt} = corresponding bending resistances with respect to the extreme tensile fibre.

Additionally, if at any section within the middle third of the length of the member the maximum compressive stress due to bending exceeds the tensile stress due to axial load the design should be such that:

$$\frac{M_{xmax}}{M_{Dxc}} + \frac{M_{ymax}}{M_{Dyc}} < 1 + \frac{P}{P_D}$$

where M_{xmax}, M_{ymax} = maximum moments anywhere within the middle third,
M_{Dxc}, M_{Dyc} = corresponding bending resistances with respect to the extreme compression fibres.

For compact sections the bending resistance is given in **Clause 9.9.1.2**. This is:

$$M_D = \frac{Z_{pe} \sigma_{lc}}{\gamma_m \gamma_{f3}}$$

where σ_{lc} = limiting compressive stress (see **Clause 9.8**)
Z_{pe} = plastic modulus of the effective section (see **Clause 9.4.2**).

Refer to **Clause 9.9.1.3** for the resistance of non-compact sections.

17.7.3 Eccentricity of end-connections

The bending moment resulting from an eccentricity of the end-connections should be taken into account. For single angles connected by one leg these provisions may be considered to be met if the effective area of the unconnected leg is taken as:

$$\left(\frac{3A_1}{3A_1 + A_2}\right) A_2$$

where A_1 = net area of the connected leg,
A_2 = net area of the unconnected leg.

17.8 Concluding summary

1. The tension member is theoretically the simplest and most efficient structural element. End-connections often cause problems in design.
2. Many types of sections are used for tension members. These include cables, bars, rolled and built-up sections. In buildings, rolled sections such as angles, channels and structural hollow sections are the commonest members used as ties, both for trusses and for bracing.
3. Specific rules are given in the codes to eliminate the need to consider bending separately when designing tension members with eccentric end-connections such as single angles connected through one leg. Methods are given for checking members subjected to moment in addition to direct tension.
4. Practical considerations are important in design. For example, minimum sizes are often specified for tension members to prevent damage during transport and erection.
5. Provision must be made in fabrication and in the erection method to ensure that the behaviour of tension members in the finished structure is satisfactory.

Background reading

BRESLER, B., LIN, T.Y. and SCALZI, J.B. (1969) *Design of Steel Structures,* Wiley, Chichester.
Pages 244–284: design of tension members
Pages 334–341: combined bending and axial load – elastic range.
Pages 341–345: combined bending and axial load – plastic range.
GORDON, J.E. (1978) *Structures or Why Things Don't Fall Down,* Pelican, Harmondsworth.
Pages 113–131: this chapter gives a general treatment of the behaviour of tension structures and pressure vessels
HORNE, M.R. (1971) *Plastic Theory of Structures,* Nelson, Walton-on-Thames.
Pages 62–71: the effect of axial load on plastic moment
Threaded Bars (1981) Reinforced Steel Services, McCalls's Special Products
TIMOSHENKO, S.P. and GOODIER, J.N. (1970) *Theory of Elasticity,* McGraw-Hill, Maidenhead.
Pages 78–81: the effect of circular holes on stress distributions
TRAHAIR, N.S. (1977) *The Behaviour and Design of Steel Structures,* Chapman and Hall, London.
Pages 40–51: the behaviour and design of tension members

18

Introduction to buckling: 1

Objective To introduce the basic concepts of buckling in the context of column buckling.

Prior reading Simple bending theory;
Euler buckling theory;
Ideal and real elastic-plastic behaviour of materials.

Summary A general definition of buckling is provided, together with a brief outline of buckling phenomena. The basic concepts of buckling behaviour are introduced and the main parameters that govern buckling behaviour are identified in the context of axially compressed struts. The elastic buckling of an idealized strut having perfect geometry is first considered. Then the effects of assuming either ideal rigid-plastic or ideal elastic-plastic material behaviour are examined in the absence of residual stresses. Finally, the influence of imperfect geometry (initial out-of-straightness and eccentricity of loading), residual stresses and more general elastic-plastic material behaviour are studied, both separately and in combination.

18.1 Introduction and basic definition of buckling

Buckling is a class of structural behaviour in which a mode of deformation develops in a direction or plane normal to that of the loading which produces it, such deformation changing rapidly with variations in the magnitude of the applied loading. It occurs mainly in members and elements that are in a state of compression.

The simplest type of buckling to visualize (and analyse) is that of an initially straight strut compressed by equal and opposite axial forces, as shown in Figure 18.1. As the applied forces are increased in magnitude the undeflected state of the member becomes unstable and the member prefers to adopt a curved (i.e. buckled) shape.

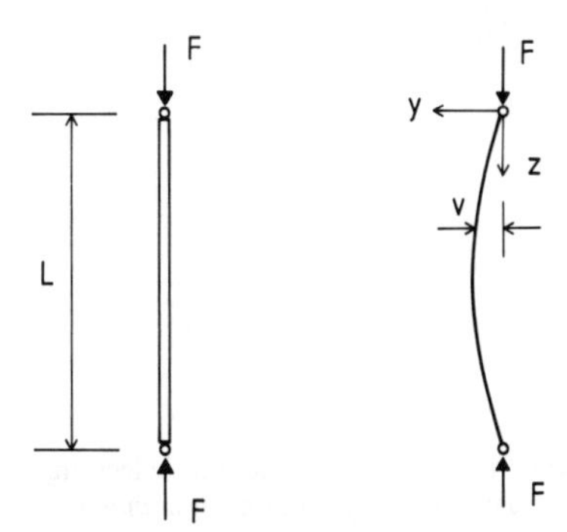

Figure 18.1 Axially loaded, initially straight, pin-ended strut

The member's buckling resistance will increase with the bending stiffness of the member, and hence with the thickness or depth of its section measured in the plane of buckling deformation. Also, it decreases as the member length is increased. Thus buckling resistance is low if a member is slender and high if it is stocky.

Buckling is of particular interest with steel members because they tend to be of slender form compared, for example, with concrete members. However, it is not only slenderness of a member as a whole that leads to buckling. The thin elements of steel members which are fabricated or cold formed from thin plate or sheet may individually experience localized buckling effects when subjected to compressive stress. Also, some members of thin, open cross-section have a low torsional stiffness and experience buckling phenomena involving torsional as well as bending deformation.

This chapter introduces buckling in the context of compressed struts and identifies the main parameters that govern buckling behaviour. First, we shall consider the elastic behaviour of an idealized strut having perfect geometry, i.e. no initial out-of-straightness or eccentricity of loading. Then we shall examine the effect of assuming either ideal rigid-plastic or ideal elastic-plastic material behaviour in the absence of residual stresses. Finally, we shall study in turn the influences of imperfect geometry, residual stresses and more general elastic-plastic material behaviour.

18.2 Elastic buckling of an ideal column or strut having pinned ends

18.2.1 Introduction

We consider flexural buckling of the idealized, pin-ended, uniform strut or column shown in Figure 18.1, and we make the following assumptions:

1. The material is linear-elastic;
2. The member is initially perfectly straight;
3. The applied loading is perfectly axial, i.e. at the centroid of the cross-section.

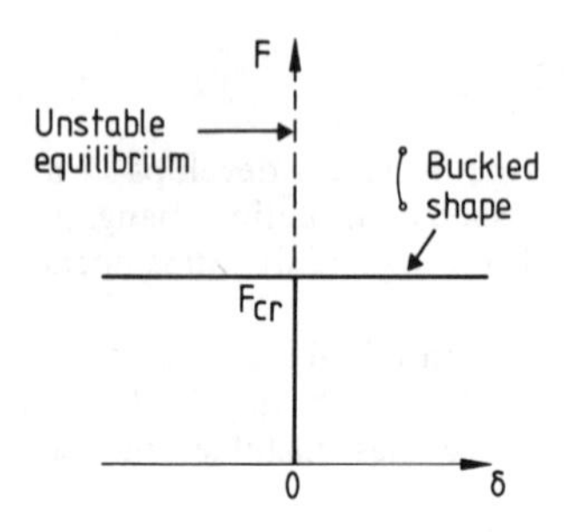

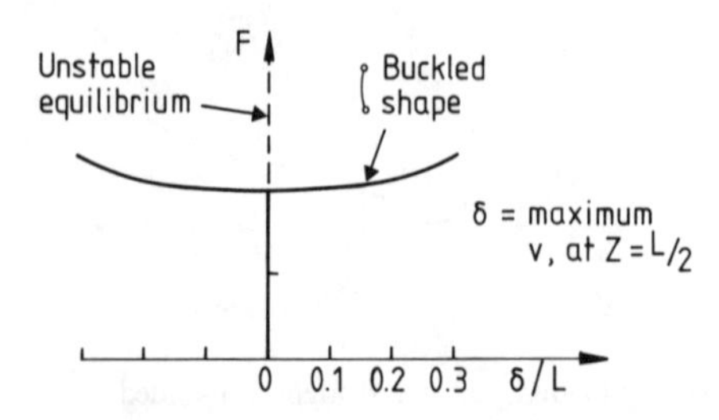

Figure 18.2 Axial load, F, against lateral deflection, δ, for a linear-elastic, initially straight, pin-ended strut

We also assume that the member is able to bend only about one principal axis. The classical Euler analysis of this problem makes an approximation which we shall discuss later, and it indicates the following behaviour:

1. The strut can apparently remain straight at all values of load (F).
2. If F has the particular value $F_{cr} = \pi^2 EI/L^2$, the strut can buckle in the shape of a half sine-wave, the amplitude of this deflection being indeterminate.
3. Other sinusoidal buckled shapes having n half wavelengths are possible at higher values of load given by $F = n^2\pi^2 EI/L^2$, where n is any integer.

Figure 18.2(a) shows a plot of successive states of static equilibrium of the member, drawn as a graph of applied axial force (F) against lateral deflection (δ). It is possible to show that all the equlibrium states at values of F greater than $F_{cr} = \pi^2 EL/L^2$, whether straight or buckled, are states of unstable equilibrium. This means that the slightest disturbance will cause the member to deflect away from that position towards a grossly deformed state. Unstable equilibrium states are unattainable by real systems, so that for practical purposes the load-deflection diagram stops at $F = F_{cr}$.

The deflected shape that the member assumes on buckling is referred to as the buckling mode. For the pin-ended strut the fundamental mode into which the member prefers to buckle is a half sine-wave:

$$v = a \sin \pi z/L \tag{18.1}$$

It is often convenient to work in terms of the mean applied compressive stress ($\sigma_{av} = F/A$) rather than a force. Elastic buckling occurs when this reaches the critical value:

$$\sigma_{cr} = \frac{F_{cr}}{A} = \frac{\pi^2 EI}{AL^2} \tag{18.2}$$

If r is the radius of gyration of the cross-section given by :

$$I = Ar^2 \tag{18.3}$$

then

$$\sigma_{cr} = \frac{\pi^2 Er^2}{L^2} = \frac{\pi^2 E}{\lambda^2} \tag{18.4}$$

where λ is the slenderness ratio defined by:

$$\lambda = L/r \tag{18.5}$$

Critical loads and critical stresses can be similarly found for struts having other end-restraint conditions. For example, if both ends are prevented from rotation as well as from lateral movement the critical load is increased to $4\pi^2 EI/L^2$. In such cases the slenderness ratio is defined as L_E/r, where L_E is the effective length of the member. This concept is developed in Chapter 22.

18.2.2 Large-deflection behaviour: post-buckling

The Euler analysis referred to above is based on the governing equation

Bending moment $M = Fv = EI \times$ curvature

where the curvature is expressed in the approximate form $-\mathrm{d}^2v/\mathrm{d}z^2$. This expression for curvature is accurate only if the deflections are small. The exact expression applicable to large as well as small deflections is:

$$-\frac{(\mathrm{d}^2v/\mathrm{d}z^2)}{[1 + (\mathrm{d}v/\mathrm{d}z)^2]^{3/2}} \tag{18.6}$$

Use of this expression leads to a much more complex analysis, known as the Theory of the Elastica. The buckling behaviour which it predicts is slightly different from that given by Euler's theory, and is represented in Figure 18.2(b). In a buckled condition the strut can withstand applied loads above the critical load (F_{cr}). As F is increased above F_{cr} the lateral deflection increases, rapidly at first and then less so. These buckled states of the member are ones of stable equilibrium, while the undeflected state is unstable for $F > F_{cr}$.

In practice, for a strut or column the deflection grows so rapidly when F exceeds F_{cr}, i.e. the post-buckling curve is so nearly flat that the behaviour is adequately represented by the Euler theory (Figure 18.2(a)). When the load is 1% above F_{cr} the lateral deflection at the midpoint of the strut is already almost one-tenth of the length of the member. Thus F_{cr} represents the maximum load that the strut can usefully support. (For certain other types of structural element the true behaviour is very different from that predicted by an Euler-type theory (see Chapter 19)).

18.2.3 Relationship between load and end-shortening

In place of the lateral deflection it is often useful to consider as a measure of deformation the end-shortening (δ_e) of the member, i.e. the decrease in the length of the straight line joining its ends. A graph of F against δ_e shows how the axial stiffness of the member varies during buckling.

The graph predicted by the simple Euler theory is shown in Figure 18.3. For $F < F_{cr}$ it is a straight line, given by:

$$F = (AE/L)\,\delta_e \tag{18.7}$$

When F reaches F_{cr} the lateral deflection (δ) – and hence also the end-shortening (δ_e) – can attain an arbitrarily large value. Thus the member's axial stiffness has been destroyed. The accurate large-deflection theory indicates that the member retains a very small axial stiffness at $F > F_{cr}$, but in practice this can generally be neglected.

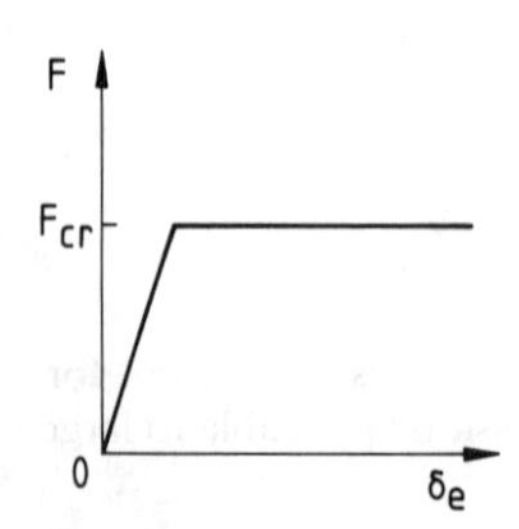

Figure 18.3 Axial load, F, against end-shortening, δ_e, for a linear-elastic, initially straight, pin-ended strut

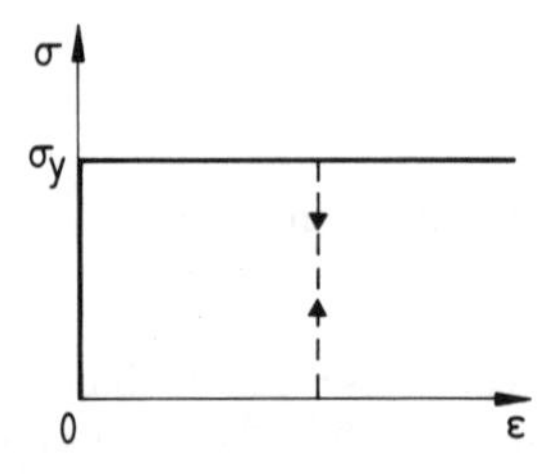

(a) Ideal rigid – plastic

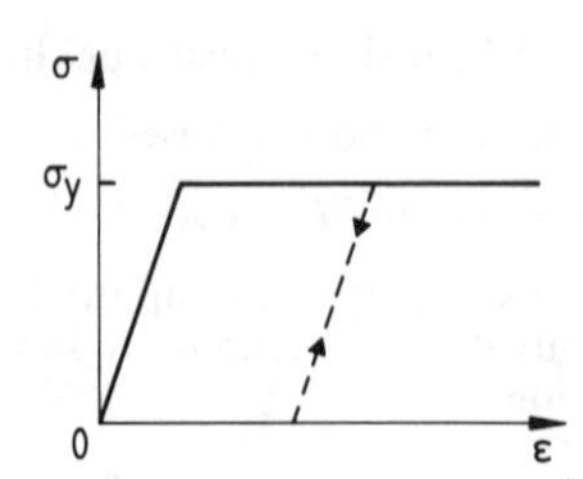

(b) Ideal elastic – plastic

Figure 18.4 Idealized material behaviour

18.3 Effect of material plasticity

18.3.1 Behaviour of an ideal strut having rigid-plastic material

In order to introduce the effects of inelastic behaviour in the simplest possible way we now consider an axially compressed member which is initially straight and made of a material having the ideal rigid-plastic stress–strain relationship shown in Figure 18.4(a). This may be expected to give an adequate description of elastic-plastic material behaviour for situations in which there is little elastic deformation.

Clearly, at low values of F there will be no deformation whatever – neither lateral nor axial. Since F is applied at the centroid of the section, apart from possible localized effects at the ends of the member all parts of the member will experience the same value of compressive stress $\sigma = F/A$. Deformation is possible only when σ reaches the yield stress (σ_y). At this stage the member is able to deform axially. The value of the axial force at which this happens (F_p) is termed the 'squash load', and is given by:

$$F_p = \sigma_y A \tag{18.8}$$

In principle, the member could deform axially and remain perfectly straight, but unless it is very stocky any slight disturbance will cause it to displace laterally and thus experience a bending moment. In the ensuing collapse of the member deformation is confined to a plastic hinge at some point within its length, as shown in Figure 18.5(a). Here, plastic deformation occurs in compression on the concave side of the member and in tension on the convex side, with the distribution of stress shown in Figure 18.5(b). F is now lower than the squash load, since the compressive yield stress is no longer attained over the entire cross-section. As the lateral displacement increases so does the bending moment at the hinge. Thus the part of the section that is in tension increases and that in compression decreases, so that the axial force continues

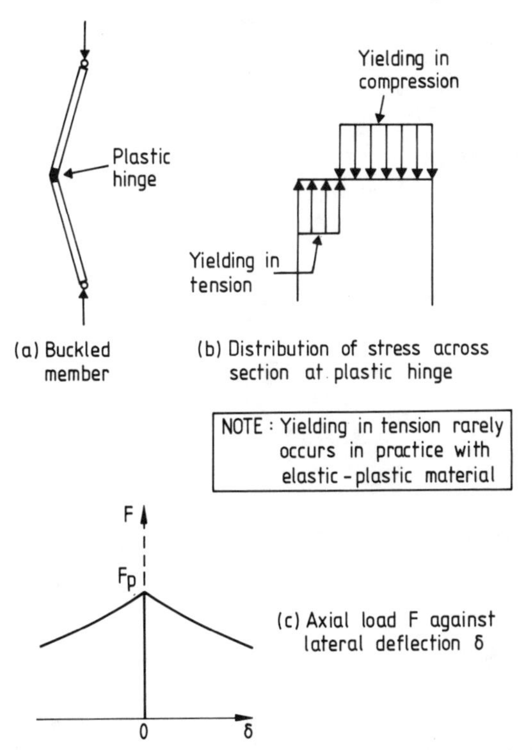

Figure 18.5 Buckling of an initially straight, pin-ended strut of ideal rigid-plastic material

to decrease. The relationship between F and the lateral displacement (δ) at the hinge is of the form shown in Figure 18.5(c).

18.3.2 Behaviour of an ideal strut having elastic-plastic material

If the member has the ideal elastic-plastic material represented by Figure 18.4(b) the behaviour under increasing axial load can be of two types according to whether F_{cr} or F_p is the smaller.

If $F_{cr} < F_p$:
At first the member experiences elastic axial deformation. Failure occurs by elastic buckling when $F = F_{cr}$. Yielding follows because bending moments are induced which increase the stress at the concave side of the member.

If $F_{cr} > F_p$:
The member at first experiences elastic axial deformation, but this time it reaches the plastic squash load first. If the member is very stocky it may squash plastically with little or no lateral displacement. Otherwise its behaviour will be similar to that of the rigid-plastic strut described earlier (see Figure 18.5(c)), but with some additional elastic axial and bending deformations. In either case the maximum axial force attained, i.e. the failure load, is F_p.

18.4 Strength curve for an ideal strut

A useful plot, both for understanding the relationship between the two types of behaviour just described and for design calculations, is one of the mean compressive stress at failure (σ_f) (or the failure load itself) against the slenderness ratio (λ). Such a plot is shown in Figure 18.6(a).

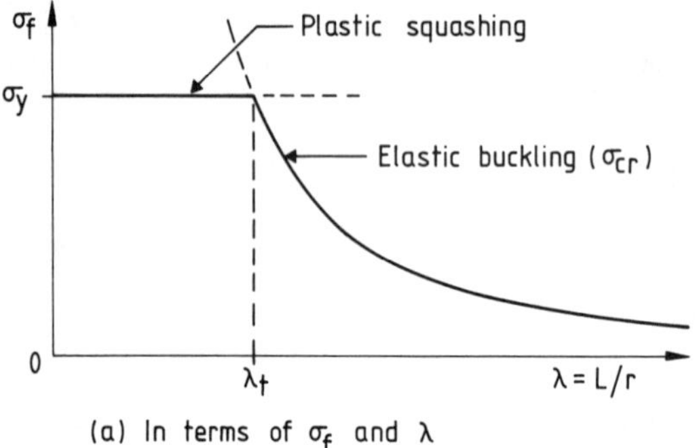

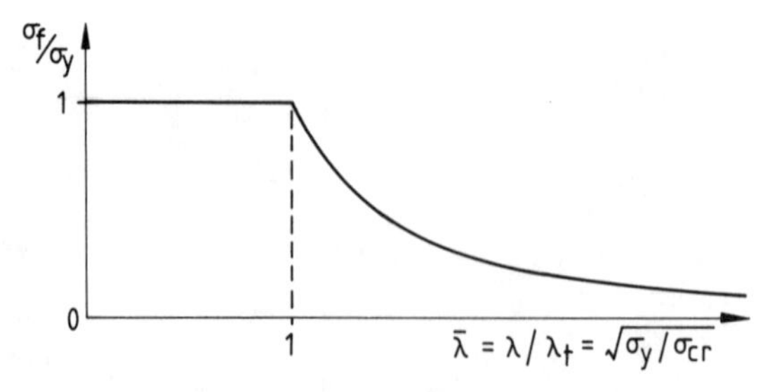

Figure 18.6 Strength curve for initially straight, pin-ended struts of ideal elastic–plastic material

Plastic squashing failure is given by the line $\sigma_f = \sigma_y$ and elastic buckling failure by the falling curve $\sigma_f = \sigma_{cr} = \pi^2 E/\lambda^2$. The two intersect at a slenderness value (λ_t) at which:

$$\sigma_y = \pi^2 E/\lambda^2 \tag{18.9}$$

so that

$$\lambda_t = \pi\sqrt{(E/\sigma_y)} \tag{18.10}$$

For struts having $\lambda < \lambda_t$, failure is by plastic squashing while for struts having $\lambda > \lambda_t$, failure is by elastic buckling. For steels having $\sigma_y = 245$ and $355\,\text{N/mm}^2$, $\lambda_t \approx 91$ and 76, respectively.

Plots of this kind are sometimes put in a special non-dimensional form, as shown in Figure 18.6(b). Here σ_f/σ_y is plotted against a generalized slenderness parameter [$\bar{\lambda} = \lambda/\lambda_t = \sqrt{(F_p/F_{cr})} = \sqrt{(\sigma_y/\sigma_{cr})}$].

This single plot now applies to all axially loaded, initially straight struts, irrespective of their E and σ_y values. A variant on this, which is used in some design codes, has on the horizontal axis the quantity $\lambda\sqrt{(\sigma_y/\sigma_{yo})}$, where σ_{yo} is an arbitrarily chosen reference value of yield stress. This modified slenderness is simply $\bar{\lambda}$ multiplied by the constant factor $\pi\sqrt{(E/\sigma_{yo})}$. BS 5400 uses this type of slenderness with $\sigma_{yo} = 355\,\text{N/mm}^2$.

18.5 Strength of real compression members

18.5.1 General

Hitherto, the struts we have considered have been of a highly idealized form that can never be achieved in practice. We have assumed perfect geometry and ideal rigid-plastic or elastic-plastic material behaviour. Deviations from perfect geometry are of two types: members are never perfectly straight and they are never loaded exactly at the centroid of the cross-section. Deviations from ideal elastic-plastic behaviour include strain-hardening at high strains and curving of the stress–strain relation so that there is no clearly defined yield point. We have also neglected the residual stresses that are inevitable in structural members.

It is difficult to deduce all at once the effects of these deviations on the member behaviour and strength. We shall therefore consider them one at a time.

18.5.2 Effect of imperfect geometry

18.5.2.1 Initial out-of-straightness

Figure 18.7(a) shows a pin-ended strut having an initial out-of-straightness. As soon as any axial load is applied the member experiences a bending moment, which in turn leads to further bending deformation and a growth in the amplitude of deflection. If the initial deflection (v_o) is in the shape:

$$v_o = a_o \sin\frac{\pi z}{L} \tag{18.11}$$

it is possible to show that, provided the material remains elastic, F magnifies the deflection at every point by the factor:

$$\frac{1}{1 - F/F_{cr}} \tag{18.12}$$

Thus the deflection tends to infinity as F approaches the elastic critical load (F_{cr}), as shown by curve 0′AB in Figure 18.8(a). (Other initial shapes are, of course, possible, but this assumption of a half sine-wave is analytically convenient and generally gives an adequate representation of the behaviour.)

As the deflection increases, the bending moments within the member also increase. The bending moment (Fv) is greatest mid way along the member (where v is greatest) and so are

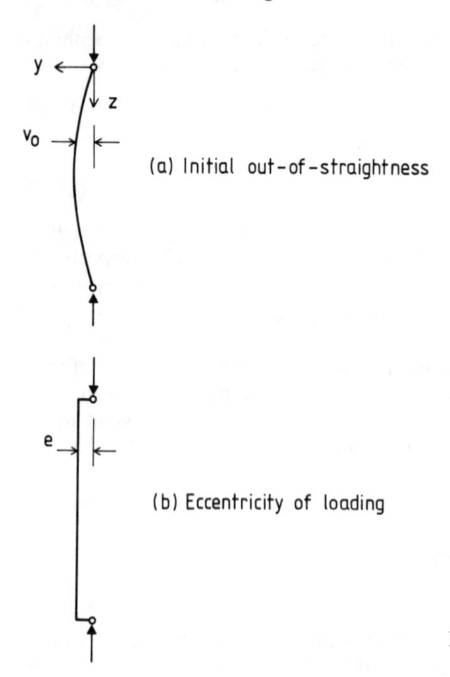

Figure 18.7 Pin-ended struts having imperfect geometry

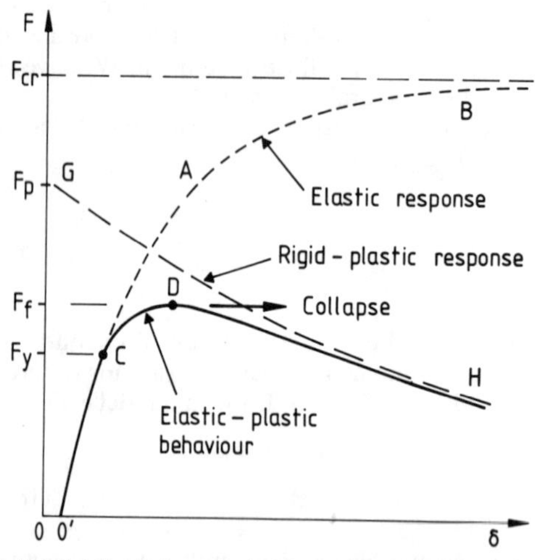

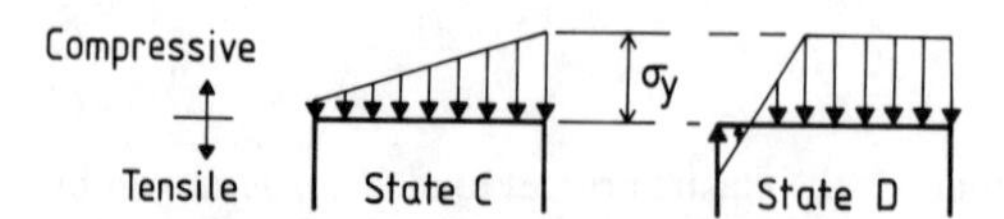

Figure 18.8 Strut with initial out-of-straightness (shown for the case when $F_p < F_{cr}$)

the resulting bending stresses. The greatest stress in the member occurs at the concave face; here the bending stress is compressive and is added to the compressive axial stress (F/A). As F is steadily increased, first yield occurs when the stress at this point reaches σ_y (point C in Figure 18.8(a)). As F is increased further, the zone of yielding spreads, both along the member and deeper into its section. This causes a deterioration in the stiffness of the member and eventually a maximum load is reached at which the strut collapses (point D). The stress distributions across the member cross-section at stages C and D are shown in Figure 18.8(b). Only exceptionally does the concave side of the member yield in tension at collapse.

Figure 18.8(a) also shows the rigid-plastic response curve GH, drawn here assuming that $F_{cr} > F_p$. Clearly, F_{cr} and F_p are both upper bounds to the loads F_y and F_f at first yield and at collapse. How close F_y and F_f are to F_{cr} and F_p depends on the amplitude a_o of the initial out-of-straightness and also on how slender or stocky the member is.

If a large number of struts, perhaps restricted to a given sectional type, were tested to failure and somehow all other deviations from ideal behaviour were eliminated it would be possible to construct a lower bound column strength curve that took into account initial out-of-straightness. Alternatively, it is possible to measure initial deflections and construct a curve from calculated collapse loads. Such a strength curve is of the form shown in Figure 18.9(a). Here, the failure stress (σ_f) is taken as the mean stress at collapse (F_f/A); that, at first yield, would give a lower curve of similar shape.

For a very stocky member, initial out-of-straightness, which is more a function of length than of cross-sectional dimensions, is negligible and failure is still at the plastic squash load. For a very slender member, F_p is much greater that F_{cr}, so that in Figure 18.8(a) point C is well up the elastic response curve, giving a failure load close to the elastic critical one. However, at intermediate values of slenderness the effect of initial out-of-straightness is very marked, the greatest loss of strength being at $\lambda \approx \lambda_t$.

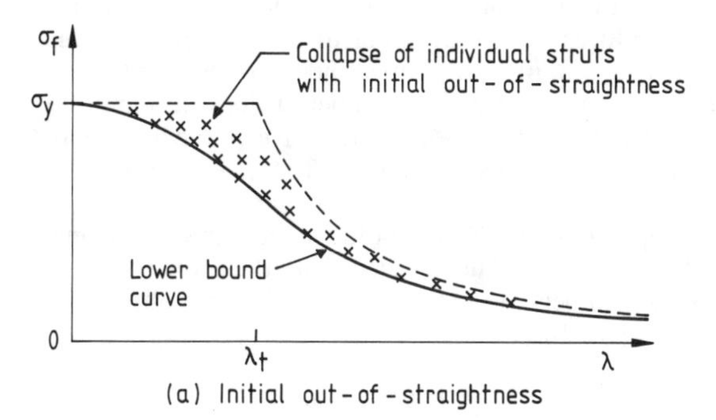

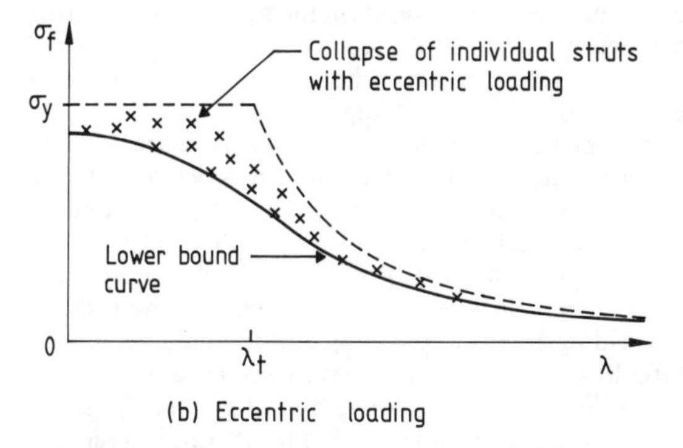

Figure 18.9 Forms of strength curves for struts of imperfect geometry

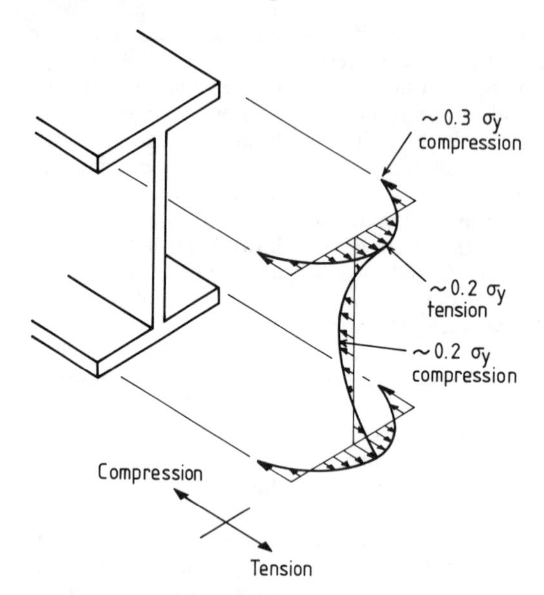

Figure 18.10 Residual stress in a rolled I-section

18.5.2.2 Eccentricity of applied loading

Figure 18.7(b) shows an initially straight, pin-ended column to which the loading is applied with an eccentricity (e). Consequently, F induces a bending moment in the member, which begins to deflect laterally in a way similar to the initially deformed member described earlier. As before, the greatest stress occurs at the concave face of the member at the section midway along it, and yielding begins at this point. The load–deflection relations for purely elastic and elastic-plastic material behaviour are similar to those in Figure 18.8(a), except that the deflection is zero at zero load.

In practical situations e is likely to be proportional to the size of the section and not strongly dependent of the length of the member. Bending effects due to eccentricity therefore reduce the load-carrying capacity even for very stocky members. The form of strength curve obtained by allowing for eccentricity is shown in Figure 18.9(b).

18.5.3 Effect of residual stresses

In practice, both rolled and fabricated sections have residual stresses in them, due to the heating and cooling processes and the large plastic deformations that have occurred during forming. Figure 18.10 shows a typical distribution of longitudinal stress in a hot-rolled universal column section. Note that compressive stresses exist at the tips of the flanges.

Consider a stub column, i.e. a very short compression member, in which there are residual stresses of this type, as shown in Figure 18.11(a), the maximum residual stress being σ_r. When a small value of axial load is applied, the uniform applied stress (σ_{av}) is added to the residual stress. This produces a total stress distribution which is non-uniform, with greatest stress ($\sigma_{av} + \sigma_r$) near the flange tips, as shown in Figure 18.11(b). Provided that the total stress nowhere reaches the yield stress, the member deforms elastically, just as if there were no residual stress. However, once $\sigma_{av} + \sigma_r$ reaches σ_y, yielding begins at the flange tips and, under further increasing load, spreads inwards and eventually develops also in the web (see Figure 18.11(c)). When $\sigma_{av} = \sigma_y$ the entire section yields, as in Figure 18.11(d). The graph of mean axial stress against mean axial strain (shortening/length) is of the form shown in Figure 18.12(a). Averaged thus, the material appears to have a limit of proportionality at a stress $\sigma_p = \sigma_y - \sigma_r$, the curve becoming horizontal at the yield stress.

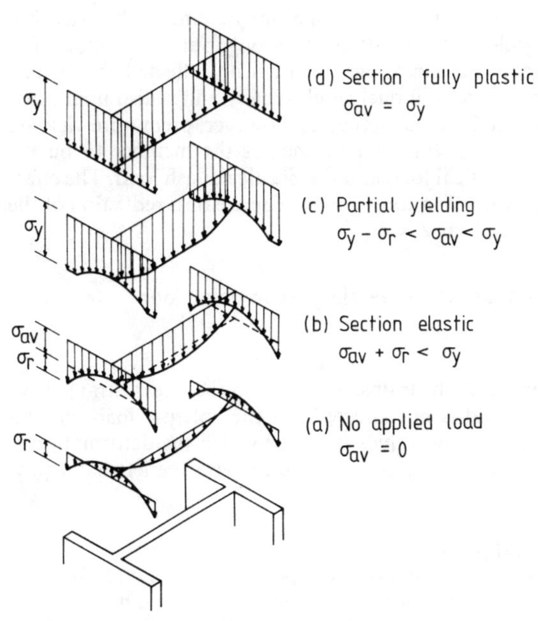

Figure 18.11 Stress distributions during axial loading of stub column with residual stresses

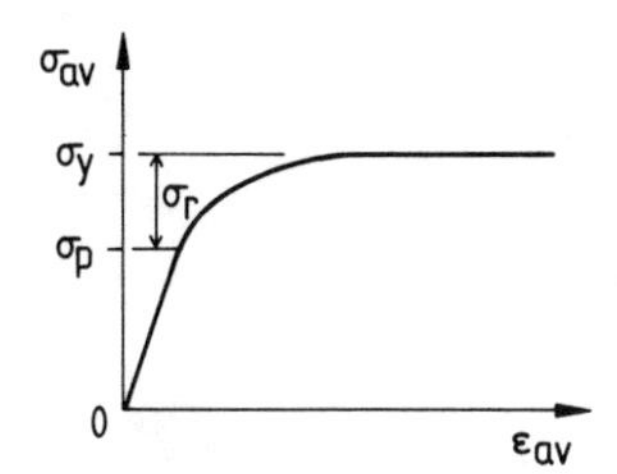

(a) Mean axial stress against mean axial strain from stub column test

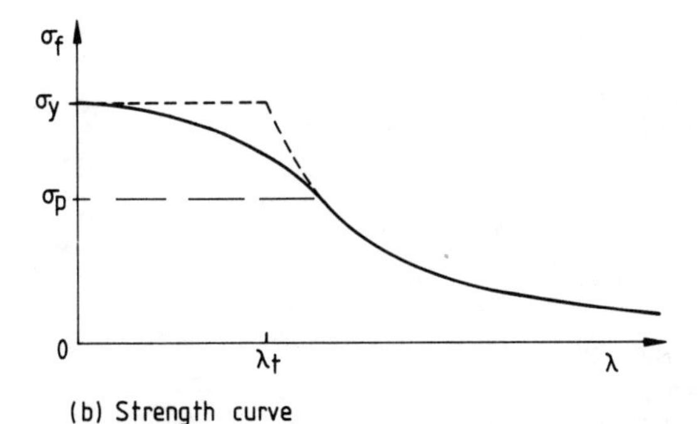

(b) Strength curve

Figure 18.12 Buckling of initially straight struts having residual stresses

In a very stocky column the residual stresses may cause premature yielding in this way, but the member remains straight until the yield stress is attained throughout the section. The member still fails at the plastic squash load, i.e. the collapse load is not affected by residual stresses. A very slender strut, for which $\sigma_{cr} < \sigma_p$, still buckles elastically and is also unaffected by residual stresses. For struts having intermediate slendernesses, however, premature yielding at the flange tips reduces the effective bending stiffness and enables the member to buckle inelastically at a load below both the elastic critical load and the plastic squash load. The effect on the column strength curve is thus as shown in Figure 18.12(b). The greatest reduction of the failure load from the ideal-plastic case is again at $\lambda \approx \lambda_t$.

18.5.4 Effect of material having other than ideal elastic-plastic behaviour

18.5.4.1 Strain-hardening

If the material experiences strain-hardening at high strains, as shown in the stress–strain curve of Figure 18.13(a), the condition for first yield is not affected but the collapse load may be increased. This will clearly be more marked at low values of λ, for which the deformation is predominantly plastic. The effect of strain-hardening is therefore to raise the left-hand (low slenderness) end of the column strength curve.

18.5.4.2 Absence of clearly defined yield point

High-strength steels generally have stress–strain curves of the type shown in Figures 18.13(b) or 18.13(c). At stresses above σ_p the material behaviour is non-linear and plastic deformation occurs, unloading and reloading being linear-elastic. Most high-strength structural steels have a yield stress at which the curve becomes more or less horizontal, as in Figure 18.13(b), this plateau being possibly followed by strain-hardening. Some steels do not have a plateau but experience strain-hardening throughout the inelastic range, as in Figure 18.13(c); in such cases the yield stress in generally taken as the 0.2% proof stress.

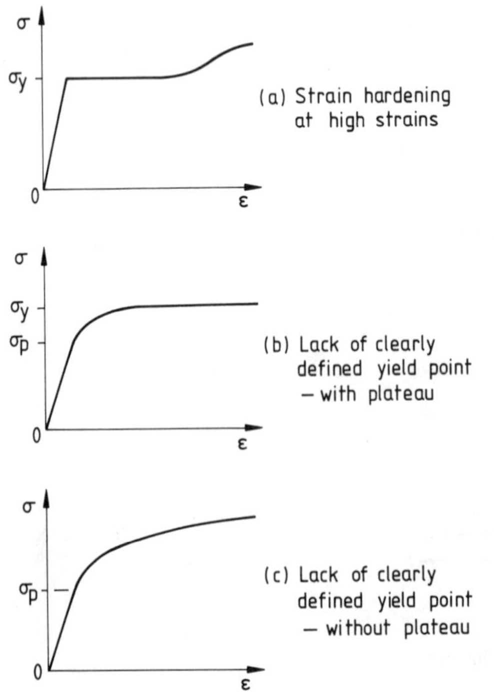

Figure 18.13 Non-ideal elastic–plastic behaviour

We consider first a strut having the material behaviour shown in Figure 18.13(b), which is similar to the effective stress–strain curve for a stub column having ideal elastic-plastic material combined with residual stresses (Figure 18.12(a)). If the strut is sufficiently slender for σ_{cr} to be below σ_p then failure is by elastic buckling at the elastic critical load. More stocky members remain straight while the applied stress rises above σ_p, and inelastic deformation begins to develop uniformly across the section. If it is very stocky the member fails by axial squashing without lateral deflection; the stress remains uniform across the section and as the plastic strain increases the yield stress is approached everywhere in the member. If the strut has an intermediate slenderness ratio, then, as the stress increases above σ_p, the stiffness of the material in resisting further increase of stress, as characterized by the current slope of the stress–strain curve, progressively decreases. A form of inelastic buckling is then possible at a load below both the squash load and the elastic critical one. The strength curve for struts having this type of material behaviour is thus of a form similar to Figure 18.12(b), which applies to struts of ideal elastic-plastic material having residual stresses.

If the material experiences strain-hardening, whether as in Figure 18.13(c) or at high strains following initial behaviour as in Figure 18.13(b), this has a beneficial effect at low slenderness ratios, as noted earlier.

18.5.5 Effect of all features taken together

In practice, all these effects occur simultaneously, i.e. out-of-straightness, eccentricity of loading, residual stresses, strain-hardening and, depending on the material, loss of a clearly defined yield point. Only strain-hardening tends to raise the column strength curve, and this happens only at low slenderness values. All the other effects lower the column strength curve for all or part of the slenderness range. When all the effects are put together the resulting column strength curve is generally of the form shown in Figure 18.14. At very low slenderness the beneficial effect of strain-hardening is generally more than sufficient to compensate for any loss of strength due to small, accidental eccentricities (though larger eccentricity must be allowed for separately). Although the member strength (σ_f) rises above the yield strength σ_y at this end, for structural design purposes the column strength curve is generally considered as having a cut-off at σ_y.

Since geometric imperfections, residual stresses and material properties are subject to statistical variations it is impossible to predict with certainty the greatest reduction of strength that they might produce in practice. Thus for design purposes it is impossible to draw a true lower bound column strength curve. One approach is to construct a curve on the basis of a specified survival probability, i.e. such that a specified percentage of the columns to which the curve relates can be expected, on a statistical basis, to fail at applied loads equal to or above those given by the curve. The column strength data in BS 5400 and BS 5950 are derived from curves that were based originally on a 97.7% survival probability.

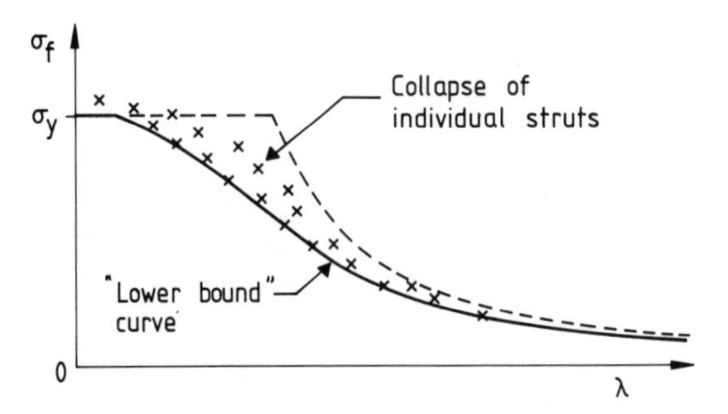

Figure 18.14 Form of strength curve for struts with initial out-of-straightness, eccentricity, strain-hardening and residual stresses

18.6 Concluding summary

The following features have been noted in the context of compressed struts but are typical of a wide range of members that are subject to buckling failure.

1. An idealized, initially straight, axially loaded strut of ideal elastic-plastic material can fail by elastic buckling or by plastic squashing.
2. In practice, the strength is influenced by initial out-of-straightness, eccentricity of loading, residual stresses, strain-hardening and, in some cases, lack of a clearly defined yield point.
3. The greatest reduction of strength caused by these effects occurs at intermediate values of slenderness, where there is a transition between plastic squashing failure and predominantly elastic buckling failure.
4. The slenderness ratio is an important geometric parameter which determines the nature of the behaviour and, consequently, the strength.

Background reading

ALLEN, H.G. and BULSON, P.S. (1980) *Background to Buckling*, McGraw-Hill, Maidenhead
HORNE, M.R. and MERCHANT, W. (1965) *The Stability of Frames*, Pergamon, Oxford
TIMOSHENKO, S.P. and GERE, J.M. (1961) *Theory of Elastic Stability*, 2nd edn, McGraw-Hill, New York

19

Introduction to buckling: 2

Objective To describe and illustrate a wide range of buckling phenomena relevant to steel structures.

Prior reading Chapter 18 (Introduction to buckling: 1); Simple torsion theory and shear centre.

Summary The chapter includes descriptions of a wide range of buckling phenomena relevant to steel structures and identifies the appropriate geometric parameters. First, the local buckling of thin, plate-like elements of members is introduced with a description of the buckling, post-buckling and collapse behaviour of compressed rectangular plates. Buckling and collapse of web plates in shear are also briefly described. Forms of buckling involving torsional deformation are then discussed, i.e. torsional and flexural–torsional buckling of compressed struts and lateral–torsional buckling of beams bent about their major axes. Finally, the means available for preventing buckling failure are briefly considered.

19.1 Buckling of plates in compression: local buckling

19.1.1 Introduction

Figure 19.1 shows two cases of local buckling. Case (a) is a box girder consisting of four flat plates welded together. The applied sagging bending moment induces in the top flange plate a compressive longitudinal stress. This may cause the flange plate to develop bulges of the type shown. Similar effects are possible if a box member is used as a column, but then all four plates are in compression and liable to buckle.

Case (b) is a plate girder of I-section, fabricated also from flat plates. Under a sagging bending moment the top flange is again in compression. The flange plate projecting to one side of the web is referred to as an outstand; it is supported and kept virtually straight along one of its longitudinal edges by the junction with the web but is free at its other longitudinal edge. Each compressed outstand may thus develop bulges, which would be visible at the free edges. This contrasts with the box girder flange, which is supported by webs at both longitudinal edges.

Local buckling occurs where thin elements of a member are placed in compression. This is similar to the buckling of a compressed strut but involves individual parts of the member cross-section rather than displacement of the section as a whole. It occurs not only in compression flanges but also in web plates, where compressive stresses arise from bending of the member and from concentrated loadings applied to the flanges; a form of local buckling also occurs in web plates subjected to shear stresses. Local buckling is mainly of concern with thin-walled members, i.e. cold-formed sections and sections fabricated from thin plate. Similar problems can arise with thin-walled tubular members.

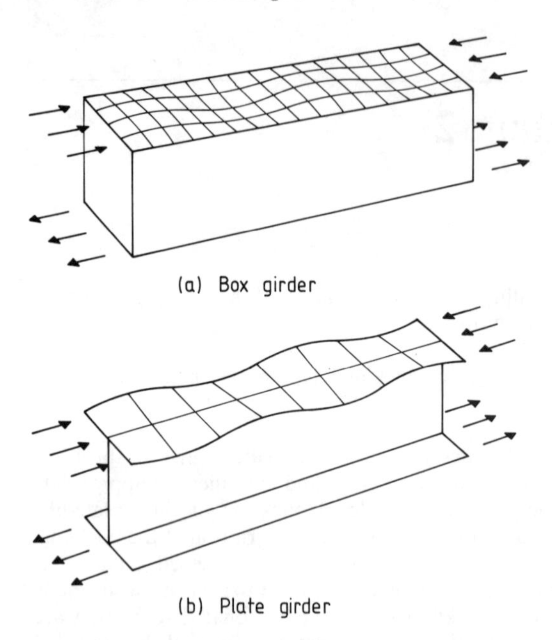

Figure 19.1 Local buckling in compression flanges of box girder and plate girder

19.1.2 Elastic buckling modes and critical stresses for compressed plates

We consider first the elastic buckling behaviour of a rectangular plate having equal compressive loadings applied to two opposite edges as shown in Figure 19.2. The loading might be applied in a variety of ways; two limiting cases are (1) a uniform stress applied across the entire width of plate and (2) a uniform displacement applied, for example, through rigid platens. In Figure 19.2 case (b) models the loading on a flange panel of a long box girder, where plane sections

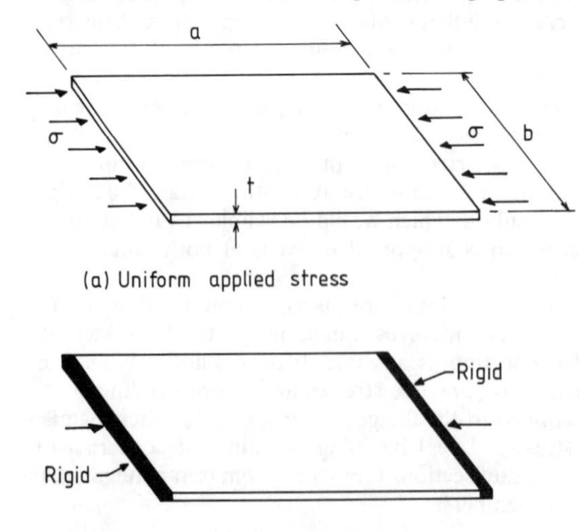

Figure 19.2 Rectangular plates in compression

of the girder remain approximately plane. Provided that the plate remains flat, the longitudinal stress (σ) is uniform throughout the plate whether the loading is case (a) or (b); only after buckling do the two cases differ.

In a way analogous to the buckling of a compressed strut the plate can apparently remain flat at all values of applied stress, but at a critical stress σ_{cr} a lateral deformation becomes possible. The shape of this lateral deformation is the buckling mode. The buckling mode and critical stress depend on two major features–the conditions of support at the edges of the plate and the plate aspect ratio (a/b). Clearly, the critical stress also depends on the plate thickness.

Suppose the two loaded edges of the plate are simply supported (prevented from displacing out-of-plane but free to rotate) and the other two edges are free (totally unsupported). This is like a wide, pin-ended strut of length a having a rectangular cross-section $b \times t$. The critical stress given by the Euler formula is:

$$\sigma_{cr} = \frac{\pi^2 EI}{a^2 A} = \frac{\pi^2 Ebt^3}{12a^2 bt} \qquad (19.1)$$

However, it is necessary to insert an extra $(1 - \nu^2)$ in the denominator, where ν is Poisson's ratio, to allow for the suppression of anticlastic curvature (transverse curvature caused by thinning of the longitudinal fibres that are in tension and thickening of those in compression). Thus:

$$\sigma_{cr} = \frac{\pi^2 Et^2}{12(1 - \nu^2)a^2} \qquad (19.2)$$

A case of much greater practical importance is that when all four edges are simply supported. This approximates to the situation in the top flange of the box girder in Figure 19.1(a), though in the box girder the webs usually offer a degree of rotational restraint to the flange edges. Provided that the aspect ratio (a/b) is less than $\sqrt{2}$, the mode of buckling is a single bulge, as shown in Figure 19.3. However, for somewhat higher values of a/b (i.e. longer plates) the plate buckles into a mode having two bulges in opposite directions, as in Figure 19.3(b). For aspect ratios in the region of 3 the plate develops three bulges in alternate directions, and so on. Thus for simply supported plates having $a > b$ the buckling mode has bulges with approximately square perimeters.

Some insight into this behaviour can be gained by considering a long plate which has been replaced by a system of strips, as in Figure 19.4. A longitudinal strip such as AB is able to buckle under the compressive loading but is partially restrained by the transverse strips. If the

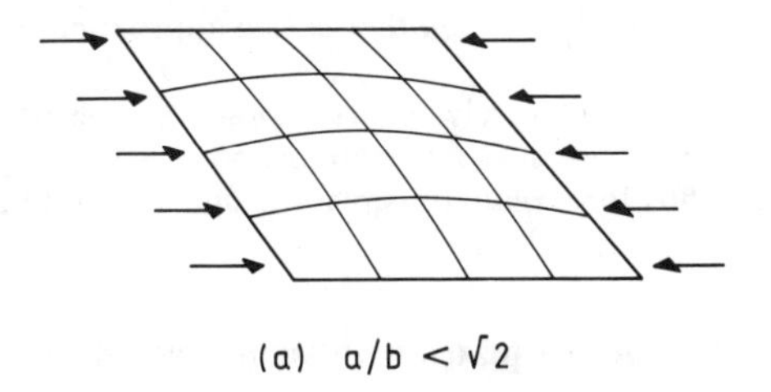

(a) $a/b < \sqrt{2}$

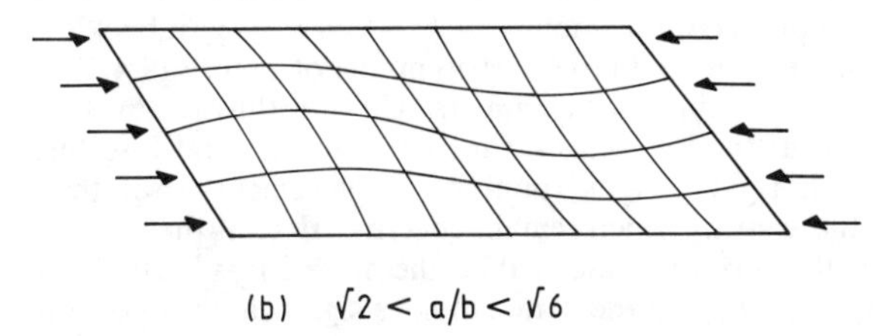

(b) $\sqrt{2} < a/b < \sqrt{6}$

Figure 19.3 Elastic buckling models for simply supported plates

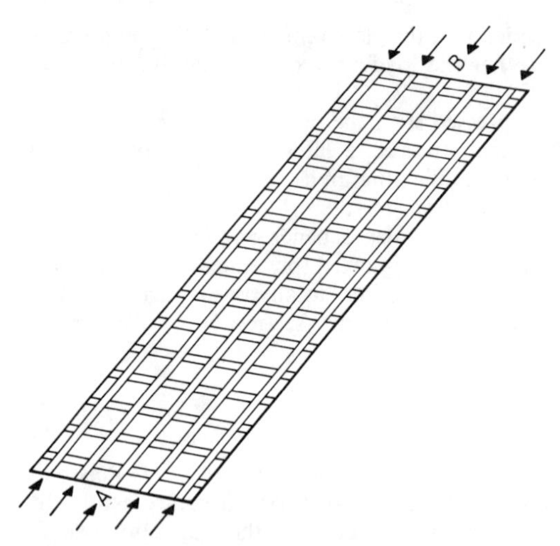

Figure 19.4 Strip model of long plate

plate formed only one buckle the curvature of the longitudinal strips would be very much less than that of the transverse ones. The plate prefers instead to take up a position such that the curvatures of the longitudinal and transverse strips are as near equal as possible; this represents an optimum form of deformation giving the minimum strain energy. In such a state the buckles are as near to square as possible.

The critical stress is given by:

$$\sigma_{cr} = k \frac{\pi^2 E t^2}{12(1 - \nu^2) b^2} \tag{19.3}$$

where k is a dimensionless buckling coefficient. If the aspect ratio (a/b) is an integer the plate can form exactly square buckles, and for this case $k = 4$. A long plate can always buckle into very nearly square panels so that $k \approx 4$. In general, k is a function of a/b, but $k = 4$ gives an adequate, conservative approximation provided that a/b is greater than about 0.8.

Figure 19.5 shows buckling coefficients for long plates having various edge conditions. The case in which one longitudinal edge is simply supported and the other is free approximates to a flange outstand.

Except in the case when both long edges are free (which rarely occurs in practice), the elastic critical stress of a long plate is proportional to t^2/b^2, i.e. it is inversely proportional to $(b/t)^2$ rather than $(a/t)^2$. The width-to-thickness ratio (b/t) for a plate thus plays a similar role to the slenderness ratio (L/r) for a strut or column.

19.1.3 Post-buckling behaviour and ultimate loads of plates with supported edges

Consider an initially flat plate, having all edges either simply supported or clamped, and loaded as shown in Figures 19.2(a) or 19.2(b). Below the elastic critical stress the plate remains flat, with uniform longitudinal stress (σ), but shortens by an amount $\sigma a/E$. When $\sigma = \sigma_{cr}$ it buckles. To describe the post-buckling behaviour we refer again to the strip model of Figure 19.4. The longitudinal strips close to the edges are constrained to remain straight and thus retain their axial stiffness. Only those strips away from the long edges are free to buckle and lose this stiffness. If the load is applied through rigid platens (uniform displacement case) then shortening continues to be resisted by those strips which remain effective. Even if the loading is a uniform stress, the in-plane shear stiffness of the plate enables the outer strips partially to restrain the inner ones from shortening, so that again the plate retains a significant proportion of its original stiffness.

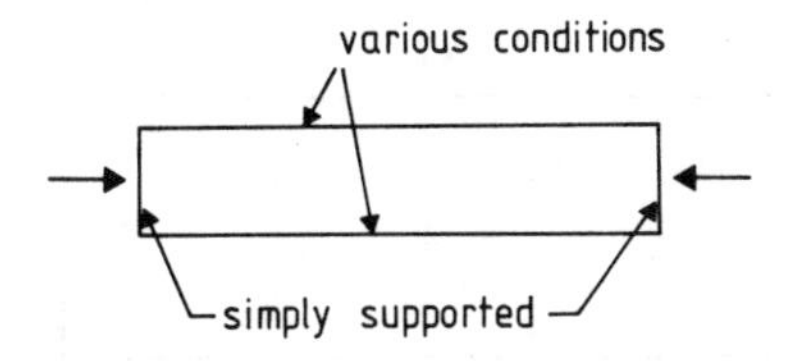

Support conditions at long edges	Buckling coefficient k
Clamped + clamped	6.97
Clamped + simply supported	5.41
Simply supported + simply supported	4.00
Clamped + free	1.25
Simply supported + free	0.43
Free + free	$(b/a)^2$ $(\rightarrow 0)$

Figure 19.5 Elastic buckling coefficients for long plates in compression

Thus in either case, in order to buckle, the plate must not only bend but also undergo deformation in its own plane (which is clearly necessary, since the buckled plate no longer forms a developable surface). The plate's resistance to this in-plane deformation enables it to support an applied stress considerably greater than σ_{cr}, provided that the material does not yield; this contrasts with the behaviour of a strut. The plot of mean applied stress (σ_{av}) against lateral deflection (δ) is of the form shown in Figure 19.6(a). At twice the critical stress the lateral deflection is typically about twice the plate thickness. The post-buckling axial stiffness is readily seen in the plot of applied stress against end-shortening (δ_e), (Figure 19.6(b)). The value of this stiffness depends on the in-plane restraints offered by the longitudinal edges as well as by the loaded ones; generally, it is between 40% and 75% of the initial stiffness.

The effect of an initial out-of-flatness is also demonstrated in Figure 19.6. As the applied stress is increased the lateral displacement grows steadily, accelerating in the region of the critical stress and then slowing down again; at stresses significantly higher than σ_{cr} the deflection is close to that for an initially flat plate. In the end-shortening plot the effect is to round off the corner at the critical stress.

The feature that eventually limits the load capacity of the plate is yielding. When the plate has buckled the compressive stress is no longer distributed evenly across the width of the plate, being greatest at the edges, as shown in Figure 19.7. Once the longitudinal stress at the edge has reached the yield stress, plastic deformation spreads rapidly and a collapse condition is very soon reached. For a slender plate the mean applied stress at failure may be as much as twice or three times the elastic critical stress.

19.1.4 Plate strength curves

Figure 19.8(a) shows how a strength curve is constructed for ideal, initially flat plates of a given aspect ratio. The applied compressive stress at failure (σ_f) is plotted against the slenderness parameter $\lambda' = b/t$. If the plate is very thick it fails by plastic squashing at a failure stress $\sigma_f = \sigma_y$, similar to a stocky strut. If it is slender then failure involves buckling, the critical stress being given by:

$$\sigma_{cr} = k \frac{\pi^2 E}{12(1 - \upsilon^2)} \frac{1}{(\lambda')^2} \tag{19.4}$$

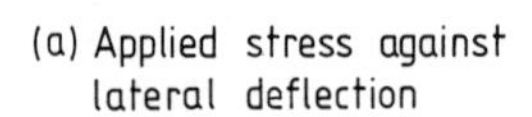

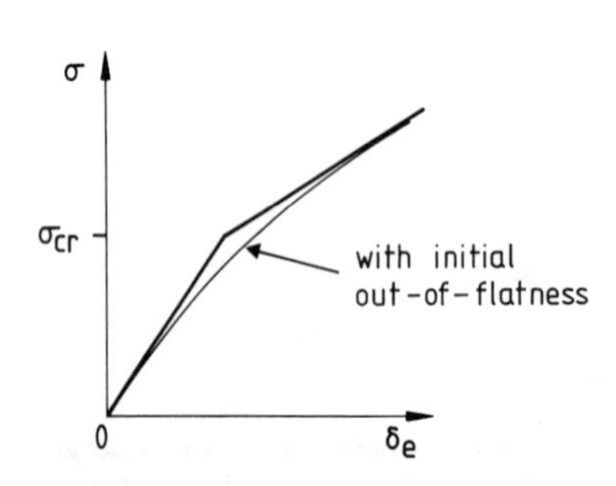

Figure 19.6 Elastic post-buckling behaviour of plates

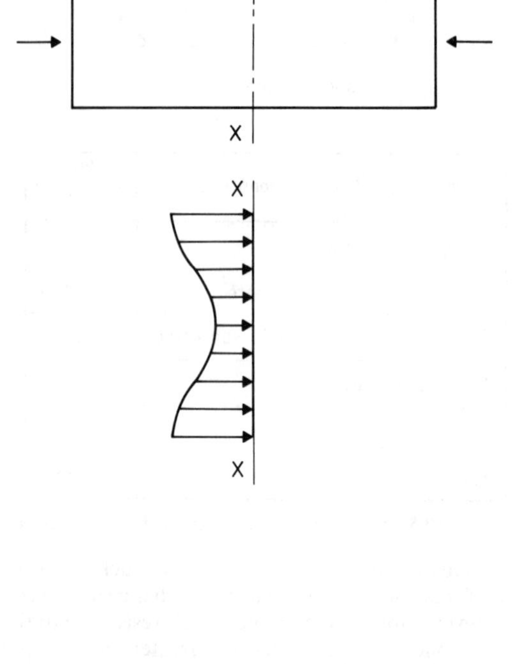

Figure 19.7 Distribution of mid-plane longitudinal stress across width of compressed plate after buckling

This relationship is similar to that for elastic buckling of a column. However, unlike the column, the ultimate strength of a slender plate is above the elastic critical stress and follows a higher curve as shown.

The transition from squashing behaviour to buckling occurs where $\sigma_{cr} = \sigma_y$ at a slenderness:

$$\lambda_t^1 = \pi\sqrt{\left[\frac{kE}{12(1-\upsilon^2)\sigma_y}\right]} \tag{19.5}$$

For simply supported plates having a yield stress of 355 N/mm^2 and aspect ratios greater than about 0.8, $k \approx 4$ and the transition occurs when $b/t \approx 46$.

As with columns, compressed plates have, in reality, initial out-of-flatness, residual stresses, strain-hardening, etc. which modify their behaviour and influence their strength, as shown in Figure 19.8(b). The effect on the plate strength curve is similar to that for columns, but at high slenderness the modified curve approaches the ideal collapse (ultimate strength) curve rather than that for elastic buckling.

19.2 Buckling of web plates in shear

Fagure 19.9(a) shows a rectangular plate whose edges are subjected not to applied compressive stresses but to shear ones (τ). A square element, whose edges are orientated at 45° to the plate edges, experiences tensile stresses on two opposing edges and compressive ones on the other two. These compressive stresses can induce a form of local buckling; the mode involves

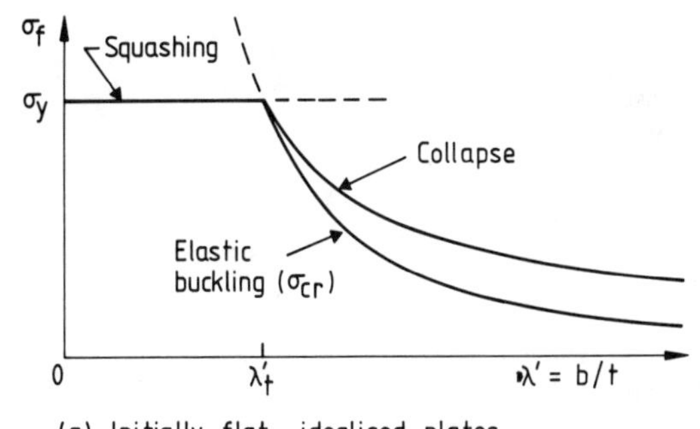

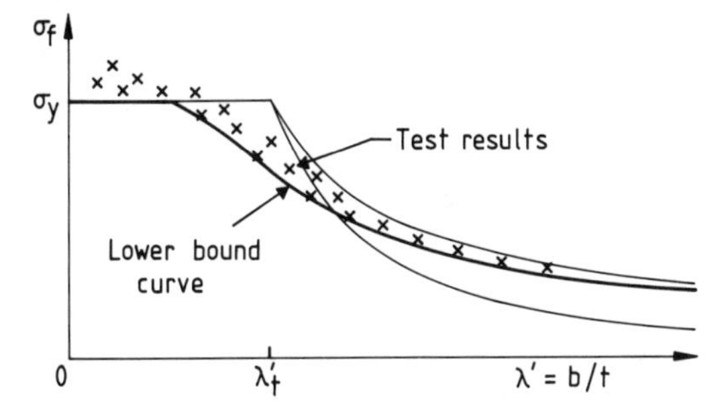

Figure 19.8 Plate strength curves

elongated bulges orientated at about 45° to the plate edges, as indicated in Figure 19.9(b). The critical shear stress (τ_{cr}) is given by an expression similar to that for uniform compressive loading; again, k depends on the aspect ratio and edge support conditions.

As with compressive loading, a thin plate loaded in shear can support an applied stress well in excess of the elastic critical one. This is due again to the resistance to in-plane deformation. As the applied shear stress is increased beyond τ_{cr} the plate buckles elastically and retains little stiffness in the direction in which the compressive component acts. However, the inclined tensile component is still resisted fully by the plate. The inclined buckles become progressively narrower and the plate acts like a series of bars in the tension direction, developing a so-called tension field. Further increase of applied stress causes plastic deformation in part of the tension field, which rotates to line up more closely with the plate diagonal, as in Figure 19.9(c).

Tension field action is particularly important in plate and box girders, in which the function of the web plates is primarily to resist shear. The extent to which it can develop depends greatly on the way in which the edges of the web panels are restrained.

19.3 Torsional buckling of a column

Consider a compression member having the cruciform section shown in Figure 19.10. Such sections are not in common use but permit ready visualization of the torsional buckling phenomenon. Each of the four equal outstands can be considered as a flat plate which is in longitudinal compression and therefore liable to buckle. If the inner edge of each plate were simply supported each plate would buckle independently at the same value of applied stress. A line 0A drawn across an outstand would remain approximately straight for this mode of buckling but would rotate about 0. The same would happen to lines 0B, 0C and 0D drawn

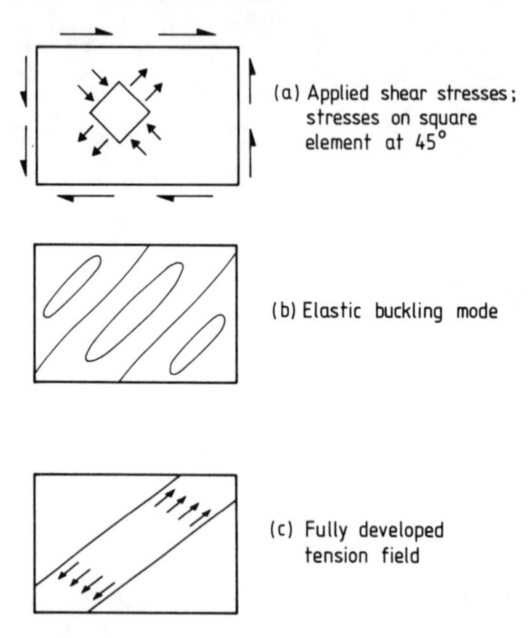

Figure 19.9 Web panels in shear: buckling and tension field action

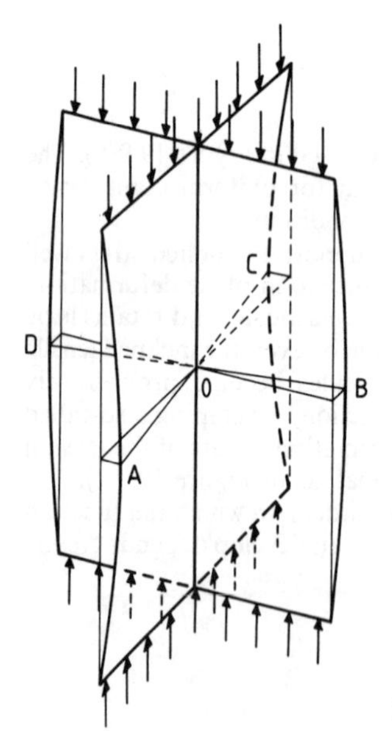

Figure 19.10 Torsional buckling of a cruciform strut

across the other outstands at the same section; these could all rotate by the same amount and in the same direction. Since these four lines would continue to meet at right angles at 0 the same behaviour would occur if the plates were rigidly joined along their common edge, as is actually the case in the cruciform member. Thus a form of buckling is possible in which the axis remains straight but sections of the member rotate, i.e. the member twists. This is termed 'torsional buckling'.

The analysis of torsional buckling is quite complex. The mode of buckling and corresponding critical axial force or stress clearly depend on the restraints provided at the ends. The critical stress depends on the torsional stiffness of the member as well as on the resistance to warping deformations provided by the member and by the restraints at its ends. If the ends are prevented from torsional rotation but are not otherwise restrained the critical stress for torsional buckling is given by the equation:

$$\sigma_{cr} = \frac{1}{I_p}\left(GJ + EH\frac{4\pi^2}{L^2}\right) \qquad (19.6)$$

where I_p is the polar second moment of area of the section, J is the St Venant torsional constant and H is the warping constant.

Pure torsional buckling is possible only if the centroid and shear centre of the cross-section are coincident. This limits it to sections having either double or point symmetry. It is important to recognize that purely flexural buckling of the member as a strut is also possible, and in many cases this will happen at a lower critical stress than torsional buckling.

19.4 Flexural–torsional buckling

If the shear centre and centroid of the section do not coincide, as happens with an angle or channel section, coupling between flexural and torsional deformations complicates the behaviour considerably. This is readily seen in the case of a channel section having broad flanges, as in Figure 19.11. This member could buckle in a purely flexural mode in the x-direction. Suppose, however, that the flanges are so wide that $I_{xx} > I_{yy}$ and it prefers to buckle in the y-direction. If the member ends are subjected to a uniform applied stress, the resultant axial force (F) acts at the centroid C. Suppose that, as a consequence of flexural buckling in the y-direction, a given section deflects by an amount (v). A bending moment equal to Fv is then induced at this section. However, this bending moment and the associated shear forces are acting in the Y–Y plane containing C and not in the Y′–Y′ plane containing the shear

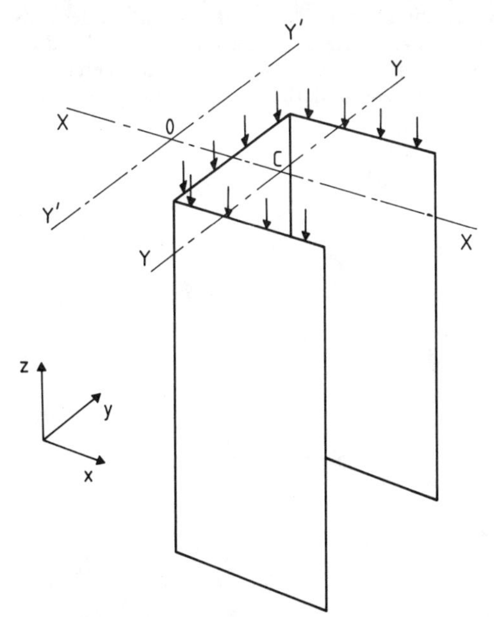

Figure 19.11 Axially compressed member of channel section with broad flanges

centre 0. They do not, therefore, induce purely bending deformations but also cause the section to twist. Thus purely flexural buckling in this direction is impossible, and bending and torsional deformations are inseparable. The resulting form of buckling is termed 'flexural–torsional buckling'.

There is no single, physically obvious slenderness ratio for torsional and flexural–torsional buckling. Strength curves can, however, be plotted using either the parameter $\bar{\lambda} = \sqrt{(\sigma_y/\sigma_{cr})}$ or a slenderness ratio based on an effective length that allows for torsional effects. Torsional and flexural–torsional buckling are associated mainly with compression members with low torsional stiffness, i.e. open, thin-walled sections such as cold-formed sections. Purely torsional buckling can occur with Z-sections and also with I-sections having broad flanges. Flexural–torsional buckling can occur with certain angle sections and with channels having broad flanges.

19.5 Lateral–torsional buckling

Lateral–torsional buckling (also called lateral buckling) is a buckling phenomenon which occurs with certain types of beams and girders. Figure 19.12 shows a cantilever beam having a deep I-section and loaded vertically, thus producing bending about its major axis. A hogging bending moment is induced in the beam, resulting in tensile stresses in the top flange and compressive stresses in the bottom one. If it were not restrained by the web the bottom flange would buckle either horizontally or vertically as a strut. The web completely prevents buckling in the vertical plane but only partly restrains the section from buckling horizontally. This causes the section to twist, so that the principal axes are no longer horizontal and vertical. The vertically applied loading on the beam now has a component in the weak direction of bending, i.e. in the direction of the rotated X–X axis, and induces a displacement in that direction. The deformation thus consists of a mixture of torsional and lateral flexure. Similar types of behaviour occur in beams or girders supported at both ends, but under downward loading the top flange is now the one in compression.

The condition for buckling is generally expressed in terms of a critical value of either the maximum bending moment or the compressive bending stress. Clearly, the condition depends on the pattern of loading applied to the beam, as characterized by the shape of the bending moment diagram. The buckling resistance depends also on several geometric parameters–the beam length and support conditions, its lateral stiffness and the torsional properties and

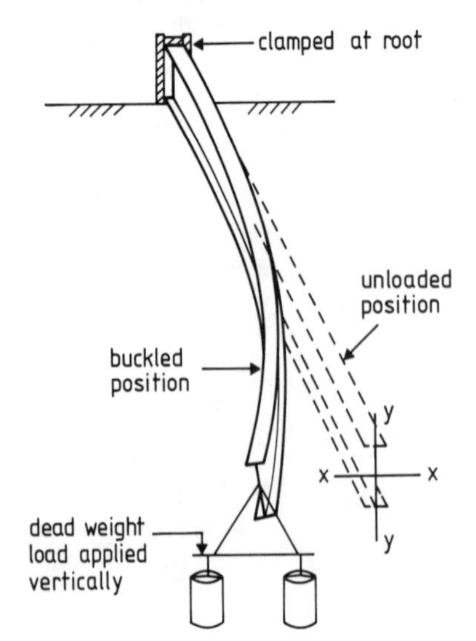

Figure 19.12 Lateral–torsional buckling of a cantilever beam

warping resistance of its section. There is no single, physically obvious slenderness ratio for lateral–torsional buckling. However, for a given loading type a beam strength curve can be constructed by plotting the failure moment against the parameter $\bar{\lambda}_{LT} = \sqrt{(M_p/M_{cr})}$ or alternatively the 'slenderness ratio' $\lambda_{LT} = \bar{\lambda}_{LT}\pi\sqrt{(E/\sigma_y)}$.

19.6 Prevention of buckling failure

An obvious way to increase the buckling resistance of a member is to increase its cross-sectional dimensions. With a strut, for example, this gives a larger cross-sectional area and can also reduce the slenderness ratio, giving a higher failure stress.

An alternative, and often more effective, approach is to modify the buckling mode. This can sometimes be achieved by improving the end-restraint on a member – for example, by preventing rotation at the end of a strut by fixing it rigidly to a support or another member. Sometimes it is possible to suppress altogether the particular mode that is governing the behaviour. For example, restraining the midpoint of a pin-ended column suppresses the first buckling mode and causes it to buckle instead in a mode having two half sine-waves. This increases the elastic critical load by a factor of 4, and does not necessarily require a very substantial form of bracing.

Similar possibilities exist with compressed plates. Stiffening the edge of an outstand by the formation of a lip, for example, can greatly increase its buckling resistance. Attaching a longitudinal stiffener along the middle of a simply supported plate is analogous to centrally bracing a strut; by doubling the number of buckles required this can quadruple the elastic critical stress. Similarly, attaching transverse and/or longitudinal stiffeners to the webs of plate and box girders can greatly increase their resistance to buckling under shear stress.

Restraining a beam from lateral and/or torsional displacements can, if properly carried out, totally eliminate lateral–torsional buckling and enable the beam to develop its full plastic resistance.

In the context of overall buckling of frames, restraining sway displacements by appropriate diagonal bracing, lift/service shafts or shear walls can eliminate the most damaging modes of buckling and greatly increase the strength of the frame.

Buckling Phenomenon	Member type	Loading	Stiffness Parameters	Slenderness Ratio	Generalised Slenderness
Flexural	Struts & Columns	Axial Compression	EI_{xx} or EI_{yy}	L/r_x or L/r_y	$\sqrt{F_p/F_{cr}} = \sqrt{\sigma_y/\sigma_{cr}}$
Lateral – torsional	Beams	Major axis bending	GJ, EH, EI_{yy}	—	$\sqrt{M_p/M_{cr}}$
Local	Thin flat plate or element	Compression or shear	$\dfrac{Et^3}{12(1-\nu^2)}$	b/t	$\sqrt{\sigma_y/\sigma_{cr}}$
Torsional	Struts (open, thin-walled sections with double or point symmetry)	Axial Compression	GJ, EH	—	$\sqrt{F_p/F_{cr}} = \sqrt{\sigma_y/\sigma_{cr}}$
Flexural – torsional	Struts (open thin-walled sections without double or point symmetry	Axial Compression	GJ, EH EI_{xx} &/or EI_{yy}	—	$\sqrt{F_p/F_{cr}} = \sqrt{\sigma_y/\sigma_{cr}}$

Figure 19.13 Summary of buckling phenomena and parameters

19.7 Concluding summary

1. The buckling phenomena that have been described are summarized in Figure 19.13.
2. The most effective way of designing against buckling is often to modify or eliminate the most damaging mode of buckling.

Background reading

Buckling of plates: local buckling

ALLEN, H.G. and BULSON, P.S (1980) *Background to Buckling*, McGraw-Hill, Maidenhead. Pages 378–393: good description of large-deflection behaviour, including shear loading.

TIMOSHENKO, S.P. and GERE, J.M. (1961) *Theory of Elastic Stability*, 2nd edn, McGraw-Hill, New York. Chapter 9, especially pp. 348–356, 379–383 and 408–423. Plate buckling and post-buckling (including shear loading) with emphasis on theory.

TRAHAIR, N.S. (1977) *The Behaviour and Design of Steel Structures*, Chapman and Hall, London. Chapter 4, especially pp. 98–109 and 114–119. A good general introduction.

WALKER, A.C. (ed.) (1975) *Design and Analysis of Cold-formed Sections*, Intertext Books, London. Pages 34–53: description of local buckling and post-buckling for compressive loading.

Torsional and flexural–torsional buckling

ALLEN, H.G. and BULSON P.S., *op. cit.* Pages 429–445: description and theory.

Lateral–torsional buckling

TRAHAIR, N.S., *op. cit.* Chapter 6, especially pp.178–181. Brief description followed by much theory.

20

Restrained compact beams

Objective To derive, discuss and present the processes used to design restrained compact beams for bending, shear and deflection according to BS 5400: part 3 and BS 5950.

Prior reading Elastic theory for uniaxial and biaxial bending; Elastic theory for shear stress due to bending; Simple plastic theory, including shear and normal forces; Simple torsion theory; Chapter 18 (Introduction to buckling: 1).

Summary This chapter is restricted to beams whose design may be based upon simple strength of materials considerations. Behaviour in simple bending is discussed, leading to the concept of section modulus as the basis for strength design. Subsidiary considerations of shear strength and adequate stiffness against deflection are also mentioned. More complex behaviour, involving combined bending and torsion and torsion or bending about both principal axes, is introduced. For worked examples, see Appendix, p. 368.

20.1 Introduction

Probably the most basic structural component is the beam intended to span between two supports and to transmit loads, principally by bending action. Steel beams, which may be drawn from a wide variety of structural types and shapes, can often be designed using little more than the simple theory of bending. However, situations will arise in which the beam's response to its loading will be more complex, with the result that other forms of behaviour must also be considered. The main purpose of this chapter is to concentrate on the design of that class of steel beam for which basic strength of materials provides the foundations of the design approach. These are termed 'restrained compact beams'; in order to come within this category the beam must not be susceptible to either local or lateral–torsional instability.

The first requirement will be met if the individual plate elements of the cross-section are not allowed to be too wide relative to their thickness. Sections for which the ratios of flange width/flange thickness, web depth/web thickness, etc. have been limited, so that their full moment capacity may be achieved, are said to be compact. The majority of hot-rolled sections (for example, UB) meet these requirements. However, care is necessary when using fabricated sections, for which suitable rules are provided in codes of practice.

Lateral–torsional instability will not occur if one or more of the following apply:

1. The section is bent about its minor axis;
2. Full lateral restraint is provided: for example, by positive attachment of the top flange of a simply supported beam to a concrete slab;
3. Closely spaced, discrete bracing is provided so that the weak axis slenderness (L/r_y) of the beam is low;
4. The section has high torsional and lateral bending stiffness – for example, rectangular box sections bent about their major axis are unlikely to fail in this way.

For the purpose of this chapter beams within any of these categories will be classed as 'restrained'.

20.2 Beam types

Several factors, some of them tending to conflict with one another, influence the designer's choice of beam type for a given application. Clearly, the beam must possess adequate strength but it should also not deflect too much. It must be capable of being connected to adjacent parts of the structure; this will often involve the use of site connections which should be simple and quick to effect, with the minimum requirements for skilled labour or special equipment. Particular features, such as the passage of services beneath the floor of a building, may dictate the use of a section with openings in the web, while architectural requirements may require the profile to be varied to improve the line. Table 20.1 summarizes the main types of steel beam, indicates the range of spans for which each is most appropriate and gives some idea of any special features.

20.3 Design of beams for simple bending

For a doubly symmetrical section or a monosymmetric section bent about the axis of symmetry the basic theory of bending, assuming elastic behaviour, gives the distribution of bending stress shown in Figure 20.1. Since the maximum stress f_m is given by:

$$f_m = M \frac{D/2}{I_x} \tag{20.1}$$

where M = moment at cross-section under consideration, D = depth of section, I_x = second

Table 20.1

Beam type	*Span range (m)*	*Notes*
(0) Angles	3–6	Used for roof purlins, sheeting rails, etc. where only light loads have to be carried
(1) Cold-formed sections	4–8	Used for roof purlins, sheeting rails, etc. where only light loads have to be carried
(2) Rolled sections UB, UC, RSJ, RSC	1–30	Most frequently used type of section, proportions selected to eliminate several possible types of failure
(3) Open-web joists	4–40	Prefabricated, using angles or tubes as chords and round bar for web diagonals, used in place of rolled sections
(4) Castellated beams	6–60	Used for long spans and/or light loads, depth of UB increased by 50%, web openings may be used for services, etc.
(5) Compound sections (e.g. UB + RSC)	5–15	Used when a single-rolled section used would not provide sufficient capacity, often arranged to provide enhanced horizontal bending strength as well
(6) Plate girders	10–100	Made by welding together three plates sometimes automatically, web depths up to 3–4 m sometimes need stiffening
(7) Trusses	10–100	Heavier version of (3), may be made from tubes, angles or if spanning large distances, rolled sections
(8) Box girders	15–200	Fabricated from plate, usually stiffened, used for OHT cranes and bridges due to good torsional and transverse stiffness properties

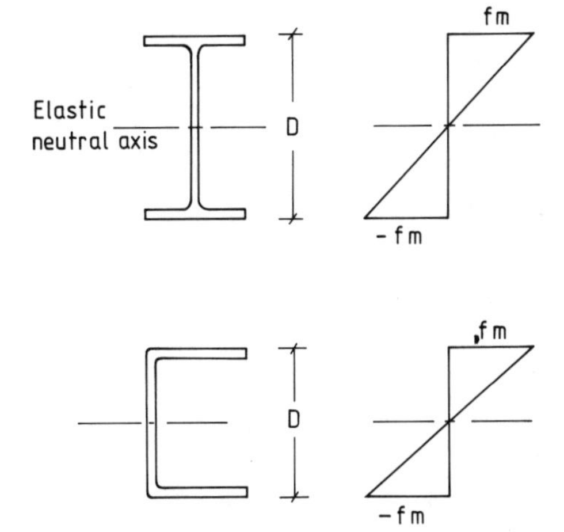

Figure 20.1 Distribution of bending stresses for bending about an axis of symmetry according to elastic theory

moment of area about the neutral axis (line of zero strain), it follows that limiting this to a fraction of the material yield stress gives a design condition of the form:

$$Z \nless M/p_b \qquad (20.2)$$

where $Z = I_x/D/2$ is the section modulus, p_b is the maximum permissible bending stress. This approach to design is termed 'elastic' or 'permissible stress' design. Values of Z for the standard range of sections are available in tables of section properties.

Selection of a suitable beam therefore involves:

1. Determination of the maximum moment in the beam;
2. Extraction of the appropriate value of p_b;
3. Selection of a section with an adequate value of Z subject to considerations of minimum weight, depth of section, rationalization of sizes throughout the structure, etc.

Clearly, sections for which the majority of the material is located as far away as possible from the neutral axis will tend to be the most efficient in bending. Figure 20.2 gives some quantitative idea of this for some of the more common structural shapes. I-sections are most often chosen for beams because of their structural efficiency; being open sections they can also be connected to adjacent parts of the structure without undue difficulty, and Figure 20.3 gives some typical beam-to-column connections.

Utilization of the plastic part of the stress–strain curve for steel enables moments in excess of those which just cause yield to be carried. At full plasticity the distribution of bending stress

Section type	Rect-angle	UC (typical)	UB (typical)	RHS (typical)
Section properties				
A area	1	1	1	1
Z_x elastic section modulus	1	3.5	6.2	2.3
S_x plastic section modulus	1.5	3.9	7.2	2.9

Figure 20.2 Relative section properties for bending of different cross-sectional shapes

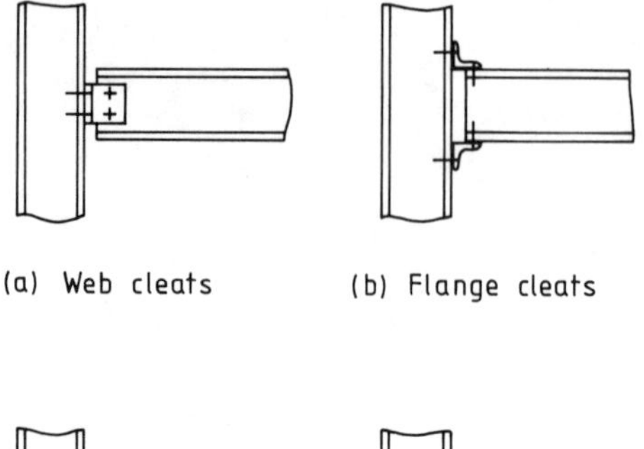

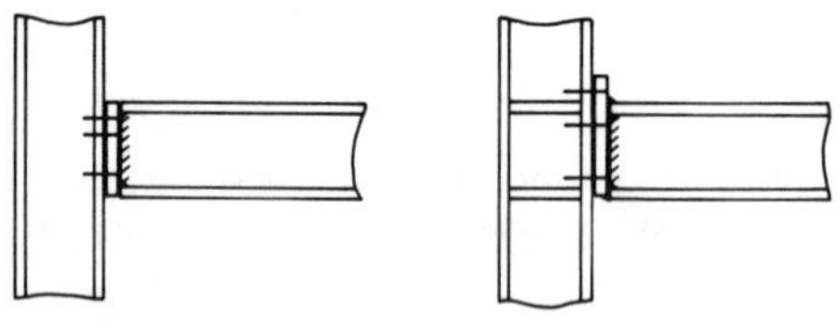

Figure 20.3 Examples of beam-to-column connections

will be as illustrated in Figure 20.4, with half the section being yielded in compression and half in tension. The corresponding moment is termed the fully plastic moment, M_p. This may be calculated by taking moments of the stress diagram about the neutral axis to give:

$$M_p = \sigma_y S \qquad (20.3)$$

where σ_y = material yield stress (assumed identical in tension and compression), S = plastic section modulus.

Basing design on Equation 20.3 means that the full strength of the cross-section in bending is now being used, with the design condition being given by:

$$S \nless M/p_y$$

where M = moment at cross-section under consideration, p_y = the design strength. In BS 5950 and BS 5400: Part 3 the value of p_y is taken as the material yield strength reduced slightly, so as to cover possible variations from the expected value.

For continuous (statically indeterminate) structures, attainment of M_p at the point of maximum moment will not normally imply collapse. Provided that nothing triggers unloading at this point (for example, local buckling does not occur), then the local rotational stiffness will virtually disappear (i.e. the cross-section will behave as if it were a hinge), and the pattern of moments within the structure will alter from the original elastic distribution as successive plastic hinges form. This redistribution of moments will enable the structure to withstand loads beyond that which produces the first plastic hinge, until eventually collapse will occur when sufficient hinges have formed to convert the structure into a mechanism, as shown in Figure 20.5. Utilization of this property of redistribution is termed 'plastic design'. It can only be used

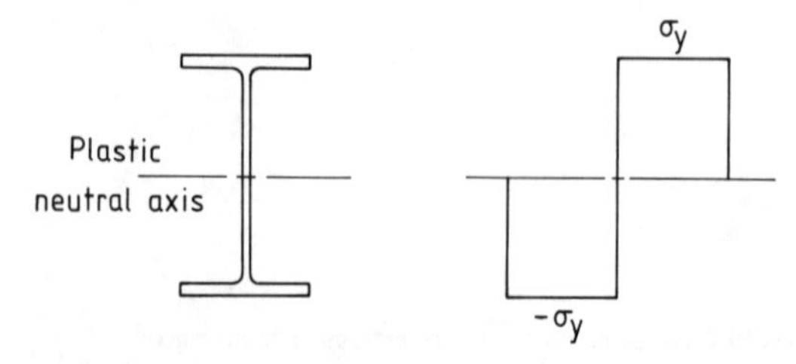

Figure 20.4 Plastic stress distribution for a doubly symmetrical section

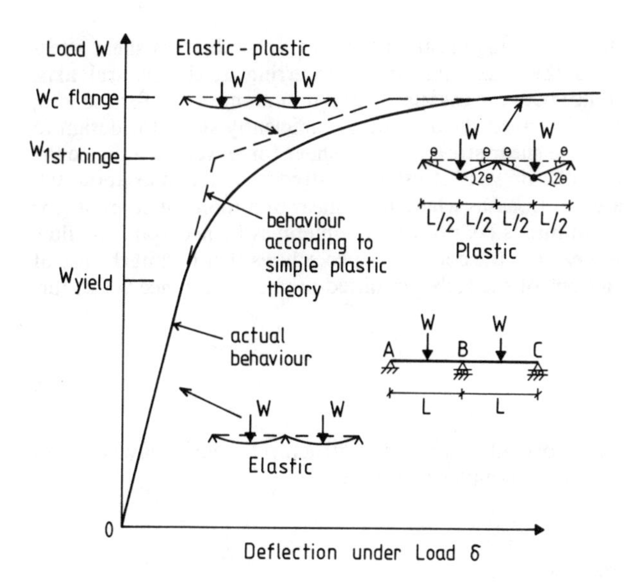

Figure 20.5 Load–deflection curve for a statically indeterminate steel beam

for continuous structures and then only when certain restrictions on cross-sectional geometry, member slenderness, etc. are observed.

20.4 Design of beams for shear

Although bending will govern the design of most steel beams, situations will arise (for example, short beams carrying heavy concentrated loads) in which shear forces are sufficiently high for them to be the controlling factor. Figure 20.6 illustrates the pattern of shear stress found in a

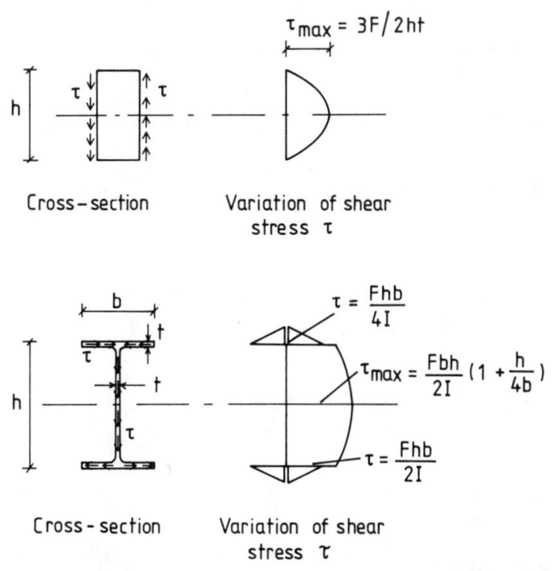

Figure 20.6 Distribution of shear stress in beams

rectangular section and in an I-section assuming elastic behaviour. In both cases shear stress varies parabolically with depth, with the maximum value occurring at the neutral axis. However, for the I-section the difference between maximum and minimum values for the web, which carries virtually the whole of the vertical shear force, is sufficiently small for design to be simplified by working with the average shear stress = total shear force/web area. Since the shear yield stress of steel is $1/\sqrt{3}$ of its tensile yield stress shear strength is taken as about 0.6 times the design strength of the material. In cases where high shear and moment co-exist (for example, the internal support of a continuous beam) it may sometimes be necessary to allow for interaction effects. However, since each of the codes uses procedures that permit the use of full shear capacity at quite large fractions of the fully permitted moment, and vice versa, this will not often be required.

20.5 Deflections

Beams that deflect too much, while not normally leading to a structural failure, may nonetheless impair the operation of the structure. Examples might include:

1. Cracking of plaster ceilings;
2. Allowing crane rails to become misaligned;
3. Causing difficulty in opening large doors.

Since these affect the performance of the structure in its working conditions, it is usual to conduct this type of serviceability check at working load levels. BS 5950 and BS 5400: Part 3 both draw the engineer's attention to the need to check deflections at serviceability and give advice on suitable limits for use when more detailed information – for example, from crane manufacturers – is not available.

20.6 Bending of unsymmetrical sections

For sections with a single axis of symmetry bent about the other rectangular axis the elastic distribution of bending stresses will be as shown in Figure 20.7. Due to its lack of symmetry

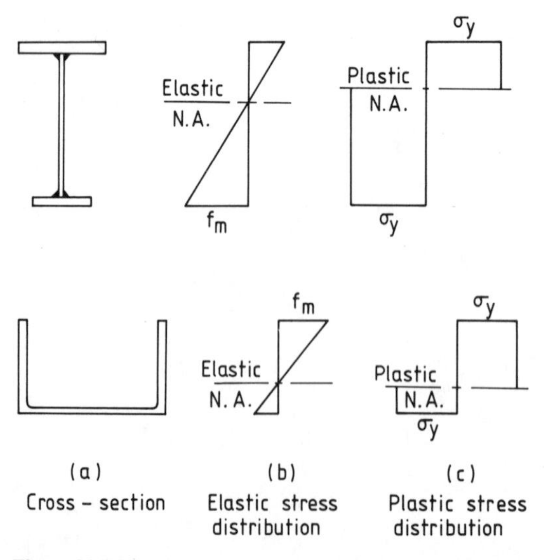

Figure 20.7 Bending in singly symmetrical sections in the plane of symmetry

about the neutral axis the stress on one face (for example, the outer face of the smaller flange in the case of an unequal flanged I-beam) will exceed that on the other. Therefore in elastic design it is necessary to use the smaller of the two elastic section moduli. If values are available for the type of section under consideration (for example, a standard fee is being used) design can proceed as before.

At full plasticity the line of zero stress must be positioned so that it divides the area of the section into two equal parts, as shown in Figure 20.7(c). Since the plastic section modulus S is defined from Equation 20.3 as the ratio of M_p/σ_y, there will be a single value for both faces of the section. Thus when using BS 5950 or BS 5400: Part 3 the design condition will remain as Equation 20.4, i.e.:

$$S \nless M/p_y \tag{20.4}$$

20.7 Biaxial bending

Doubly or singly symmetrical sections subject to bending about both principal planes may be treated as the sum of two uniaxial problems. Thus elastic design requires the satisfaction of the linear interaction equation:

$$f_x + f_y \ngtr p_b \tag{20.5}$$

in which f_x and f_y are the maximum bending stresses at the cross-section in question,

p_b = permissible stress

This is based on the idea of limiting the maximum combined stress in the section to the design value. Care is necessary when dealing with sections such as angles for which the principal axes are not the rectangular ones.

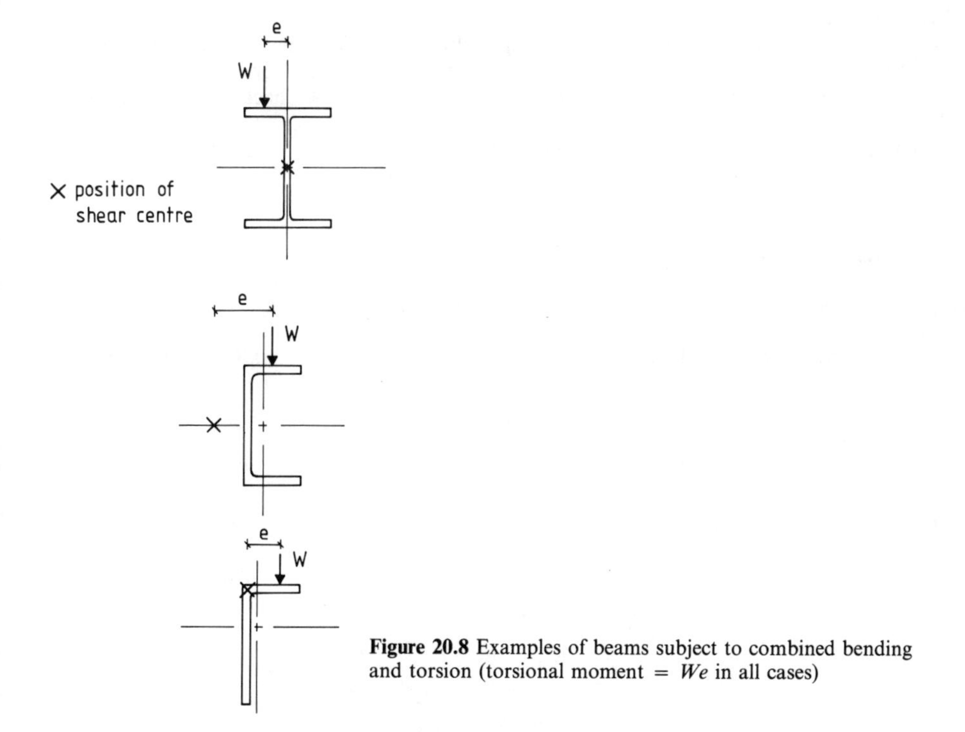

Figure 20.8 Examples of beams subject to combined bending and torsion (torsional moment = *We* in all cases)

When using the plastic strength of the cross-section, analysis (which entails the location of the plastic neutral axis) shows that the shape of the interaction between moments will depend upon the geometry of the cross-section. Safe approximations to these interaction diagrams have therefore been provided in BS 5950 and BS 5400: Part 3. As an example, the former uses:

$$\left(\frac{M_x}{M_{cx}}\right)^{z_1} + \left(\frac{M_y}{M_{cy}}\right)^{z_2} \not> 1 \tag{20.6}$$

in which M_{cx} and M_{cy} are moment capacities about the x and y axes, respectively, and the values of z_1 and z_2 depend on the particular section under consideration; safe values are $z_1 = z_2 = 1.0$.

20.8 Bending and torsion

Loads which do not act through the shear centre of the section (see Figure 20.8) will also cause twisting. (The shear centre coincides with the centroid for doubly symmetrical sections and lies on the axis of symmetry for singly symmetrical sections.) This will induce shear stresses due to torsion and, in the case of open sections, may also produce significant, additional, direct, longitudinal stresses due to the structural action known as warping. Proper consideration of this complex topic requires an appreciation of the theory of torsion. In many cases torsional effects may be minimized by careful detailing which arranges the transfer of load into members in such a way that twisting would not occur.

20.9 Concluding summary

1. The main design requirement for beams is the provision of adequate bending strength, and the types of member normally employed to fulfil this have been indicated. Means of recognizing beams whose design can be based on relatively simple structural principles have been presented.
2. Section modulus, either elastic or plastic, depending upon the design philosophy adopted, is the most appropriate property for use in selecting a suitable section.
3. Adequate shear strength and deflection limitation must also be checked in simple beam design.
4. Situations in which a more complex structural response will occur, resulting in the need to consider biaxial bending, torsion, etc., have been indicated.

Background reading

STEEL CONSTRUCTION INSTITUTE (1988) *Design of Members Subject to Combined Bending and Torsion*, SCI, Ascot

STEEL CONSTRUCTION INSTITUTE (1989) *The Steel Designer's Manual* (new edition), Blackwell, Oxford

21

Local buckling

Objective To describe the concepts and demonstrate the design implications of local plate buckling.

Prior reading Chapter 18 (Introduction to buckling: 1); Chapter 19 (Introduction to buckling: 2).

Summary The chapter discusses the distinctions between overall and local buckling, slender and compact sections, post-buckling reserve of strength and effective-width concepts.

21.1 Introduction

When a slender plate is subjected to in-plane compressive or shear loads, or to a combination of compression and shear as in Figure 21.1(a), it is prone to buckle.

Many structural sections consist of assemblages of plates of slender proportions. Cold-formed, thin-walled sections of various shapes and profiled steel sheeting are typical examples of such structures; fabricated sections, such as plate and box girders, may provide more.

Any one of the slender component plates of such a cross-section is prone to local buckling. For example, in the case of the box column in Figure 21.2(b) each plate may buckle locally before the load required to cause the overall cross-section to buckle as a strut is attained; premature failure of the member will then occur.

Such local buckling can develop whenever an individual plate is subjected to one (or a combination) of the load components shown in Figure 21.1(a). For example, Figure 21.1(c) shows a box girder cross-section subjected to pure bending which places the upper flange plate in a state of uniaxial compression; this flange may then buckle. Figure 21.1(d) illustrates a plate girder subjected to a typical combination of bending moment and shearing force. This again places the upper flange in compression, whereas the web sustains a combination of direct and shear stresses that may also cause buckling. Figure 21.1(e) shows another possible load combination where a multi-cellular structure, such as the double-bottom of a ship's hull, is supported on all four edges to develop a two-way bending action. A component plate of the upper flange is then liable to buckle under the biaxial compression imposed upon it.

21.2 Slender and compact sections

The buckling capacity of a plate is inversely proportional to the square of its width/thickness ratio (b/t or d/t). One way of avoiding local buckling is to ensure that the width/thickness ratio of each component plate does not exceed a certain value.

In BS 5950: Part 1 a comprehensive system is introduced for classifying the cross-sections of members that are subjected to compression due to moment and/or axial loading. The classification is primarily dependent upon the geometry of the cross-section, with four classes being identified:

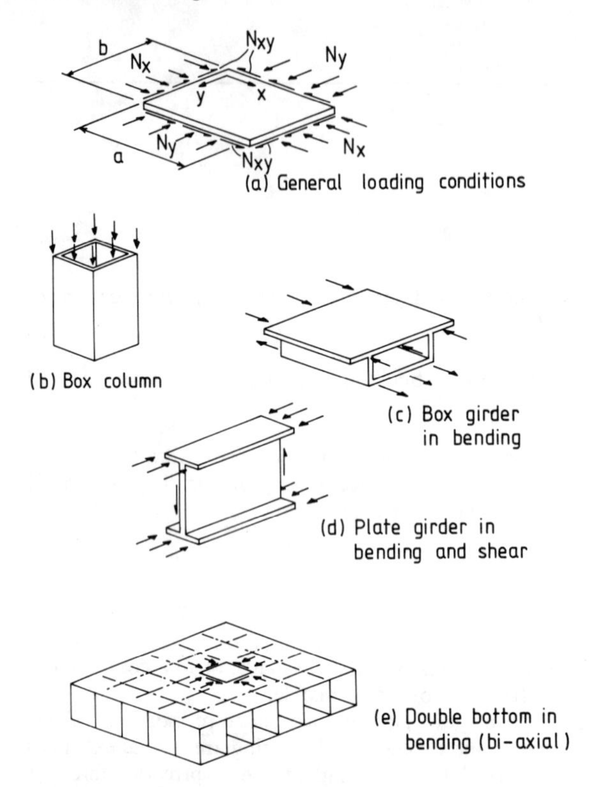

Figure 21.1 Examples of plates susceptible to buckling

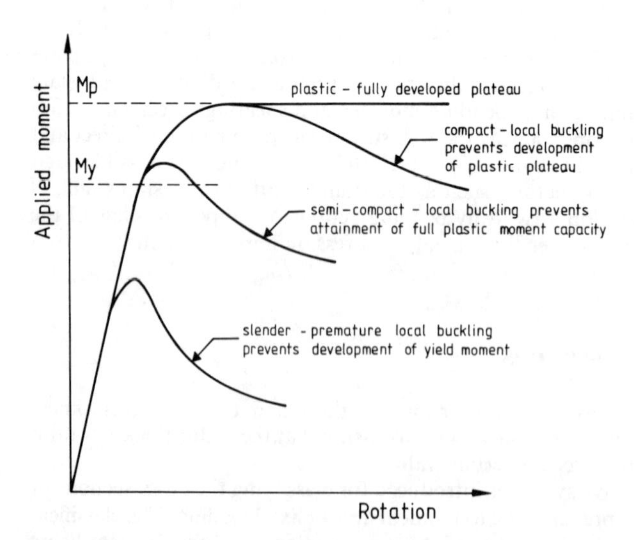

Figure 21.2 Moment/rotation curves for different types of cross-section

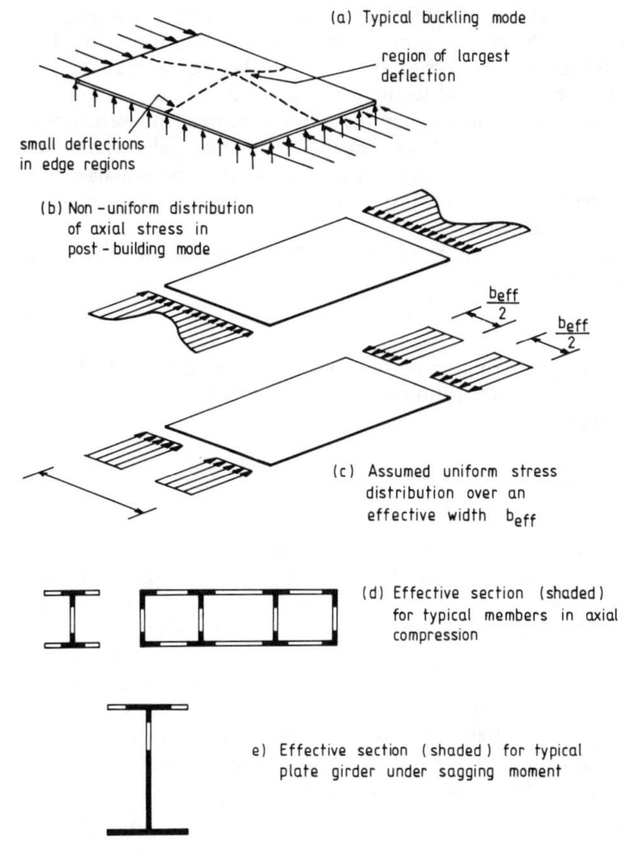

Figure 21.3 Effective width approximation in post-buckling analysis

1. *Plastic cross-sections.* Those which can develop a plastic hinge with sufficient rotation capacity to allow redistribution of bending moments in the structure.
2. *Compact cross-sections.* Those which can develop the plastic moment capacity of the section but where local buckling prevents rotation at constant moment in the structure.
3. *Semi-compact cross-sections.* Those in which the stress in the extreme fibres should be limited to yield because local buckling would prevent development of the plastic moment capacity of the section.
4. *Slender cross-sections.* Those in which yield in the extreme fibres cannot be attained because of premature local buckling.

In Figure 21.2 typical moment/rotation curves are plotted for various classes of cross-section to illustrate the differences in the assumed behaviour. BS 5950 specifies the limiting proportions of the elements of a cross-section which enable the appropriate classifications to be made. (Other codes of practice also seek to control local buckling. BS 5400: Part 3 uses the classifications compact and non-compact for cross-sections.)

21.3 Simplified approach to post-buckling behaviour and 'effective width'

When a plate buckles under compressive loading the central regions of the plate will undergo large out-of-plane deformations (see Figure 21.3(a)), whereas the edge regions, adjacent to the supports, remain almost straight. The central region thus hardly participates in carrying any further loading beyond buckling, while the edge regions continue to resist an increasing load with hardly any growth in lateral deflection. The compressive stress is then redistributed across the flange, the stresses being shed from the central to the edge regions, as shown.

Rigorous analysis of the post-buckling action is complex, but a simple and satisfactory solution can be obtained by using an 'effective-width' approach. In this approach it is assumed that in the post-buckling range the true, variable, stress distribution may be replaced by a simplified distribution. The central region of the plate is ignored and the remaining or 'effective' width (b_{eff}) is assumed to be uniformly stressed.

The ultimate load of the plate (P_{ult}) may then be equated with the attainment of the yield stress on this reduced effective section:

$$P_{ult} = \sigma_y b_{eff} t \tag{21.1}$$

The effective width (b_{eff}) may be calculated from an empirical formula established by Winter (1947) from experimental observations. The derivation of this formula is discussed by Winter and its use is described in Trahair (1977):

$$\frac{b_{eff}}{b} = \sqrt{\left(\frac{\sigma_{cr}}{\sigma_y}\right)}\left[1 - 0.22\sqrt{\left(\frac{\sigma_{cr}}{\sigma_y}\right)}\right] \not> 1.0 \tag{21.2}$$

In this formula σ_{cr} represents the critical buckling stress of the plate, i.e. $\sigma_{cr} = N_{x,cr}/t$.

When a complete section is analysed as an assembly of individual plates, as suggested earlier, the effective-width formula can be applied to each plate. Figure 21.1(4) shows the possible loss of effective section that can result for different members under axial compression. In the case of a plate girder subjected to bending only the upper part of the web is subjected to compression in a sagging-moment region. Thus if the idea of an effective width (or depth) is applied material is lost from the upper regions of the web and the resulting girder cross-section becomes unsymmetrical, as in Figure 21.3(e). The consequent shift of the neutral axis away from the mid-depth position is rather inconvenient from the point of view of design calculations, so that an empirical expression to calculate an effective web thickness is sometimes adopted instead.

21.4 Concluding summary

1. The overall load-carrying capacity of a member composed of slender plates can be reduced because of local buckling of the component plates.
2. This local buckling action can be analysed approximately by considering each plate in isolation.
3. Plates can carry loads in excess of the buckling load.
4. The post-buckling reserve of strength can be analysed approximately by assuming an empirical effective width.

Background reading

ALLEN, H.G. and BULSON, P.S. (1980) *Background to Buckling*, McGraw-Hill, Maidenhead
TRAHAIR, N.S. (1977) *The Behaviour and Design of Steel Structures*, Chapter 4, Chapman and Hall, London
WINTER, G. (1947) Strength of thin steel compression flanges. *Trans. Am. Soc. Civ. Engrs*, **112**, 527

Columns

Objective To explain the design method for axially loaded columns.

Prior reading Chapter 18 (Introduction to buckling: 1); Euler buckling theory; Basic plasticity.

Summary The elastic analysis of columns with initial curvature is presented and extended to cover inelastic behaviour. The performance of slender, intermediate and stocky columns is considered. The basis of the multiple column design curves of BS 5400 and BS 5950 is explained. The concept of effective length is introduced and its application to practical columns described, with guidance on estimating values in design. Finally, the differences between no-sway and sway columns are emphasized. For worked examples see Appendix. p. 373.

22.1 Introduction

Column behaviour is extensively discussed in Chapter 18, from which the conclusions to be drawn concerning practical column behaviour are as follows:

1. All practical columns fail by bending out of line in an Euler type bending mode. Local buckling can occur with thin flanges, but this does not affect load capacity except for extremely thin flanges, such as occur in thin-walled sections.
2. Practical columns generally fail by inelastic buckling and do not conform to the idealized assumptions of the Euler theory; most importantly, they do not normally remain linearly elastic up to failure unless they are extremely slender.
3. Slenderness (L/r_y) and material strength (σ_y) are the dominant factors affecting the ultimate strength of columns and they give an upper bound to column strength, as shown in Figure 22.1.
4. Ultimate load tests on pin-ended, axially loaded columns give the scatter band of results shown in Figure 22.1.
5. The ultimate strength of practical columns is often significantly affected by the column imperfections; these are of many forms including some, such as variation of yield stress over the cross-section, not mentioned in Chapter 18. The most important are:

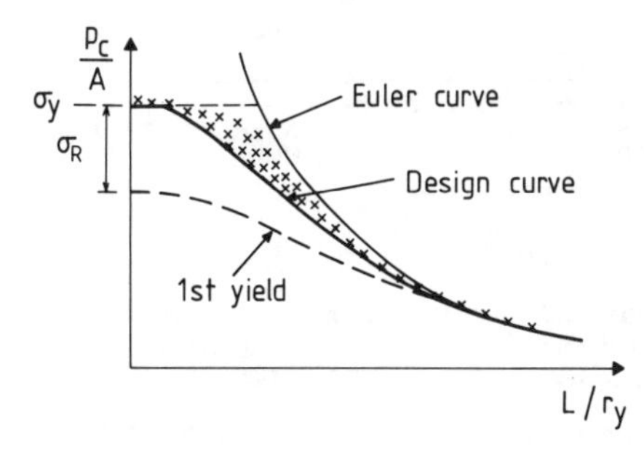

Figure 22.1 Typical column design curve

(a) Geometrical imperfections; and
(b) Residual stresses.

22.2 Analysis

22.2.1 General

Considering an axially loaded column with a sinusoidal initial imperfection [$v_o = \delta_o \sin(\pi z/L)$], as in Figure 22.2, the theoretical solution illustrates the effect of typical geometric imperfections on column behaviour in the elastic range. Under axial load (P) there is an increase in deflection (v) and equilibrium gives a modified Euler type equation:

$$EI\frac{d^2v}{dz^2} = -P(v + v_0) \tag{22.1}$$

with a solution for the total central deflection ($\delta_c = \delta + \delta_0$):

$$\delta_c = \delta_0\left(\frac{1}{1 - P/P_E}\right) \tag{22.2}$$

where P_E is the Euler critical load.

The curve of P against δ_c is as shown in Figure 22.2(b) and is asymptotic to the Euler load at large deflections, provided that the column remains elastic.

22.2.2 Slender columns

Practical columns of extreme slenderness behave in this way and their ultimate failure loads will approach P_E. Figure 22.3 shows the curve given by the Euler critical stress [$\delta_{cr} = (P_E/A)$] plotted against L/r_y from:

$$\sigma_{cr} = \frac{\pi^2 EI}{AL^2} = \frac{\pi^2 E}{(L/r_y)^2}$$

For slender columns the Euler stress is low, so the columns remain elastic until quite large bending deflections have developed; thus P_E is approached, as in Figure 22.2(b).

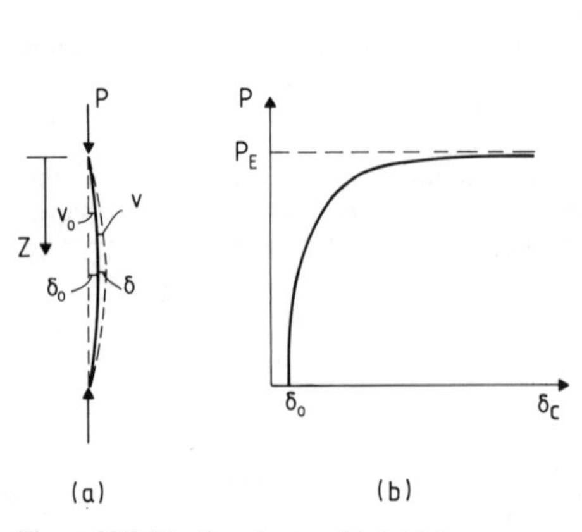

Figure 22.2 Elastic column with initial curvature

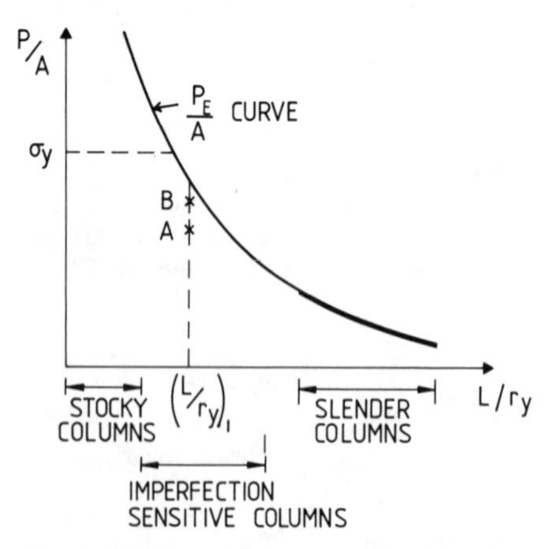

Figure 22.3 Euler stress/slenderness (showing classification of imperfection sensitivity by slenderness)

22.2.3 Columns of medium slenderness

The majority of practical columns are not extremely slender, and therefore fail as the result of becoming inelastic at loads much lower than P_E. In the elastic range the stresses at the mid-height of such a column are as shown in Figure 22.4. Considering the column shown in Figure 22.2, there will be a bending moment:

$$M_c = P\delta_c = P\delta_0\left[\frac{1}{1 - P/P_E}\right] \tag{22.3}$$

giving a maximum combined stress (δ_{max}) of:

$$\sigma_{max} = \frac{P}{A} + \sigma_R + \frac{M_c}{Z} \tag{22.4}$$

and first yield will occur when:

$$\frac{P}{A} + \frac{P}{Z}\delta_0\left(\frac{1}{1 - P/P_E}\right) = (\sigma_y - \sigma_R) \tag{22.5}$$

The shape of the section will influence the value of Z, so members with a high ratio of y/r will yield earlier than shapes with a lower y/r.

At first yield there will be no observable change in behaviour, but under further load plasticity will spread into the column from one side, as in Figure 22.5. This will decrease the column-bending stiffness from:

$$EI_y \approx 2(\tfrac{1}{12}TB^3)$$

to approximately:

$$(EI_y\alpha_D) \approx 2[\tfrac{1}{12}T(B - c)^3]$$

where $(EI_y\alpha_D)$ is the reduced bending stiffness of the column resulting from inelastic behaviour; α_D is always < 1. This deterioration of bending stiffness as the result of plasticity is similar to that occurring in a beam, where the cross-sectional stiffness falls to zero as a plastic hinge forms.

With the section inelastic, Equation 22.1 becomes modified to:

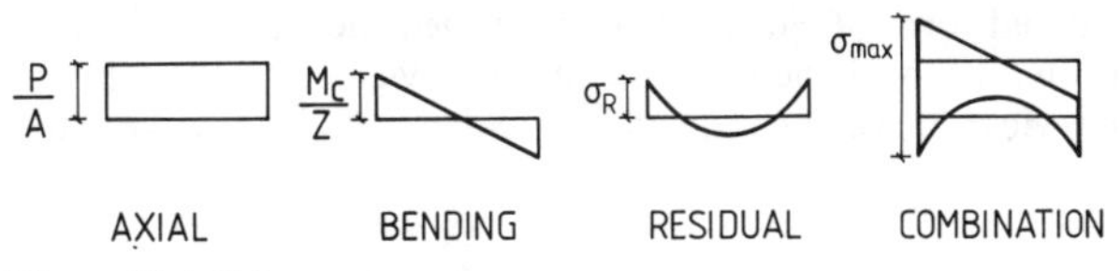

Figure 22.4 Column stresses

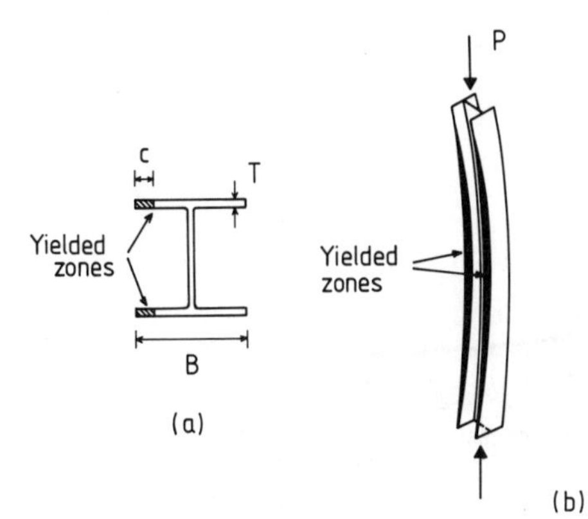

Figure 22.5 A partially yielded column

$$(EI_y \alpha_D) \frac{d^2 v}{dz^2} = -P(v + v_0) \tag{22.6}$$

where α_D varies along the column as the depth of plasticity changes (see Figure 22.5(b)). A numerical solution of this equation gives a failure load for inelastic buckling that is significantly below P_E. At failure, the plasticity is such that the buckling load has reduced (deteriorated) below P_E to equal the applied load. Typical experimental results, when non-dimensionalized with respect to the yield stress, give the scatter shown in Figure 22.1. It can be seen that columns of medium slenderness are very imperfection-sensitive.

22.2.4 Stocky columns

For stocky columns the yield stress becomes the dominant factor limiting column strength. Obviously, at the squash load ($P_y = A\sigma_y$) the column would fail, with the whole cross-section becoming plastic or inelastic. Figure 22.3 shows that P_E is high for stocky columns, and extensive plasticity has to occur before the column stiffness has deteriorated sufficiently to cause failure. This requires a load approaching P_y, and the columns are not particularly sensitive to variations in the imperfections. At extremely low slenderness experimental results tend to be above, rather than below, P_y; this is because of strain-hardening and explains the plateau shown on the ECCS curves.

22.3 Column design curves

Statistical analysis of the results of an extensive international test programme, coordinated by the ECCS, has shown that the behaviour of standard column sections can be classified by a family of column curves, as in Figure 22.6. For weak axis buckling, RHS, UB and UC sections fall on curves A, B and C, respectively. However, columns do not always fail by buckling about the YY axis, as in a design situation they may be supported to prevent this, so that they may buckle by bending about the major axis. UB and UC sections are less imperfection-sensitive about the major axis; for example, UC strengths are given by curve B for major axis buckling and curve C for minor axis buckling.

For computational convenience, formulae are required for the column design curves; British codes have traditionally used a modified form of Equation 22.5, termed the Perry-Robertson formula. Solving for P from Equation 22.5 gives the load to cause first yield, which, with high residual stresses, would give a first yield curve, as indicated in Figure 22.7. However, this

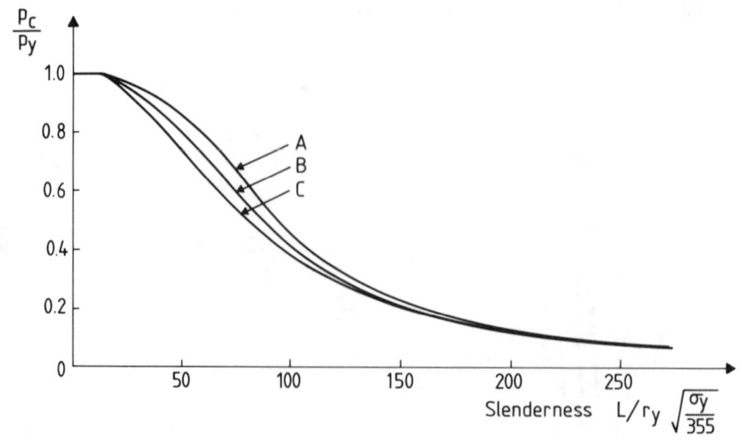

Figure 22.6 ECCS column design curves

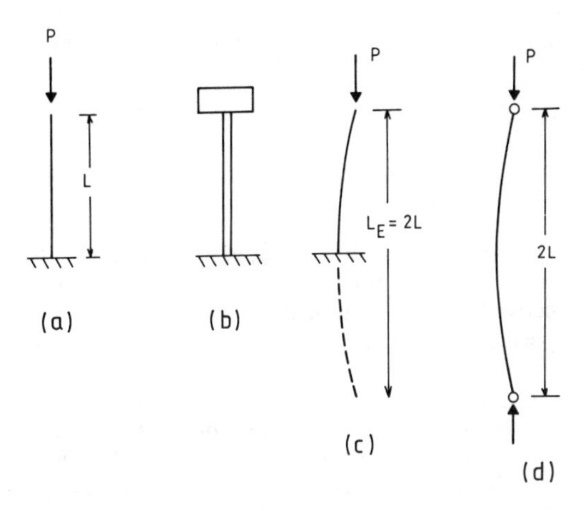

Figure 22.7 Cantilever columns

formula can be used to give a reasonable fit to any column curve based on test results by putting the residual stress term $\sigma_R = 0$ and introducing an exaggerated value for the initial curvature δ_0. The expression for the compressive design stress (p_c), given in Appendix C of BS 5950, is:

$$p_c = \frac{p_E p_y}{\phi + \sqrt{(\phi^2 - P_E P_y)}} \quad \text{where} \quad \phi = \frac{p_y + (\eta + 1)p_E}{2} \qquad (22.7)$$

with $p_E = P_E/A$ and $p_y = \sigma_y/\gamma_m$, the material strength reduced by the appropriate safety margin.

The imperfection factor is taken as:

$$\eta = 0.001a(\lambda - \lambda_0)$$

which is obtained approximately from $\eta = (A/Z)\delta_0$ by assuming that δ_0 is dependent on column length, $L/1000$. A slightly reduced slenderness $(\lambda - \lambda_0)$ is used to give the plateau required on the curves at low slenderness. The imperfection constant (a) is varied to give the desired shape of curve: for example, $a = 2.0$ for RHS and 5.5 for a UC buckling about its minor axis.

22.4 Types of sections used as columns

Universal Column (UC) sections were designed to have suitable properties for compression members. Their broad flanges and square shape give a relatively high r_y that maximizes the column design stress (p_c) for a given length; their relatively thick flanges avoid problems of local flange buckling. The open shape can be produced by traditional rolling techniques and it facilitates beam-to-column and other connections.

However, UC sections do not represent the optimum theoretical shape, which is in fact a circular hollow section (CHS). Such members have been used in many contexts, particularly as compression members in roof trusses, and recently and most spectacularly in the legs and other members of offshore oil platforms. There is no weak bending axis and r $\approx 0.35D$ (diameter), compared with $r \approx 0.29$B for a UC section. In many situations, however, tubular members are unsuitable, often because of high connection costs; rectangular hollow sections (RHS) have good properties and are widely used, but the use of larger-size RHS is also restricted by the difficulty of providing suitable connections.

For relatively light loads angle sections are often most convenient and they can be connected simply through one leg. This results in non-axial loading so that the permitted design stresses (p_c) have to be significantly reduced.

Columns are frequently subjected to bending actions in addition to axial load, and in these conditions universal beam (UB) sections are widely used.

22.5 Effective lengths

22.5.1 Introduction

Pin-ended, axially loaded columns are relatively rare in practice. However, experimental results obtained for this particular case are used as a basis for predicting strength under other end-conditions, by using the concept of an effective length (L_E). L_E is the length of a pin-ended column of the same section which would be expected to have the same strength as the case being designed.

The simple example of a free-standing column that is shown in Figure 22.7(a) with height L, fully restrained at the base and unrestrained at the top, illustrates the approach. A practical example of this would be a tubular steel column of uniform section supporting a cylindrical water tank (Figure 22.7(b)). A column as in Figure 22.7(a) would fail by weak axis bending, as shown in Figure 22.7(c); it is obvious, from the mirror image of the deflected shape shown dotted, that its buckled shape is the same as half a pin-ended column of length $2L$ (Figure 22.7(d)). Therefore the critical failure loads for cases Figure 22.7(c) and 22.7(d) would be the same, assuming identical column imperfections. This must be the case, as the Euler expressions for bending moment in the buckled shape at a distance x from the top are the same in each case, with the boundary conditions $M = 0$ at $x = 0$ and $(\mathrm{d}y/\mathrm{d}x) = 0$ at $x = L$ valid for both cases. This result would apply whether the failure was elastic or inelastic.

The design procedure for cantilever columns would be the same, but in establishing the design strength (p_c) from the column design curve the slenderness (L_E/r_y), i.e. $2L/r_y$ in this case, would be used instead of L/r_y. Other obvious cases where the effective length can easily be established are shown in Figure 22.8.

22.5.2 Effective lengths in different planes

The restraint against buckling may be different for buckling about the two column axes. For example, Figure 22.8(a) could be a pin-ended UC section braced about the minor axis against lateral movement (but not rotationally restrained) at spacing $L/3$. The minor axis buckling mode would be as indicated, with an effective pin-ended column length $(L_E)_{yy}$ of $L/3$. If there was no major axis bracing the effective length for buckling about the major axis $(L_E)_{xx}$ would remain as L. Therefore the design slenderness about the major and minor axes would be L/r_x and $(L/3)/r_y$, respectively. As $r_x < 3r_y$ for all UC sections, the major axis slenderness (L/r_x) would be greater, giving the lower value of p_c, and failure would occur by major axis buckling.

22.5.3 Points of contraflexure and effective lengths

Figure 22.8(b) shows a column with both ends fully restrained; the buckled shape has points of contraflexure, equivalent to pin ends, at $L/4$ from either end. The central length is clearly equivalent to a pin-ended column of length $L/2$ and the ends, similar to Figure 22.7(c), have the same effective length of $L/2$. This must be the case for the column to buckle in this mode, if it has full rotational restraint; Figure 22.9(a) shows the effect of partial end-restraints.

Sometimes columns are free to sway laterally, as in Figure 22.7(c), but are restrained against rotation at both ends, as in Figure 22.8(c). A water tank supported on corner columns, as in Figure 22.8(d), with rigid joints at the top could be a practical example. In this case the point of contraflexure is at mid-height of the column and the effective length ($L_E = 2 \times L/2$) remains L.

22.5.4 Effective lengths recommended for design

Approximate values for effective lengths which can be used in design are given in the design

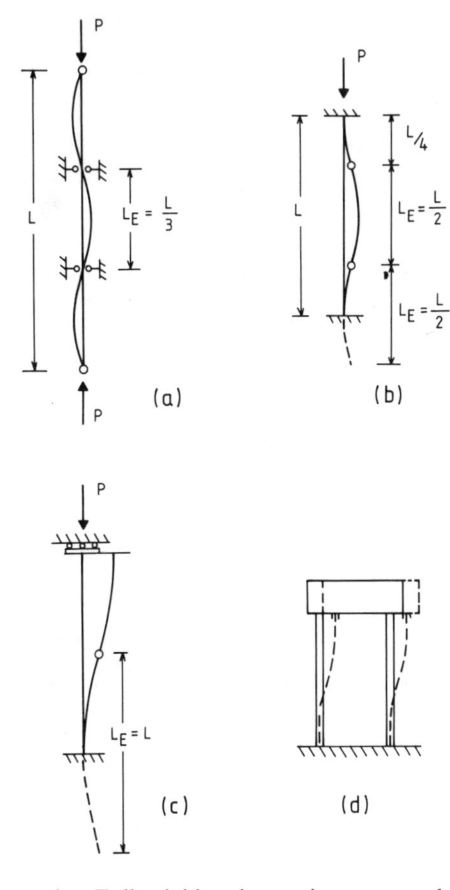

Figure 22.8 Columns with differing effective lengths

codes. Fully rigid end-restraints can rarely be achieved in practice; partial end-restraints are much more common in practice. This partial fixity can result from flexibility in the end-connections and/or flexibility of the restraining members, as illustrated by the simple frame of Figure 22.10(a). Under the rather unusual, but possible, case of nodal loading the in-plane buckling mode for this frame would be as shown in Figure 22.10(b). With the top beam bent in an S-shape the rotational end-restraint stiffness is given by:

$$K_\theta = \frac{M}{\theta} = \frac{6EI_B}{L_B}$$

For rigid beam-to-column joints this stiffness of the beam (K_θ) will control the position of the point of contraflexure in the column and thus the column effective length. These columns are represented by the model of Figure 22.10(c) for which an effective length of $1.5L$ is suggested.

22.5.5 No-sway and sway columns

Figures 22.9(a) and 22.9(b) represent the general cases of no-sway and sway columns with partial end-restraint. The buckled shapes will be of the form shown if the top restraint stiffness ($K_{\theta T}$) and the bottom restraint stiffness ($K_{\theta B}$) are equal. For the no-sway case of Figure 22.9(a) the position of the points of contraflexure will move within the column length as $K_{\theta T}$ and $K_{\theta B}$ vary. For example, Figure 22.9(c) illustrates the situation for low $K_{\theta T}$ and high $K_{\theta B}$. However, for **no-sway columns L_E is always less than or equal to L**. By contrast, for **sway columns L_E is always greater than or equal to L**. As K_θ decreases, the column end-joint rotations increase and

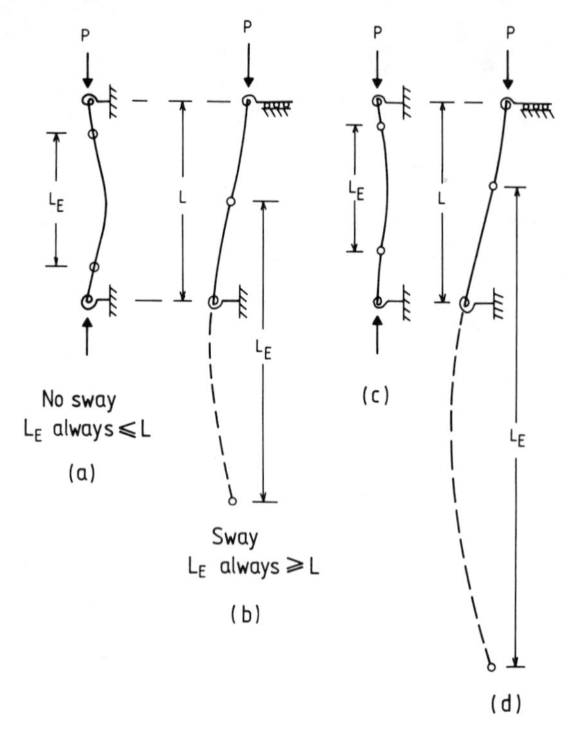

Figure 22.9 Columns with partial rotational restraint

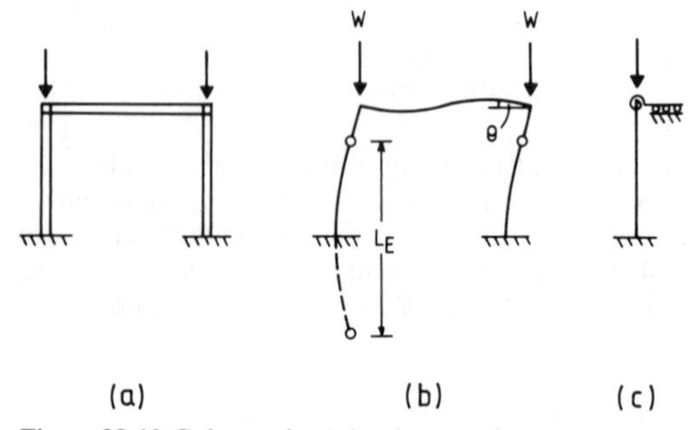

Figure 22.10 Columns in a simple sway frame

L_E can easily become 2 or $3L$ (Figure 22.9(d)). The limiting case of $K_{\theta T}$ and $K_{\theta B} = 0$ gives $L_E = \infty$.

The column design stress may be written as:

$$P_c = \frac{\pi^2 E}{(L_E/r_y)^2} \alpha_D$$

Therefore column strength is largely dependent on $(1/L_E)^2$ except for stocky columns, where α_D becomes dominant. Thus sway columns, i.e. with $L_E > L$, are much weaker than no-sway ones, and establishing the sway conditions for a column is often the single most important factor dictating column strength. For example, if the frame of Figure 22.10(a) had been braced against sway, as in Figure 22.11 (as can be done conveniently in many structures), L_E for the columns would be less than $0.85L$, compared with $1.5L$ for the sway case. Comparing $(0.85)^2$ with $(1.5)^2$

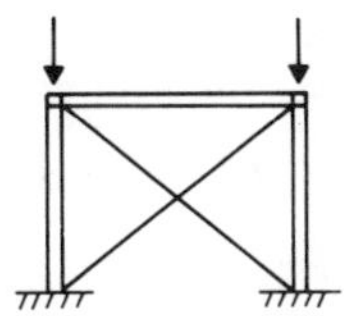

Figure 22.11 No-sway frame

gives the strength of the columns in the sway case as only one third of that of the no-sway columns.

This significant difference in strength between the sway and no-sway conditions has led to spectacular failures during construction, when a structure in which the columns have been designed as no-sway has been temporarily left unbraced. Sway instability failures occur suddenly and therefore frequently result in injury and loss of life. The collapse of scaffolding inadequately tied to the facade of a building is a common example of sway instability.

22.5.6 Accuracy using effective lengths

Since column strength is approximately dependent on $(L_E/r_y)^2$ the accuracy of predictions based on tabular data is clearly not high. On the other hand, the complexity and variability of practical column behaviour are such that it is unreasonable to expect highly accurate strength predictions. However, where the column conditions are reliably defined, more accurate estimates of effective lengths can be made assuming elastic column behaviour to failure. Elastic analysis, similar to Euler's analysis for pin-ended columns, can be extended to columns with elastic end-restraints, and to complete structures, and the results interpreted as effective lengths. These data are available in many forms; as an example, Figure 22.12 is reproduced from BS 5950. It presents effective lengths for different end-restraint factors (k_1 and k_2) at the two ends of no-sway and sway columns. k_1 and k_2 are dependent on the relative magnitudes of the beam stiffness (k_b) and column stiffness (k_c):

$$k = \frac{K_c}{K_c + K_b}$$

It is noted that k decreases with increased beam restraint. However, this formula does not cover all possible conditions, as many columns are really partially restrained against sway, as shown in Figure 22.13.

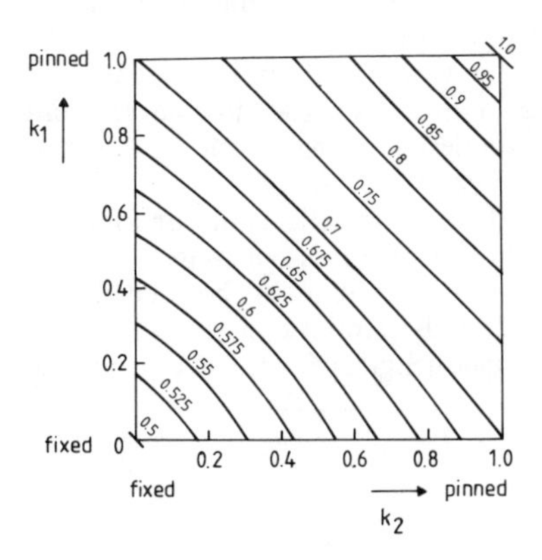

(a) Column in a rigid-jointed frame braced against sidesway

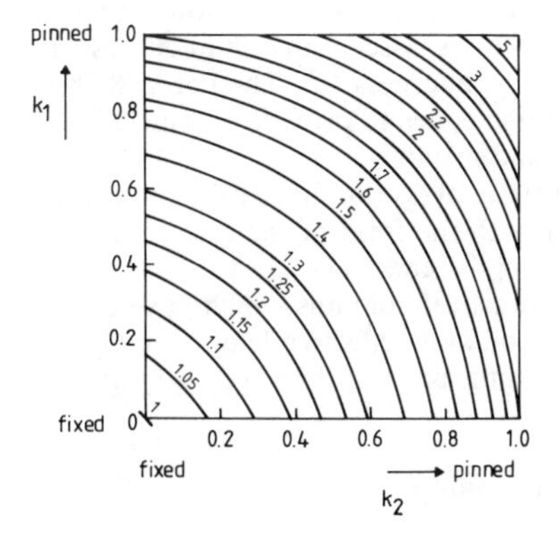

(b) Column in a rigid-jointed frame with unrestricted sidesway

Figure 22.12 Effective length ratios L_E/L

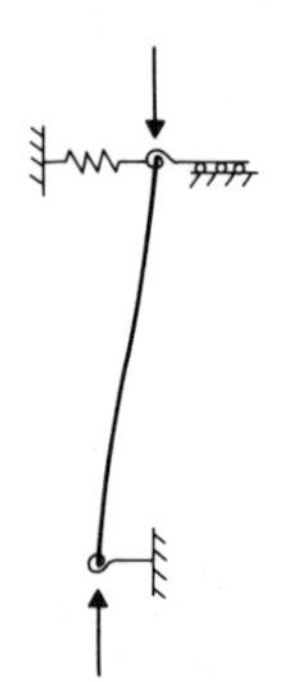

Figure 22.13 Elastically restrained column

22.5.7 Design practice

The general approach that is adopted is to use effective lengths to design columns which can clearly be considered as no-sway. This is usually the case for compression members in triangulated trusses and space frames, and in building frames if a stiff core is provided by lift shafts and staircases, or the frame is braced. L_E will always be less than L, and if the elastic data of Figure 22.12(a) are used, with realistic restraint stiffness including allowance for connection flexibility, the effective length should be conservative. This is because the column becomes inelastic and its stiffness deteriorates prior to failure, i.e. α_D reduces from unity, and the restraint stiffnesses become more effective, i.e. $k = K_c/(K_c + K_b)$ decreases.

Sway columns present a more serious problem, and while effective lengths are used in design it is becoming accepted that this approach is extremely inaccurate. Design considering the whole structure, based on approximate methods for elastic critical load analysis, is being recognized as more reliable. These approximate methods consider the effect of horizontal sway forces on the structure which subject the columns to bending actions as well as axial load; the behaviour of such beam columns is discussed in Chapter 24.

22.6 Concluding summary

1. Most practical columns fail by inelastic buckling at loads significantly below the elastic buckling load (P_E). Elastic and inelastic theory have to be considered to explain their behaviour. Both geometrical imperfections and residual stresses reduce column strength by initiating yielding at loads much lower than the squash load ($A\sigma_y$).
2. Column design is based on column curves which give the variation of mean axial stress at failure with slenderness L/r. The Perry–Robertson elastic formula is used in a curve-fitting role in British codes to provide formulae for the column design curves which are based on ultimate load results.
3. Different types of section have different column curves because the effects of imperfections vary with both the form of the section and the axis about which it is buckling.
4. Effective lengths enable column design curves for pin-ended columns to be used for the design of practical columns which have a wide range of end-restraint conditions.
5. For sway columns, effective lengths are greater than true lengths; for no-sway columns, effective lengths are less than true lengths.

Background reading

BRESLER, B., LIN, T.Y. and SCALZI, J.B. (1968) *Design of Steel Structures*, Chapter 9, McGraw-Hill, New York

COATES, R.C., COUTIE, M.G. and KONG, F.K. (1988) *Structural Analysis*, Chapter 9, Van Nostrand Reinhold, Wokingham

GAYLORD, E.H. and GAYLORD, C.N. (1968) *Design of Steel Structures*, Chapter 4, McGraw-Hill, New York
NARAYANAN, R. (ed.) (1982) *Axially Compressed Structures, Stability and Strength*, Applied Science, London and New York. Chapter 1 on Centrally compressed members by Tall, L
SALMON, C.G. and JOHNSON, J.E. (1980) *Steel Structures – Design and Behaviour*, Chapter 6 (part), Harper and Row, New York
TRAHAIR, N.S. (1980) *The Behaviour and Design of Steel Structures*, Chapter 3, Chapman and Hall, London

23

Unrestrained beams

Objective

To develop an understanding of lateral–torsional instability, identify the controlling parameters and illustrate the practical design of unrestrained beams.

Prior reading

Chapter 18 (Introduction to buckling: 1); Simple bending theory; Simple torsion theory.

Summary

The chapter commences with a simple introduction to the response of slender beams to vertical loading and then presents the elastic analysis of lateral–torsional buckling. Variations in load pattern, load level and degree of end-restraint are discussed. The background to the practical design of unrestrained beams is presented and referred to the specific requirements of BS 5950 and BS 5400: Part 3. For worked examples see Appendix, p. 375.

23.1 Structural properties of sections used as beams

Sections normally used as beams have the majority of their material concentrated in the flanges, which are themselves made relatively narrow so as to prevent local buckling. The need to connect beams to adjacent members with ease normally suggests the use of an open section, for which the torsional stiffness will be comparatively low. Figure 23.1, which compares section properties for four different shapes of equal area, shows that the high vertical bending stiffness of universal beam sections is obtained at the expense of both horizontal bending and torsional stiffness.

Section type	Flat	UC (typical)	UB (typical)	RHS (typical)
Section properties				
A	1	1	1	1
I_x (vertical loading)	1	0.35	1	0.2
I_y (horizontal loading)	0.2	3.5	1	3.5
J (twisting)	1	1	1	100

Figure 23.1 Types of cross-section used as beams showing relative values of section properties

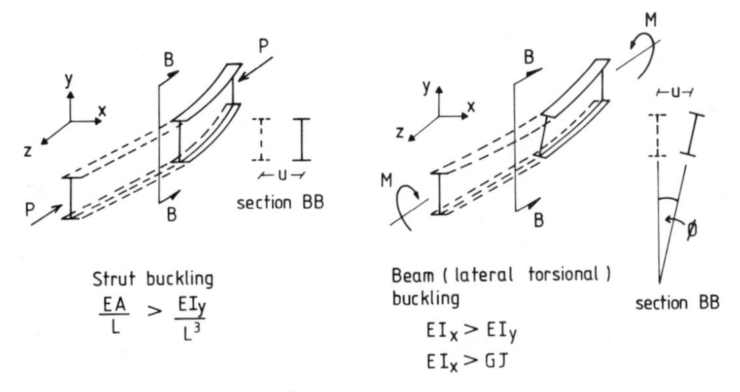

Figure 23.2 Similarity between strut buckling and beam buckling

23.2 Response of slender beams to vertical loading

Whenever a slender structural element is loaded in its stiff plane (axially in the case of the strut) there exists a tendency for it to fail by buckling in a more flexible plane (by deflecting sideways in the case of the strut). The phenomenon is termed lateral–torsional buckling. Although it involves both a lateral deflection (u) and twisting about a vertical axis through the web (ϕ) as shown in Figure 23.2, this type of instability is quite similar to the simpler flexural buckling of an axially loaded strut. Loading the beam in its stiffer plane (the plane of the web) has induced a failure by buckling in a less stiff direction (by deflecting sideways and twisting).

Many types of construction effectively prevent this form of buckling, thereby enabling the beam to be designed, with greater efficiency, for performance in the vertical direction only. In this context it is important to realize that during erection of the structure certain beams may well receive far less lateral support than will be the case when floors, decks, bracings, etc. are present, so that stability checks at this stage are also necessary.

Lateral–torsional instability influences the design of laterally unrestrained beams in much the same way that flexural buckling affects that of columns. Thus the bending strength will now be a function of the beam's slenderness, as indicated in Figure 23.3, requiring the use in design of an iterative procedure similar to the use of column curves in strut design. However, because

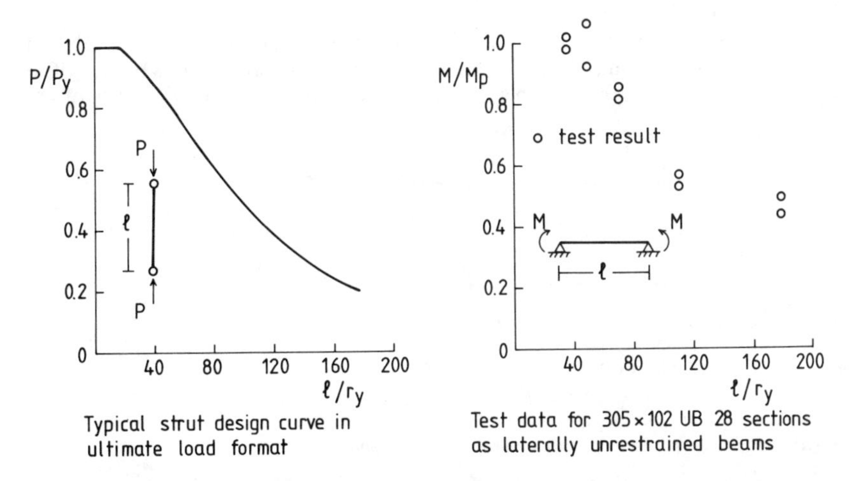

Figure 23.3 Dependence of beam strength on unrestrained length and analogy with column strength

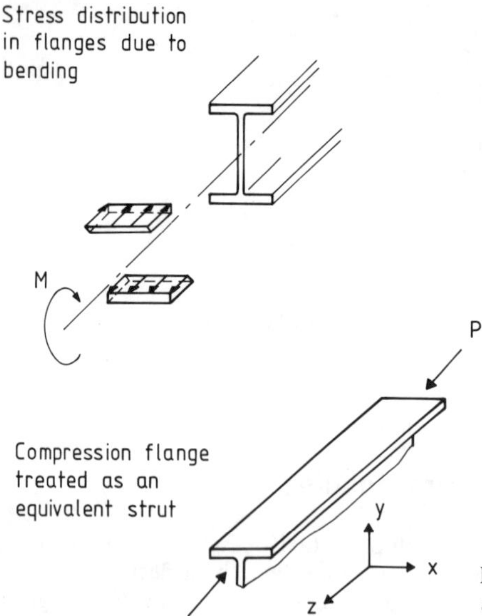

Figure 23.4 Approximation of beam-buckling problem as a strut problem

of the type of structural actions involved the analysis of lateral–torsional buckling is considerably more complex, and this is reflected in a design approach which requires a rather greater degree of calculation.

23.3 Factors influencing lateral stability

Since bending of an I-section beam is resisted principally by the tensile and compressive forces developed in two flanges, as shown in Figure 23.4, the compression flange may be regarded as a strut. This simplification permits an appreciation of a number of important features:

1. The buckling load of the beam is likely to be dependent on its unbraced span, i.e. the distance between points at which lateral deflection is prevented, and on its lateral bending stiffness (EI_y) because strut capacity $\propto EI_y/l^2$.
2. The shape of the cross-section may be expected to have some influence, with the web and the tension flange being more important for relatively shallow sections (for example, UCs) than for deep slender sections (for example, UBs). (In the former case the proximity of the

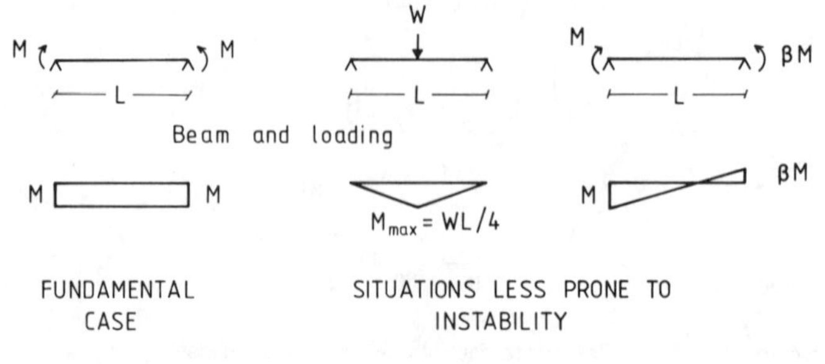

Figure 23.5 Effect of non-uniform moment on lateral–torsional buckling

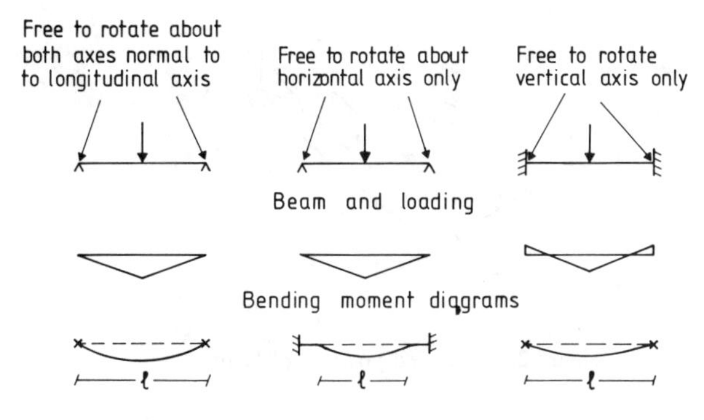

Figure 23.6 Effect of end-restraint in plan or elevation on lateral–torsional buckling

stable tension flange to the unstable compression flange increases stability and also produces a greater twisting of the cross-section: thus torsional behaviour becomes more important.)

3. For beams under non-uniform moment the force in the compression flange will no longer be constant, as shown in Figure 23.5, and therefore we might reasonably expect such members to be more stable than similar members under more uniform pattern of moment.
4. End-restraint which inhibits development of the buckling shape of Figure 23.2 is likely to increase the stability of the beam. Consideration of the buckling deformations (u and ϕ) should make it clear that this refers to rotational restraint in plan, i.e. about the y-axis (refer back to Figures 23.4 and 23.2). Rotational restraint in the vertical plane affects the pattern of moments in the beam (and may thus also lead to increased stability) but does not directly alter the buckled shape as shown in Figure 23.6. Two significant factors not illustrated by this simplified model are the warping of the cross-section and the influence of level of application of load on stability. Both of these are discussed later in the chapter.
5. Bracing may be used to improve the strength of a beam that is liable to lateral–torsional instability. Two requirement are necessary:
   (a) The bracing must be sufficiently stiff to hold the braced point effectively against lateral movement (this can normally be achieved without difficulty).
   (b) The bracing must be sufficiently strong to withstand the forces transmitted to it by the main member (these are normally a few per cent of the compressive force in the main member).

23.4 Lateral–torsional buckling

23.4.1 Assumptions

The basic problem used to illustrate the theory of lateral–torsional buckling is shown in Figure 23.7. It embraces the following assumptions:

1. Beam is initially straight;
2. Elastic behaviour;
3. Uniform equal flanged I-section;
4. Ends simply supported in the lateral plane (twist and lateral deflection prevented, no rotational restraint in plan);
5. Loaded by equal and opposite end moments in the plane of the web.

This problem may be regarded as being analogous to the basic pin-ended Euler strut.

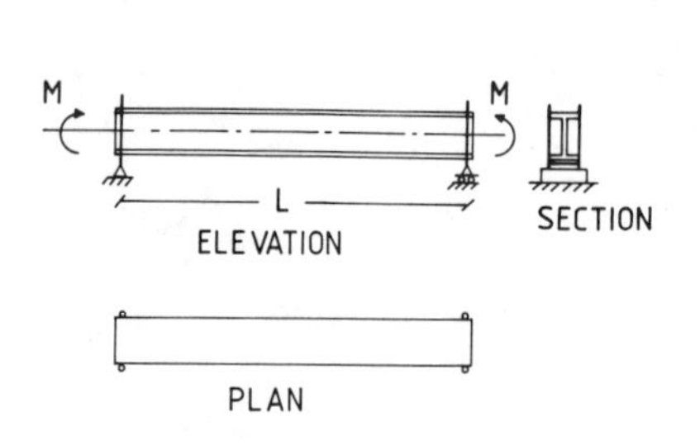

Figure 23.7 Definition of basic lateral–torsional buckling problem

Figure 23.8 Buckled position of beam

23.4.2 Derivation of governing equation

The beam is placed in its buckled position, as in Figure 23.8, and the magnitude of the applied load necessary to hold it there determined by equating the disturbing effect of the end-moments, acting through the buckling deformations, to the internal (bending and torsional) resistance of the section.

The derivation and solution of the equations leading to the critical value of applied end moments (M_{cr}) at which the beam of Figure 23.7 just becomes unstable is provided in standard texts.

23.4.3 Physical significance of the solution

Comparing the buckled shape of the beam, Figure 23.8, with the expression for elastic critical moment of equation 23.1 (page 199), it is clear that the presence of the flexural (EI_y) and torsional (GJ and EH) stiffnesses of the member in the latter is a direct consequence of the

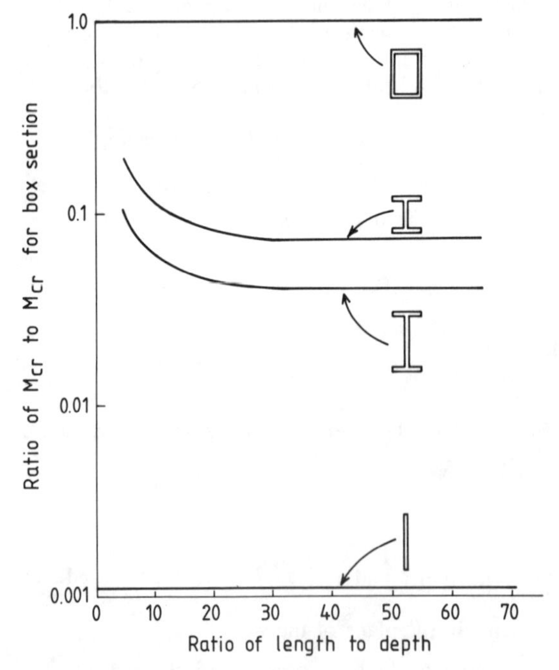

Figure 23.9 Effect of type of cross-section on theoretical elastic critical moment

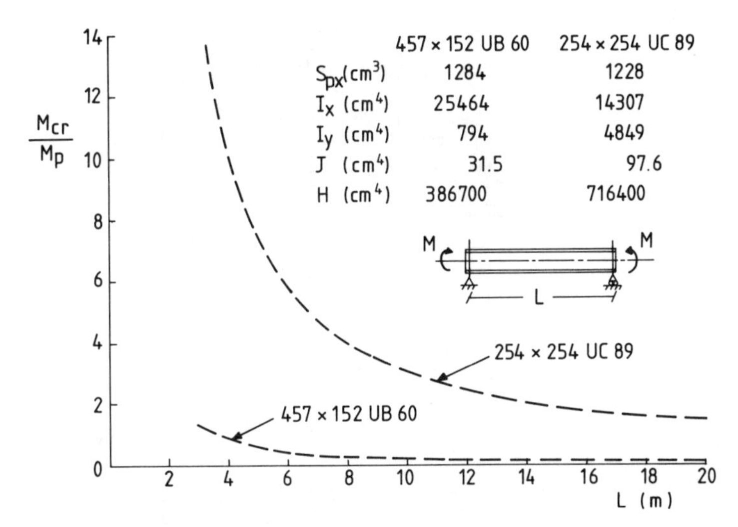

Figure 23.10 Comparison of elastic critical moments for a UB and UC section of similar in-plane bending strengths

lateral and torsional components of the buckling deformations. The relative importance of the two mechanisms for resisting twisting is reflected by the second square-root term. Length is also important, entering both directly and indirectly via the $\pi^2 EH/L^2 GJ$ term. It is not possible to simplify Equation 23.1 by omitting terms without imposing limits on the range of application of the resulting approximate solution. Figure 23.9 gives some quantitative idea of the application of Equation 23.1 to the different types of beam sections. The region of the curves for both I-sections of low length/depth ratios corresponds to the situation in which the value of the second square-root term in Equation 23.1 adopts a value significantly in excess of unity. Since warping effects will be most important for deep sections composed of thin plates, it follows that the $\pi^2 EH/L^2 GJ$ term will, in general, tend to be large for short, deep girders and small for long, shallow beams.

Figure 23.10 gives some quantitative indication of the effect of shape of cross-section for structural steel I-beams by comparing values of M_{cr} for a UB and a UC section having approximately equal in-plane (plastic) moment capacities. Clearly, lateral–torsional buckling is a potentially more significant design consideration for the laterally much less stiff UB type of section.

23.4.4 Extension to other cases

23.4.4.1 Load pattern

As an example, consider the beam subjected to a central load acting at the level of the centroidal axis shown in Figure 23.11. The solution for this example may conveniently be compared with the basic case in terms of the critical moments for each, i.e. maximum moment when the beam is on the point of buckling:

$$\text{Basic case: } M_{cr} = \frac{\pi}{L}\sqrt{(EI_y GJ)}\sqrt{\left(1 + \frac{\pi^2 EH}{L^2 GJ}\right)} \tag{23.1}$$

$$\text{Central load: } M_{cr} = \frac{4.24}{L}\sqrt{(EI_y GJ)}\sqrt{\left(1 + \frac{\pi^2 EH}{L^2 GJ}\right)} \tag{23.2}$$

The ratio of the two constants $\pi/4.24 = 0.74$ is often termed the 'equivalent uniform moment factor' m. Its value is a direct measure of the severity of a particular pattern of moments relative to the basic case. Figure 23.12, which gives a few examples in the form of m-factors, shows how lateral stability generally increases as the moment pattern becomes less uniform.

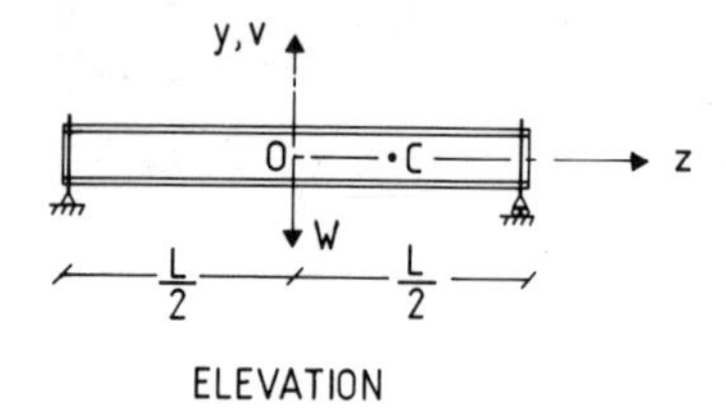

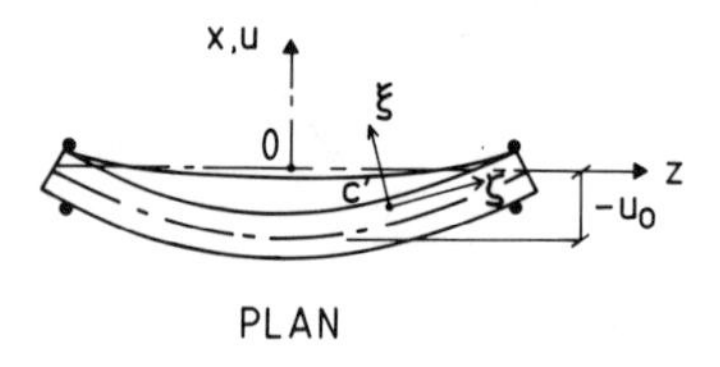

Figure 23.11 Buckling of a beam with a central transverse load

$$M_{cr} = \frac{1}{m}\frac{\pi}{L}\sqrt{EI_y GJ}\sqrt{1+\frac{\pi^2 EH}{L^2 GJ}}$$

Beam and loads	Bending moment	M_{max}	m
M M		M	1.00
M		M	0.60
M −M		M	0.40
W		$\frac{WL}{4}$	0.74
w		$\frac{wL^2}{8}$	0.89
W W		$\frac{WL}{4}$	0.94
W		$\frac{3WL}{16}$	0.68

Figure 23.12 Equivalent uniform moment factors, *m*, for simply supported beams

23.4.4.2 Load level

For transverse loads free to move sideways with the beam on buckling, the level (relative to the centroid) of application is important. When the load is applied to either the top or the bottom flanges (for example, by a crane trolley) the solution of Equation 23.2 may still be used, providing the numerical constant is replaced by a variable, the value of which depends upon the ratio L^2GJ/EH, as shown in Figure 23.13. The reason why top- and bottom-flange loading are respectively more or less severe than centroidal loading may be appreciated from the sketches in Figure 23.13, which show the destablizing and stabilizing effects. Clearly, this would be expected to become more significant as the depth of the section increases and/or the span reduces, i.e. as L^2GJ/EH becomes smaller.

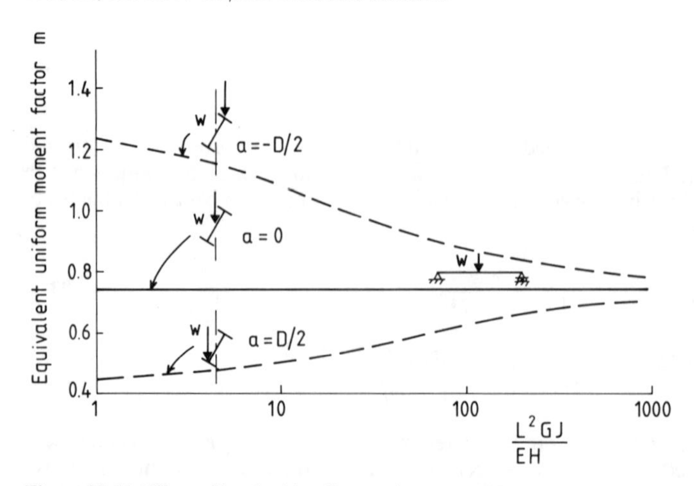

Figure 23.13 Effect of level of loading on beam stability

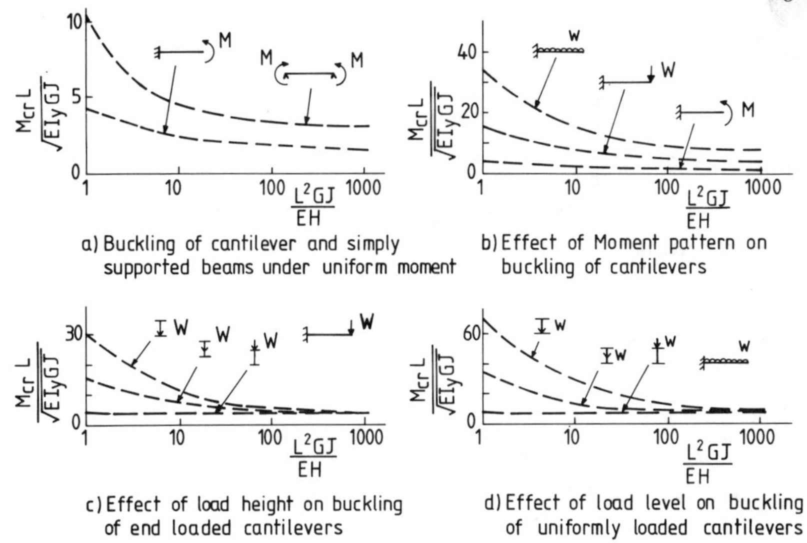

Figure 23.14 Buckling of cantilever beams

23.4.3 Conditions of lateral support

Lateral support arrangements which inhibit the growth of the buckling deformations will improve a beam's lateral stability. Equally, less effective conditions will reduce stability. Providing that the appropriate boundary conditions can be incorporated into the analysis, any arrangement can be dealt with, although recourse to a numerical solution may well be necessary.

One case of particular practical interest is the cantilever, for which some results are presented in Figure 23.14. These show that:

1. Cantilevers under end-moment are less stable than similar, simply supported, beams;
2. Concentrating the moment adjacent to the support, as happens when the applied loading changes from pure moment to an end-load or to a distributed load, improves lateral stability;
3. The effect of load height is even more significant for cantilevers than for simply supported beams.

23.5 Design approach

23.5.1 General introduction

Direct use of the theory of the previous section for design is inappropriate because:

1. The formulae are too complex for routine use;
2. Significant differences exist between the assumptions which form the basis of the theory and the characteristics of real beams. Since the theory assumes elastic behaviour it provides an upper bound on the true strength.

Figure 23.15 compares a typical set of lateral–torsional buckling test data obtained using actual hot-rolled sections with the theoretical elastic critical moments given by Equation 23.1. In Figure 23.15(a) only one set of data for a 305 × 102 × 28 UB section is shown, while the use of the $\bar{\lambda}_{LT}$ non-dimensional format in Figure 23.15(b) has the advantage of permitting results from different test series (using different cross-section and material strengths) to be compared directly. In both figures three distinct regions of behaviour may be observed:

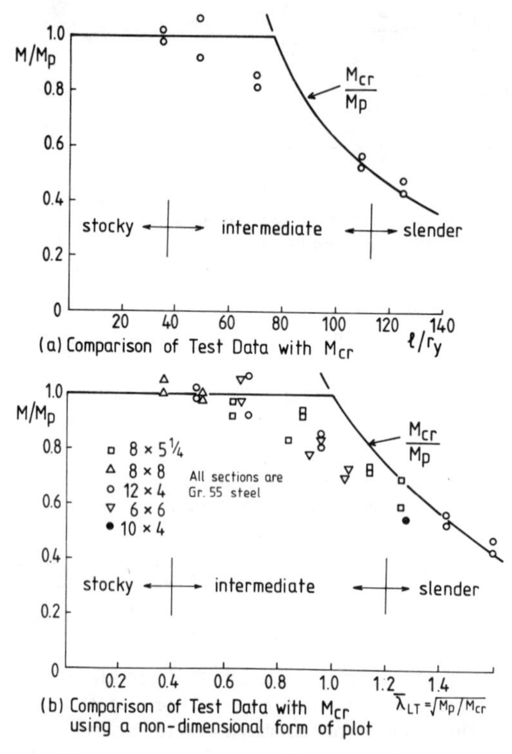

Figure 23.15 Comparison of test data with theoretical elastic critical moments

1. Stocky beams which are able to attain M_p, with values of $\bar{\lambda}_{LT}$ below about 0.4 in Figure 23.15(b);
2. Slender beams which fail at moments close to M_{cr}: for example, values of $\bar{\lambda}_{LT}$ above 1.2 in Figure 23.15(b);
3. Beams of intermediate slenderness which fail to reach either M_p or M_{cr}, i.e. $0.4 < \bar{\lambda}_{LT} < 1.2$ in Figure 23.15(b).

Only in the case of beams in region 1 does lateral stability not influence design; such beams form one class for which the methods of Chapter 20 are sufficient. For beams in region 2, which covers much of the practical range of beams without lateral restraint, design must be based on considerations of inelastic buckling suitably modified to allow for geometrical imperfections, residual stresses, etc. Thus both theory and tests must play a part, with the inherent complexity

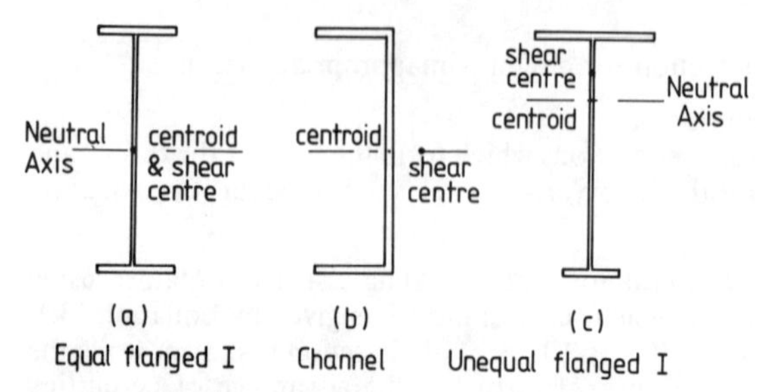

Figure 23.16 Basic equal-flanged section and examples of sections with one axis of symmetry

of the problem being such that the final design rules are likely to involve some degree of empiricism. The ways in which the provisions of BS 5950 and BS 5400: Part 3 meet the designer's requirement for an adequate representation of the three regions depicted in Figure 23.15, preferably in the form of a simple direct relationship between beam strength and beam geometry, are now outlined.

The basic treatment of each of the three design codes for the sections shown in Figures 23.16(a) and 23.16(b) is presented first. It should be noted that sections of the type illustrated in Figure 23.16(b), with one axis of symmetry (for example, channels) may only be included if the section is bent about the axis of symmetry (i.e. loads are applied through the shear centre parallel to the web for the channel). Monosymmetric sections bent in the other plane (for example, an unequal flanged I-section bent about its major-axis as shown in Figure 23.16(c)) may only be treated by an extended version of the theory, principally because the section's shear centre will no longer lie on the neutral axis. These and other special cases are discussed subsequently.

23.5.2 Method of BS 5950 – elementary cases

Moment capacity M_b is:

$$M_b \doteq p_b S_x \tag{23.3}$$

in which p_b = bending strength allowing for susceptibility to lateral–torsional buckling, S_x = plastic section modulus for compact sections.*NB*: $S_x \not> 1.2Z_x$(for other than compact sections S_x must be reduced accordingly). The slenderness used, λ_{LT}, is defined as:

$$\lambda_{LT} = \sqrt{(\pi^2 E/p_y)}\,\bar{\lambda}_{LT} \tag{23.4}$$

where $\bar{\lambda}_{LT} = \sqrt{(M_p/M_{cr})}$ (see Figure 23.15(b)). This adopts values more readily associated with the slenderness for compressive buckling, as shown in Figure 23.17.

Thus in the absence of instability, Equation 23.3 permits the full plastic moment capacity, i.e. beam strength is controlled by the development of full plasticity at the most heavily stressed cross-section. This is not plastic design (since no redistribution of moments is involved), merely a clear recognition of the limiting condition. Referring to Figure 23.15(b), p_b adopts the value p_y for $\lambda_{LT} < 0.4$, which corresponds for Grade 43 steel to an upper limit of $\lambda_{LT} = 37$ as the maximum slenderness for which strength is unaffected by lateral instability.

For more slender beams p_b is a function of λ_{LT}; determination of λ_{LT} for a given beam is facilitated by the use of some approximation of cross-sectional properties to give:

$$\lambda_{LT} = uvL/r_y \tag{23.5}$$

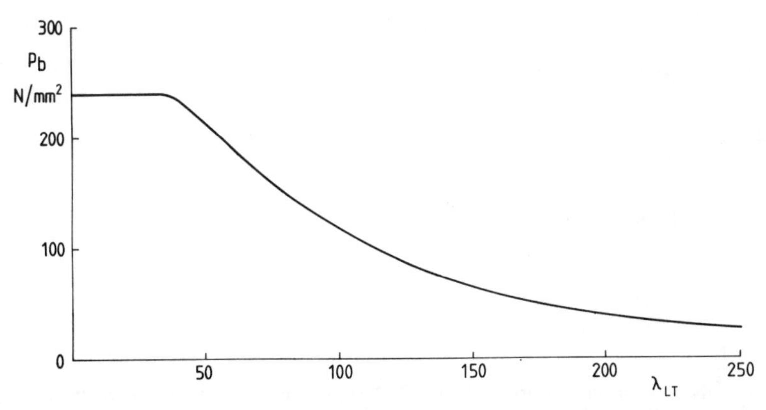

Figure 23.17 Bending strength for rolled sections of design strength 240 N/mm² according to BS 5950

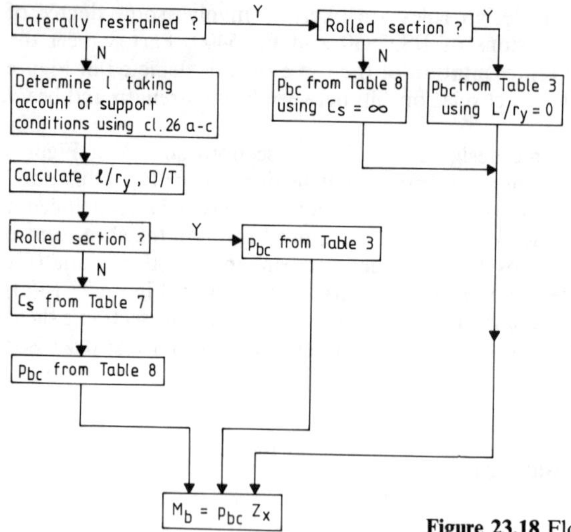

Figure 23.18 Flowchart representation of the basic procedure for checking lateral buckling strength of equal-flange compact section beams according to BS 5950

where $u = 0.9$ for rolled UBs, UCs, RSJs and channels ($= 1.0$ for all other sections), $v = f(L/r_y, x)$ is given in Table 14 of BS 5950, but for a preliminary assessment it may safely be taken as 1.0 where $x = D/T$ providing the above values of u are used.

The accuracy with which λ_{LT} (and hence p_b) is determined is thus within the control of the designer; reducting the degree of approximation will lead to lower value of λ_{LT} and hence higher design strengths. For very accurate work, tables of u and x, together with the appropriate formulae, are provided. Figure 23.18 illustrates the process in flowchart form.

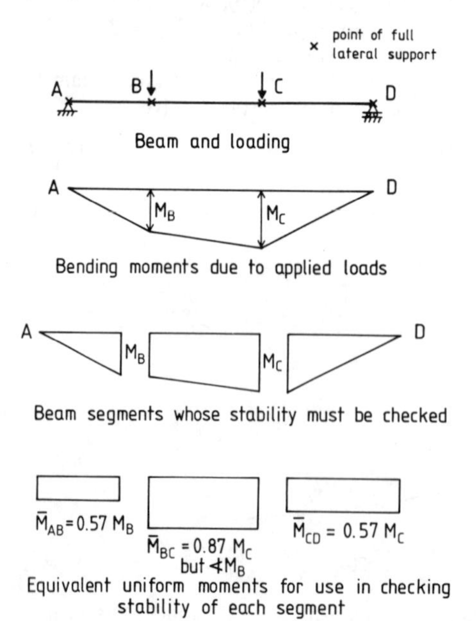

Figure 23.19 Use of equivalent uniform moment concept of BS 5950

Guidance on the choice of suitable effective length (l) values as a means of allowing for different lateral support conditions is given. In the case of beams, or segments of beams between points of full lateral support, subject to non-uniform moments as illustrated in Figure 23.19, direct use of the equivalent uniform moment introduced in Section 23.4.4.1 as a means of comparing the relative severity of different moment patterns in elastic lateral–torsional buckling is permitted. Provided that the point of maximum moment occurs at a braced cross-section, i.e. one end of the segment, the values given in Figure 23.12 are suitable for use with design capacities. Thus lateral stability is checked for an equivalent moment $\bar{M}$ given by:

$$\bar{M} = mM_{max} \tag{23.6}$$

in which $m = 0.57 + 0.33\beta + 0.10\beta^2 \nless 0.43$

$$\beta = M_{min}/M_{max}(1.0 \geqslant \beta \geqslant -1.0) \tag{23.7}$$

Of course, it is also necessary to check local strength at the more heavily stressed end against the development of M_p, i.e.:

$$M_{max} \ngtr M_p \tag{23.8}$$

23.5.3 Method of BS 5950 – special cases

Since moment capacity is normally expressed in terms of the plastic section modulus (S_x), of which there is only one value for an unequal flanged section, Equation 23.3 may still be used provided that p_b is determined accordingly. This requires that λ_{LT} be evaluated from Equation 23.5 using the appropriate section properties. Of these, u may safely be taken as 1.0 (although many sections will actually possess significantly lower values – 0.6 is not uncommon), while v now includes an allowance for the degree of monosymmetry through the parameter $N = I_{cf}/(I_{cf} + I_{cf})$. For sections with a high degree of monosymmetry for which the smaller flange is in tension, i.e. high values of N, design will be controlled by the prior attainment of yield in the tension flange.

23.5.4 Method of BS 5400: Part 3 – elementary cases

The format of Figure 23.15(b) forms the basis, with beam slenderness λ_{LT} being effectively defined by:

$$\lambda_{LT} = \sqrt{(\pi^2 p_y/355)}\,\bar{\lambda}_{LT} \tag{23.9}$$

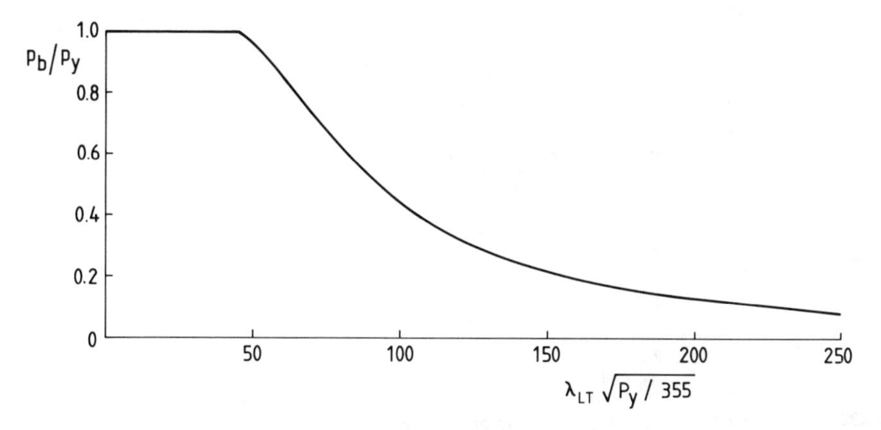

Figure 23.20 Bending strength according to BS 5400: Part 3

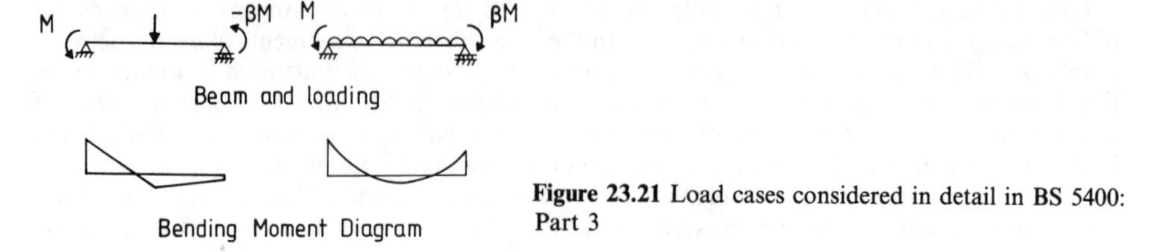

Figure 23.21 Load cases considered in detail in BS 5400: Part 3

Evaluation of λ_{LT} is facilitated by some degree of approximation of cross-sectional properties, leading to:

$$\lambda_{LT} = uvL/r_y \qquad (23.10)$$

in which u = 0.9 for rolled UBs, UCs, RSJs and channels (= 1.0 for all other sections), $v = f(L/r_y, D/T)$ is given in Table 9 of BS 5400, but for a preliminary assessment it may safely be taken as 1.0.

Knowing λ_{LT}, the limiting compressive stress (p_b) is obtained as a proportion of the yield stress of the compressive flange (p_y) from the single non-dimensional curve (Figure 23.20). Moment capacity (M_b) is then determined using:

$$M_b = p_b S_x \qquad (23.11)$$

For non-compact sections, i.e. those whose full cross-sectional strength cannot be realized because of considerations of local buckling, S_x must be replaced by the elastic section modulus (Z).

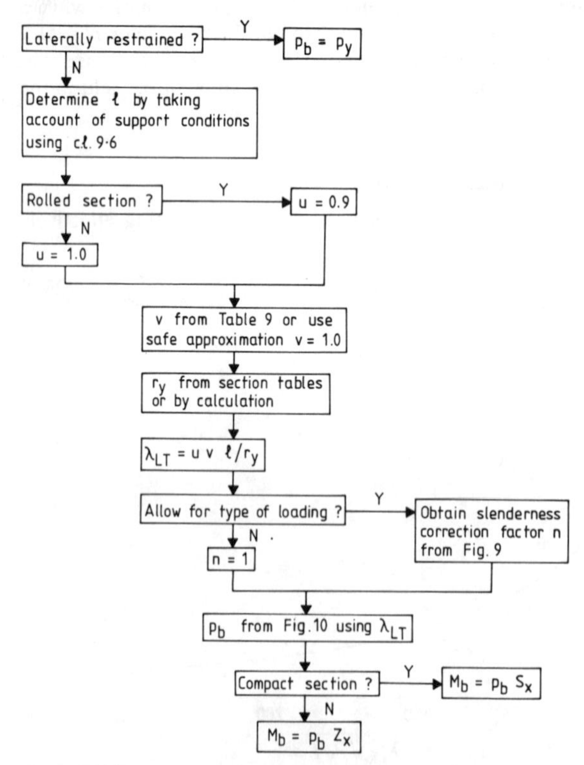

Figure 23.22 Flowchart representation of the basic procedure for checking lateral buckling strength of equal-flange beams according to BS 5400: Part 3

The effect of different types of support conditions in the lateral plane, including various forms of cantilever, is allowed for by providing a series of effective length (l) factors. Similarly, variations in the pattern of moments of the type shown in Figure 23.21 that lead to more stable arrangements than the uniform moment used to derive the basic design curve are recognized by the provision of a slenderness reduction factor (n). Figure 23.22 illustrates the process in flowchart form.

23.5.5 Method of BS 5400: Part 3 – special cases

For compact sections with unequal flanges Equation 23.11 may still be used to determine moment capacity, provided that p_b is determined using a value of λ_{LT} which properly reflects the rather more complex behaviour of such sections. Approximation of the relevant section properties in the expression for M_{cr} enables λ_{LT} to be determined from Equation 23.10, provided that the value of v now also allows for the degree of monosymmetry through the parameter $\varepsilon = I_c/(I_c + I_t)$. Although the safe value for u is 1.0, significantly lower values may be obtained from sections with very unequal size flanges. When approximating x as D/T the mean value of the flange thicknesses should be used for T.

A similar process is used for non-compact sections, which occur more frequently in bridges, although care is necessary when introducing the elastic section moduli (Z_{xc} and Z_{xt}) in place of the plastic section modulus (S_x) since one of these may be numerically greater than S_x. Thus the limiting compressive stress must be taken as the lesser of p_y or:

$$\frac{D}{2Y_t} p_b$$

in which Y_t is the distance from the axis of zero stress to the extreme tension fibre, so that the moment capacity will be the lesser of:

$$M_b = Z_{xc} \frac{D}{2Y_t} p_b$$

$$M_b = Z_{xt} p_y$$

23.6 Bracing

Faced with a situation in which calculations show his trial beam to possess insufficient moment capacity, and given that this moment capacity is limited by considerations of lateral instability (i.e. the beam's slenderness is sufficiently high that less than its full in-plane strength is available) the designer usually has two options:

1. Replace the trial beam with a potentially more stable section; or
2. Increase the strength of the original section by improving its lateral stability.

The first alternative will almost certainly involve the use of a heavier section since greater lateral stiffness must be provided without reducing vertical bending stiffness (for example, by using an I-section with wider flanges). In certain circumstances this may well be the most appropriate and economic solution.

The other possibility is to retain the original section but to improve its performance by eliminating, or at least reducing, the loss of strength through instability. If the beam's effective slenderness can be reduced sufficiently, design can be based on its full in-plane capacity. Intermediate bracing, which subdivides the original unbraced span into shorter lengths as shown in Figure 23.23, permits this. In order to be effective, such bracing must have sufficient stiffness and strength to restrain either the lateral deflection (u) or the twist (ϕ) at the braced cross-section. In many circumstances it is appropriate to restrain both displacements by bracing directly to the compression flange. Although bracing which is only partially effective will normally provide some benefit, it is often difficult to quantify this. Moreover, since relatively light bracing is normally able to provide full support, there is little point in providing less.

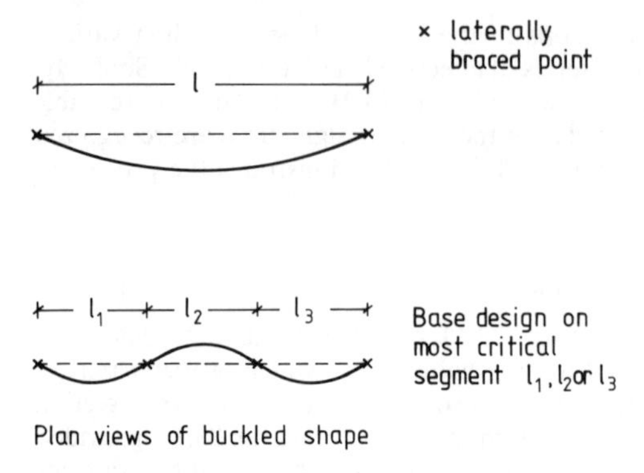

Figure 23.23 Reduction of effective length by using intermediate bracing

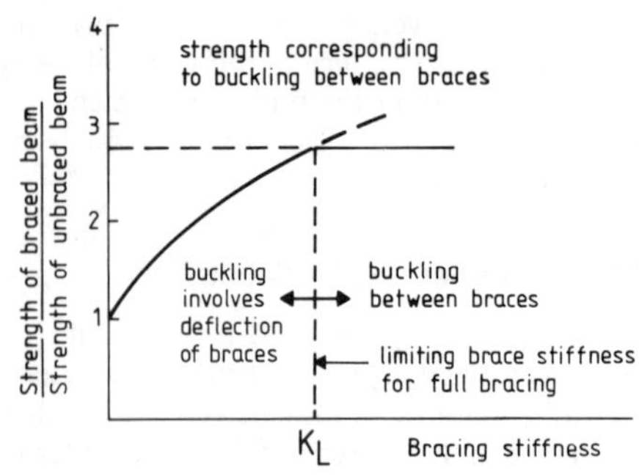

Figure 23.24 Typical relationship between beam strength and bracing stiffness

Figure 23.24 shows the typical relationship between increased stability and bracing stiffness; it is usual to provide bracing of a stiffness just in excess of the limiting value K_L.

Since the beam is generally neither straight nor free from lateral deformation or twist, some forces will be induced in the bracing. These will normally be quite small, particularly at working load levels. For design purposes it is usual to relate the strength of the bracing to the level of compressive force in the main beam, i.e. the force in the compression flange at the point of maximum moment. In many codes of practice the bracing is required to be designed for $2\frac{1}{2}\%$ of this force.

Although this is a notional force, i.e. it is introduced into the system as a means of quantifying the bracing requirement, the actual behaviour of laterally braced beams confirms that forces of approximately this magnitude are induced in the bracing members as the collapse load of the main beam is approached. Therefore the anchorage for the bracing as well as the connections to the main beam must be capable of withstanding it. The development of lateral deformations and twist in a beam will also mean that the end-supports, which are normally required to prevent both actions at the beam ends, will be subjected to a lateral force and a twisting moment.

23.7 Concluding summary

1. Beams that are not restrained along their length and are bent about their strong axis may be subject to lateral–torsional buckling.
2. Unbraced span, lateral slenderness (l/r_y), cross-sectional shape (torsional and warping rigidities), moment distribution and end-restraint are the primary influences on buckling resistance.
3. Although elastic critical load theory provides a background for understanding the behaviour of laterally unrestrained beams it requires both simplification and empirical modification if it is to form a suitable basis for a design approach.
4. Extensions to the basic theory permit the effects of load pattern, end-restraint and level of application of transverse loads to be quantified.
5. Bracing of sufficient stiffness and strength that restrains either torsional or lateral deformations may be used to increase strength.

Background reading

GALAMBOS, T.V. (1968) *Structural Members and Frames*, Prentice-Hall, Englewood Cliffs, NJ
KIRBY, P.A. and NETHERCOT, D.A. (1979) *Design for Structural Stability*, Granada, St Albans
NETHERCOT, D.A. (1982) Design of beams and plate girders: treatment of overall and local flange buckling, Chapter 13 of *The Design of Steel Bridges* (eds. R.C. Rockey and H.R. Evans,) Granada, St Albans
NETHERCOT, D.A. and TRAHAIR, N.S. (1983) Design of laterally unsupported beams, Chapter 3 of *Beams and Beam Columns: Stability and Strength* (ed. R. Narayanan), Applied Science, London
TIMOSHENKO, S.P. and GERE, J.M. (1961) *Theory of Elastic Stability*, 2nd edn, McGraw-Hill, New York

24

Beam columns

Objective

To introduce beam columns and the essential features of their behaviour, explaining why and how design is based on both strength and buckling checks.

Prior reading

Chapter 20 (Restrained compact beams); Chapter 22 (Columns); Chapter 23 (Unrestrained beams).

Summary

Examples of beam columns are introduced, together with the concept of their interaction with the structure. Single and double curvatures are defined and behaviour at low and high axial loads is described. The general loading case of biaxial bending is then considered. Biaxial interaction formulae for the strength of short columns are discussed and related to interaction surfaces. The significance of moment magnification under sway and no-sway conditions is expounded. Sketches of typical buckling interaction curves and surfaces are shown. Finally, all these theoretical considerations are related to the practical beam column design problem with discussion of the specific strength and buckling requirements of BS 5400 and BS 5950. For worked examples, see Appendix, p. 378.

24.1 Introduction

Most columns are subject to bending actions applied at the ends in addition to axial load: considerable care would have to be taken in a practical situation to load a column under axial load only. Where significant bending actions are present as well as axial load the member is termed a 'beam column'.

A common example is a portal frame, shown in Figure 24.1(a), where the columns and rafters normally have relatively light axial loads combined with severe bending. Another example is a typical column in a steel building frame (see Figure 24.1(b)); in addition to carrying the axial load from floors above, the column will generally be subject to bending actions at both ends from the floor beams. These bending actions are dependent on the interaction between the column and its adjacent members; sometimes at failure the column will support the beams, as in Figure 24.1(c), and in other situations the beams will support the columns, as shown in Figure 24.1(d).

Most building frames are braced against sway, as in Figure 24.2(a), by staircases, etc., but in some the horizontal wind forces have to be resisted by bending actions in the columns, as indicated in Figure 24.2(b). The S-shape of the latter is termed 'double-curvature' bending, in contrast to the 'single-curvature' bending of Figure 24.1(c). Figure 24.3 shows these cases diagramatically. Wind forces can also produce lateral loading on a column, as in Figure 24.2(c), giving beam-type bending moment distributions.

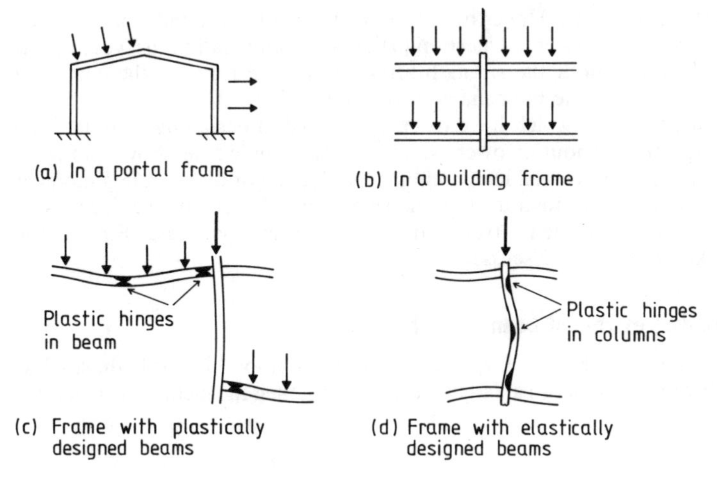

Figure 24.1 Practical beam columns

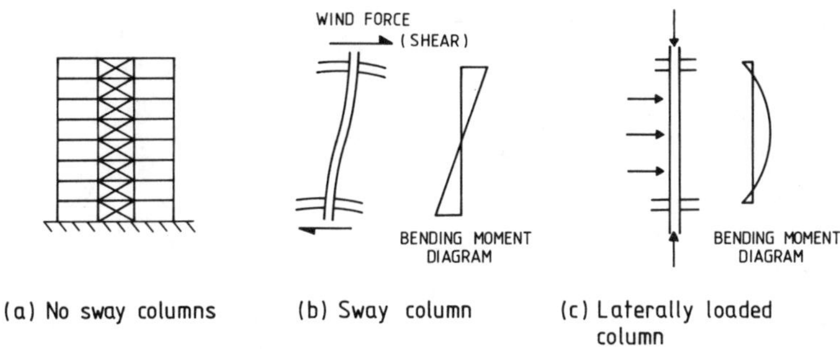

Figure 24.2 Classes of beam column

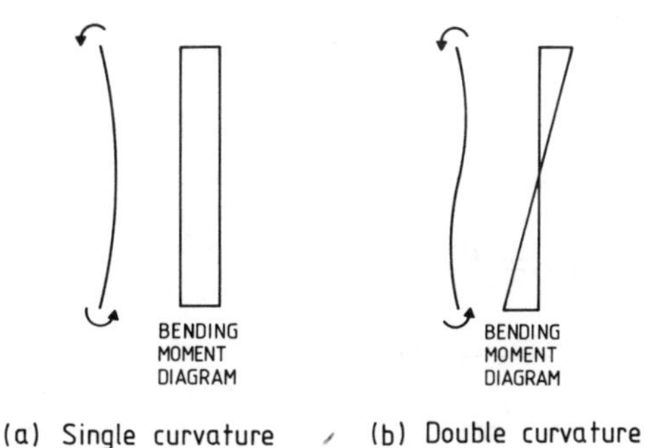

Figure 24.3 Bending moment distributions

24.2 Beam column behaviour

24.2.1 General

Beam columns are aptly named, as sometimes they can behave essentially like restrained beams, forming plastic hinges, and under other conditions fail by buckling in a similar way to axially loaded columns or unrestrained beams. Under both bending moments (M) and axial load (P) the response of a typical column for lateral deflection (δ) or end-joint rotation (θ) would be as shown in Figure 24.4. However, both the strength attained and the form of the curve are dependent on which features dominate the behaviour of the member.

The two basic types of behaviour which have to be considered in designing a column are introduced by studying typical behaviour at lower axial loads. The simple case, shown in Figure 24.5, of a UC section under moderate axial load combined with a major axis bending moment applied at one end only is used to illustrate typical behaviour. The distinctive features of behaviour at higher axial loads are then introduced using the same load case. Finally, the general load case of biaxial bending is discussed.

24.2.2 Lower axial loads: restrained beam type behaviour

If the column is not very slender, perhaps $L/r_y < 50$, and the axial load is not high, say less than a third of the member's strength as an axially loaded column, then the member will deform

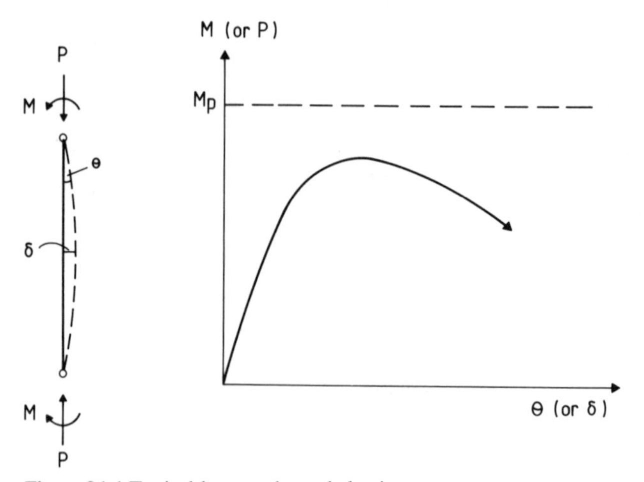

Figure 24.4 Typical beam column behaviour

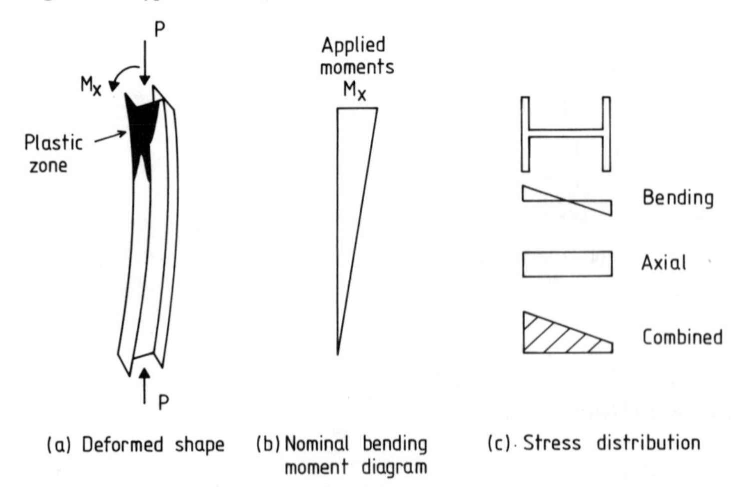

Figure 24.5 Simple beam column subject to axial load and major axis moment at one end only

in plane to failure, forming a plastic hinge near one end. Many beam columns behave in this simple way (i.e. similar to restrained beam behaviour), and their performance is relatively easy to predict. The fully plastic moment strength is reduced by axial load, and this interaction can be calculated for any cross-section. The member stiffness, i.e. the relation between the column end-moment and the column end-rotation (θ), will depend on the sequence in which M and P are applied; if the column-bending strength under constant axial load is considered a deformation curve of the type shown in Figure 24.6 (full line) would be obtained.

When the column is loaded under these conditions, i.e. with P applied first and then M_x increased, it would initially develop an elastic curvature and the elastic stresses shown in Figure 24.5(c). The most highly stressed section would be at the top and yielding would start on the flange that is subject to the greater compression, under the combination of bending stress, axial stress and any residual stresses in the section. The stress in the other flange would be much lower, a significant difference between plastic hinges in beams and columns; in columns one flange will normally remain elastic except when a high hinge rotation is combined with a very small axial load. Under increasing moment, plasticity would spread into the section, as in Figure 24.5(a), and a local hinge rotation would be developed. The hinge would spread a short way down the column and this, for reasons discussed in Section 24.2.4, causes the slight downward slope on the $M - \theta$ curve of Figure 24.6.

24.2.3 Lower axial loads: buckling behaviour

A more slender column under the same loading (perhaps $L/r_y > 60$) could fail by buckling out of plane before the full in-plane strength and behaviour of Figure 24.6 can develop. If the member were predominantly a beam under a single end-moment the failure mode would be similar to the lateral–torsional buckling of an unrestrained beam. However, the axial load produces minor axis moments ($P\delta_y$) which tend to exaggerate the minor axis bending deformations in the buckling mode. Therefore in beam columns the buckling mode is predominantly minor axis bending with a small torsional component.

The effect of this mode of failure on the end-moment/rotation performance of the column is that, at some point on the $M_x - \theta_x$ curve of Figure 24.6, out-of-plane buckling will occur and the column will develop no further strength. Possible failure curves are shown dotted in Figure 24.6; the buckling will be inelastic except for very slender members. The plasticity over the central length of the column will be caused by minor axis imperfection moments ($P\delta_y$), as in the case of an axially loaded column, combined with the major axis bending and axial load to give biaxial yielding, as discussed later. Like most inelastic buckling failures, there will be

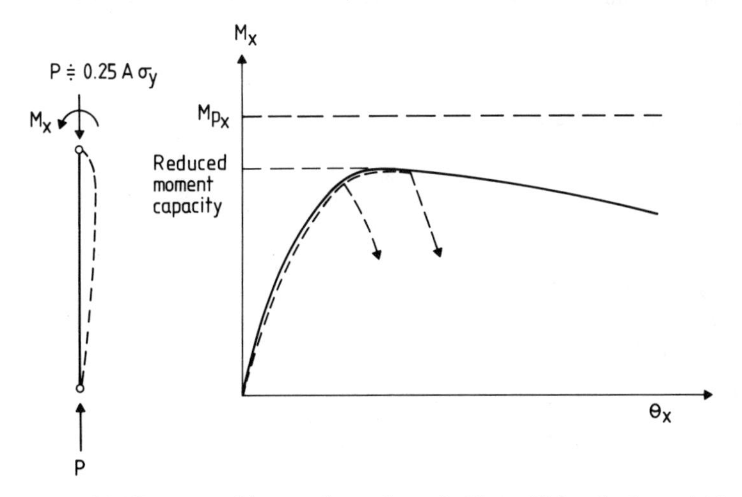

Figure 24.6 Response of beam column shown in Figure 24.5 under low axial load

a sudden loss of strength; obviously, a column failing in this way could not participate in a plastic hinge failure mechanism.

Beam column design is therefore primarily concerned with:

1. Checking the cross-sectional strength (capacity) of the member at the ends; and
2. Making an overall buckling strength check for the member.

24.2.4 Behaviour of beam columns at higher axial loads

At higher axial loads, say above half the squash load ($P > 0.5A\sigma_y$), the behaviour, as shown in Figure 24.7, is essentially similar to that at lower ones. However, there will be much more plasticity throughout the length of the column. The high axial stress present at all sections of the column, plus the residual stresses, means that very little bending stress is required to initiate yielding at any cross-section. Localized hinges of the type indicated in Figure 24.5 do not form. Instead, plasticity will spread over a considerable proportion of the column length, as is the case at the centre of a simple beam under uniformly distributed load. The inelastic curvatures (previously associated with an end-hinge rotation) will also be distributed along the column, giving deflected shapes of the type shown in Figure 24.8(a). With plasticity starting at the end under maximum in-plane moment and spreading into the central length of the column, overall out-of-plane buckling becomes affected by the associated deterioration of stiffness, and the two effects interact. The effect of out-of-plane buckling during hinge-formation behaviour will still be as indicated by the dotted lines in Figure 24.7. The only distinct change in behaviour as axial load increases is that the in-plane moment/rotation curve shows severe offloading once the maximum moment has been exceeded; this can be seen by comparing Figures 24.6 and 24.7.

Figure 24.7 shows the response of the member of Figure 24.5 under high axial load with M_x increasing, as previously considered. The variation of stresses and forces along the member when maximum M_x was attained (point A), could be as shown diagrammatically in Figure 24.8(a). The applied bending moment (M_x) will vary linearly along the member but, as a result of the major axis bending deflections (δ_x), additional significant major axis moments ($P\delta_x$) will develop, giving a combined moment distribution as shown. Thus the most highly stressed section will no longer be confined to the end of the column. With further increase of end-rotation (θ_x), causing increased curvatures and $P\delta_x$ moments along the member, the end-moment capacity could be reduced, as indicated in Figure 24.8(b). This corresponds to point B in Figure 24.7. This downward slope of the $M - \theta$ curve represents in-plane inelastic buckling similar to the failure of an axially loaded column; a simple rigid plastic analysis of the post-buckling response of a column will show the same phenomenon. If the member were part

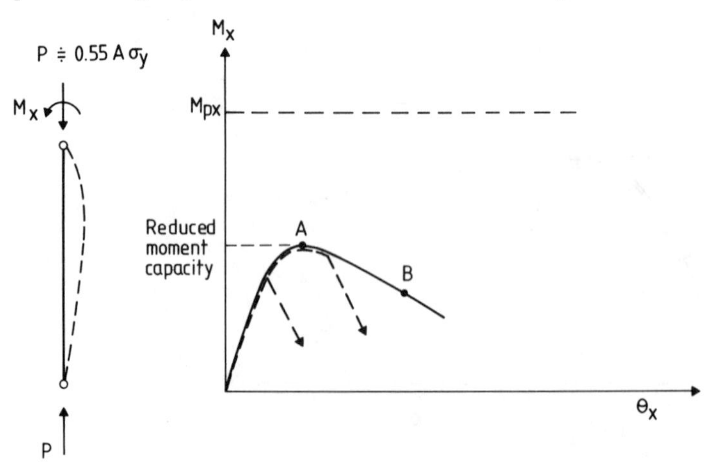

Figure 24.7 Response of beam column shown in Figure 24.5 under high axial load

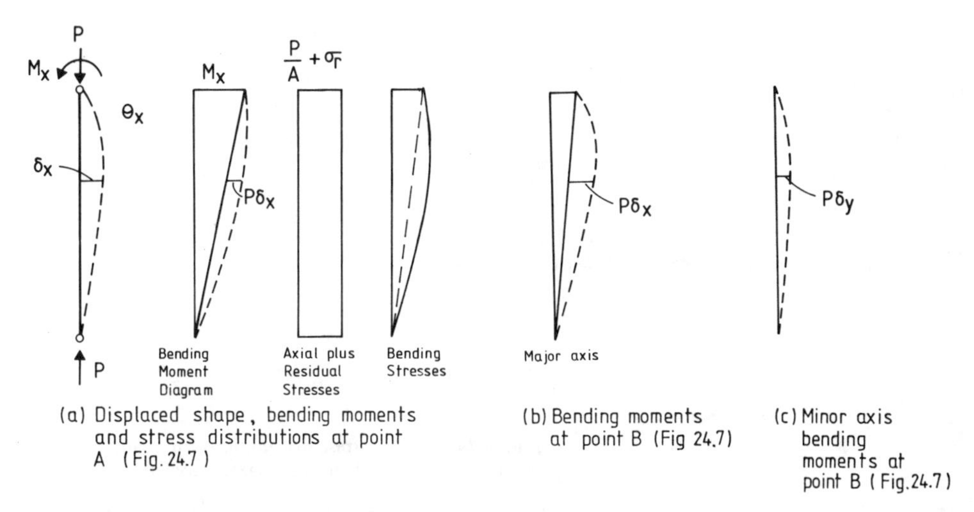

Figure 24.8 Detailed behaviour of beam column shown in Figure 24.5 under high axial load

of a structure failing by the formation of a plastic mechanism this would represent a column where the plastic moment capacity reduced with rotation.

The above discussion has ignored the significant $P\delta_y$ moments and stresses shown in Figure 24.8(c), which would also have developed, from initial curvatures about the minor axis. At any point, either before or after the maximum in-plane moment had been attained, out-of-plane buckling could occur, as indicated by the dotted lines in Figure 24.7. As a result of these δ_y moments behaviour is partly biaxial, as if the member were under biaxial end-moments; this case is discussed in the following section.

Thus, at higher axial loads, design again requires a check on strength and an overall buckling check, with the possibility of either in-plane (i.e. the downward-sloping line of Figure 24.7) or out-of-plane buckling.

24.2.5 Columns under biaxial bending

Practical columns are often subjected to more complex loading combinations than the case of Figure 24.5, with moments applied at both ends and about both axes. When both major and minor axis moments are present the column is said to be under 'biaxial bending'. A typical elastic biaxial bending stress distribution is indicated in Figure 24.9(a), resulting from combining the major axis bending plus axial stress wedge of Figure 24.9(b) with the minor axis bending stresses of Figure 24.9(c). The maximum stress occurs at one corner and is the sum of the axial stress plus the major and minor axis-bending stresses. The overall bending deformations of the column will be a combination of the curvatures about both axes. With the possibility of single-curvature bending about one axis and double-curvature about the other and any ratio of end-moments, complicated three-dimensional shapes can be developed. However, the possible modes of behaviour, i.e. hinge formation at one or both ends of the column or overall buckling, are the same as in the simple uniaxial case.

If the member is not very slender and the axial load is low, a biaxial hinge will form at one end, giving a failure curve of the type shown in Figure 24.6, but with end-joint rotations (θ_x and θ_y) about both axes. A more slender member would behave in this way until at some point overall buckling would occur and the member would appear to fail like an axially loaded column bending about the minor axis. However, there would be some torsional component in the mode and this would affect the buckling load, bringing in the M_x (du/dz) and $M_x\phi$ terms from the lateral–torsional buckling analysis of unrestrained beams. Minor axis-bending affects minor axis buckling strength in a way similar to the effect of minor axis curvature imperfections on axially loaded columns, so the buckling load will, in general, be dependent on P, M_x and

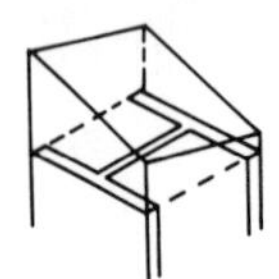

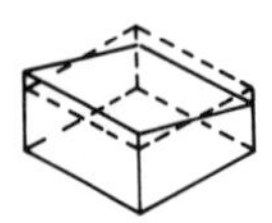

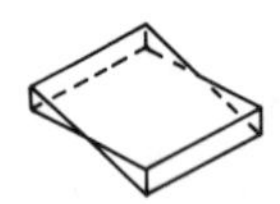

Figure 24.9 Elastic stress distributions for beam column under axial load and biaxial moments

M_y. It will also be dependent on the ratio of the end-moments, with those cases that approach uniform single-curvature bending being the most severe (for example, beams). As with axially loaded columns, only extremely slender members will fail by elastic buckling; the buckling will generally be inelastic, occurring after some inelastic deterioration (α_D) of the member's stiffness.

At higher axial loads plasticity will spread along the member; if biaxial hinges form, they will be less localized. There will still be the two possible modes of behaviour; either the column will attain the maximum strength of its cross-section or it will fail by buckling. Therefore in all beam column design a strength and an overall buckling check are required.

For the rare special case of uniform bending under equal end-moments, either uniaxial or biaxial, the maximum moment will be acting at the centre of the member, so the buckling check will then also cover strength.

24.3 Interaction between column and structure

The buckling of beam columns is controlled by the behaviour of the structure of which they are a part, and this has to be taken into account in their design. For example, a column in a no-sway building frame, as shown in Figure 24.1(b), would have an effective length less than L, as discussed in Chapter 22. It would be much more stable than a similar column in a sway frame, as, for example, in Figure 24.2(b), which would have an effective length greater than L. Effective lengths are often used in design, but this can lead to difficulties and contradictions if the designer thinks in terms of linear elastic structural behaviour when columns are failing by buckling.

Let us consider the no-sway column shown in Figure 24.10(a), which is being bent in single curvature about its minor axis by loading on adjacent beams. These moments, which would be of the type predicted by a linear analysis, will be considered positive. Paradoxically, the same beams can restrain the column against minor axis buckling, i.e. giving the column an effective length less than L; however, for this to occur at failure the column end-moments must be of opposite sign, as indicated in Figure 24.10(b). Both conditions cannot occur at the same time, and what actually happens is as follows.

When only part of the axial load is applied the column restrains the beams, as predicted by linear elastic analysis, with positive end-moments as in Figure 24.10(a). Then, as the axial load approaches failure, the column loses stiffness, usually with the section going partially inelastic and reducing its elastic stiffness (EI_y) by a factor α_D, as discussed in Chapter 22. The Euler load [$P = (\pi^2/L^2)\,EI_y$] for the equivalent pin-ended column of length L becomes approximately $(\pi^2/L^2)(EI_y)\alpha_D$; when this is equal to the applied load the column is incapable of restraining the beams and the end-moments must have been relaxed to zero. Under further increase of axial

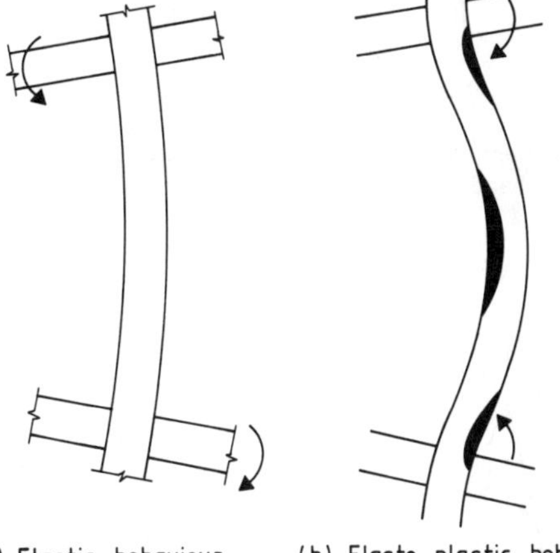

Figure 24.10 Interactions between beam column and surrounding frame

load the column must be supported by the beams, with negative end-moments and an effective length of less than L, until failure occurs in the configuration of Figure 24.10(b).

Column stiffnesses can also alter directly, without stability effects, as the result of plasticity; thus columns can often significantly reduce their major axis applied end-moments before failure by deforming plastically at the ends, as shown in Figure 24.5(a). Most practical beam columns behave in this way.

It is difficult to develop accurate design procedures to deal with these aspects of the behaviour of beam columns, and all current design procedures are approximate. They attempt to deal with the essential features of behaviour, sometimes at the expense of introducing contradictions, such as the one illustrated in Figure 24.10, where a positive design moment is combined with an effective length less than L. Arbitrary assumptions about design moments (for example, those made in 'simple design') are thus as valid as those obtained from an idealized linear elastic frame analysis. In fact, slavish adherence to linear analysis often restricts the designer's choice to a type of structure in which heavier beam columns support the beams, whereas 'simple design' gives slightly heavier beams which support lighter columns, often the more desirable solution.

24.4 Short-column strength

One obvious limit to the capacity of a beam column is the strength of a short length of section under uniform bending and axial load; this is often referred to as the short-column strength. The only buckling which could occur would be local flange or web buckling; if the section is compact the capacity will be the full plastic strength of the section. The ultimate strength of a section under uniaxial bending combined with axial load is given in the interaction curve shown in Figure 24.11. The dotted line shows the nominal first yield limit, ignoring residual stresses. Under biaxial bending, i.e. P, M_x and M_y, the condition for nominal first yield becomes (for sections where the extreme compression fibres for bending about both axes coincide, as shown in Figure 24.12(a):

$$\frac{P}{A} + \frac{M_x}{Z_x} + \frac{M_y}{Z_y} = \sigma_y$$

This can be plotted as the plane surface shown in Figure 24.12(b), which intersects the axes at

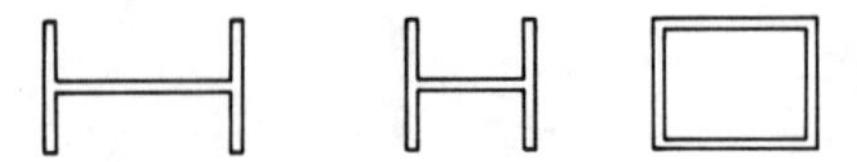

(a) Cross-sections where extreme compression fibres for bending about both axes coincide

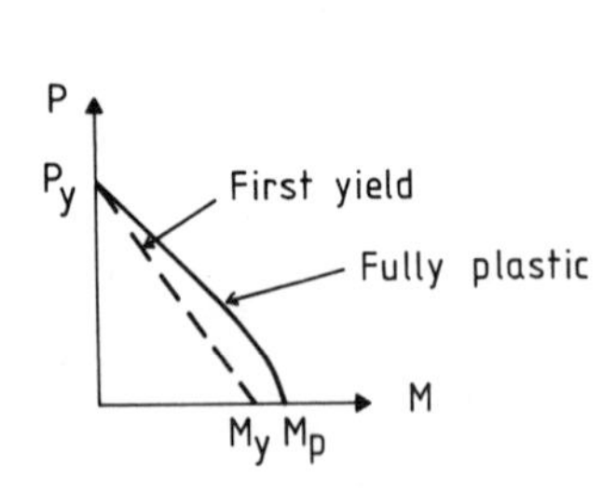

Figure 24.11 Interaction diagram for UC section under axial load and major axis moment

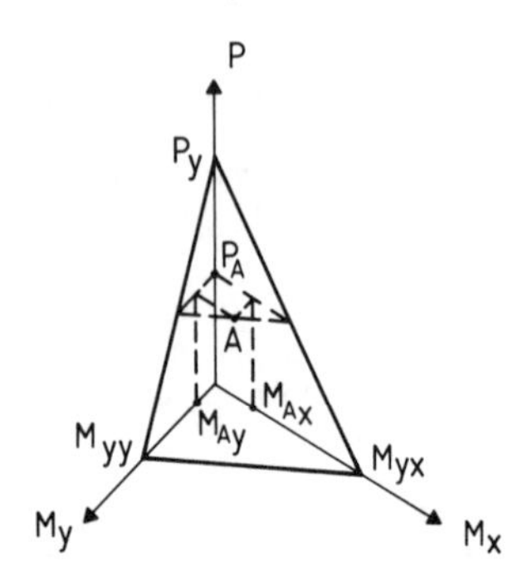

(b) Interaction diagram for sections in (a)

Figure 24.12 Elastic interaction of cross-section

the nominal first-yield strengths M_{yx}, M_{yy} and P_y. Any typical point A on the surface would correspond to yielding under the combination of forces P_A, M_{Ax} and M_{Ay}. Sometimes, when it is desired to limit the amount of plasticity occurring in a column, this nominal first-yield interaction surface is used as a strength limit in the design code (for example, BS 5400: Part 3).

Under biaxial bending and axial load the ultimate strength for a RHS gives a smooth convex surface, as in Figure 24.13. UB and UC sections give a convex surface of a less regular shape because of discontinuities in the cross-section. However, ignoring such complexity, an equation of the following convenient form gives conservative results:

$$\left(\frac{M_{ux}}{M_{rx}}\right)^2 + \left(\frac{M_{uy}}{M_{ry}}\right)^2 = 1 \qquad (24.1)$$

where M_{rx} and M_{ry} are the plastic moment strengths about the X and Y axes, respectively, as

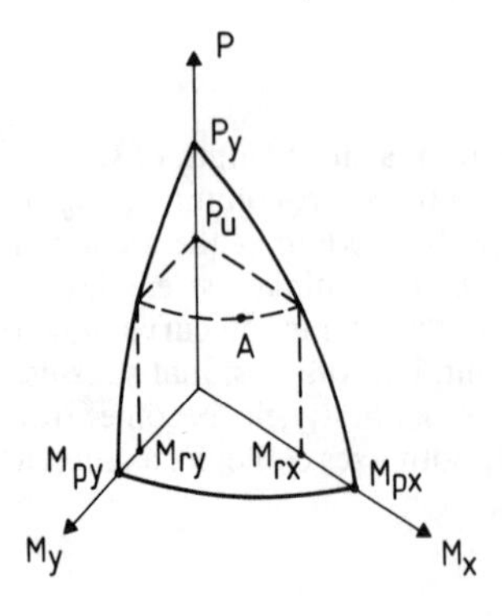

Figure 24.13 Fully elastic interaction of RHS section under axial load and biaxial moments

reduced by the axial load P_u in Figure 24.13. The equation gives a convex line as indicated: section tables give data for calculating M_{rx} and M_{ry}.

24.5 Moment magnification

With columns of normal length the strength can be significantly below the short-column strength. The analysis of a column under uniaxial bending with equal end-moments illustrates the aspect of behaviour termed 'moment magnification'. Let us consider an elastic column, as in Figure 24.14, bent about its minor axis so that there is no possibility of out-of-plane buckling. The end-moments will produce a curvature in the member, and P acting on that deformed shape will produce extra Pv moments, as indicated. The theoretical solution gives the maximum moment at the centre of the column as:

$$M_{max} = M \sec k \frac{L}{2}$$

It is interesting to note that this secant formula was used in America, instead of the similar solution for a column with initial curvatures, to give formulae for axially loaded columns; it is found that, very approximately:

$$\sec k \frac{L}{2} \approx \frac{1}{[1 - (P/P_E)]}$$

Therefore columns are often required to resist forces on a cross-section in excess of those applied at the ends. This effect also occurs in beam columns which are free to sway and which have to resist wind shears, as in Figure 24.15. The end-moments M are increased by $P\Delta/2$ for the symmetrical case shown and a theoretical analysis again gives the maximum moment:

$$M_{max} \approx M \frac{1}{[1 - (P/P_E)]} \tag{24.2}$$

Moment magnification will not always be significant. For example, in a column under a single end-moment as shown in Figure 24.16(a), the end-moments will only be exceeded within the length of the column at high axial loads or when the column is about to buckle in-plane. This is also the case for a no-sway column under end-moments of opposite sign, as shown in Figure 24.16(b). This figure should be compared with Figure 24.15, for a sway column, where any moment magnification within the column has to be added to the overall magnification from the sway displacements over the height of the column; this reinforces the point concerning the weakness of sway columns.

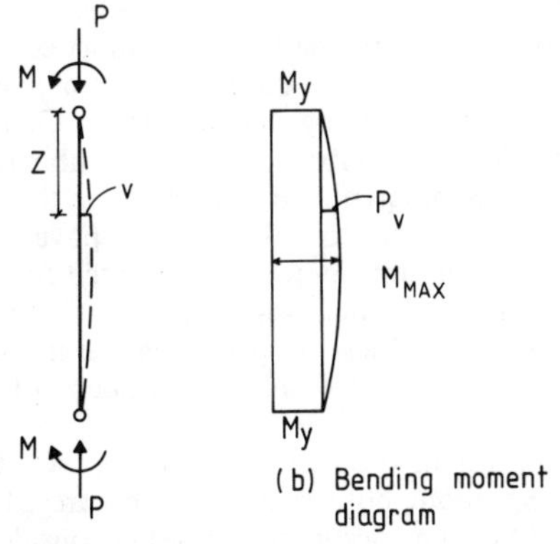

Figure 24.14 Elastic behaviour of no-sway beam column under equal end-moments

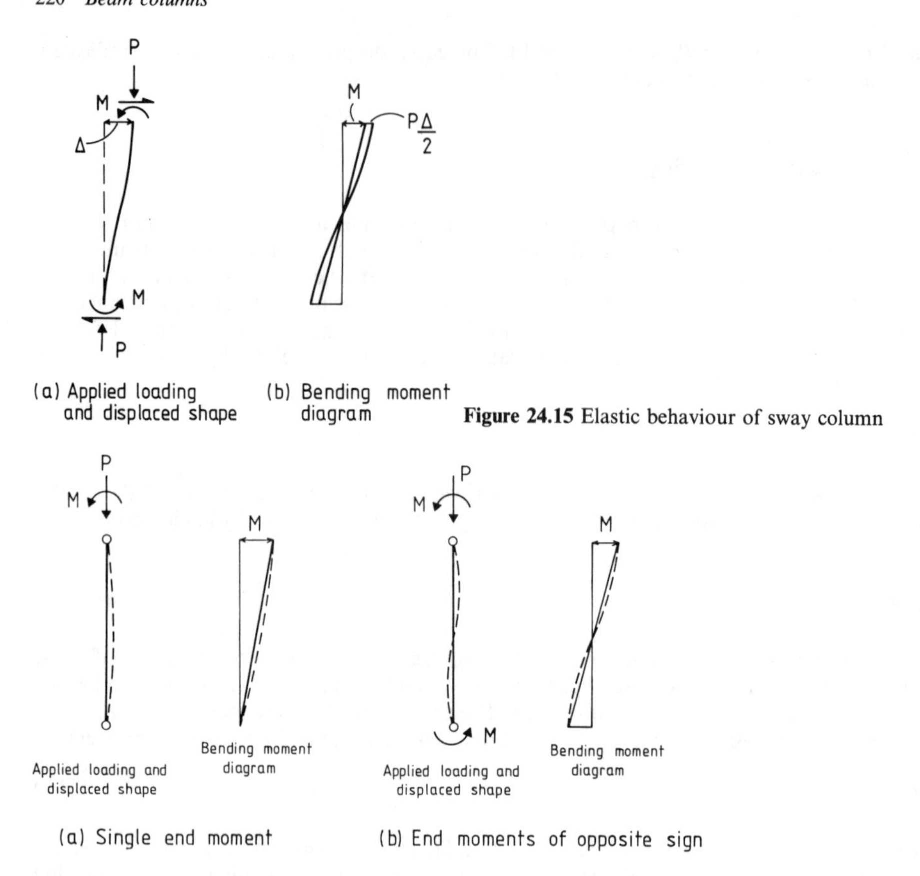

Figure 24.15 Elastic behaviour of sway column

Figure 24.16 Elastic behaviour of no-sway beam column under non-uniform moments

24.6 Buckling

24.6.1 Introduction

Many beam columns will fail by buckling under end-forces that are less than their short-column strength. The failure can be elastic instability for very slender columns, as in the case of axially loaded members, but for columns of medium slenderness the buckling will be inelastic. General equations can be established for the buckling of axially loaded members under any combination of non-uniform biaxial bending; this approach gives three differential equations, as in the case of the lateral–torsional buckling of beams. Additional terms for axial load effects have to be included and the effects of minor axis bending and non-uniform moments accounted for. All cases of member buckling are special cases of these general equations. They can be solved numerically, taking account of deterioration of sectional stiffness properties in the inelastic range. However, as the computations are extensive, solutions have only been obtained for a limited range of cases.

There is a very large range of possible bending conditions occurring in practice, including biaxial bending combinations with moments varying about both axes, as in Figure 24.17. However, for certain loading cases particular types of buckling occur, and these are considered first.

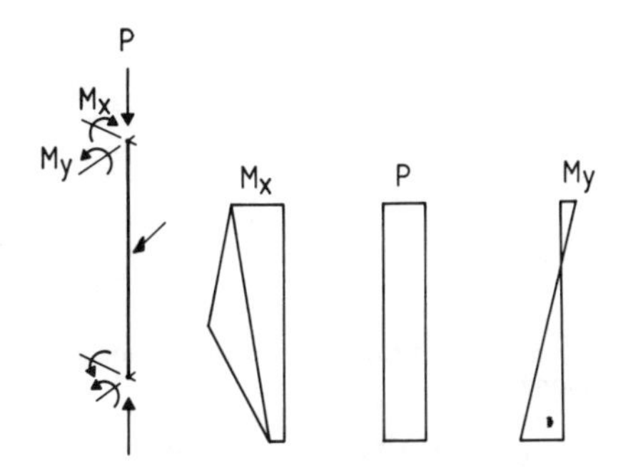

Figure 24.17 Typical loadings for biaxial beam columns

24.6.2 In-plane buckling

The simplest case of buckling is in-plane, which can only occur when the member is under uniaxial bending and failure is by deformations in the plane of bending. The failure of the member shown in Figure 24.14 is an example. Elastic theory gives deflections v, tending to ∞ as P approaches P_E, but a real beam column would buckle inelastically at a much lower load. The magnified moment combined with the axial load would cause plasticity, and for minor axis bending reduce the stiffness (EI_y) of the member by a factor $\alpha_D < 1$ in a way similar to that seen in an axially loaded column, giving inelastic in-plane minor axis buckling.

Columns under major axis bending which are restrained against out-of-plane buckling, as is sometimes the case in practice, can buckle in-plane about the major axis. Under certain conditions it is also possible to get in-plane inelastic buckling under M_x bending without out-of-plane restraints. (This explains the two formulae for M_x buckling in the second part of Clause 4.8.3.2 of BS 5950. The first formula takes account of in-plane buckling; the other covers the more obvious case of lateral–torsional buckling, discussed below.)

24.6.3 Lateral–torsional buckling

This is the type of buckling out of plane that was introduced in considering unrestrained beams (Chapter 23). It can only occur in columns under axial load and uniaxial bending about the major axis; under such loading the column buckles in a mode involving twisting and minor axis bending. The twisting, which distinguishes it from minor axis buckling and reduces the buckling load, is significant for I and H shapes buckling at lower axial loads. RHS shapes are stiff in torsion, so the mode then effectively becomes minor axis bending.

24.6.4 Biaxial buckling

Under biaxial bending buckling will most often be in a mode dominated by minor axis bending; however, some twisting and major axis deformations in the mode will also affect the failure load. Inelastic deterioration of the member's stiffness under biaxial bending will start at one corner of the cross-section, producing a section without axes of symmetry. In effect, this rotates the bending axes to give a situation similar to the bending of an unequal angle section. Bending about the inclined weakest axis can then involve both major and minor axis deformations. However, for I and H shapes minor axis displacements will be dominant, unless one flange goes completely plastic, thus considerably reducing the major axis stiffness.

24.7 Buckling interaction curves

Lateral–torsional buckling will reduce the strength of members under axial load and major axis bending, as shown in Figure 24.18. As indicated by the dotted lines, the buckling strength will decrease below the short-column strength, shown as a full line, as the slenderness (L/r_y)

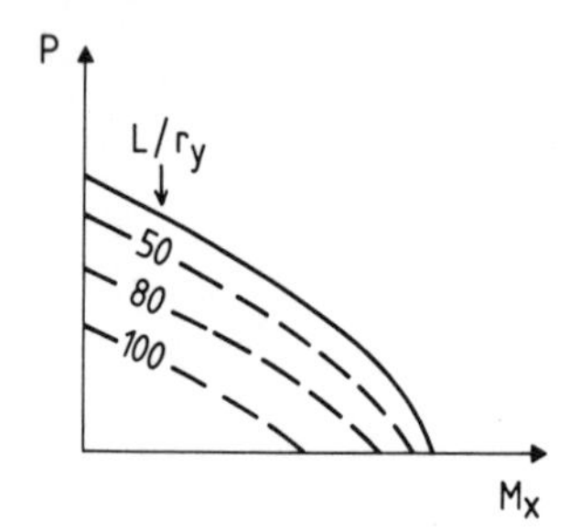

Figure 24.18 Typical interaction diagrams for beam columns under axial load and major axis moments

increases. The values where the dotted interaction curves intersect the M_x axis, i.e. with $P = 0$, are given by the lateral–torsional buckling strength of unrestrained beams. Similarly, the values on the P axis correspond to the limiting case where there is no torsion in the buckling mode and are given by the minor axis buckling strength of axially loaded columns. The shape of the dotted interaction curves will depend on the member cross-section and other variables, but most available data suggest that they are convex.

Results for the general case of buckling under major and minor axis bending would give interaction surfaces of the form sketched in Figure 24.19. The intersection with the P, M_y plane represents the case of a member under minor axis bending combined with axial load that is shown in Figure 24.20. Such a member would buckle in-plane by bending about the minor axis with no torsion. The curves of Figure 24.20 differ from those of Figure 24.18 in that, at zero axial load, minor axis buckling cannot occur and the strength of the member will always be M_{py}, except where this is reduced by local buckling for non-compact sections. The coordinates on the P axis are, of course, the same as in Figure 24.18. The interaction curves will again generally be convex, but some work has suggested that they can become slightly concave at high slendernesses.

The precise shape of the interaction surfaces will depend on:

1. The cross-sectional shape and the column imperfections;
2. The variation of moments along the column, as in Figure 24.17;
3. The end-restraint conditions of the column.

All these variables can only be dealt with on an approximate basis and so various interaction formulae are introduced in the design codes which attempt to allow for the above effects. End-restraint is normally allowed for by using effective lengths: this introduces uncertainty concerning the values of the end-moments. Therefore design under these conditions cannot be exact.

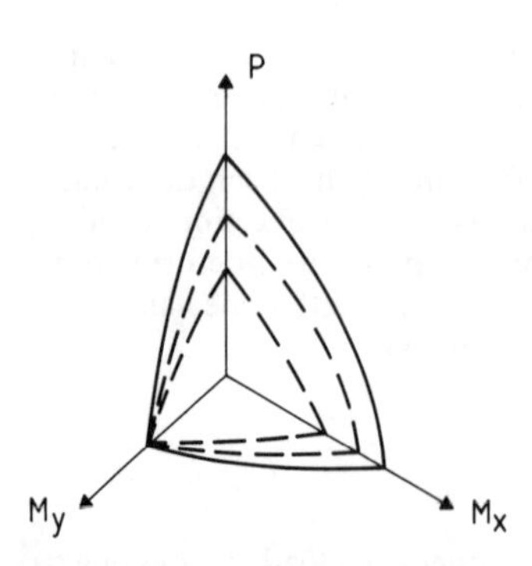

Figure 24.19 Typical interaction diagrams for beam columns under axial load and biaxial moments

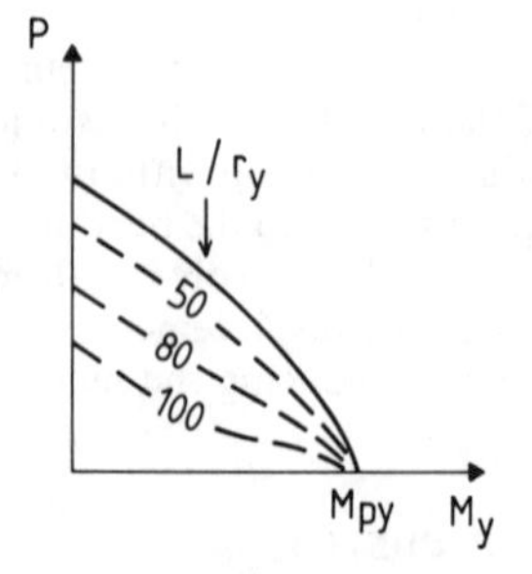

Figure 24.20 Typical interaction diagrams for beam columns under axial load and minor axis moments

24.8 Choice of section

In selecting the type of section to be used as a beam column it must be appreciated that the magnitude of the moments will often be dependent on the section chosen. An I-section shape will be stiff and often the most efficient choice for resisting M_x bending, as, for example, in the eccentrically loaded column of Figure 24.21(a) and the portal frame of Figure 24.21(b). However, using a stiff I-section in a no-sway building frame would develop higher major axis moments by restraining the beams than using a square H shape. Thus UC sections have been designed to be used where the primary function of the column is to carry axial load. RHS are efficient shapes to resist both major and minor axis bending. Also they have a high torsional resistance which effectively prevents lateral–torsional buckling. Except for the limitations of the available weight of the RHS sections, their costs and the difficulties with connections, they would be the ideal section in many applications.

Some members become beam columns as the result of their sectional shape in places where they would otherwise be predominantly axially loaded. For example, a channel section loaded through a connection to the web would be subject to minor axis bending. Such members should only be used where they are appropriate for some special reason.

24.9 Design forces on the column

In some structures the forces acting on a beam column can be unambiguously established; examples are a column supporting an eccentric load, as in Figure 24.21(a), and one loaded by a beam carrying a fully plastic end-moment. More often, however, the column is interacting inelastically with adjacent structural members, and the forces acting on it cannot be established exactly. In such cases the nominal assumed design forces are dependent on the design method being used for the structure. For example, in BS 5950 there are code recommendations for the moments to be assumed using 'simple design' for a frame structure with flexible joints; there are other assumptions for 'semi-rigid design'. A linear elastic analysis is used for structures with 'rigid' connections.

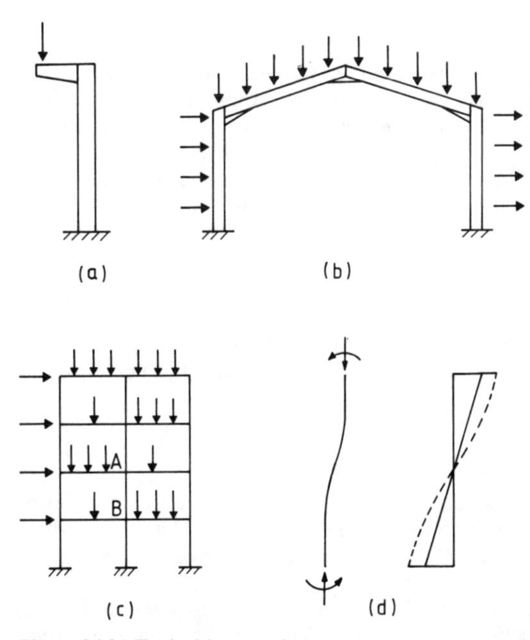

Figure 24.21 Typical beam columns

Consideration of the case of the sway column AB in the building frame of Figure 24.21(b), under wind and vertical loads, illustrates how the first stage of beam column design is to establish the nominal design end-moments. A linear analysis program, assuming elastic behaviour and rigid joints, could be used to estimate a worst combination of axial load and primary sway moments on the column, as shown by the full line in Figure 24.12(d). These moments would then be increased by the moment magnification effect of the axial load, as indicated by the dotted line in Figure 24.21(d). Such sway columns can either be designed:

1. Using effective lengths greater than L and taking the lower nominal moments, shown by the full line in Figure 24.21(d), as the design end-moments; or
2. As columns of effective length L, using the magnified moments, denoted by the dotted line in Figure 24.21(d), as the design end-moments to deal with sway instability effects.

With the latter procedure, instead of using the individual column moment magnification factors, $1/[1 - (P/P_E)]$, the effect of sway interaction with other columns can be allowed for by using a factor, $1/[1 - (1/\lambda_{cr})]$, where λ_{cr} is the critical load factor for the whole frame.

24.10 Strength assessment for design

Having established the nominal design forces to which the column is considered to be subjected, including Υ_{fl} factors, the design procedure requires two checks, which are discussed below.

24.10.1 A strength check

This is to ensure that, at any cross-section of the column, the worst combination of forces for which the column is being designed does not exceed the short-column strength of the section. The worst combination of forces will occur at the ends of the member, unless there is lateral loading within the column length. The types of code formulae used in BS 5950 and 5400 have been discussed under short-column strength in Chapter 22.

24.10.2 A buckling check

Buckling strength may control the load capacity of the member; the wide range of possible cases is dealt with by using interaction formulae. In assessing the buckling strengths under axial load and major axis moment that are used in these formulae effective lengths must be used to make allowance for restraint conditions.

The traditional approach to design has been to assume a linear interaction surface between the ordinates on the P, M_x and M_y axes, as shown in Figure 24.22. This gives design formulae of the type:

$$\frac{F}{P_c} + \frac{M_x}{M_{bx}} + \frac{M_y}{M_{py}} = 1 \qquad (24.3)$$

where M_{bx} and P_c are the buckling strengths of the column under M_x and P acting separately. Non-linear interaction surfaces have been introduced in BS 5950. Details of the buckling check under each code now follow.

24.10.2.1 BS 5400

Clause 10.6.2.1 uses the linear interaction of Equation 24.3 but allows for the effect of variation of moment along the column by only considering the maximum moments that occur within the middle third of the column length (M_{xmax} and M_{ymax}). These moments need not occur at the same cross-section.

The variation of M_x along the member is also allowed for in calculating the major axis bending resistance (M_{Dxc}) by using a reduction factor (η) in calculating the beam slenderness

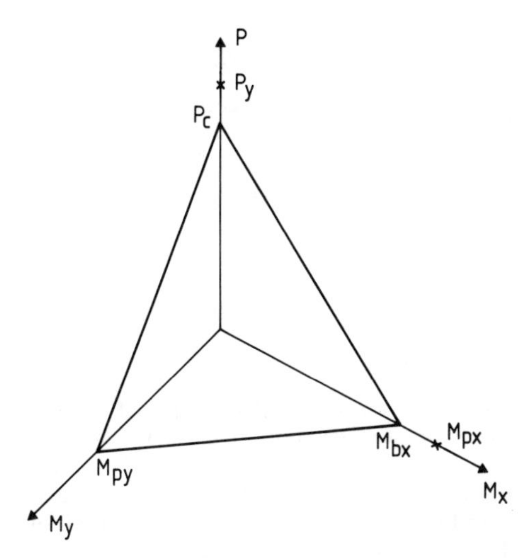

Figure 24.22 Beam column design strength with linear interaction between P, M_x and M_y

(λ_{LT}) in **Clause 9.7.2.** End-restraint is recognized by the use of reduced effective lengths which, if desired, can be calculated more accurately using **Figure 7** of **Clause 9.6.**

In calculating the axial buckling stress, effective lengths are given for no-sway columns in **Table 10**. Alternatively, more accurate values can be obtained using **Figure 7a**. Sway columns, fully restrained at one end, are also included in **Table 10**; for other sway columns the use of an elastic critical load analysis is permitted but no data are given.

24.10.2.2 BS 5950

Clause 4.8.3.2 presents the simplified approach which is based on the linear interaction of Equation 24.3. In **Clause 4.3.7.2** allowance for the variation of moments along the column is made by using equivalent uniform moments mM_{ax} and mM_{ay}, where m is a function of moment gradient.

The variation of M_x along the member is also allowed for in calculating M_{bx} by using a reduction factor (η) on the beam slenderness (λ_{LT}) in **Clause 4.3.7.5**. An effective length (**Clause 4.3.5**) is used to cover end-restraint.

The limiting moment under minor axis bending (M_{py} in Equation 24.3) is restricted to the nominal first-yield moment $Z_y p_y$.

For calculating the axial buckling stress (p_c) effective length data are given for both sway and no-sway columns. **Table 24** of **Clause 4.7.2** presents standard cases; more accurate data, covering a wide range of cases, are given in **Appendix E, Figures 23–26**.

The simplified formula has to be used for the simple design of no-sway frames (**Clause 4.7.7**) with no reduction factors for the variation of moments along the column, i.e. $m = n = 1$. The buckling check then also acts as the strength check.

Clause 4.8.3.3.2 gives a more exact approach, using a non-linear interaction surface. It allows for concave interaction for in-plane buckling using a formula:

$$M_{ay} = \frac{M_{cy}(1 - F/P_{cy})}{[1 + (0.5/P_{cy})F]}$$

This gives a curve as shown in Figure 24.23 for M_{ay}. Linear interaction is assumed for lateral–torsional buckling under M_x and P, using the formula:

$$M_{ax} = M_b\left(1 - \frac{F}{P_{cy}}\right)$$

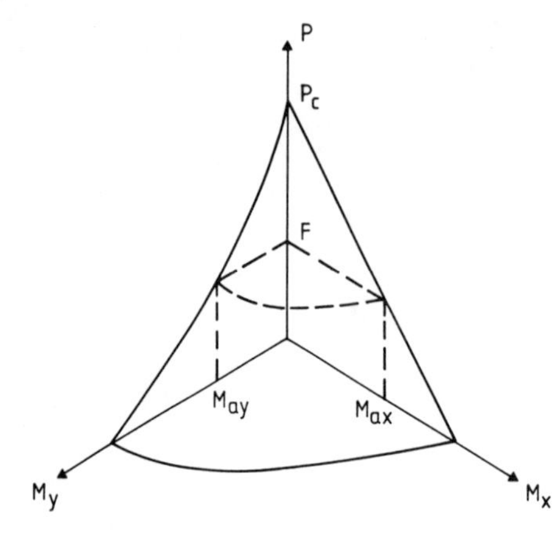

Figure 24.23 Beam column design strength with non-linear interaction between P, M_x and M_y

which gives the line for M_{ax}. The possibility of in-plane buckling under M_x is also covered by using the same formula as for M_{ay} as an alternative for M_{ax}. Then a convex curve is used for interaction between M_{ax} and M_{ay}, as shown in Figure 24.23 by using an equation:

$$\frac{M_x}{M_{ax}} + \frac{M_y}{M_{ay}} \leqslant 1$$

24.11 Concluding summary

1. Beam columns can either fail by reaching the ultimate strength of the cross-section under combined bending and axial load or by buckling, generally inelastically.
2. At lower axial loads beam columns can form localized plastic hinges, provided that they do not buckle. With I shapes under predominantly major axis bending the neutral axis will be near one flange, and this will normally remain elastic.
3. At higher axial loads plasticity tends to spread over considerable lengths of the column and is not concentrated at localized hinge positions.
4. Linear and non-linear interaction formulae are used in design to give conservative predictions of short-column strength. Moment-magnification effects can reduce beam column strength below short-column strength, particularly with sway columns.
5. Beam columns interact with the structure of which they are a part, and this must be considered in establishing their buckling loads. It is particularly important to determine sway restraint conditions.

Background reading

GALAMBOS, T.V. (1968) *Structural Members and Frames*, Prentice-Hall, Englewood Cliffs, NJ. Chapter 5: introduction to behaviour of beam columns and introduction to research and theory

JOHNSON, B.G. (ED.) (1976) *Guide to Stability Design Criteria for Metal Structures*, 3rd edn, Wiley Chichester. Chapter 8: introduction to beam column interaction curves

MARCUS, S.H. (1977) *Basics of Structural Steel Design*, Restor Publishing (Prentice–Hall). Chapter 9: brief introduction to beam columns and American design

TIMOSHENKO, S.P. and GERE, J.M. (1961) *Theory of Elastic Stability*, McGraw-Hill, Maidenhead

25

Plate girders

Objective To develop an understanding of the phenomena of web buckling and of the post-buckling reserve of strength derived from tension field action.

Prior reading Chapter 18 (Introduction to buckling: 1); Chapter 20 (Restrained compact beams); Chapter 21 (Local buckling).

Summary The chapter begins by outlining the primary design considerations for plate girders. It describes the behaviour of unstiffened webs and indicates how the strength of a slender web may be increased by the addition of transverse stiffeners. It outlines the analysis of post-buckling tension field action and explains how this is used as the basis for modern codes of practice. It briefly discusses stiffener design, longitudinal stiffening and patch loading on webs. For worked examples, see Appendix, p. 381

25.1 Structural action of plate girders

When large loads have to be supported over long spans the high bending moments and shearing forces developed may well exceed the capacity of universal beam sections. To carry such loads a plate girder may be fabricated with its proportions designed specifically for each situation, to reduce self-weight and provide a high strength/weight ratio. A typical girder is shown in Figure 25.1, where the components of the cross-section are identified as the top and bottom flange plates, the web plate and the longitudinal fillet welds connecting the web to the flanges.

In contrast to universal beam sections the component plates of a fabricated plate girder are often of slender proportions; this is particularly true of the web plate. The full plastic moment capacity of a typical plate girder cross-section cannot then be achieved because of local buckling of the web, but there is a post-buckling reserve of strength possessed by plate girder webs which enables them to carry shear loads considerably in excess of the load at which web buckling first occurs.

25.2 Outline of design procedure

In the fabricated plate girder the primary function of the top and bottom flange is to resist the axial compressive and tensile forces arising from the bending moment. In a symmetrical girder, the approximate magnitude of these flange forces may be obtained simply by dividing the moment applied to the cross-section by the girder depth. The major function of the web plate is to resist the shearing force. The average value of the shear stress is obtained by dividing the applied shear by the web area. The fillet welds connecting the component plates ensure the transfer of longitudinal shear from the web to the flange.

As indicated in Figure 25.1, any cross-section of a plate girder is normally subjected to a combination of shear force and bending moment. For example, panels 1 and 6, close to the

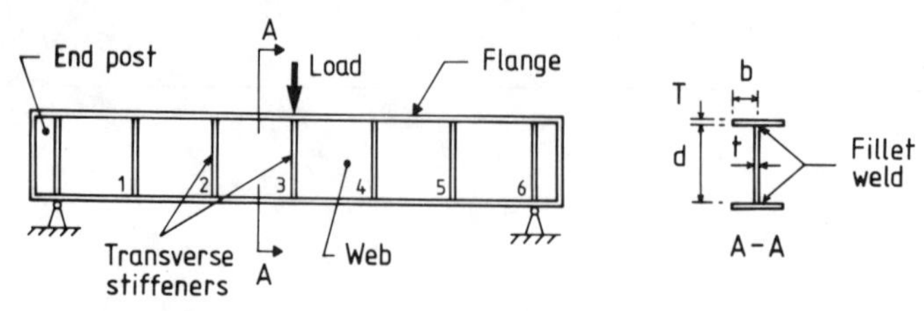

Figure 25.1 Elevation of a typical plate girder

supports, will be subjected primarily to shear, whereas in the central panels, 3 and 4, the effects of bending moment are dominant.

A simple and satisfactory design is achieved by assuming that the bending moment is resisted by the flanges alone and that the shear is resisted by the web. The design code for steel buildings, BS 5950, allows the separation of moment and shear effects in this way for simplification; this chapter shows the link between the basic theory and the relevant clauses of BS 5950. The design code for steel bridges, BS 5400, takes the interaction between moment and shear effects into account by means of a simplified interaction relationship which can only be discussed briefly here.

25.3 Proportions of girder cross-section

The proportions of the component flange and web plates may be chosen independently to ensure the optimum distribution of material within the cross-section. To avoid local buckling of the compression flange and consequent loss of effectiveness BS 5950 specifies that the outstand (b) should be limited to:

$$\frac{b}{T} \leqslant 13\sqrt{\frac{275}{p_{yf}}}$$

The flange then becomes 'semi-compact' and there is rarely a good reason for exceeding this limit in plate girder design.

The situation is, however, different in the case of the web plate. For an efficient girder design, the web depth (d) should be increased as far as possible, so as to give the lowest flange forces for a given bending moment. Also, to reduce self-weight, the web thickness (t) should be reduced to a minimum. The consequence of these two conflicting requirements is that webs are frequently slender. For example, BS 5950 permits the use of unstiffened webs, provided that the web slenderness ratio d/t does not exceed 250.

Such slender webs are prone to buckling at relatively low levels of applied shear. BS 5950 specifies that if the web slenderness ratio (d/t) exceeds $63\sqrt{(275/p_{yf})}$, as will normally be the case, then the effects of web buckling must be taken into account.

A plate girder of proportions appropriate to the majority of design situations may, therefore, be assumed to consist of semi-compact flange plates (which alone resist the bending moment) and a slender web (which carries the shear).

25.4 Introduction to design

25.4.1 General

The design of the flange plates to resist bending moments is similar to the procedure described for beams in Chapter 20 and will not be considered further; it is assumed here that sufficient restraints are provided to prevent lateral buckling of the compression flange. Attention will now be concentrated upon the more complex design of the slender web plate to resist the effects of shear.

The web of the typical plate girder shown is Figure 25.1 is reinforced by intermediate transverse stiffeners. As will be shown later, this transverse reinforcement can increase the ultimate load capacity of a girder significantly, but the introduction of stiffeners will obviously increase fabrication cost. Where the saving of self-weight is not of prime importance a designer may choose to use a thicker web plate rather than stiffeners. We shall consider the design of unstiffened webs first, before proceeding to discuss the post-buckling action of stiffened webs.

25.4.2 Unstiffened webs

The strength of a stocky web is, of course, limited to its yield strength. An unstiffened girder with a slender web may be designed simply by neglecting any post-buckling reserve of strength and assuming the shear capacity of the web to be equal to the shear buckling resistance (V_{cr}), given by:

$$V_{cr} = \tau_{cr} dt \qquad (25.1)$$

where the web has length (a), depth (d) and thickness (t).

The critical shear strength (τ_{cr}) may be taken as the elastic critical shear buckling stress ($N_{xy,cr}/t$) of the web panel, i.e:

$$\tau_{cr} = \frac{N_{xy,cr}}{t} = \frac{k\pi^2 E}{12(1 - \nu^2)}\left[\frac{t}{d}\right]^2 \qquad (25.2)$$

Taking the properties of steel ($E = 205$ kN/mm^2, $\nu = 0.3$) specified in BS 5950, and the value for the buckling coefficient k for a panel where $a/d > 1$, the expression for the critical shear strength becomes:

$$\tau_{cr} = \left[1 + \frac{0.75}{(a/d)^2}\right]\left(\frac{1000}{d/t}\right)^2 \text{ N/mm}^2 \qquad (25.3)$$

Of course, for an unstiffened web, the stiffener spacing (a) is infinite, so that the expression becomes:

$$\tau_{cr} = \left(\frac{1000}{d/t}\right)^2$$

25.5 Webs with intermediate transverse stiffeners

Webs of intermediate slenderness may be designed with transverse stiffeners. BS 5950 defines the maximum permissible slenderness ratio (d/t) for a transversely stiffened web to be $250\sqrt{(275/p_{yf})}$ to satisfy serviceability requirements. Webs of higher slenderness must be reinforced by longitudinal as well as transverse stiffeners. The introduction of transverse web stiffeners, as in Figure 25.1, increases the shear-carrying capacity of a plate girder in two ways. First, the shear buckling resistance of the web is increased; second, and more importantly, the stiffeners enable the girder to withstand loads considerably in excess of the load at which buckling occurs. This 'post-buckling' reserve of strength arises from the development of 'tension field' within the web plate, and advantage should be taken of this action in order to achieve an effective design.

Figure 25.2(a) shows the development of tension field action in the individual web panels of a typical girder. Once a web panel has buckled it loses its capacity to carry significant additional compressive stresses. In this post-buckling range a new load-carrying mechanism is developed whereby any additional shear load is carried by an inclined tensile membrane stress field. The tensile membrane field pulls against the top and bottom flanges and against the transverse stiffener on either side of the web panel, as shown. Adequate rigidity of these boundary members is essential to allow the post-buckling action to develop.

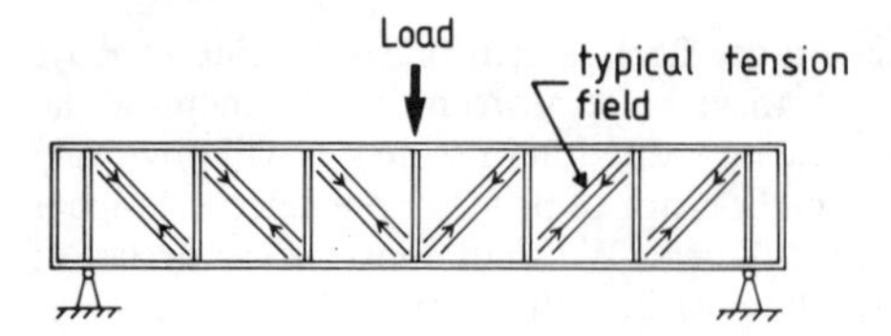

(a) Tension field action in individual sub panels as a result of transverse stiffeners

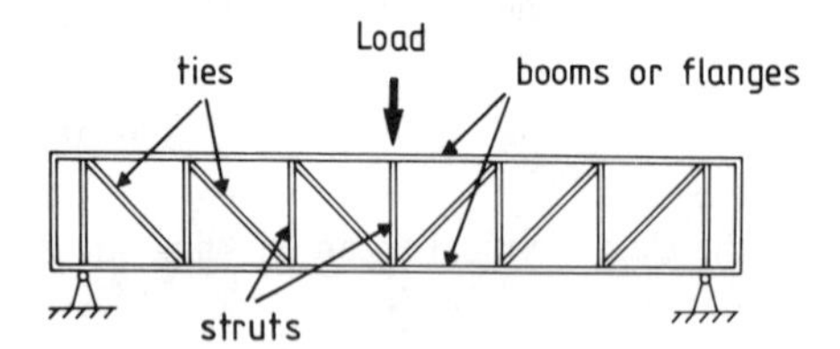

(b) Typical N-truss for comparison

Figure 25.2 Tension field action

The load-carrying action of the plate girder with buckled webs shown in Figure 25.2(a) is similar to that of the N-truss in Figure 25.2(b). In the post-buckling range the resistance offered by the web plates is analogous to that of the diagonal tie bars in the truss.

The influence of the transverse stiffeners in increasing the shear buckling resistance of the web is apparent from Equation 25.3. The introduction of more stiffeners will decrease the stiffener spacing (a), thereby increasing the elastic critical strength. If the stiffener spacing is reduced so that $a/d < 1$, then the expression for the buckling coefficient (k) changes to give the following expression for the critical shear strength:

$$\tau_{cr} = \left[0.75 + \frac{1}{(a/d)^2}\right]\left(\frac{1000}{(d/t)}\right)^2 \text{ N/mm}^2 \tag{25.4}$$

Values of τ_{cr} are tabulated in BS 5950 for different values of stiffener spacing (a/d) and web slenderness (d/t).

25.6 Analysis of post-buckling tension field action

25.6.1 Introduction

The post-buckling action that is developed as a result of the introduction of transverse stiffeners is more complex to analyse. This action will now be considered in some detail for one of the typical subpanels of the girder in Figure 25.2(a); such a panel is shown in Figure 25.3.

The behaviour of a plate girder subjected to an increasing shear load may be divided into three phases, as shown in Figure 25.3. Prior to buckling, equal tensile and compressive principal stresses are developed in the plate, as in Figure 25.3(a). In the post-buckling range an inclined tensile membrane stress field is developed, as shown in Figure 25.3(b). The magnitude of the tensile membrane stress is indicated by σ_t in Figure 25.3(b) and its inclination to the horizontal is shown as θ. Since the flanges of the girder are flexible they will begin to bend inwards under the pull exerted by the tension field. Further increase in the load will result in yield occurring in the web under the combined effect of the membrane stress field and the shear stress at buckling. The value of σ_t at which yield occurs is identified as the basic tension field strength (y_b).

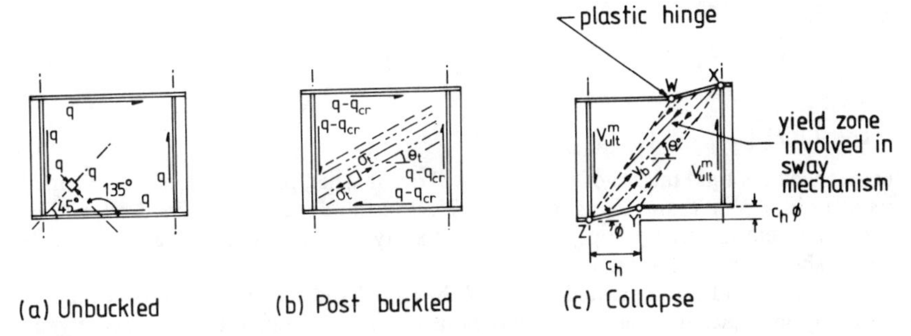

Figure 25.3 Phases in behaviour up to collapse of a typical panel in shear

Failure of the girder occurs when four plastic hinges form in the flanges of the girder, as shown in Figure 25.3(c). The resulting collapse mechanism then allows a shear displacement to occur as indicated.

25.6.2 Detailed analysis

In the post-buckling range, as shown in Figure 25.3(b), a tensile membrane stress (σ_t) will develop at an inclination (θ_t) to the horizontal. The total state of stress at this stage is obtained by superimposing these post-buckling stresses upon those developed at buckling (q_{cr}), as in Figure 25.3(a). By resolving these stresses in the direction along (θ_{cr}) and perpendicular ($\theta_t + 90$) to the inclination of the tension field the total stress state (see Figure 25.4) may be expressed as:

$$\sigma_{\theta_t} = q_{cr} \sin 2\theta_t + \sigma_t$$

$$\sigma_{(\theta_t+90)} = -q_{cr} \sin 2\theta_t$$

$$\tau_{\theta_t} = q_{cr} \cos 2\theta_t$$

Under an increasing applied load the post-buckling stress (σ_t) will increase until, eventually, the tensile membrane stress ($\sigma_{\theta t}$) will reach the design strength (p_{yw}) of the web material. The value of σ_t required to produce yield is identified as the basic tension field strength (y_b). It may be determined by applying the Von Mises–Hencky yield criterion:

$$\sigma^2_{\theta_t} + \sigma^2_{(\theta_t+90)} - \sigma_{\theta_t}\sigma_{(\theta_t+90)} + 3\tau^2_{\theta_t} = p^2_{yw}$$

to give:

$$y_b = [p^2_{yw} - 3q^2_{cr} + 2.25q^2_{cr} \sin^2 2\theta_t]^{1/2} - 1.5q_{cr} \sin 2\theta_t \qquad (25.5)$$

The inclination (θ_t) of the tension field cannot be determined other than by an iterative procedure. For simplicity, it is assumed in BS 5950 that the inclination is half that of the panel diagonal, i.e:

$$2\theta = \tan^{-1} (d/a)$$

This assumption will result in a conservative estimation of the ultimate shear capacity and the following expression for the basic tension field strength:

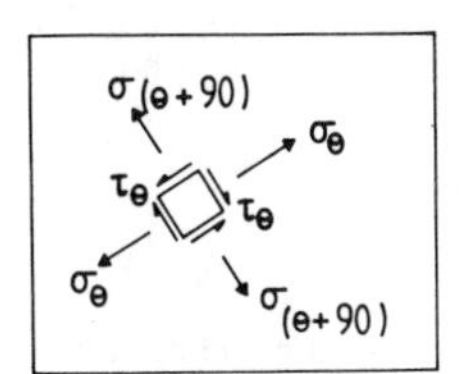

Figure 25.4 State of stress in web in post-buckling phase

$$y_b = (p_{yw}^2 - 3q_{cr}^2 + \phi_t^2)^{1/2} - \phi_t$$

where

$$\phi_t = \frac{1.5_{cr}}{\sqrt{[1 + (a/d)^2]}}$$

Once the web has yielded, final failure of the girder will occur when plastic hinges have formed in the flanges at points WXY and Z, as in Figure 25.3(c); the distance is defined by c_h. The failure load may be determined by applying a virtual sway displacement to the girder in its collapse state, as shown in the figure.

It is convenient to consider the yielded region WXYZ of the web to be removed and to replace its action upon adjacent flange and web regions by the tensile membrane stress at yield (y_b) as in Figure 25.5. These stresses have a resultant force F_{xy}, given by:

$$F_{xy} = y_b t \sin \theta_t (d \cot \theta_t - a + c_h)$$

During the imposed virtual displacement shown in Figure 25.3(c) this resultant force, together with the force (V_{ult}^m) shown in Figure 25.5(a) as the post-buckling shear load that causes the mechanism to develop, will create external work:

$$\text{external work} = (V_{ult}^m - F_{xy} \sin \theta_t) c_h \phi$$

$$= [V_{ult}^m - y_b t \sin^2 \theta_t (d \cot \theta_t - a + c_h)] c_h \phi$$

From the principle of virtual work this external work must be balanced by the internal work done at the four plastic hinges, i.e. $4 M_{pf} \phi$, so that the value of V_{ult}^m may be determined.

The total shear buckling resistance (V_b) is then obtained as the summation of the initial buckling load (V_{cr}) and the post-buckling load (V_{ult}^m), as shown below:

$$V_b = V_{cr} + V_{ult}^m = q_{cr} dt + y_b t \sin^2 \theta_t (d \cot \theta_t - a + c_h) + \frac{4M_{pf}}{c_h} \tag{25.6}$$

This equation is not yet in a suitable form, since it contains the term c_h, representing the position of the plastic hinges in the flanges; the hinge position is not yet known. It may be determined by considering the equilibrium of the flange, and a free-body diagram of the portion WX as shown in Figure 25.5(b). Remembering that the internal hinge will form at the point of maximum moment, and consequently zero shear, taking moments about X gives:

$$c_h = \frac{2}{\sin \theta_t} \sqrt{\frac{M_{pf}}{y_b t}}$$

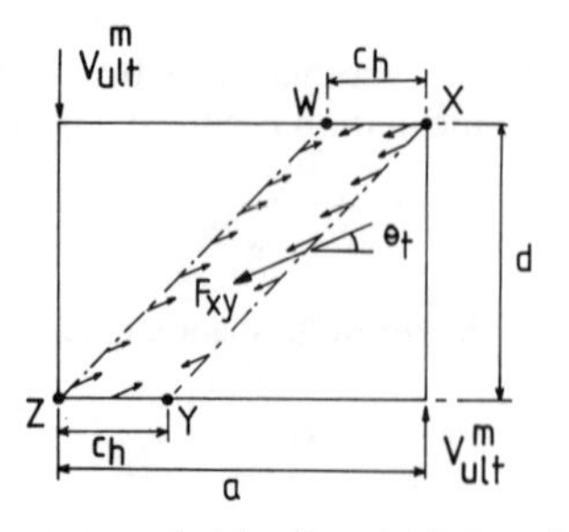

Figure 25.5 Forces in typical web panel at collapse

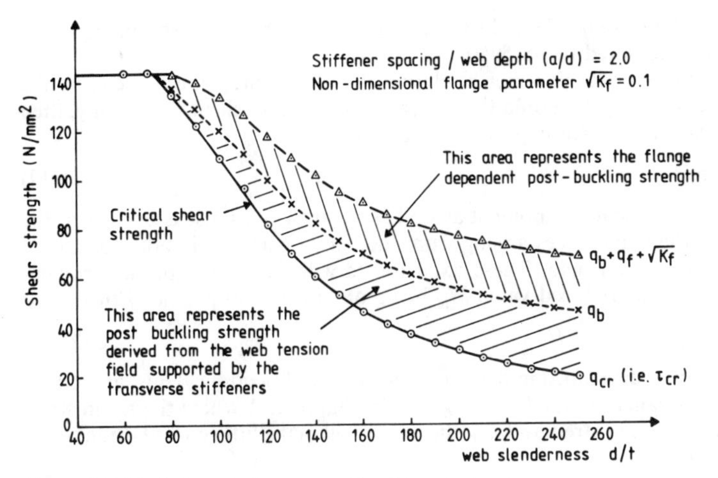

Figure 25.6 Variation of shear strength with web slenderness

Substituting into Equation 25.6 gives:

$$V_b = q_{cr}dt + y_b\sin^2\theta_t(\cot\theta_t - a/d)\,dt + 4\sin\theta_t\sqrt{(P_{yw}y_b)}\sqrt{(K_f dt)} \tag{25.7}$$

where

$$K_f = \frac{M_{pf}}{4M_{pw}}$$

Hence:

$$V_b = (q_b + q_f\sqrt{K_f})dt \tag{25.8}$$

where

$$q_b = q_{cr} + \frac{y_b}{2\{a/d + \sqrt{[1 + (a/d)^2]}\}} \tag{25.9}$$

and

$$q_f = \left[4\sqrt{3}\sin\left(\frac{\theta_t}{2}\right)\sqrt{\frac{y_b}{p_{yw}}}\right]0.6\,p_{yw} \tag{25.10}$$

25.7 Code of practice for buildings: BS 5950

The expression for the shear resistance (V_b) of a girder with transverse stiffeners is derived and given in Equation 25.7. In BS 5950 this equation is written as:

$$V_b = (q_b + q_f\sqrt{K_f})dt \tag{25.11}$$

where q_b represents the basic shear strength of the web panel as defined in Equation 25.9. This term combines the critical shear strength of the panel (τ_{cr} from Equation 25.2) and the post-buckling strength derived from that part of the web tension field supported by the transverse stiffeners. Values of the basic shear strength are tabulated in BS 5950 for different values of stiffener spacing (a/d) and web slenderness (d/t) and for different grades of steel.

The second term within the bracket in Equation 25.11 (i.e. $q_f\sqrt{K_f}$) represents the contribution made by the flanges to the post-buckling strength. The flanges support that part of the web tension field that pulls against them, as in Figure 25.3(b), and finally develop plastic hinges at collapse. The term q_f is a flange-dependent shear strength factor which is given in Equation 25.10. This factor is again tabulated in BS 5950.

The K_f term is a non-dimensional parameter relating the plastic moment capacity of the flange (M_{pf}) to that of the web (M_{pw}). Should the girder be subjected to shear forces only, then M_{pf} is taken as the full plastic moment capacity of the flange, i.e.

$$M_{pf} = 0.25\,(2b)(T^2)p_{yf} \qquad (25.12)$$

When a girder is subjected to a bending moment as well as shear then, as discussed earlier, the flanges will have to carry the axial forces arising from the bending action, in addition to the lateral pull exerted by the tension field. These axial forces will reduce the plastic moment capacity of the flange plates, the reduced value M'_{pf} being obtained from plasticity theory as:

$$M'_{pf} = M_{pf}\,(1 - f/p_{yf}) \qquad (25.13)$$

where f is the mean axial stress developed in the flange as a result of bending. Thus, at sections of high moment, such as in panels 3 and 4 of the girder in Figure 25.1 where the mean stress (f) may approach the flange design strength (p_{yf}), the flange will make little contribution to the shear buckling resistance (V_b).

The relative importance of the various terms in Equation 25.11 is illustrated for a typical case in Figure 25.6. In this the various components of shear strength are plotted against web slenderness for $a/d = 2.0$ and $\sqrt{K_f} = 0.1$.

25.8 Code of practice for steel bridges: BS 5400

For the design of steel bridges the appropriate code BS 5400 adopts a similar tension field approach to evaluate the shear resistance of transversely stiffened girders. Although the basic theory is similar, its presentation in the two codes is very different. Whereas tables are given in BS 5950, design curves are provided in BS 5400.

25.9 Design parameters

The expression for V_b, given in Equation 25.11, separates the contribution made by the web (q_b) from that made by the flange. It shows that the important non-dimensional parameters are:

1. Web-aspect ratio (a/d);
2. Web-slenderness ratio (d/t);
3. Flange strength parameter (K_f).

With increasing values of web-aspect ratio and, particularly, of web slenderness, the critical shear strength (q_{cr}) decreases so that the post-buckling action becomes more significant. This is reflected in a relative increase in the magnitude of the basic tension field strength (y_b) and of the second term of the expression given in Equation 25.9 for the basic shear strength (q_b) of the web panel, as illustrated in Figure 25.6.

As the flange strength parameter increases, the second term in the expression for V_b, given in Equation 25.11, becomes increasingly important.

25.10 Further design considerations

Only the major design considerations have been discussed: there are others which can only be mentioned now but which are dealt with more fully in Narayanan (1983, Chapter 1).

It is apparent from Figure 25.2(a) that the intermediate transverse stiffeners play an important role. They must increase the buckling resistance of the web; therefore they need adequate stiffness, and they must also be of sufficient strength to withstand the forces imposed upon them by the post-buckling tension field action. From the analogy with the N-truss in Figure 25.2(b) the behaviour of the stiffeners is seen to be similar to that of the vertical compression members. The stiffeners are, indeed, designed as struts, assuming that a certain width of web is effective in acting with the stiffener to resist the axial loading. Those stiffeners at the ends of the girder, i.e. the 'end-posts' as defined in Figure 25.1, must provide adequate anchorage for the tension field in the end-panels. Since the web plate does not usually extend beyond these stiffeners, more stringent requirements are imposed upon their design.

The web can be further strengthened by introducing longitudinal stiffeners to increase the buckling resistance. However, as in the case of transverse stiffeners, the increased strength must be balanced against the additional fabrication costs of welding the stiffeners into position. The design of all these stiffening elements must be considered very carefully (see Narayanan, 1983, Chapter 4).

A further point to consider is the possibility of buckling of the web under randomly placed point loads. There are situations (for example, when a plate girder carries a travelling crane or when a bridge girder is being launched) when loads are applied at positions between transverse stiffeners. The additional design considerations for such a situation are discussed fully in Narayanan (1983, Chapter 3). In some cases webs may also contain openings for inspection and services, and the effect of these upon the shear capacity of the girder is discussed in Narayanan (1983, Chapter 2).

The co-existent effects of bending moments and shearing forces can be conveniently taken into account by means of the simplified interaction diagram shown in Figure 25.7. It has been shown that the flanges contribute to the shear capacity; similarly, the webs contribute to the bending strength. However, tests have shown that if the shear and bending capacities are calculated assuming no contribution from the flanges and webs, respectively, then a girder can carry these magnitudes of bending and shear acting simultaneously.

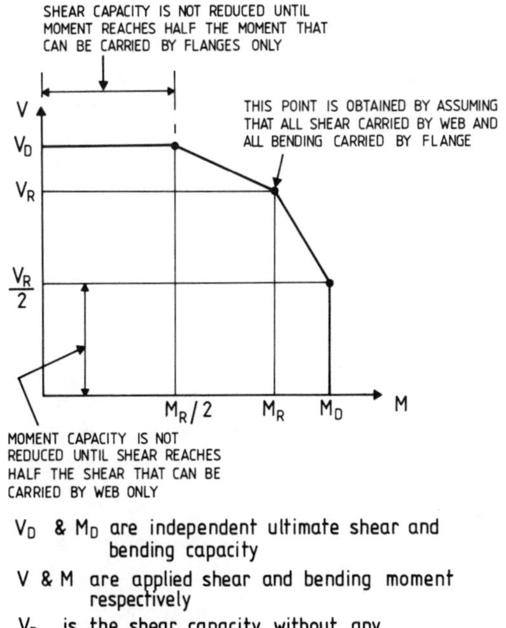

Figure 25.7 Simplified interaction diagram for shear and bending moment

25.11 Concluding summary

1. A plate girder may be designed conveniently and safely assuming that the bending moment is resisted by the flanges and that the shear force is resisted by the web.
2. Normally, effective design requires the flanges to be semi-compact and the web to be slender.
3. The procedure for flange design is similar to that for rolled beams.
4. The web may be transversely stiffened to allow the development of post-buckling tension field action.
5. The shear buckling resistance (V_b) may be obtained from Equation 25.11: this indicates the separate contributions made by the web and the flanges.

Background reading

NARAYANAN, R. (ed.) (1983) *Plated Structures, Stability and Strength*, Applied Science Publishers, London.
Chapter 1: Longitudinally and transversely reinforced plate girders (Evans, H.R.)
Chapter 4: Optimum rigidity of stiffeners of webs and flanges (Skaloud, M)
Chapter 3: Patch loading on plate girders (Roberts, T.M.)
Chapter 2: Ultimate shear capacity of plate girders with openings in their webs (Narayanan, R.)

26

Cold-formed sections and sheeting

Objective To illustrate the potential uses and advantages of cold-formed sections and sheeting and to outline the respective design procedures.

Prior reading Chapter 18 (Introduction to buckling: 1); Chapter 21 (Local buckling).

Summary This chapter describes cold-formed sections and their manufacture. It reviews behaviour and associated design procedures for overall response and local effects and goes on to describe the uses of sheeting. The design procedure for simple trapezoidal profiles is outlined.

26.1 Cold-formed sections

26.1.1 Introduction

The essential differences between cold-formed sections and their hot-rolled counterparts are that the former can be manufactured from very thin material, are generally of uniform thickness throughout the cross-section and can be held to very close dimensional tolerances. Sections can be formed to almost any required shape to obtain very strong, stiff, structural members of low weight and high efficiency. Examples of cold-formed section profiles are shown in Figure 26.1. Very often sections are formed from material that has been previously galvanized, or plastic coated, to form an attractive product without the need for additional protection against corrosion.

The benefits of using slender cross-sections are illustrated in Figure 26.2, which shows some elastic cross-sectional properties for a standard hot-rolled channel compared with those of corresponding cold-formed steel channels of equal weight and formed from different sheet thicknesses. As can be seen, decreasing the material thickness greatly increases the cross-sectional properties for a volumne of material.

26.1.2 Methods of manufacture

Cold-formed sections are manufactured using a variety of techniques, including folding, press braking and rolling. Cold-rolling accounts for the great majority of commercially produced sections. Using this technique, the strip is formed gradually by feeding it continuously through successive pairs of rolls which progressively alter the cross-section until the finished shape is obtained. The thickness of material used in cold-formed construction lies mainly in the range of 1–3.2 mm, this being the thickest material which can be galvanized before forming, but much thicker material can be formed and 8 mm is considered a reasonable upper limit to the range for most purposes. The average yield strength of steel used in cold-formed sections is 280 N/mm^2, although there is a trend towards the use of steels with higher yield strength.

26.1.3 Design considerations

In the UK the design of cold-formed steel sections is covered by Part 5 of BS 5950.

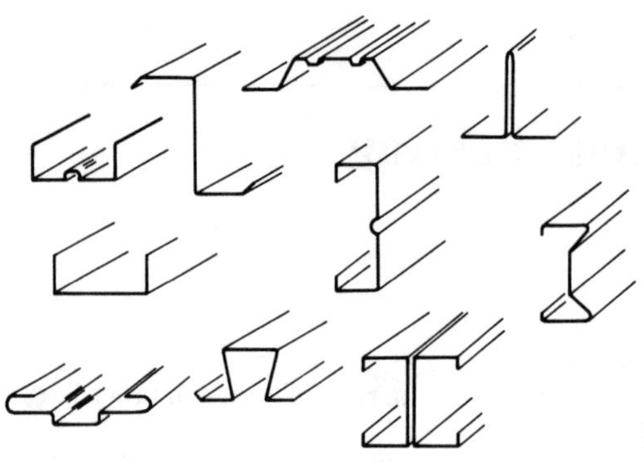

Figure 26.1 Typical cold-formed section profiles

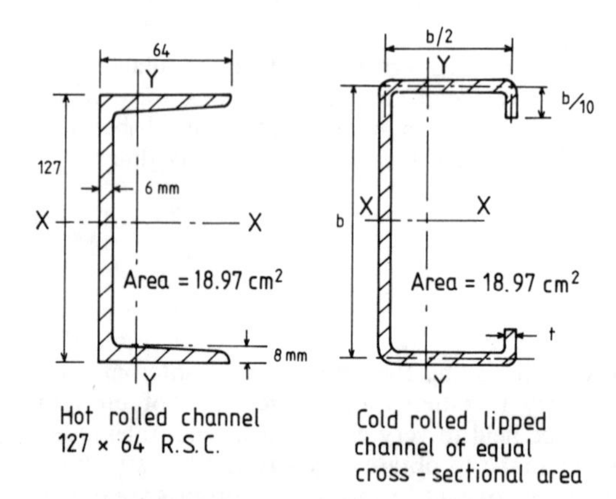

	HOT ROLLED CHANNEL	COLD ROLLED CHANNEL				
		t = 6 mm b = 143 mm	t = 5 mm b = 172 mm	t = 4 mm* b = 215 mm	t = 3 mm* b = 287 mm	t = 2 mm* b = 431 mm
I_{xx} cm^4	482.5	666	959	1498	2664	5995
Z_{xx} cm^3	76	92.7	111.2	139	185	278
I_{yy} cm^4	67.2	138.3	199.2	311	553	1245
min Z_{yy} cm^4	15.3	28.2	33.8	42.3	56.3	84.5

* THESE ARE GROSS SECTION PROPERTIES, EFFECTIVE PROPERTIES WILL BE LESS.

Figure 26.2 Comparison of hot-rolled and cold-formed steel channels

26.1.3.1 Local buckling

In design, the effects of local buckling are taken into account using the concept of effective width, as illustrated in Figure 26.3. Loads on very thin elements are assumed to be resisted only by effective portions near the supported edges, and the contribution of the heavily buckled centre portion is neglected.

The effective width of a stiffened element (an element which is supported on two edges) is obtained from the expression:

$$\frac{b_{\text{eff}}}{b} = \left\{1 + 14\left[\sqrt{\left(\frac{\sigma_{\text{cr}}}{\sigma_{\text{E}}}\right)} - 0.53\right]^4\right\}^{-1/5} \qquad (26.1)$$

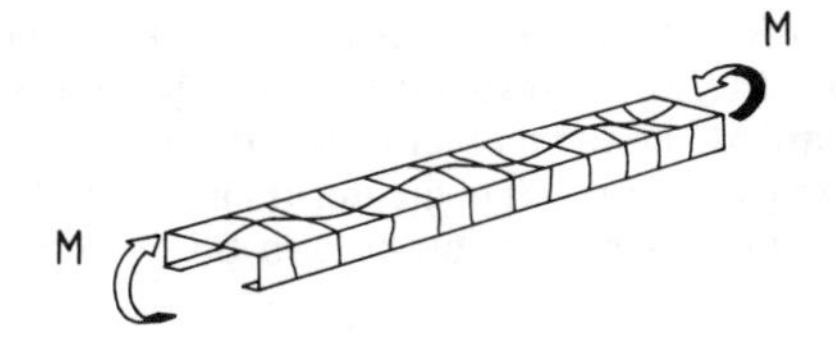

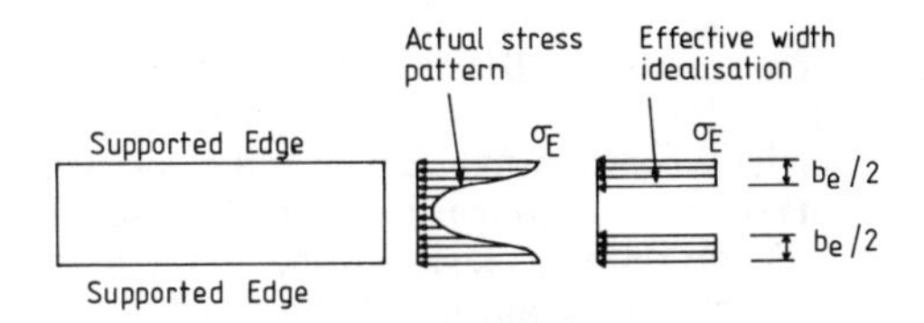

Figure 26.3 Local buckling of a beam-compression element

where b_{eff} = the effective width of the element, b = its full width, σ_{cr} = the local buckling stress, σ_E the stress at the edges of the element, assumed constant over the effective width.

Because cold-formed sections are mainly open ones they generally contain some elements which are supported on one edge only. These are termed 'unstiffened elements' and are illustrated in Figure 26.4(a).

For an unstiffened element, local buckling will become a significant problem if the width-to-thickness ratio is greater than about 12; the effective widths of such elements are much less than those of stiffened ones of the same actual width. For this reason, edge-stiffeners are often used to convert otherwise unstiffened elements into stiffened ones. The simplest form of edge stiffener is the 90° lip shown in Figure 26.4(b). For very thin-walled sections, lips can themselves be subject to local buckling as unstiffened elements, and in order to avoid this complication compound lips, also illustrated in Figure 26.4(b), are sometimes used.

Another type of stiffener is the formed intermediate stiffener (Figure 26.4(c)) for elements of high width-to-thickness ratio. This type of stiffener has the effect of making a very slender, highly ineffective, element behave as if it were two less slender elements of greater effectiveness.

26.1.4 Beam behaviour

In beams the effects of local buckling are largely confined to the compression flange or flanges.

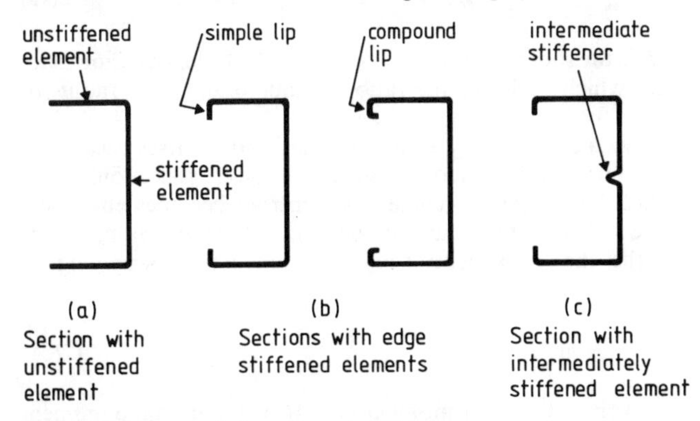

Figure 26.4 Types of elements and stiffeners

The first step in the analysis of their moment capacity is to evaluate the effective width of the compression elements at the design stress. Thereafter, the section properties of the beam are calculated with the ineffective parts completely disregarded and the moment capacity evaluated on this basis. For beams with webs of high depth-to-thickness ratio (i.e. greater than about 80) local buckling effects in the webs can become significant. The expression used in BS 5950 for the maximum compressive stress can be written:

$$\sigma_{max} = \left[1.13 - 0.0019\sqrt{\left(\frac{\sigma_y}{280}\right)} \cdot \frac{D}{t}\right]\sigma_y \quad \text{but } \sigma_{max} \not> \sigma_y \tag{26.2}$$

where D and t are the web depth and thickness, respectively. As D/t increases, the maximum stress becomes progressively less than the yield strength σ_y.

For beams at the heavier end of the range, which have compression element width-to-thickness ratios less than about 25 for stiffened elements (or 8 for unstiffened elements), design on a fully plastic basis is realistic. As the width-to-thickness ratio increases beyond about 40 (or 13 for unstiffened elements) then plasticity in the compression elements induces rapid failure, although plasticity on the tension side can be tolerated.

Other buckling problems which must be considered in cold-formed beam design are as follows:

1. *Web crushing*: at supports or points of concentrated load;
2. *Shear buckling* of webs. This type of buckling generally occurs in short deep beams with thin webs;
3. *Lateral buckling*: sometimes called lateral–torsional buckling. This type of buckling is important for long, deep, open section beams which are bent about their major axis. In BS 5950: Part 5 design formulae are given for the analysis of lateral buckling of I, T, Z and C sections.

26.1.5 Columns

26.1.5.1 Flexural buckling

In assessing the strength of cold-formed steel columns the interaction of local and overall buckling must be taken into account. The short strut failure load is first evaluated by summing the strengths of each individual element, taking account of loss of effectiveness, to obtain a 'smeared' uniformly reduced effectiveness of the cross-section. This short-strut failure load is then treated as a squash load, and a Perry–Robertson formulation is used to obtain the appropriate interaction between failure of the partially ineffective short strut and the behaviour as a fully effective long column. The interaction equation is of the form:

$$P_c = \frac{1}{2}\left[[P_s + (1 + \eta)P_E] - \sqrt{\{[P_s + (1 + \eta)P_E]^2 - 4P_sP_E\}}\right] \tag{26.3}$$

where P_c is the ultimate axial load, P_s is the short-strut failure load, P_E is the Euler buckling load, and η is an imperfection parameter which varies as the ratio of column length to radius of gyration.

For columns which have only one axis of symmetry a further complication arises because of the different degrees of loss of effectiveness in the different elements of the cross-section. When local buckling occurs a column loaded through its geometrical centroid becomes effectively eccentrically loaded, and is therefore subjected to combined bending and compression. This is taken into account by evaluating the change in neutral axis position, e_s, and applying an interaction expression of the form:

$$P_{ult} = \frac{P_c}{[1 + (P_c e_s/M_c)]} \tag{26.4}$$

where P_c is the ultimate axial load in the absence of moment and M_c is the ultimate moment in the absence of axial load.

26.1.5.2 Torsional–flexural buckling

In columns this is related to lateral buckling of beams and is characterized by a combination of bending and twisting in the column during buckling. A simple device is employed to take torsional–flexural behaviour into account for a variety of sections without requiring intricate analysis by the designer. This consists of an effective length multiplication factor (α) which sets the torsional–flexural buckling load of the member under examination with length (L) equal to the flexural buckling load of a column of length (αL). If the value of α for a particular case is known then this problem is simply treated as a flexural one. Values of α for a variety of section geometries are tabulated in the British Standard.

26.1.6 Connections

Where applicable, the design rules for connections in cold-formed sections are in broad agreement with those for hot-rolled steelwork. There are, however, some areas of divergence. For example, the strength of bolted connections in bearing is generally found for cold-formed steel to be significantly higher than would be obtained on the basis of hot-rolled steel design rules.

26.1.7 Testing

Testing is widely used by manufacturers of mass-produced components, such as purlins, as a basis for load tables which can be used by potential customers. The use of such load tables provides an alternative design approach for the non-expert who wishes to obtain the benefits of cold-formed sections without a detailed knowledge of the design complexities. To use these tables the designer simply sets the purlin centres and span and selects a suitable section on the basis of the safe loads given in the tables.

26.1.8 Concluding summary

1. Cold-formed steel sections can often offer advantages over hot-rolled ones with regard to aesthetics, functionality and strength-to-weight ratio.
2. The use of thin material produces design complexities such as local buckling, and design in the post-local buckling range is essential for the optimum use of material.
3. The design rules for structural members such as beams and columns take local buckling into account using the effective-width concept.
4. There is widespread use of testing as a basis for the production of load tables which can thereafter be employed to facilitate simplified design analysis.

26.2 Sheeting

26.2.1 Introduction

In a typical sheeted industrial building the breakdown of building costs would show that the value of the cold-rolled steel members and the steel sheeting in the walls and roof exceeds the cost of the hot-rolled steelwork. However, it is safe to assume that the latter accounts for nearly all the design effort in the office.

The design approach to mass-produced components such as sheeting is often on two levels; that of the user and that of the manufacturer. The user can simply select the correct profile from tables prepared by the manufacturer, but the latter has to carry out detailed calculations, verified by tests, in order to ensure that the design tables are as economic as possible, considering the number of spans, the safe load and the permissible deflection.

26.2.2 Types of sheeting

Profiled steel sheeting is used in quite different functions in the roofs and walls of buildings as follows:

1. *Roof sheeting*. When the sheeting is used as the outer skin of a pitched roof, with the insulation inside, it is known as roof sheeting or 'cold roof' construction. In this situation weather-tightness, durability and insulation are all-important. Common terms in sheeting terminology are shown in Figure 26.5. Insulation is usually effected by means of a glass fibre quilt held in place by a thin gauge steel liner panel fitted over the purlins. It has been found that the most economic spans in practice are of the order of 2 or 3 m.

   The simple trapezoidal profile illustrated in Figure 26.6(a) is complemented by profiles with rolled-in stiffeners (Figure 26.6(b)), which are more efficient in bending, and by profiles with a very wide bottom flange, as shown in Figure 26.6(c), which are efficient in shedding water in large-span buildings with low roof pitches.

   The steel quality of many sheeting profiles is about 250 N/mm^2, but an increasing number of manufacturers are using steel with a yield stress of up to 350 N/mm^2 and even 550 N/mm^2. Under these conditions the minimum acceptable thickness of sheeting is no longer regarded as being 0.7 mm, and thicknesses down to about 0.5 mm have been used in agricultural buildings.

   Because roof sheeting may be continuous over several spans the maximum span may be limited by strength or deflection. The usual deflection limitation is span/100 to span/200. Wind suction, particularly in local areas such as the eaves, gable or apex, is likely to be a more important design case than dead plus imposed load, and due attention must be given to the design of fixings. To allow for loads incidental to maintenance, the sheeting must be able to sustain a concentrated load.
2. *Roof decking*. If the insulation and waterproof membrane are placed on top of the sheeting then the sheeting is termed 'decking' and the method is known as 'warm roof' construction. It is usually reserved for flat or nearly flat roofs.
3. *Wall cladding*. For many applications the profiles used for roof sheeting are also suitable for vertical cladding. In many respects the loading is similar, except that the requirement for the profile to sustain a concentrated point load no longer applies. The insulation arrangements are also similar to those for roof sheeting.

26.2.3 Design considerations

26.2.3.1 Codes of practice

European Recommendations for the design of profiled sheeting have been produced by the European Convention for Constructional Steelwork (ECCS). BS 5950: Part 6, 'Code of Practice for Design of Light Gauge Sheeting, Decking and Cladding' is in preparation.

26.2.3.2 Detailed calculation procedures for trapezoidal profiles

For trapezoidal profiles in bending, a single corrugation is taken out and considered as a beam,

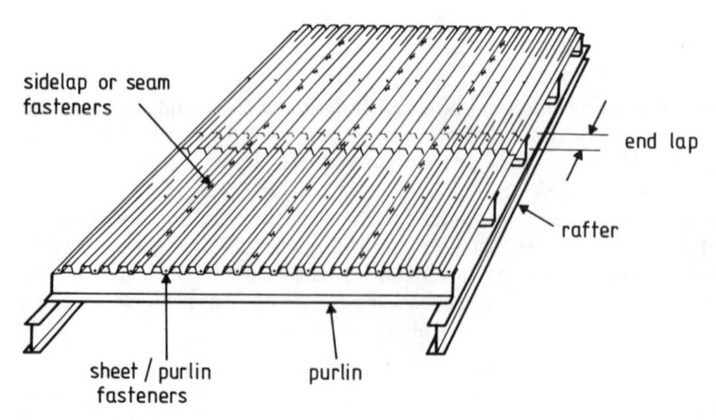

Figure 26.5 Terms used in sheeting

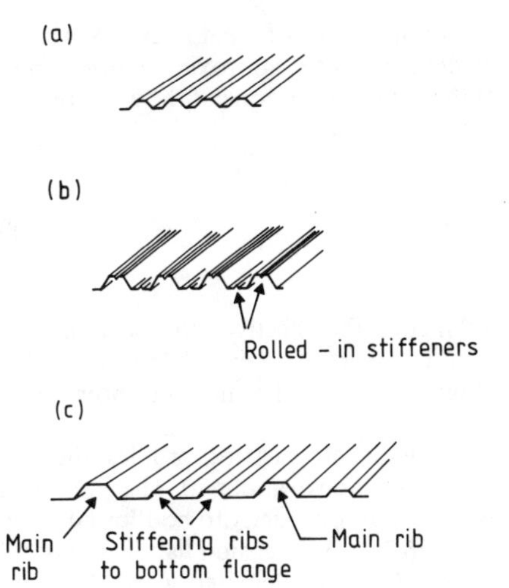

Figure 26.6 Roof-sheeting profiles

as shown in the basic analytical model (Figure 26.7). The effective width (b_{eff}) of the compression flange is calculated using the following expressions:

$$b_{eff} = b \qquad \text{when } \lambda \leqslant 1.27 \tag{26.5}$$

$$b_{eff} = \frac{1.9b}{\lambda}\left(1 - \frac{0.42}{\lambda}\right) \qquad \text{when } \lambda > 1.27 \tag{26.6}$$

where

$$\lambda = \frac{2}{\sqrt{k}} \frac{b}{t} \sqrt{\frac{p_y}{E}} \tag{26.7}$$

It may be shown that the flange is fully effective (i.e. $b_{eff} = b$) when the b/t ratio is less than about 36 for steel with a design yield stress of 250 N/mm^2 or 31 for steel with a design yield stress of 350 N/mm^2. In the above expressions b = flat width of compression flange (mm), E = modulus of elasticity (N/mm^2), k = buckling coefficient (unless otherwise given, k = 4

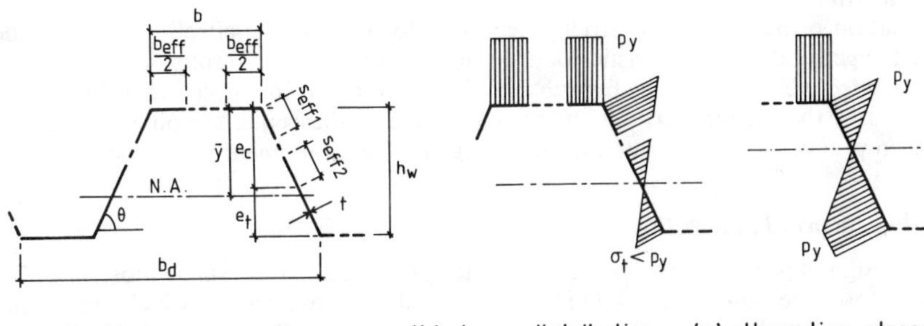

Figure 26.7 Analytical model for trapezoidal profiles in bending

may be used), t = net thickness of sheet (mm), p_y = design value of yield stress (N/mm^2). With the compression flange reduced in this way, the depth e_c to the approximate position of the neutral axis is calculated. The effective portions of the compressive zone of the web are then positioned as shown in Figure 26.7(a) and have lengths given by:

$$s_{eff1} = 0.76\sqrt{\frac{E}{f_1}} \qquad (26.8)$$

$$s_{eff2} = 1.5 s_{eff1}$$

where f_1 is the stress at the level of the compression flange in the effective cross-section.

Using the doubly reduced cross-section, the new section properties are calculated so that, for first yield in the compression flange as shown in Figure 26.7(b), the ultimate moment of resistance (M_d) may be found.

In the above calculation it can be shown that, for reasonably symmetrical profiles, the web is fully effective unless the ratio of profile height to thickness is above about 70. The European Recommendations also make provision for an increase of bending capacity to be allowed when first yield occurs in the tension flange as shown in Figure 26.7(c). The tension flange is allowed to become plastic and the ultimate moment of resistance (M_d) is calculated when the compression flange also reaches yield. If advantage is taken of this concession, f_1 is always equal to the design yield stress of the material (p_y).

For continuous sheeting, where the bending moment changes sign over the support, the above calculation has to be repeated in the case of unsymmetrical sheeting. The ultimate moment of resistance over the support will not generally be the same as that near midspan.

The European Recommendations also require that the support reaction capacity (R_d) is calculated and that account is taken of the interaction between moment and reaction when sheeting is continuous over an intermediate support.

The value of R_d at which web crippling takes place at a support or under a concentrated load has been found from tests to be:

$$R_d = 0.15t^2\sqrt{(Ef_y)}[1 - 0.1\sqrt{(r/t)}][0.5 + \sqrt{(0.02l_s/t)}][2.4 + (\theta/90)^2] \qquad (26.9)$$

where t = net sheet thickness (mm), E = modulus of elasticity (N/mm^2), f_y = yield stress (N/mm^2), r = inside radius of bends ($r \leqslant 10t$), l_s = length of bearing ($10\,\text{mm} < l_s < 200\,\text{mm}$), θ = web slope ($50° < \theta \leqslant 90°$).

The relationships between the moment (M) and reaction (R) at the design load and the above ultimate moment and reaction capacities (M_d and R_d) must satisfy the interaction relationships at an internal support (or concentrated load) given in Figure 26.8.

26.2.3.3 Deflections

For the calculation of deflection it is strictly necessary to consider the effective width of the compression flange at the characteristic load, although conservative results will always be obtained if the effective width in the strength calculation is taken. It is noted that the second moment of area of the section varies with the magnitude of the bending moment along the member, but in practice a single value for an averaged bending moment will suffice.

26.2.4 Testing of profiled sheeting

Although the design of profiled sheeting is carried out by calculation it is usually most desirable to test as well in order to observe the actual behaviour and to obtain data on which to base the design assumptions. The European Recommendations for the testing of profiled metal sheets specify tests for determining the bending capacity of the sheeting at midspan and over a support, the capacity of the sheet under concentrated line load or reaction and the capacity under a concentrated point load.

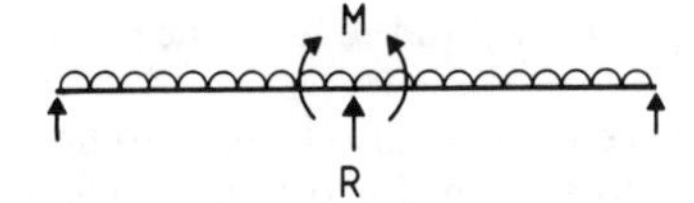

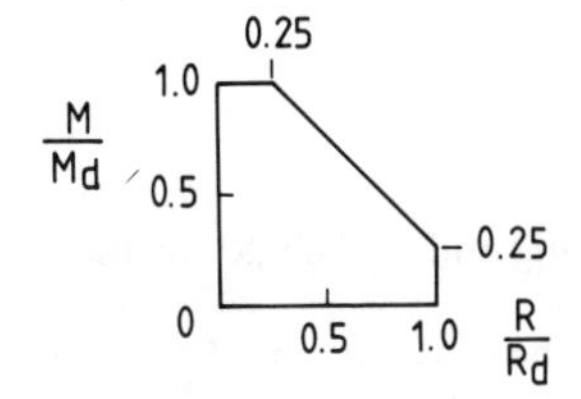

$$\frac{M}{M_d} \leqslant 1.00 \text{ when } \frac{R}{R_d} \leqslant 0.25$$

$$\frac{M}{M_d} + \frac{R}{R_d} \leqslant 1.25 \text{ when } 0.25 < \frac{R}{R_d} \leqslant 1$$

Figure 26.8 Empirical interaction diagram for permissible design values of moment M and support reaction R at an internal support

26.2.5 Structural benefits of sheeting

In addition to providing a membrane to exclude the weather, steel sheeting can produce important structural benefits to the steel framework of buildings. These are considered under the general term 'stressed skin design'. In rectangular framed buildings, as in Figure 26.9(a), the roof decking acts as the web of a deep plate girder, taking transverse wind loads back to the sheeted end-gables. It also provides lateral restraint to the main beams and can be designed to take the wind load on the end-gables so that horizontal bracing in the plane of the roof is

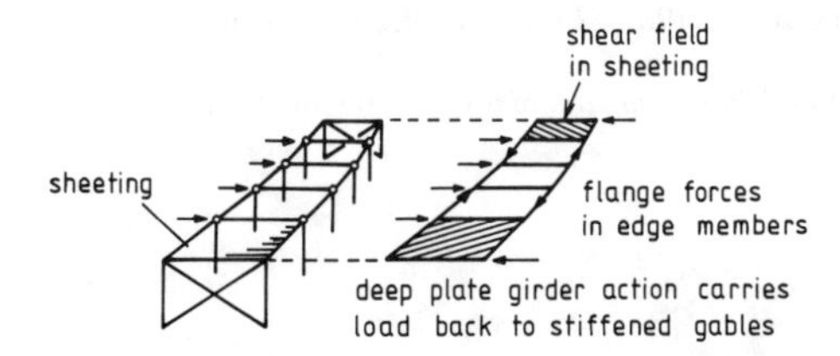

(a) Diaphragm action in a flat roofed building

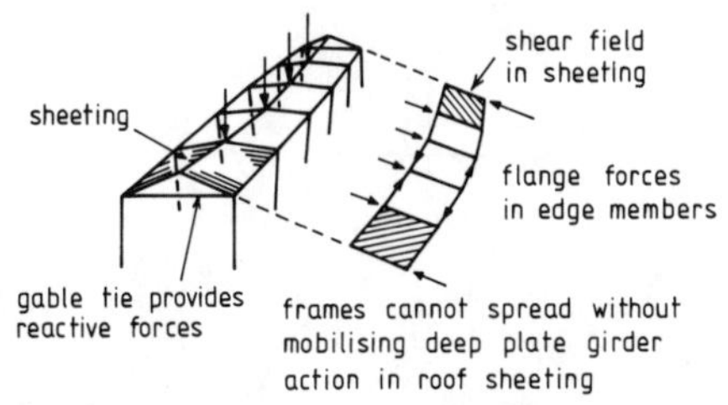

(b) Diaphragm action in a pitched roofed building

Figure 26.9 Stressed skin action in buildings

not necessary. In many cases the sheeting virtually eliminates sidesway and reduces the size of the main frame members.

In pitched roof buildings, as shown in Figure 26.9(b), the sheeting may also be used to help resist vertical loads, although this effect is not worth taking into account for roof pitches less than 12°. For steeper pitches (over 30°) folded plate roofs of steel sheeting and cold-rolled apex and valley members can be used to span from one end-gable to another without the need for intermediate frames.

26.2.6 Concluding summary

1. Different profiles have evolved for use as roof sheeting, roof decking and wall cladding; these satisfy the particular requirements of each type of sheeting.
2. Behaviour is influenced by local buckling of the thin elements of the cross-section. Effective-width concepts are widely used in analysis and design.
3. Design development of a new profile is usually supported by testing.
4. There are important structural benefits of sheeting (stressed skin design).

Background reading

Cold-rolled sections

BS 5950: Part 5. Code of Practice for the Design of Cold Formed Steel Sections

Draft European Recommendations for the Design of Light Gauge Steel Members (1983). European Convention for Constructional Steelwork, ECCS-TC7-1983

Specification for the Design of Cold-formed Steel Structural Members (1983). American Iron and Steel Institute, Washington, DC

WALKER, A.C. (ED.) (1975) *Design and Analysis of Cold-Formed Sections*, Intertext, London

WEI-WEI, YU (1973) *Cold Formed Steel Structures*, McGraw-Hill, New York

Sheeting

DAVIES, J.M. and BRYAN, E.R. (1982) *Manual of Stressed Skin Diaphragm Design*, Granada, St Albans

EUROPEAN CONVENTION FOR CONSTRUCTIONAL STEELWORK (1977a) *European Recommendations for the Stressed Skin Design of Steel Structures*, Constrado, London

EUROPEAN CONVENTION FOR CONSTRUCTIONAL STEELWORK (1977b) *European Recommendations for the Testing of Profiled Metal Sheets*, Constrado, London

EUROPEAN CONVENTION FOR CONSTRUCTIONAL STEELWORK (1983) *European Recommendations for the Design of Profiled Sheeting*, Constrado, London

Profiled Steel Cladding and Decking for Commercial and Industrial Buildings (1980) Constrado, London

27

Composite beams

Objective To introduce the fundamentals of composite action for elements in bending with particular reference to plastic and elastic analysis of beam sections and shear connection.

Prior reading Chapter 20 (Restrained compact beams); Chapter 21 (Local buckling); Elastic theory of longitudinal shear; Basic properties of concrete; Theory of transformed sections.

Summary The advantages and disadvantages of composite construction for beams are outlined and construction methods are summarized. Methods for allowing for shear lag are presented and discussed. The various failure modes for simply supported beams are identified and the ultimate moment of resistance is determined. Elastic theory is presented for the design of beams for the serviceability limit state. Recommendations are given for a reduced modulus for concrete under sustained loading. The effects of shrinkage of concrete and temperature difference through the section are outlined. The plastic design of continuous beams is introduced. The purposes of shear connection are stated and tests on connectors are described. Design of shear connection in buildings and bridges is discussed.

27.1 Introduction

In buildings and bridges it is common to find concrete slabs supported on steel beams. If slip at the interface is free to occur, each component will act independently, as shown in Figure 27.1(a). If slip at the interface is eliminated, or at least reduced, the slab and steel member will

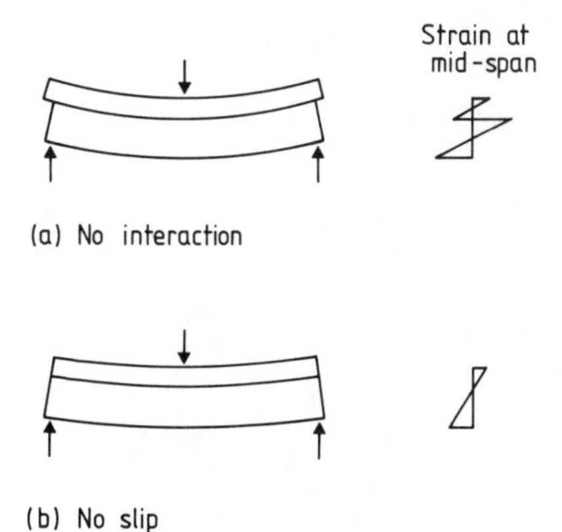

Figure 27.1 Non-composite and composite action

act together as a composite unit. The resulting increase in strength and stiffness will depend on the extent to which slip is prevented. Here it is assumed that the connection between the slab and the steel member is such that the effects of slip can be ignored, as in Figure 27.1(b). In practice, such action is achieved by headed studs or other connectors which are welded to the structural steel and embedded in the concrete slab, as shown in Figures 27.2(a)–(c). Pre-cast concrete floor or deck units can be used in place of an *in situ* slab (Figure 27.2(d)).

If profiled steel sheeting is used as permanent formwork for the underside of the slab, as shown in Figure 27.2(e), then composite action can be achieved between these two components after construction. Steel stanchions, encased or filled with concrete as shown in Figure 27.2(f) and 27.2(g), can also be treated as composite elements.

27.2 Advantages and disadvantages of composite construction for beams

A composite beam has greater stiffness and generally a higher load capacity than its non-composite counterpart. Consequently, a smaller steel section is usually required. This results in the saving of material and depth of construction. In turn, the latter leads to lower storey heights in buildings and lower embankments for bridges.

The main disadvantage is the need to provide connectors at the steel–concrete interface. There is also some increase in the complexity of design, particularly for continuous structures. However, it is clear that composite construction is particularly competitive, in terms of overall cost, for medium- or long-span structures where a concrete slab is needed for other reasons and where there is a premium on rapid construction.

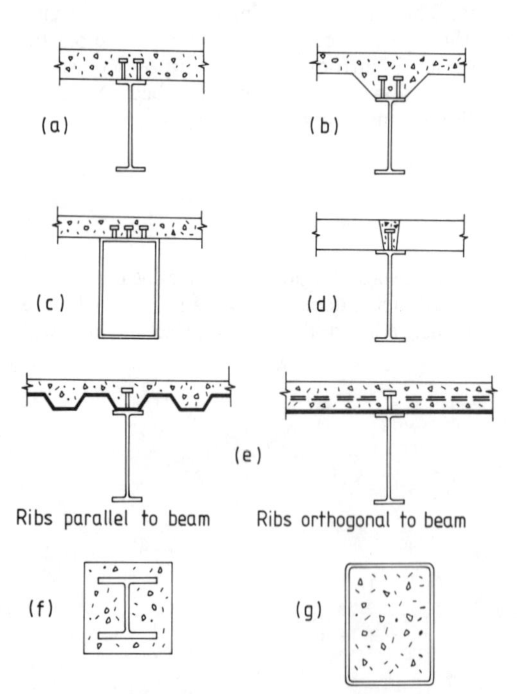

Figure 27.2 (a) Tee-beam; (b) haunched slab; (c) steel box girder; (d) pre-cast concrete units; (e) profiled steel sheeting; (f) encased column; (g) filled column

27.3 Construction methods

Because of the benefits in structural performance it would seem preferable to ensure that the concrete slab and steel member act compositely at all times. This would result in all loads, including the dead weight of the structure, being resisted by the composite section. For this to be achieved it is necessary to support the steel beam until the concrete has hardened; a method known as 'propping'. The number of temporary supports need not be high; propping at the quarter-span points and mid-span is generally adequate. The props are usually left in place until the concrete slab has developed three-quarters of its design strength.

However, in the majority of bridges 'unpropped construction' is the only practical method. This technique is also usually preferred for a building in order to reduce the time of construction. Initially the steel beam alone resists its own weight, and that of the formwork, wet concrete and placement loads. Other loads are added later and so are carried by the composite member.

27.4 Effective width

A typical form of composite construction consists of a slab connected to a series of parallel steel members; essentially a series of interconnected T-beams with wide, thin flanges, as shown in Figure 27.3(a). In such a system the flange width may not be fully effective in resisting compression due to 'shear lag'. This phenomenon will be explained by reference to a simply supported member, part of whose length is shown in plan in Figure 27.3(b).

The maximum force in the slab will be at midspan, while the force at the ends will be zero. The change in longitudinal force is associated with shear in the plane of the slab. The resulting deformation, shown in Figure 27.3(b), is inconsistent with simple bending theory in which initially plane sections are assumed to remain plane thereafter. The edge regions of the slab are

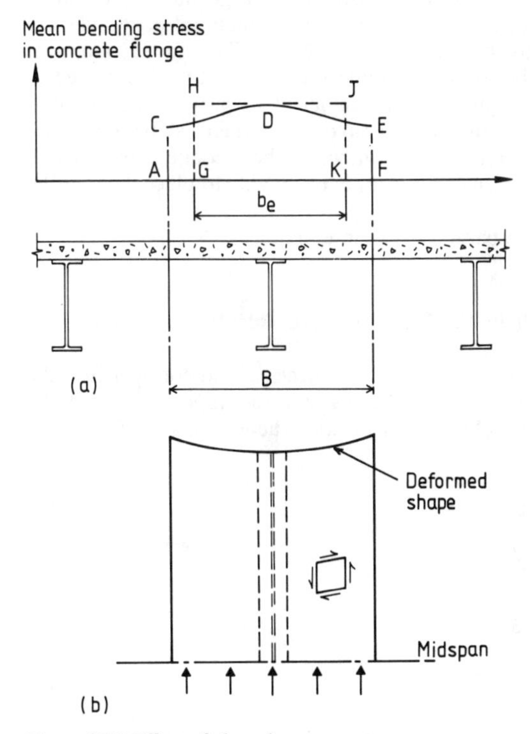

Figure 27.3 Effect of shear lag

effectively less stiff, and a non-uniform distribution of longitudinal bending stress is obtained across the section. The simple theory will give the correct value of the maximum stress if the real flange width B is replaced by an effective width, b_e, such that the area GHJK equals the area ACDEF.

The ratio b_e/B depends not only on the relative dimensions of the system but also on the type of loading, the support conditions and the cross-section considered. The effects on a simply supported span of the ratio of beam spacing to span, B/L, and the type of loading are shown in Figure 27.4.

In most codes of practice very simple formulae are given for the calculation of effective widths, though this may lead to some loss of economy. For simply supported beams the effective width on each side of the steel web should be $L/6$, but not greater than half the distance to the next adjacent web nor greater than the projection of the cantilever slab for edge beams are the limits given in the bridge code BS 5400: Part 5 for those sections which can attain the plastic moment before the onset of local buckling. For composite members with steel sections susceptible to local buckling elastic theory is used for the analysis of sections at the ultimate limit state. For these, BS 5400: Part 5 gives a more extensive treatment of effective breadth, with the ratio b_e/B depending on the support conditions, the type of loading and position along the span, as well as B/L.

27.5 Shear connection: full and partial interaction

As connection between the slab and the steel member is provided mainly to resist longitudinal shear it is referred to as the 'shear connection'. If this is inadequate the full flexural strength of the composite beam cannot be developed. If there are sufficient connectors for the strength of the beam to be not less than that given by plastic analysis the resulting method of design is known as 'full-interaction design'. However, this may result in a large number of shear connectors, particularly if the dimensions of the cross-section or the material properties are governed by factors other than the strength of the composite beam. The most economical design may then be one in which the number of shear connectors is such that the degree of interaction between the slab and the steel member is just sufficient to provide the required flexural strength. Such 'partial-interaction' design may be unavoidable when a slab is constructed with permanent formwork of profiled sheeting. The number of shear connectors attached to the steel beam may then be limited by the restriction of only being able to place them in the troughs of the sheeting profile.

In this chapter it is assumed that design is based on full interaction.

27.6 Failure modes for simply supported composite beams

In order to develop a design method for composite beams it is necessary to consider first the various possible failure modes. Only simply supported beams with the upper surface of the concrete in compression are considered here. This form of construction is widely used for

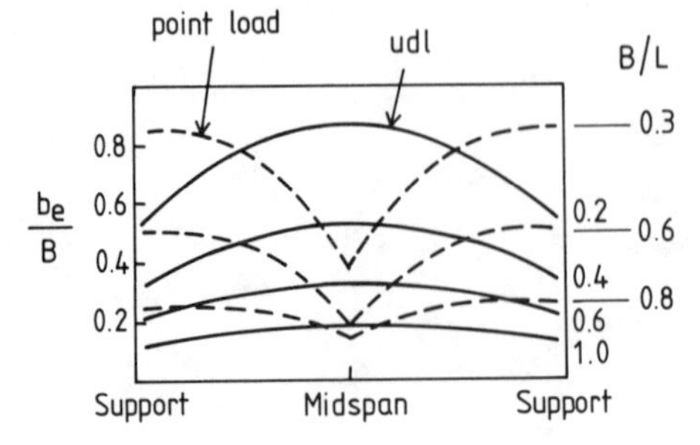

Figure 27.4 Variation of effective width

buildings and similar structures, in which horizontal forces are resisted by stiff cores or some other form of bracing. Simple spans are also used in bridges, although in multi-span structures continuity between spans will lead to a better surface finish and quality of ride.

Tests and other studies show that many modes of failure are possible in composite beams, even when simply supported:

1. Failure of the composite member in flexure by the development of a plastic hinge;
2. Reduction or complete loss of composite action, leading to collapse of the member in flexure, due to excessive slip or complete failure of the shear connection;
3. Local shear failure in the slab in regions of high stress around shear connectors;
4. Failure in vertical shear;
5. Local buckling in the steel members.

In addition, although the beam may not collapse it may become unfit for its purpose due to excessive deflection.

Failure in the shear connection is more sudden and less predictable than flexural failure of the composite member. For this reason design rules for both shear connectors and neighbouring slab reinforcement aim to ensure that a properly designed beam fails in flexure at a bending moment not less than that predicted by plastic hinge analysis of the cross-section.

When designing composite beams it is assumed that vertical shear is resisted by the structural steel member acting alone. The treatment of vertical shear therefore follows that used in the design of steel beams. The influence of vertical shear on the ultimate moment of resistance is neglected here.

In simply supported composite beams no account need usually be taken of local buckling in the steel section. The compression flange is attached to the concrete slab by the shear connectors and the depth of the web in compression is usually small. However, there remains the possibility, theoretically at least, that local buckling could occur in the web of a deep plate girder or in a flange with a wide outstand beyond the shear connectors. It is assumed that the proportions are such that plastic analysis can be applied to the cross-section of the composite beam. The calculation of the ultimate moment of resistance is therefore an application of the rectangular stress block theory used for structural steel and reinforced concrete members. Due to redistribution of stress the ultimate moment of resistance is independent of the method of construction.

27.7 Ultimate moment of resistance in positive bending

It is assumed here that concrete resists a compressive stress of $0.6f_{cu}/\gamma_m$, but no tension. Taking γ_m as the usual value of 1.5 for concrete, a design compressive stress of $0.4f_{cu}$ is obtained. Although the longitudinal reinforcement within the concrete flange is stressed to its design yield strength its contribution to the flexural strength is small, and is usually neglected. The whole of the area of the structural steel member is taken to be stressed at its design yield strength (p_y), in tension or compression.

From Figures 27.5 and 27.6 two cases must be considered for the position of the neutral axis. From equilibrium of longitudinal forces the plastic neutral axis will lie in the slab providing:

$$0.4\, b_e h_c f_{cu} \geqslant A_s p_y \qquad (27.1)$$

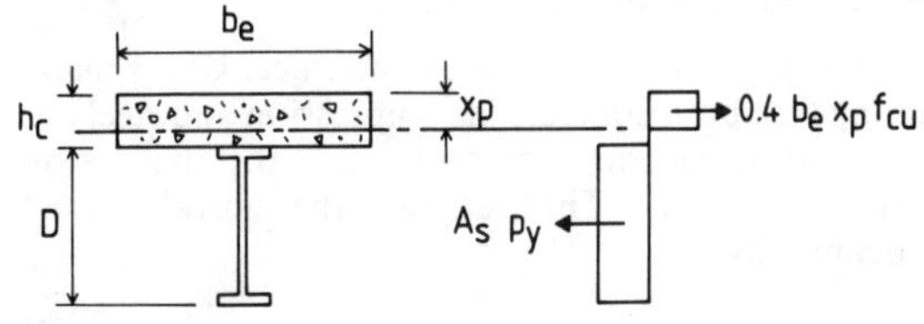

Figure 27.5 Stress distribution at flexural failure: neutral axis in slab

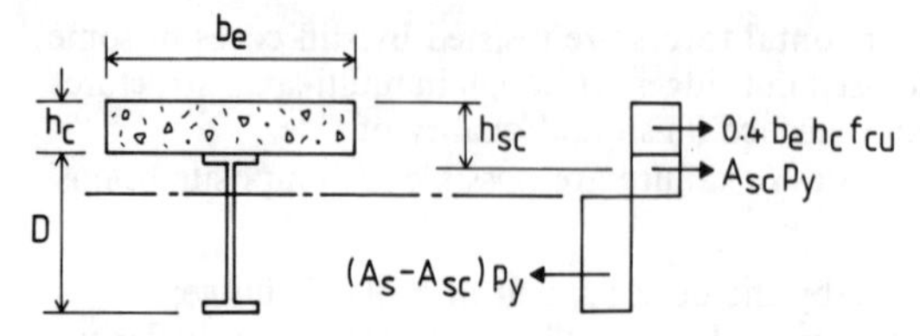

Figure 27.6 Stress distribution at flexural failure: neutral axis in steel section

In this case, the depth (x_p) of the neutral axis is given by:

$$x_p = \frac{A_s p_y}{0.4 b_e f_{cu}} \tag{27.2}$$

The assumed stress distribution of flexural failure will then be as shown in Figure 27.5 and for a symmetrical steel section of overall depth (D) the ultimate moment of resistance of the composite beam, M_u, will be given by:

$$M_u = A_s p_y \left(\frac{D}{2} + h_c - \frac{x_p}{2} \right) \tag{27.3}$$

If the concrete flange is too small to satisfy Equation 27.1 the plastic neutral axis will lie in the steel section. The stress blocks then will be as in Figure 27.6, where A_{sc} is the area of the steel section in compression. Resolving longitudinally gives:

$$0.4 b_e h_c f_{cu} + A_{sc} p_y = (A_s - A_{sc}) p_y \tag{27.4}$$

This can be solved for A_{sc} and hence the depth to the centroid of the structural steel in compression (h_{sc}) can be determined. Taking moments about the centroid of the slab:

$$M_u = A_s p_y \left(\frac{D}{2} + \frac{h_c}{2} \right) - 2 A_{sc} p_y \left(h_{sc} - \frac{h_c}{2} \right) \tag{27.5}$$

27.8 Elastic analysis

27.8.1 Introduction

It is necessary to employ elastic theory to ensure that, under service loads, both the deflection of the beam is acceptable and stresses are within the elastic range. At the ultimate limit state elastic methods may also be required when designing continuous composite beams. For these members, the unrestrained bottom flange of the steel section and the greater part of the web will be in compression in the region of an internal support. If these elements are not compact the resistance to bending will be limited by local buckling, thereby precluding plastic analysis of the cross-section.

When elastic theory is used it may be necessary to include the effects of creep and shrinkage of concrete, and of temperature, if accurate results are required.

27.8.2 Elastic analysis of cross-sections assuming full interaction

Full interaction implies that there is negligible slip at the steel–concrete interface. Use is made of the theory of transformed sections, assuming that both concrete in compression and steel are linearly elastic materials. This enables the composite section to be replaced in the analysis by an equivalent all-steel cross-section, as shown in Figure 27.7. The breadth of the equivalent steel slab depends upon the modular ratio, α_e, defined by:

$$\alpha_e = E_s / E_c \tag{27.6}$$

where E_c is the appropriate elastic modulus of concrete.

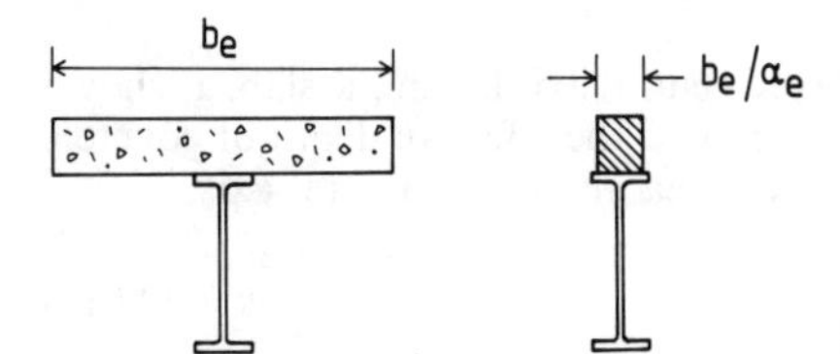

Figure 27.7 Equivalent all-steel section

For generality the steel section is shown in Figure 27.8 as asymmetrical, with a cross-sectional area (A_s), second moment of area (I_s) and centroid distance (d) below the top of the concrete slab. As with plastic analysis, two cases must be considered for the depth of the elastic neutral axis (x_e).

Case 1: Elastic neutral axis within the steel section

The whole of the concrete is in compression, as shown in Figure 27.8(a). Taking moments of area about the neutral axis:

$$A_s(d - x_e) = \frac{b_e h_c}{\alpha_e}(x_e - h_c/2) \tag{27.7}$$

Solving for x_e:

$$x_e = \frac{\alpha_e A_s d + b_e h_c^2/2}{\alpha_e A_s + b_e h_c} \tag{27.8}$$

The second moment of area of the equivalent all-steel section, I, is given by:

$$I = \frac{b_e h_c^3}{12\alpha_e} + \frac{b_e h_c}{\alpha_e}\left(x_e - \frac{h_c}{2}\right)^2 + I_s + A_s(d - x_e)^2 \tag{27.9}$$

Longitudinal stress due to bending can be calculated using the appropriate distance from the neutral axis and, for the concrete slab, the modular ratio. Thus a positive bending moment M causes the following maximum stresses in the steel and the concrete, denoted by f_{sb} and f_{cb}, respectively:

$$f_{sb} = \frac{M}{I}(h_c + D - x_e) \tag{27.10}$$

$$f_{cb} = \frac{Mx_e}{\alpha_e I} \tag{27.11}$$

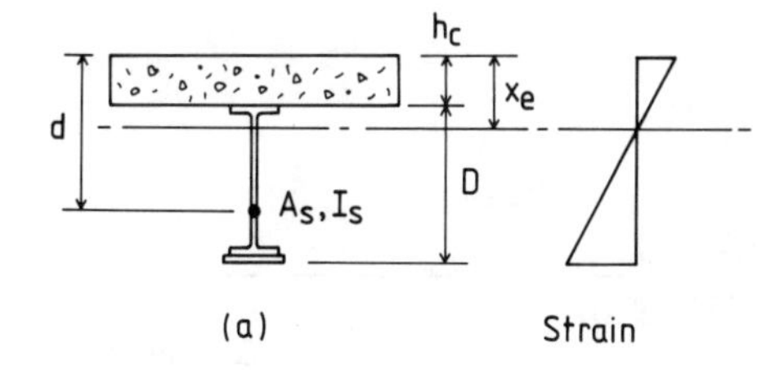

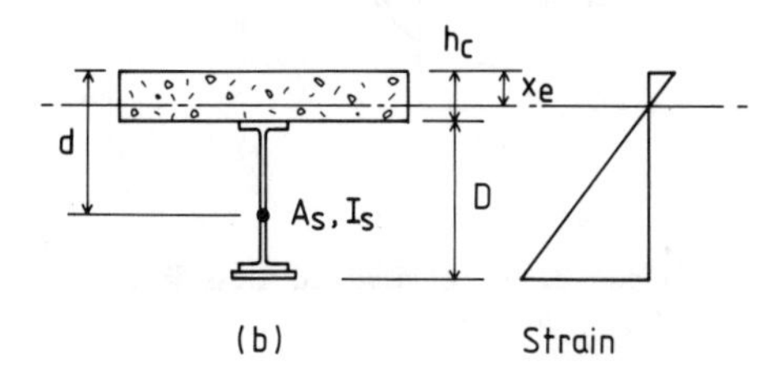

Figure 27.8 Elastic analysis of cross-sections

Case 2: Elastic neutral axis within the concrete slab
When x_e, given by Equation 27.8, is less than h_c the elastic neutral axis lies in the slab, as shown in Figure 27.8(b). If the tensile strength of concrete is ignored the effective depth of concrete reduces to the neutral axis depth x_e. Taking moments of area about the neutral axis:

$$A_s(d - x_e) = \frac{b_e x_e}{\alpha_e}\left(\frac{x_e}{2}\right) \tag{27.12}$$

Hence x_e is determined by solution of the quadratic equation:

$$b_e x_e^2 + 2\alpha_e A_s x_e - 2\alpha_e A_s d = 0 \tag{27.13}$$

The second moment of area of the equivalent all-steel section is given by:

$$I = \frac{b_e x_e^3}{3\alpha_e} + I_s + A_s(d - x_e)^2 \tag{17.14}$$

The maximum stresses are calculated from Equations 27.10 and 27.11 using the value of I given by Equation 27.14.

27.8.2.1 Creep, shrinkage and temperature

Under sustained compressive stress concrete undergoes a time-dependent increase in strain, as shown in Figure 27.9, known as 'creep'. For a given concrete specification creep is principally a function of the age at which loading is applied, its duration and intensity, and the relative humidity of the environment. In a composite beam creep of the concrete causes an increase in deflection. This is accompanied by a relaxation of stress in the concrete and an increase in stress in the steel member.

Concrete also undergoes long-term deformation due to shrinkage, thereby causing additional deflection in the composite beam. Unlike creep, shrinkage occurs irrespective of whether the concrete is subjected to stress. For plain concrete of a given specification shrinkage is mainly dependent on the relative humidity of the environment. For example, in the fairly dry environment of a building an unrestrained concrete slab could be expected to shrink by 0.03% of its length or more. In a composite beam the slab is restrained by the steel member which exerts a tensile force on it through the shear connection. The apparent shrinkage is therefore less than the unrestrained value, tensile stresses arise in the concrete, and, in a simply supported member, the shrinkage causes sagging curvature, with compressive stress in the upper part of the steel section, as shown in Figure 27.10.

Composite beams may also be subject to temperature difference through the depth of the section. As temperature difference and shrinkage both lead to stress caused by differential strain their effects are usually treated in the same way in design.

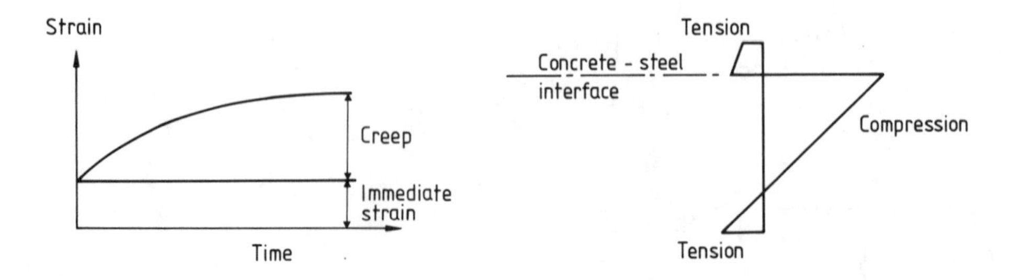

Figure 27.9 Change in strain due to creep

Figure 27.10 Longitudinal stress distribution due to shrinkage

27.9 Calculations for deflections of composite beams

If deflections are to be calculated the effects of creep should be included in the elastic analysis. This is usually done by using a reduced modulus of elasticity for concrete, E_c', under sustained loading. The composite bridge code BS 5400: Part 5 gives three methods for determining E_c' from the static secant modulus, E_c, appropriate to short-term loading. In increasing order of accuracy these are as follows:

1. $E_c' = E_c/2$ (27.15)
2. $E_c' = \phi_c E_c$ (27.16)

where the creep reduction factor ϕ_c is given by **Table 9** of the code. For humid open-air environments $\phi_c = 0.5$, while for dry interior enclosures ϕ_c falls to 0.3:

3. $E_c' = \dfrac{E_c}{1 + \phi}$ (27.17)

where ϕ is a creep coefficient determined in accordance with Part 4 of BS 5400, Appendix C. Values of ϕ lie within the range 1–3.

Values for the short-term modulus E_c are given in **Table 2** of BS 5400: Part 4, and in CP8110. In the absence of other recommendations, the above procedures may also be used for composite beams in buildings.

Curvature due to differential temperature strains can be ignored because control of deflection is rarely a main criterion for the design of bridges, and differential temperatures rarely occur in buildings.

As an alternative to calculation from basic principles deflections can be controlled by limiting span–depth ratios. Such a method is necessarily approximate, but would seem appropriate for use in design because of the many factors which influence deflection that cannot be quantified. These factors include variations in the elasticity, creep and shrinkage properties of the concrete, the stiffening effect of finishes and partial fixity at the supports. There is also difficulty in defining when a deflection becomes excessive.

27.10 Plastic analysis of continuous composite beams

In buildings composite beams usually employ rolled steel sections which are often sufficiently compact for the negative moment of resistance to be calculated by plastic analysis of the cross-section.

27.10.1 Negative moment of resistance of a compact cross-section

The cross-section considered is shown in Figure 27.11(a). Research on shear lag in negative moment regions has shown that the effective breadth of the concrete slab (b_e) may be taken as the same value used for positive moment of resistance. All properly anchored longitudinal reinforcement within this breadth is assumed to be effective. Let the area of this reinforcement

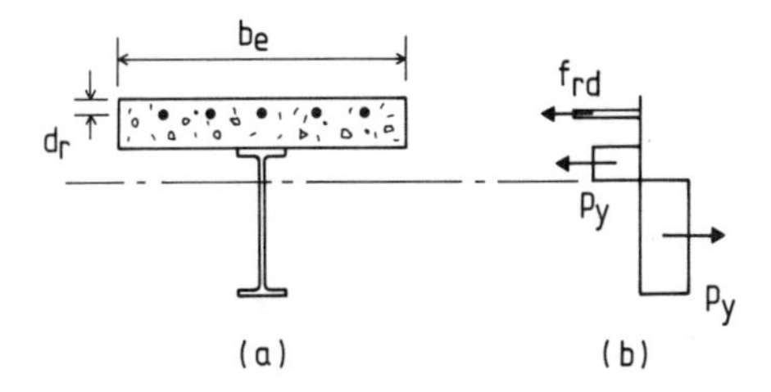

Figure 27.11 Negative moment of resistance

be A_r, located a distance d_r below the top of the slab. The design yield strength is denoted by f_{rd}.

At flexural failure the whole of the concrete slab may be assumed to be cracked, while all the structural steel is at its design strength (p_y), either in tension or compression. The plastic neutral axis may be in the top flange or in the web. For the latter case the stresses are as shown in Figure 27.11(b). The position of the neutral axis is determined by considering longitudinal equilibrium, and the ultimate moment of resistance (M_u') can then be calculated.

27.10.2 Distribution of bending moments

To design a suitable cross-section against flexure it remains to determine the distribution of bending moments due to the applied load. The simplest calculation results from the use of simple plastic hinge theory for the structural analysis.

Let the ratio of the negative to the positive moments of resistance in a proposed section be ψ. Therefore:

$$\psi = M_u'/M_u \qquad (27.18)$$

Consider, for example, the end-span of a continuous composite beam, subjected to a uniformly distributed design load of w_f per unit length. The bending moment diagram at collapse is as shown in Figure 27.12. It can easily be shown that:

$$\beta = \left(\frac{1}{\psi}\right)[(1 + \psi)^{1/2} - 1] \qquad (27.19)$$

and the required value of M_u is:

$$M_u = \tfrac{1}{2}w_f\beta^2 L^2 \qquad (27.20)$$

27.10.3 The applicability of simple plastic theory

For the above structural analysis to be valid critical cross-sections must be capable of developing the calculated moments of resistance and sustaining them during rotation until a plastic hinge mechanism is complete. In negative bending the effective section consists of the steel member and the longitudinal reinforcement in the slab. This section can be assumed to have sufficient ductility to maintain the calculated plastic moment M_u' as rotation takes place, provided that the reinforcement is suitably anchored and buckling of the steel section is prevented. The former is achieved by correct detailing and the latter by satisfying limits on the slenderness of the section.

27.11 Shear connection

27.11.1 Introduction

The performance of composite beams depends on an effective transfer of shear at the interface between the concrete slab and the steel member. Under increasing load the natural bond at the

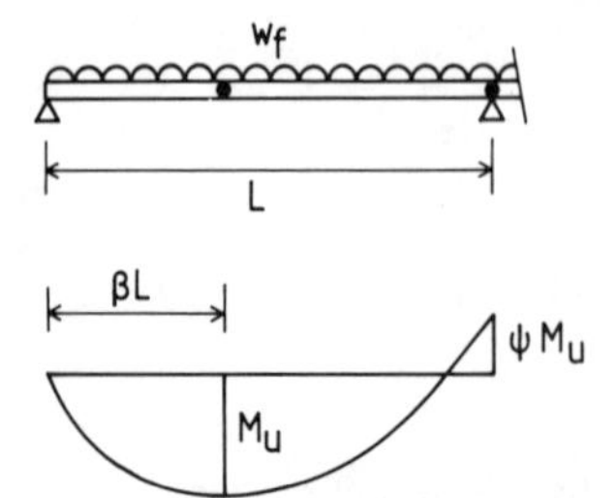

Figure 27.12 End-span of continuous beam

interface, once broken, cannot be restored. The use of mechanical shear connectors is therefore essential for the satisfactory performance of composite beams. Shear connectors also serve the function of holding the slab down onto the steel member. Although loads acting downwards on the steel member are the most obvious possible cause of vertical separation there are other situations in which this can occur. These arise from more complex effects, such as torsion or triaxial stresses in the vicinity of the connectors. All connectors in common use are therefore shaped to provide adequate resistance to uplift as well as slip. In normal design calculation related to uplift is not required.

27.11.2 Stud connectors

The most widely used type of connector is the headed stud, shown in Figure 27.13. Such studs range in diameter from 13 to 25 mm, and from 65 to 100 mm or more in length. The popularity of the headed stud lies mainly in the speed of the welding process, carried out using a portable hand-tool connected via a control unit to mains or a generator. Highly skilled operatives are not required and welding of the stud takes only 10–15 s. The semi-automatic process also results in uniformity of quality and involves only very low heat input to the parent metal. Other advantages of studs are that they provide little obstruction to the reinforcement in the slab and are equally strong and stiff in all lateral directions. When pre-cast units are used for the concrete slab, stud connectors may project through holes in the slab which are later filled in.

A slab reinforced by profiled steel sheeting can be attached to the steel beam by shear connectors welded through the decking. The sheeting may be up to 1.25 mm thick and galvanized to a maximum of 0.02 mm per face, but not plastic coated. Through-deck welding is not recommended through sheet overlaps. As an alternative, the sheeting may be slotted over studs already welded to the steel beam.

27.11.3 Behaviour of shear connectors

27.11.3.1 Push-out tests

The property of a shear connector that is most relevant to design is the relationship between the shear force transmitted and the slip at the interface. This load slip curve should ideally be found from tests on composite beams. In practice, most of the data on the strength of connectors under static loading have been obtained from 'push-out' tests, as shown in Figure 27.14. The specimen consists of a short length of steel beam connected to two small *in situ* concrete slabs. The natural bond at the interface is deliberately destroyed by greasing the steel flanges before the concrete is cast (except when high-strength friction-grip bolts are used). The slabs are bedded in mortar on the lower platen of a standard compression testing machine and load is applied concentrically at the top cover plate of the vertical specimen.

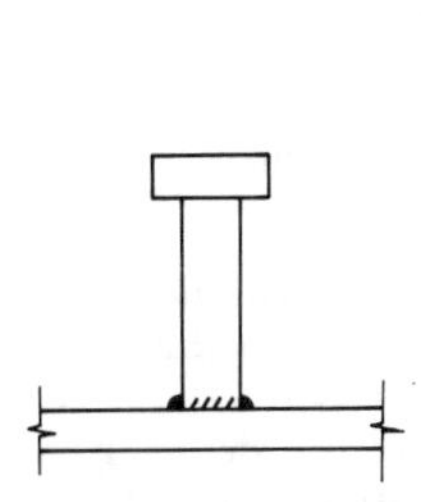

Figure 27.13 Headed stud shear connector

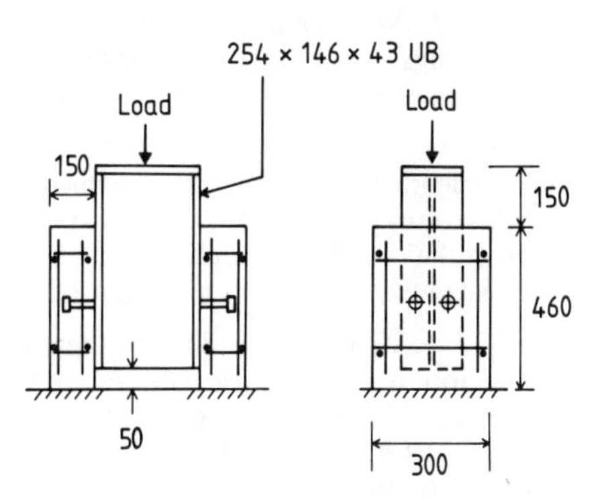

Figure 27.14 Push-out test on shear connectors

The push-out test cannot exactly simulate the actual conditions occurring in a composite beam. Despite this, it does enable the behaviour of different types and sizes of connector to be compared, provided that the test is standardized.

A typical load slip curve obtained from a test using headed studs is shown in Figure 27.15. The distribution of bearing stress on the shank is likely to have the form shown in Figure 27.16, the majority of the shear being resisted at the base of the stud. At ultimate load the concrete at the base may be subjected to a local pressure of two or three times its cube strength. Such high stresses are possible only because the concrete bearing on the connector is restrained laterally by the surrounding concrete, the reinforcement and the steel flange. Possible modes of failure are crushing or splitting of the concrete and shearing of the stud. For studs embedded in strong concrete it is found that their strength depends on the ultimate tensile strength of the stud material rather than the yield stress.

Values of the nominal static strength for headed studs are given in Table 7 of BS 5400: Part 5. The values show that the strength is roughly proportional to the square of the diameter and to the square root of the cube strength.

27.11.3.2 Behaviour in beams

Although the load slip curve for a connector in a beam cannot be assumed to be that found in a push-out test it has been found that the ultimate strength is usually about the same. However, there are four situations in which a connector strength found from a push-out test may be too high for use in design:

1. When the beam is subjected to repeated loading, as in a bridge:
2. When the lateral restraint to the concrete in contact with the connector is less than that provided in a push-out test, as may be the case when the slab is haunched. Restrictions on the shape of the haunch are given in design recommendations;
3. When lightweight concrete is to be used. Tests show that the ability of such concrete to resist the very high local stresses at shear connectors is slightly less than for normal density concrete of the same cube strength. Provided that the density of the lightweight concrete exceeds 1400 kg/m^3, it is usual to take the connector strength as 85% of the value applicable to normal density material;
4. When the slab is in tension, as in the negative-moment regions of a continuous composite beam. It has been proposed that the ultimate strength of connectors in these regions should be taken as 20% less than the value used for positive-moment regions. However, this reduction is not necessary for bridge girders designed to BS 5400, for reasons described below.

27.11.4 Design of shear connection

27.11.4.1 Number and spacing

In buildings the number of shear connectors required in beams under static loading is based on ultimate strength behaviour, and is determined by two criteria:

1. Since shear failure is more sudden and less predictable than flexural failure there must be sufficient connectors for the beam to fail in flexure, not in shear.
2. There must be sufficient connectors for the resistance of the beam in bending to be not less than the calculated value. For a beam in positive bending designed for full interaction the resistance is calculated using plastic analysis of the cross-section.

Ultimate strength design of the shear connection is based on the assumption that individual connectors possess sufficient plasticity to redistribute load between them until all those in a shear span fail as a group. In determining the number of connectors it is usual to base calculations on a strength of $0.8\,P_u$ to ensure that the criteria stated above will definitely be satisfied. In limit state design it will also be necessary to reduce further the strength by a factor $\gamma_m \geqslant 1.0$.

As failure of some connectors, such as large studs, occurs mainly in the surrounding concrete,

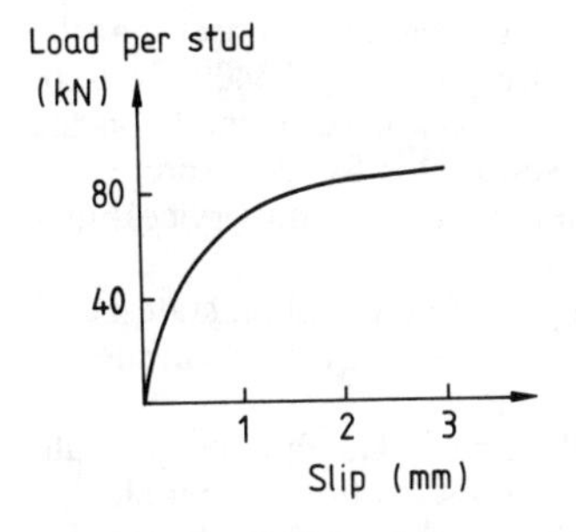

Figure 27.15 Load–slip curve

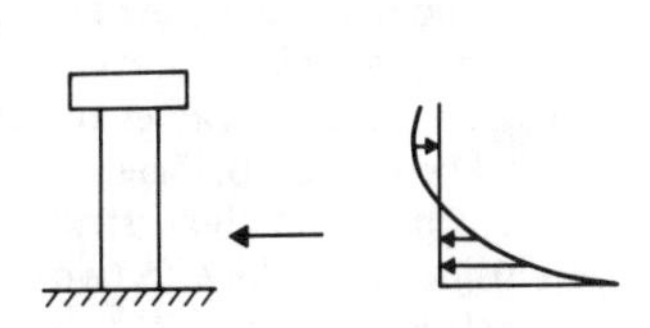

Figure 27.16 Distribution of bearing stress

while small-diameter studs may shear off, γ_m should lie between the values for concrete and structural steel. Hence the design load per stud (P_d) is given by:

$$P_d = \frac{0.8P_u}{1.1} \tag{27.21}$$

At or near ultimate load there is no simple way of calculating the longitudinal shear at particular points along the steel–concrete interface in a composite beam. In buildings not subjected to heavy point loads connectors can be spaced uniformly between supports and sections of maximum positive bending moment. Their number is determined by the change in longitudinal force in the slab between these cross-sections. Figure 27.17 shows separately the steel and concrete components of one half of a simply supported beam of span (L) that is at flexural failure under loading symmetrical about mid-span. If the plastic neutral axis is in the concrete, then the maximum longitudinal force in the steel member is $A_s p_y$. Hence the required number of connectors (N) between a support and midspan is given by:

$$N = \frac{A_s p_y}{0.8P_u/\gamma_m} \tag{27.22}$$

The number of connectors may also be influenced by empirical limits placed by design recommendations on the maximum spacing of connectors. The limits aim to control uplift between connectors and to avoid excessive stress concentrations in the slab.

If redistribution of load is to occur between connectors, then some longitudinal slip must occur. In beams in buildings at ultimate load this will be of the order of 1–2 mm. The span of the composite bridge girder may be ten times that of a beam in a building, yet the connectors may be of the same size. From geometric similarity it is evident that connectors would have to tolerate a slip of 10–20 mm for redistribution to occur. They cannot do this and to avoid progressive failure the bridge code, BS 5400: Part 5, requires the connectors to be spaced in accordance with elastic theory.

The design of the shear connection for a composite bridge will be influenced by the following additional considerations:

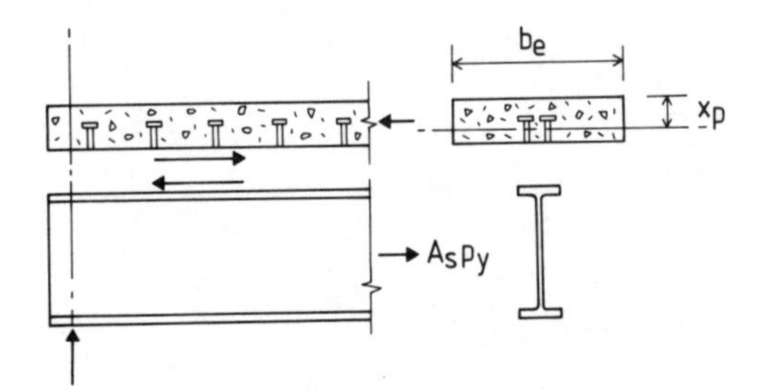

Figure 27.17 Action of shear connection

1. The possibility of fatigue failure. Part 5 of BS 5400 limits the static load per connector at the serviceability limit state to $0.55P_u$, and further requirements are given in BS 5400: Part 10.
2. Unless plastic analysis is used for analysis of cross-sections no check need be made on the static strength of the shear connection at the ultimate limit state. This follows from:
   (a) The relative value of the design strengths for connectors at the ultimate and serviceability limit states;
   (b) The relative values of the partial safety factors on loading at these two limit states; and
   (c) The need in bridges to relate the spacing of connectors to the envelope of vertical shears, rather than the distribution of shear due to any one load case.
3. Variations of the effective cross-section along the length of the girder. The familiar formula for shear flow, $q = V\bar{A}y/I$, is only valid for members of uniform cross-section. To enable this to be used where the steel cross-section and overall breadth of the concrete flange are constant within each span two approximations are made:
   (a) The effective breadth of the concrete flange is assumed to be constant throughout any span;
   (b) The deck slab is assumed to be uncracked and unreinforced along its length.
   The second approximation also enables the designer to ignore the reduced strength of shear connectors in the cracked concrete of the negative moment regions of continuous girders.

27.11.4.2 Transverse reinforcement

In order to achieve the composite action assumed in design force must be transferred from the shear connection to the full effective width of the slab. It follows that shear failure of concrete surrounding the connectors must be prevented. When sufficient connectors are provided, the critical planes for shear failure pass round the connectors or through the slab. In Figure 27.18 the critical shear planes are 1–1 and 2–3–3–2. The situation is complicated due to transverse bending of the slab in the vicinity of the steel section. In buildings the transverse moment will be negative. This will cause cracks on the top surface along the line of the steel members and transverse compression in the region of the connectors. The latter effect enhances the resistance to shear. The design recommendations used to check the resistance of the slab to longitudinal shear are based on research into the behaviour of reinforced concrete slabs.

27.11.5 Control of cracking in concrete

Extensive cracking of concrete may adversely affect the appearance and durability of a structure, in which case provision should be made to minimize its extent. The problem arises mainly at the top surface of the concrete slab, near the internal support for a continuous beam. This surface may be in the non-corrosive atmosphere of a heated building and covered from view by a floor finish. In this situation the designer may decide that no check on crack width is necessary. However, in a multi-storey car park or bridge, water containing de-icing salt may enter the cracks, and severe corrosion may result if their width is not controlled.

An easy way of satisfying serviceability criteria in relation to cracking is to adopt bar spacings recommended by design guides such as BS 5400: Part 5. Such rules are inevitably conservative because of the simplifying assumptions that have to be made, and may be restricted in their application. If it is necessary to calculate the width of cracks, use is made of a crack-width formula such as that given in Appendix B of BS 5400: Part 5.

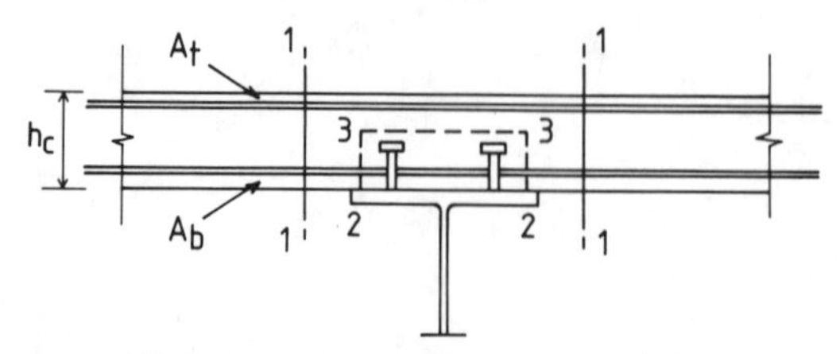

Figure 27.18 Shear planes and transverse reinforcement

27.12 Concluding summary

1. Simple composite beams utilize the compressive resistance of concrete slabs in conjunction with steel members, leading to an increase in strength and stiffness. Such members can be constructed as 'propped' or 'unpropped'.
2. Design can be based on full or partial interaction between the steel and concrete components. With full interaction it is assumed that sufficient shear connectors are provided for the strength of the beam to be not less than that given by plastic analysis. The resulting expressions for the ultimate moment of resistance in positive bending have been developed.
3. Elastic theory is needed in the design of composite beams to ensure that, under service loads, deflection is not excessive, and that stresses are within the elastic range.
4. If accurate results are required from elastic analysis, account should be taken of the effects of creep, shrinkage and differential temperature, as appropriate.
5. The most widely used type of shear connector is the headed stud. Ultimate strengths, determined from push-out tests, are tabulated in design codes. The values are mainly dependent on the ultimate tensile strength of the stud material and the grade of the concrete.
6. In buildings, the number of shear connectors is based on the assumption that individual connectors possess sufficient ductility to redistribute load until all those in a shear span fail as a group. In the absence of heavy point loads, connectors are spaced uniformly between supports and sections of maximum positive bending moment. For bridges, the possibility of fatigue failure must be considered, and the spacing of connectors is related to the envelope of vertical shears.

Background reading

European Convention for Constructional Steelwork (1981) *Composite Structures*, The Construction Press, London

JOHNSON, R.P. (1975) *Composite Structures of Steel and Concrete*, Vol. 1, Crosby Lockwood Staples, London

JOHNSON, R.P. and BUCKBY, R.J. (1979) *Composite Structures of Steel and Concrete*, Vol. 2, Granada, St Albans. Chapter 7: Design philosophy, materials and loading; Chapter 8: Design of superstructure for beam-and-slab bridges. Pages 154–171 discuss the design of shear connection for static loading for beam-and-slab bridges.

YAM, L.C.P. (1981) *Design of Composite Steel–concrete Structures*, Surrey University Press, Brighton. Chapter 4: Continuous composite beams; Chapter 6: Miscellaneous problems (includes shrinkage, creep and temperature effects)

28

Composite floors

Objective To present the design requirements for composite slabs using profiled steel sheeting in terms of the construction, in-service and fire conditions.

Prior reading Chapter 27 (Composite beams).

Summary This chapter reviews the range of sheeting profiles used in composite floors and their use both as composite slabs and as permanent formwork. Construction and fire aspects are discussed together with the interaction between the design of composite slabs and beams.

28.1 Introduction

Composite construction means utilising the compressive resistance of concrete slabs in conjunction with steel beams to increase the beams' strength and stiffness. Profiled steel sheeting has been designed to act as permanent formwork for the concrete slab and also to behave compositely with the slab for subsequent in-service loading. Once the concrete has achieved adequate strength the sheeting can act as slab reinforcement to resist live loading. This is achieved by some form of interlock or anchorage, as well as bond, between the slab and sheeting.

A summary of the different sheeting profiles marketed for use in composite slabs is given in Figure 28.1. Profile heights are usually in the range 38–75 mm and sheet thicknesses are between 0.8 and 1.5 mm. There are two well-known generic types: the dovetail profile (Holorib) and the trapezoidal profile with web indentations (for example, Hibond). Spans of the order of 3–3.5 m between beams are common.

Slab thicknesses are controlled by insulation requirements in fire: those above the profile of between 65 and 120 mm are sufficient to give a fire rating of 1–2 h, depending on the profile shape and concrete type. For simply supported slabs with nominal reinforcement ratios of span to overall depth of 35 for normal weight concrete or 30 for lightweight concrete are usually acceptable in limiting deflections. Lightweight concrete is popular, despite its slightly higher initial cost, because of economies in weight and enhanced fire-insulation properties.

28.2 Behaviour of profiled sheeting as permanent formwork

The sheeting must support the weight of concrete in the finished slab, excess weight from concrete placement and ponding, the weight of the operatives and any impact loads during construction. The design of single-span sheeting is usually controlled by deflection and ponding of concrete. Continuous sheeting is generally designed on strength criteria and longer unpropped spans are permissible. The construction load is applied in design as a pattern load even though only one span is likely to be loaded to this extent, because of the progressive nature of concreting. In BS 5950 a limit on residual deflection of the soffit after concreting of the smaller of span/180 or slab depth/10 is specified. Greater deflections are likely to be experienced during concreting.

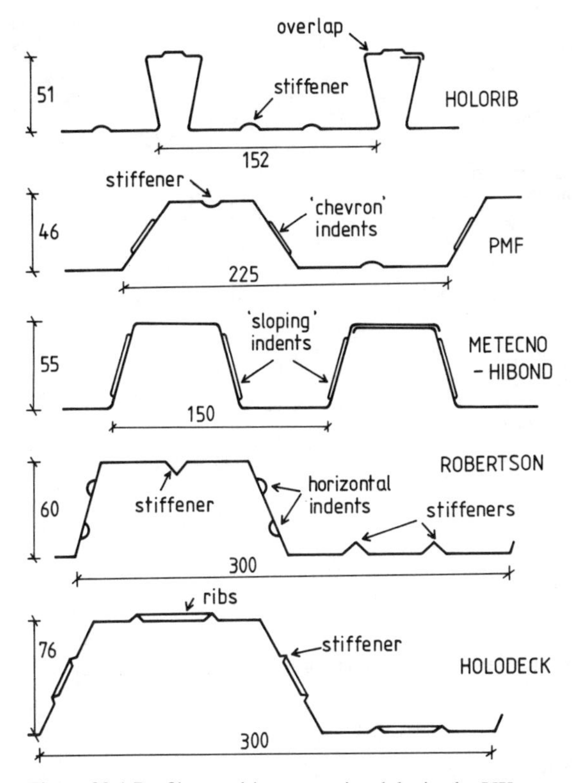

Figure 28.1 Profiles used in composite slabs in the UK

The design of the sheeting in bending is dependent on the properties of the profile, and particularly the thin plate elements in compression. Where profiles are stiffened by one or two folds in the compression plate the section is more efficient in bending. The design of continuous sheeting is to be based, according to code requirements, on an elastic distribution of moment, as a safe lower bound to the collapse strength. Many profiles with wide troughs are weaker under negative moment than positive, and the effect of the reaction at the internal support is to reduce further the negative moment capacity. In design to code requirements, therefore, continuous sheeting may be weaker than simply supported sheeting even though in reality it must be considerably stronger. The real behaviour of continuous sheeting is presented below.

As the stiffness of the profile changes with increased stress so some elastic redistribution of moment occurs from the highly stressed region of the support to midspan. Beyond the point at which yield takes place, plate collapse rather than 'plasticity' tends to occur with increasing deformation. If the sheeting were able to deform in an ideal plastic way as in Figure 28.2(a) then its failure load would be approximately:

$$W_u = \frac{8}{L^2}(M_p + 0.46KM_n)$$

where M_p and M_n are the elastic moments of resistance of the profile under positive and negative moment, respectively, at the design yield stress of the steel. The factor K (less than unity) is introduced because only a proportion of M_n can be mobilized at failure. Spans 10–15% in excess of those predicted by elastic design are possible while still maintaining an adequate factor of safety against failure and satisfactory performance at working load.

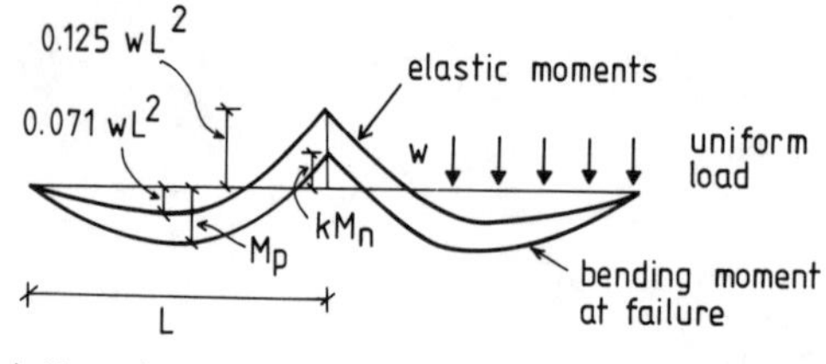

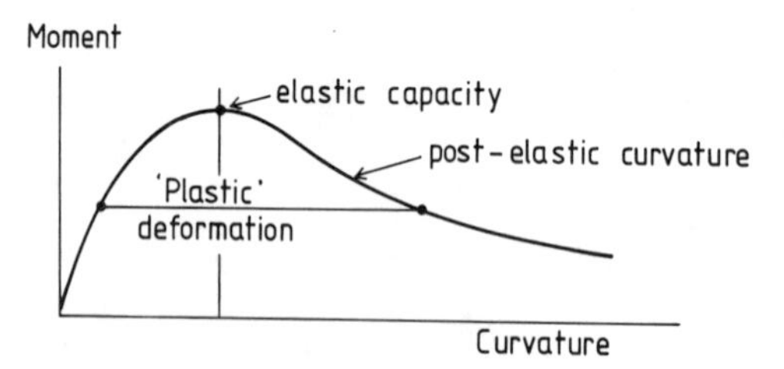

Figure 28.2 Ultimate strength behaviour of sheeting in construction conditions

23.3 Behaviour of composite slabs

The cross-sectional area of the sheeting A_p acts as conventional reinforcement at a lever arm determined by the centroid of the profile, as shown in Figure 28.3. Tensile forces are developed as a result of the bond between the sheeting and the concrete, enhanced by some form of mechanical connection such as by indenting the profile. If the beam-to-slab shear connectors are included in the design then the full flexural strength of the composite slab can usually be mobilized because of the end-anchorages provided. The degree of composite action developed is such that, for all but very heavy imposed loads, permissible spans are controlled by the construction condition.

The ultimate behaviour of composite slabs is controlled by a combination of friction, bond and mechanical interlock after initial slippage. Because of this, the performance of a particular sheeting system can only be assessed by testing.

If design were to be carried out on elastic design principles permissible bond strengths would be of the order of 0.05 N/mm² for plain profiles rising to about 0.3 N/mm² for certain types of

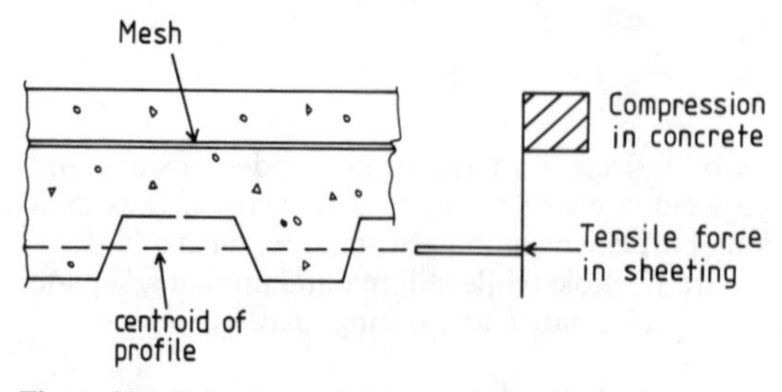

Figure 28.3 Behaviour of slab as reinforced section

indented profile. However, the modern method of determining the ultimate shear-bond strength of composite slabs with a particular profile shape is based on a relationship derived empirically, of the following form:

$$V_u = 0.8A_c(m_d p d_e/L_v + k_d\sqrt{f_{cu}})$$

where V_u is the vertical shear strength per unit width of a simply supported slab, p is the ratio of the profile area to the concrete area A_c per unit slab width, f_{cu} is the cube strength of concrete, d_e is the effective slab depth to the centroid of the profile. This shear strength is equated to the applied ultimate shear force on the slab using load factors of 1.6 for imposed load and 1.45 for self-weight and finishes.

The empirical constants m_d, k_d are determined from prototype slab tests to failure and are calculated from the slope and intercept, respectively, of the reduced regression line as defined in BS 5950: Part 4. The term L_v is the shear span length which is taken as one quarter of the span (L).

Deflection is important because composite slabs are usually designed to be simply supported, and deflection calculations should be based on the average of the cracked and uncracked inertias. Values of the modular ratio between steel and concrete of 10 for normal weight and 15 for lightweight concrete are usually acceptable when assessing the deflection from normal office loading.

28.4 Construction aspects

The sheeting acts as an efficient working platform. It is fastened to the support beams, usually by self-tapping screws or by the welded stud shear-connectors used in developing the composite action of the slab and beam. During concreting the sheeting is assumed to stabilize the beams. Therefore to enable the sheeting to act as a shear diaphragm seam fasteners should connect the individual sheet lengths at about 1 m spacing.

Concrete grades of 25 or 30 are usually specified, higher grades not being structurally necessary. Volume or area limits on the placing of concrete are not as stringent as for reinforced concrete slabs because minor cracking is distributed by the restraining action of the sheeting and additional reinforcement.

28.5 Interaction between composite floor and beam

Composite floors may be designed independently of, or in conjunction with, the supporting structure. The composite action of the beams and slab is developed by use of shear-connectors, normally of the welded stud form. The efficiency of this shear-connection is strongly dependent on the area of concrete around the stud, particularly where the stud is placed within the trough of the profile.

The sheeting may cross or run parallel to the composite beams under consideration, and it is normal practice for it to be placed over a number of beams with the shear studs welded directly through the deck. The following effective strength formula for studs has now been adopted by most codes:

$$P_{d,eff} = \frac{0.85}{\sqrt{n}}\frac{\omega}{e}\frac{(h-e)}{e}P_d$$

where P_d is the design strength of the stud in a solid slab and n is the number of studs per trough ($n < 3$). The profile shape is represented by the profile height (e) and the average trough width (ω). The stud of height (h) should project at least 35 mm above the profile. This formula applies where the sheeting crosses the beam; where the sheeting is placed parallel to it there is no reduction for the number of studs and the coefficient is reduced to 0.6. There are various limitations on the spacing of the studs. Relatively large in-plane forces resultings from wind action can be developed in the floor and transferred to vertically braced elements or walls. In such cases, the resulting forces in the shear-connectors may need to be calculated.

Longitudinal shear transfer is particularly important where composite floors are used because of the relatively shallow depth of concrete above the sheeting. Where the sheeting is continuous, or rigidly connected to the support beams by welded studs, then the sheeting may be taken to perform the function of bottom transverse reinforcement. The mesh in the concrete topping may need to be supplemented close to the beam, particularly at edge beams where shear is transferred to one slab only.

28.6 Concluding summary

The design of composite slabs using profiled steel sheeting must take into acount:

1. The behaviour of the sheeting in supporting concrete and other loads during construction;
2. The in-service composite behaviour based on empirical formulae using test parameters;
3. The requirements for additional reinforcement for adequate fire resistance;
4. The interaction between the slab and the beam in the design of a composite frame.

Background reading

British Standards Institution (1983) The Design of Composite Floors with Profiled Steel Sheeting, BS 5950: Part 4

Design of Multi-span Profiled Sheeting (1984) CIRIA Technical Note 116

European Convention for Constructional Steelwork (1981) *Composite Construction*, The Construction Press, London

Fire Resistance of Composite Slabs with Profiled Steel Decking (1988) The Steel Construction Institute

LAWSON, R.M. (1983) *Composite Beams and Slabs with Profiled Steel Sheeting*, CIRIA Report 99

29

Trusses and lattice girders

Objective To discuss the types and uses of trusses and lattice girders and set out the methods of analysis and design for these elements.

Prior reading Chapter 17 (Tension members); Chapter 22 (Columns).

Summary This chapter presents the types and uses of trusses and lattice girders and indicates the members that are most often used in their construction. A discussion of overall truss design considers primary analysis, secondary stresses, rigorous elastic analysis, cross-braced trusses and truss deflections. The practical design of truss members is discussed with a design example.

29.1 Types and uses

29.1.1 General

The truss or lattice girder is a triangulated framework of members in which lateral loads are resisted by axial forces in the individual members. The terms are generally applied to the planar truss. A 'space frame' is formed when the members lie in three dimensions. The main uses are:

1. In buildings, to support roofs and floors, to span large distances and carry relatively light loads (see Figures 4.5 and 4.6);
2. In road and rail bridges, for short and intermediate spans and in footbridges, as shown in Figure 5.7;
3. As bracing in buildings and bridges, to provide stability where the bracing members form a truss with other structural members such as the columns in a building; examples are shown in Figure 6.5.

Trusses and lattice girders are classified in accordance with their overall form and internal member arrangement. Pitched trusses are used for roofs and parallel chord lattice girders are employed to support roofs and floors and for bridges. In the past, proper names were given to the various types of trusses such as the Fink truss, Warren girder, etc. The most commonly used truss is single span, simply supported and statically determinate with joints assumed to act as pins.

The Vierendeel girder should also be mentioned. This consists of rigid jointed rectangular panels as shown in Figure 4.5. This truss is statically indeterminate and so will not be further considered here, although it has a pleasing appearance and is often used in footbridges.

The lattice girder acts as a beam, but gives a large saving in weight of steel when compared with a universal beam designed for the same service. The saving arises from the much greater depth used for the lattice girder. Some of the saving in material costs is offset by fabrication costs.

29.1.2 Typical members

Truss, lattice girder and bracing members for buildings are selected from:

1. Open sections, primarily angles, channels, tees and joists;
2. Compound sections, i.e. double angle and channels;
3. Closed sections – in practice, structural hollow sections;

For bridges, members are selected from:

1. Rolled sections;
2. Compound sections;
3. Built-up H and box sections;

The selection of members depends on the location, use, span, type of connection to be used and appearance required. Hollow sections are more expensive than open ones but are cheaper to maintain and have a better appearance. Angles are the sections traditionally used for truss construction.

29.2 Loads on trusses and lattice girders in buildings

The main types of loads are shown in Figure 29.1:

1. Dead loads. These are caused by self-weight, sheeting, decking, floor or roof slabs, purlins, beams, insulation, ceilings, services and finishes. Dead loads must be carefully estimated for the construction to be used in any particular case from material weights given in handbooks and manufacturers' literature.
2. Imposed loads. These are given in BS 6399: Part 1 for floors in various types of building and for roofs with or without access. The imposed load may cover the whole or part of the member and should be applied in such a way as to cause the most severe effect.
3. Wind loads. These are given in CP3: Chapter V: Part II and can be estimated from the location of the building, its dimensions and the sizes of openings on its faces. Wind generally causes uplift on roofs; this may lead to reversal of load in truss members in light construc-

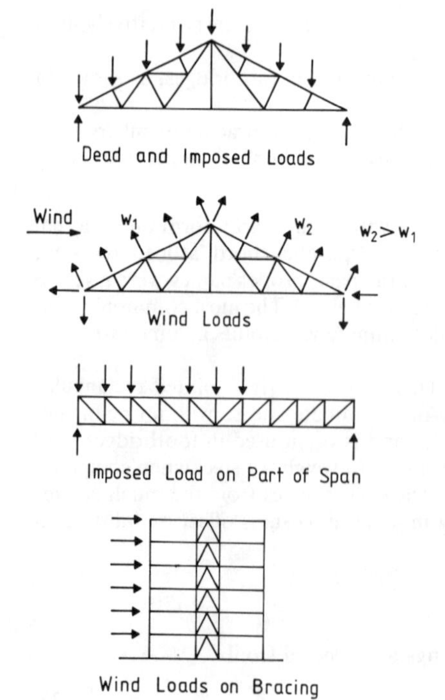

Figure 29.1 Loads on trusses and bracing

tion. In multi-storey buildings wind gives rise to horizontal loads that must be resisted by the bracing.

In special cases trusses resist dynamic, seismic and wave loads. A careful watch should be kept for unusual loads applied during erection. Failures may occur at this stage when the final lateral support system is not fully installed.

29.3 Analysis of trusses

29.3.1 Primary forces

Trusses may be single span, statically determinate or indeterminate, or may be continuous over two or more spans. Manual methods of analysis for trusses, where the loads are applied at the nodes, are joint resolution, the method of sections and the force diagram. Joint resolution is the quickest method for analysing parallel chord lattice girders when all the forces are required. The method of sections is useful where the values of the forces in only a few critical members are needed. The force diagram is the best general manual method. Computer programs are also available for truss analysis.

29.3.2 Secondary stresses in trusses

In many cases in the design of trusses and lattice girders it is not necessary to consider secondary stresses. These should, however, be calculated for heavy trusses used in industrial buildings and bridges. Secondary stresses are caused by:

1. Eccentricity at connections;
2. Loads applied between the truss nodes;
3. Moments resulting from rigid joints and truss deflection.

These are discussed in detail below.

29.3.2.1 Eccentricity at connections

BS 5950 states that trusses should be so detailed that either the centroidal axes of the members or the bolt gauge lines meet at a point at the nodes. Otherwise, members and connections should be designed to resist the moments due to eccentricity. These moments should be divided between members meeting at joints in proportion to their rotational stiffnesses. Stresses due to small eccentricities are often neglected.

29.3.2.2 Loads applied between the nodes

Moments due to these loads must be calculated and the stresses arising combined with those due to the primary axial loads, i.e. the members concerned must be designed as beam columns. This situation often occurs in roof trusses where the loads are applied to the top chord through purlins which may not be located at the nodes, as shown in Figure 29.2. The manual method of calculation is first to analyse the truss for the loads applied at the nodes which gives the axial forces in the members. Then a separate analysis is made for bending in the top chord which is considered as a continuous beam. The ridge joint E is fixed because of symmetry, but the eaves joint A should be taken as pinned; otherwise, moment will be transferred into the bottom chord if the joint between the truss and column is assumed to be pinned. The top chord is designed for axial load and bending. Computer analysis is mentioned below.

29.3.2.3 Moment, resulting from rigid joints and truss deflection

Stresses resulting from secondary moments are important in trusses with short, thick members. Approximate rules specify when such an analysis should be made. BS 5950 states in Clause 4.10 that secondary stresses will be insignificant if the slenderness of the chord members in the plane of the truss is greater than 50 and that of most of the web members is greater than 100. In trusses

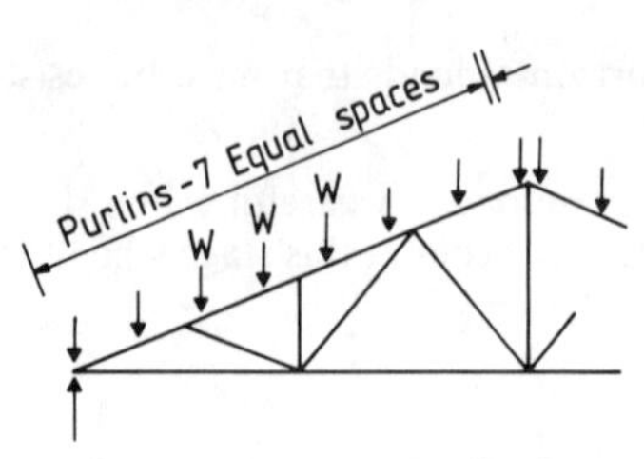

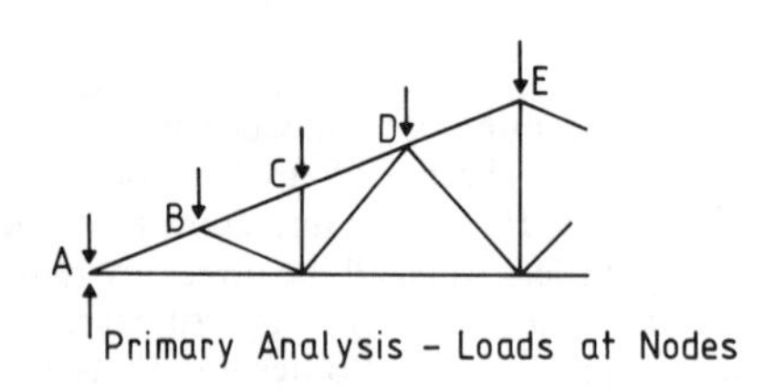

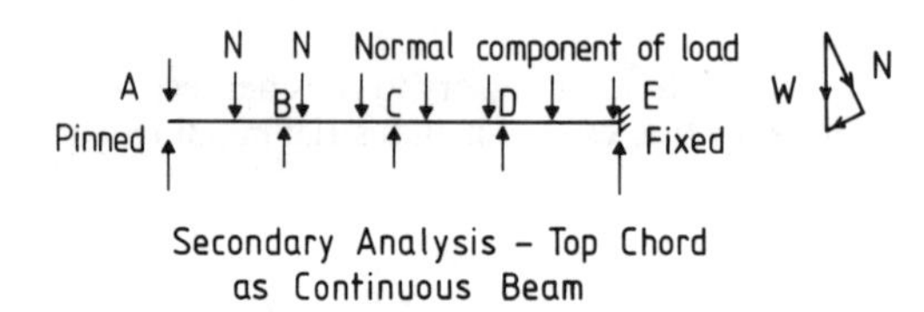

Figure 29.2 Loads applied between truss nodes

in buildings the loads are predominantly static and it is not necessary to calculate these stresses. The maximum stresses from secondary moments occur at the ends of members and are not likely to cause collapse. The method of analysis for secondary moments is set out below.

29.3.3 Rigorous elastic analysis

Rigid-jointed, redundant or continuous trusses or trusses with loads applied between the nodes can be analysed using a plane frame program based on the matrix-stiffness method of frame analysis. The truss can also be modelled taking account of joint eccentricity. Member sizes must be determined in advance using a manual analysis. All information required for design is output including joint deflections.

It is important that a consistent approach is adopted for analysis and design. This means that if secondary moments are to be ignored then the primary axial forces to be used in design must be obtained from the simple analysis of the truss as a pin-jointed frame. The axial forces obtained from a rigid frame computer analysis may be modified considerably by the joint moments.

29.4 Secondary considerations

29.4.1 Cross-braced trusses

In the bracing provided to stabilize multi-storey buildings the panels often have cross-diagonals, as shown in Figure 29.3. It is customary to consider the truss as statically determinate,

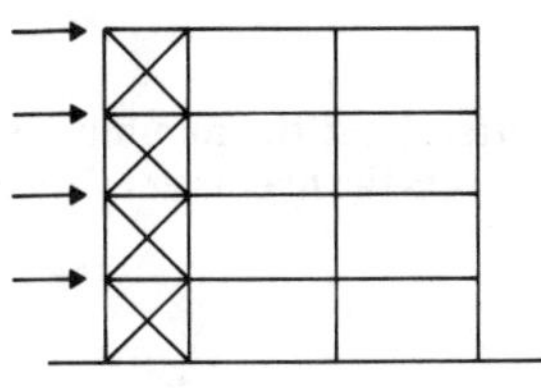

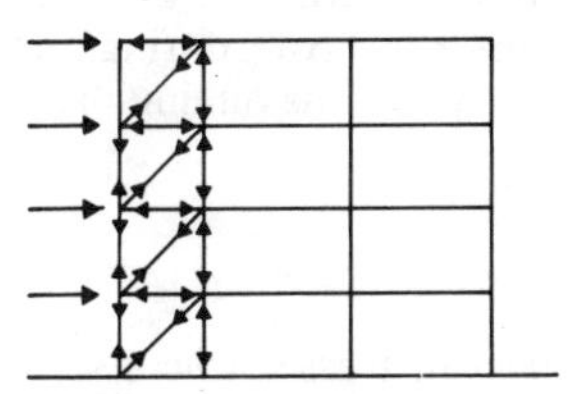

a) Bracing in a Multi–Storey Building

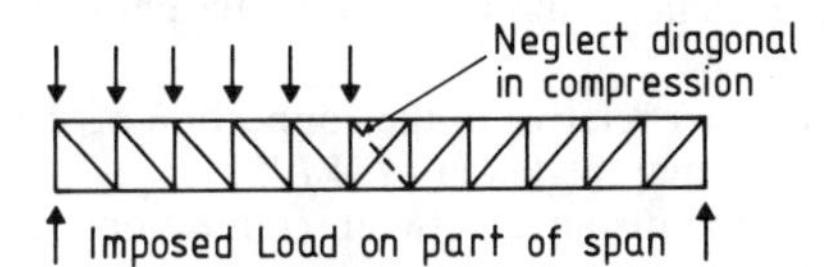

b) Lattice Girder – Odd No. of Panels

Figure 29.3 Cross-braced trusses

with only the set of diagonals in tension assumed to be effective. When the wind reverses the other set becomes active.

Another common case is the lattice girder with an odd number of panels. The centre panel is cross-braced as shown in Figure 29.3 and under symmetrical loading there are no forces in these diagonals. If imposed load is placed over part of the span only the diagonal in tension is assumed to be effective.

29.4.2 Deflection of trusses

The deflection for a pin-jointed truss can be calculated using either the strain energy or virtual work method. The deflection using the strain energy method is given by:

$$\delta = \Sigma FuL/EA$$

The Williot–Mohr graphical method can also be used to determine truss deflections. If a computer analysis is carried out joint deflections are given as part of the output.

A truss may be cambered during fabrication to offset deflections due to applied loads. The term 'cambering' means that a given upward deflection can be built into, say, a nominally horizontal truss during fabrication by adjusting the member lengths slightly to cause the truss to bow upwards.

29.5 Design of truss members

The truss should be analysed for the separate load cases. These cases are combined to give the most severe conditions for design of each element. Some important aspects of design are set out below.

29.5.1 Compression members

Maximum slenderness ratios are defined in codes and these often limit the minimum size of members that can be used in light trusses. BS 5950: Clause 4.7.3.2 sets the maximum slenderness values as:

Members resisting dead and imposed load	180
Members resisting wind load	250
Any member normally acting as a tie but subject to reversal of stress due to wind	350

These limits ensure that reasonably robust members are selected when only light loads are involved. Wind loads are transient and larger slenderness values are permitted than for dead and imposed loads. These rules also reduce the likelihood of damage occurring during transport and erection. In this regard it has been common practice to specify that the minimum sizes for angles should be as follows:

1. Equal angles – 50 × 50 × 6
2. Unequal angles – 65 × 50 × 6

BS 5950: Clause 4.10 gives empirical design rules for members in trusses where secondary bending stresses are insignificant. The following assumptions are made:

1. For the purpose of analysis the joints are taken as pinned.
2. For calculating effective lengths the fixity of connections and rigidity of adjacent members may be taken into account.
3. Where the exact position of point loads on the rafter relative to the connection of the web members is not known the local bending moment may be taken as $WL/6$.
4. The length of chord members may be taken as the distance between connections to web members in the plane and the distance between purlins or ties out of the plane of the truss.

Figure 29.4 shows roof trusses in place in a building with the purlins providing lateral support to the top chord.

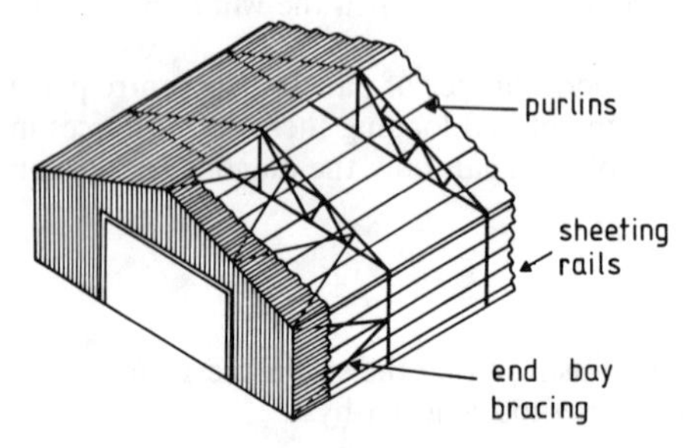

a) Purlins provide Lateral Support to Top Chord of Truss

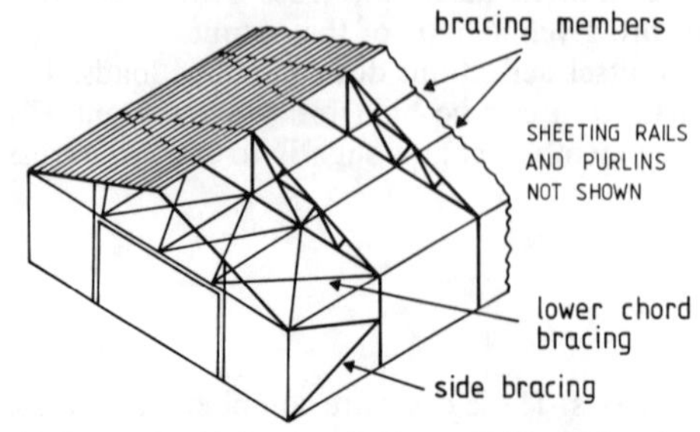

b) Lower Chord Bracing System provides Lateral Support to Bottom Chord of Truss

Figure 29.4 Lateral support for roof truss

The effective lengths for the continuous top chord of a truss are shown in Figure 29.5. The web joints reduce the in-plane effective lengths while the purlins only hold the chord in position laterally. Truss chords are often designed for axial load and moment.

Two common internal truss members are the single-angle discontinuous strut connected to a gusset or another member and the double-angle discontinuous strut connected to both sides of a gusset or another member. These should be connected by at least two bolts or the equivalent in welding. BS 5950: Clause 4.7.10 states that the end-eccentricity may be ignored and the struts designed as axially loaded members. Slenderness values for these two cases are shown in Figure 29.6. Values for other sections and member arrangements are given in the codes.

29.5.2 Tension members

Structural hollow sections connected by welding may be fully effective. The 'effective area' is to be used for angles connected through one leg. Theoretically, rounds or cables could be used, but these are unsuitable for practical reasons, because they lack stiffness and are easily damaged. The same minimum sections for angle members set out above for compression members should be adopted for tension members.

29.5.3 Members subject to reversal of load

These are to be designed for the worst conditions. The bottom chord is often the most vulnerable member and will require lateral support at intervals in its length from a lower chord-bracing system. This is shown in Figure 29.4.

29.5.4 Practical design points

1. It is not economic to make every member a different size. The designer should rationalize the sizes and use only two or three different sections.

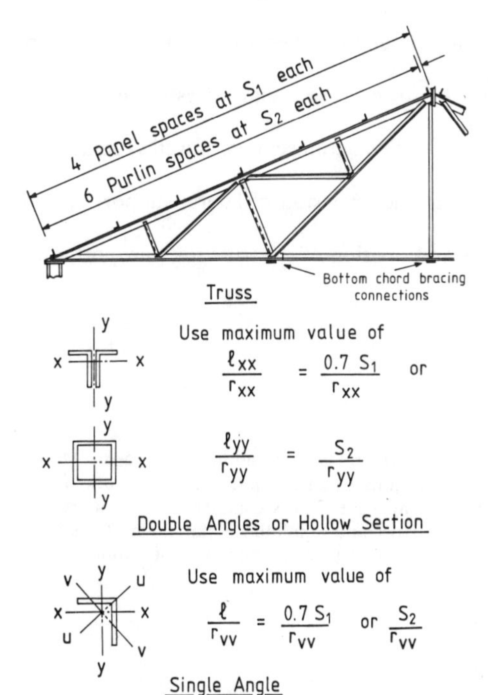

Figure 29.5 Slenderness ratios for continuous chords

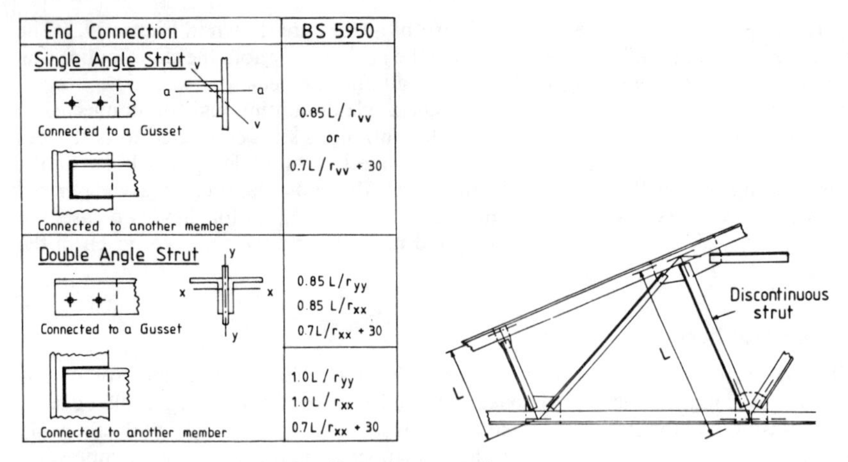

End Connection	BS 5950
Single Angle Strut: Connected to a Gusset; Connected to another member	$0.85\,L/r_{vv}$ or $0.7L/r_{vv} + 30$
Double Angle Strut: Connected to a Gusset	$0.85\,L/r_{yy}$ $0.85\,L/r_{xx}$ $0.7L/r_{xx} + 30$
Connected to another member	$1.0\,L/r_{yy}$ $1.0\,L/r_{xx}$ $0.7L/r_{xx} + 30$

Figure 29.6 Slenderness ratios for discontinuous angle struts

2. Minimum sizes should be adopted to prevent damage during transport and erection. Recommendations are set out above.
3. Safe load tables are very useful; members subjected to axial load can be selected directly. Those subjected to axial load and moment must be designed by successive trials. Select the initial size by assuming that the compression resistance is 60% of full capacity.
4. Large trusses must be subdivided for transport. Bolted site splices are used to assemble the truss on site.

29.6 Concluding summary

1. Trusses and lattice girders are important elements in building where they are used to support floors and roofs and provide bracing.
2. Statically determinate trusses are generally used. However, secondary stresses due to loads applied between the nodes must be calculated.
3. Selection of members must be based on practicality of end-connections, as well as considerations of strength and economy of weight.
4. Careful consideration must be given in design to the provision of lateral support and to possible stress reversals.
5. Practical design points such as minimum sizes and division of the truss for transport must be considered.

Background reading

GHALI, A. and NEVILLE, A.M. (1978) *Structural Analysis*, Chapman and Hall, London. Pages 45–86: Matrix stiffness method of analysis; Pages 119–122: Truss deflections using method of virtual work; Pages 177–180: Graphical determination of deflection of trusses; Pages 359–364: Secondary moments in trusses.

JENKINS, W.M. (1982) *Structural Mechanics and Analysis*, Van Nostrand Reinhold, Wokingham. Pages 5–21: Analysis of determinate trusses; Pages 204–215: Deflections of trusses and analysis of redundant pin jointed trusses.

Introduction to connection design

Objective To introduce connection design.

Prior reading Basic understanding of elastic and plastic behaviour.

Summary The chapter describes the principal components of a connection and presents some typical connections. It explains the importance of connection design to the structure as a whole and emphasizes the significance of practical considerations within connection design. The behaviour of connections is discussed and an overall approach for design and serviceability is presented.

30.1 Introduction

30.1.1 Why do we need connections?

Steel sections are linear elements that are normally manufactured in lengths of around 15 m. Steel plates usually range in size from $4 \times 1\frac{1}{2}$ m to $8 \times 2\frac{1}{2}$ m. These components of a steel structure cannot be larger because of economic constraints on manufacture and transportation. Connections are therefore necessary to overcome these size limitations. In addition, most steel structures are three-dimensional and connections are also necessary to synthesize such spatial structures from one- and two-dimensional elements. Put another way, connections are necessary to create structures with sufficient continuity from an essentially discontinuous set of components.

30.1.2 The components of a connection

Connections may be carried out either wholly or partly by welding. Figures 30.1 and 30.2 summarize the most common types of welds: fillet welds, where the weld metal is generally

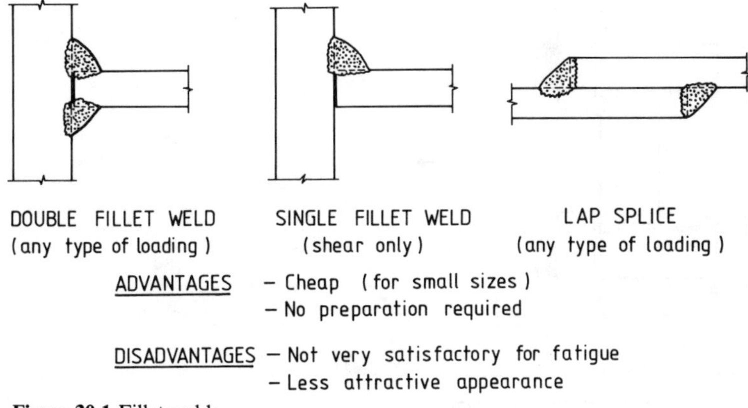

Figure 30.1 Fillet welds

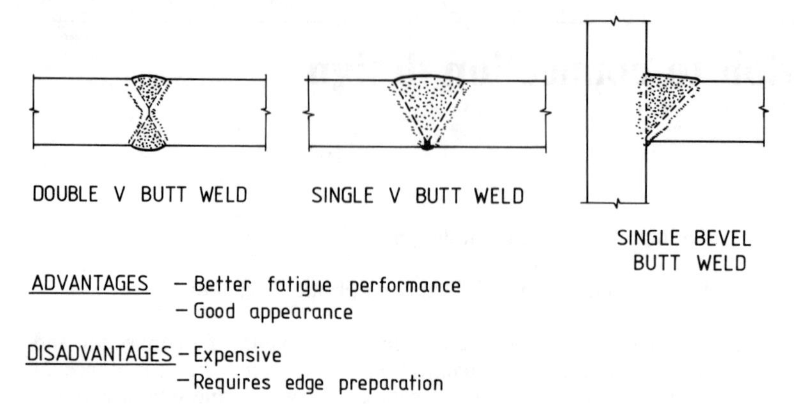

Figure 30.2 Butt welds

outside the profile of the connected elements; and butt welds, where the weld metal is generally deposited within the profile of the connected elements.

It is usual for site connections to be carried out by bolting. Figure 30.3 shows the most common type of bolt, the bearing bolt in a clearance hole, often called an 'ordinary' bolt. It is popular because of its low cost both to buy and to install, and its principal disadvantage is that it will slip into bearing at a low load in shear. This flexibility arises because it is only when the bolt is bearing against the sides of the clearance holes that load can be transferred into the bolt and be transmitted between the plies by shear and bending of the bolt. Where this flexibility

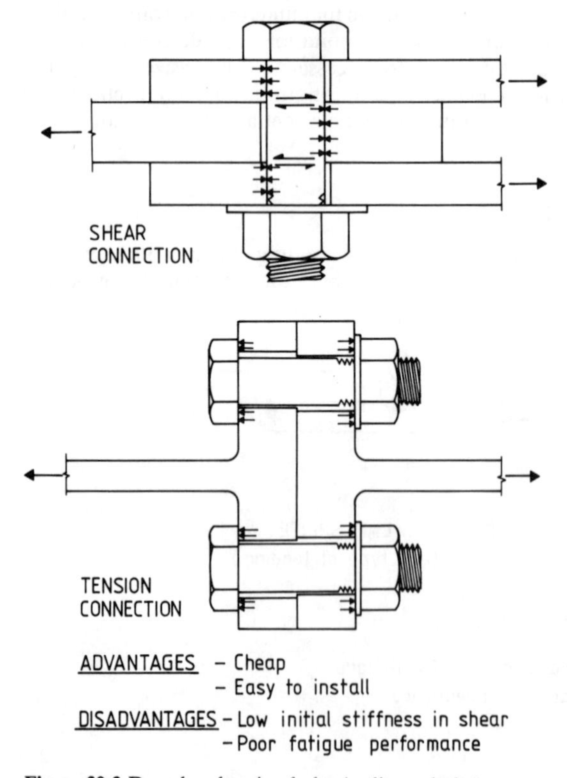

Figure 30.3 Dowel or bearing bolts (ordinary bolts)

is unacceptable the more expensive high-strength friction-grip (HSFG) bolt shown in Figure 30.4 has to be used. This achieves a high stiffness in shear because it resists such loading by friction between the plies at the interfaces.

However, connections consist of much more than just welds and bolts. Figure 30.5 gives examples of some of the other components that make up complete connections. For economy, these are cut from flats, plates, angles, tees and other rolled sections wherever possible.

30.2 The importance of connection design

30.2.1 Safety

Any structure is only as strong as its weakest links, and in many structures it is the connections that will be weaker than the components they join. Examples are shown in Figure 30.6. It is very difficult to connect angle members in tension without some local loss of effectiveness at the connection and it is expensive to achieve full continuity in a beam or column splice. The usual approach to no-sway design assumes that the beam/column connections will be so weak in flexure that their moment capacity can be discounted in analysis.

It is clearly essential that these weakest links can develop their full design strength. In addition, their ductility will be a major factor in ensuring the deformation capacity and robustness of the structure as a whole. Any unanticipated deformation of the structure from overload, settlement or accident will have to be accommodated primarily by the connections. In the particular case of the beam/column connections referred to above the entire design approach is based on the assumptions that the beams may be safely treated as simply

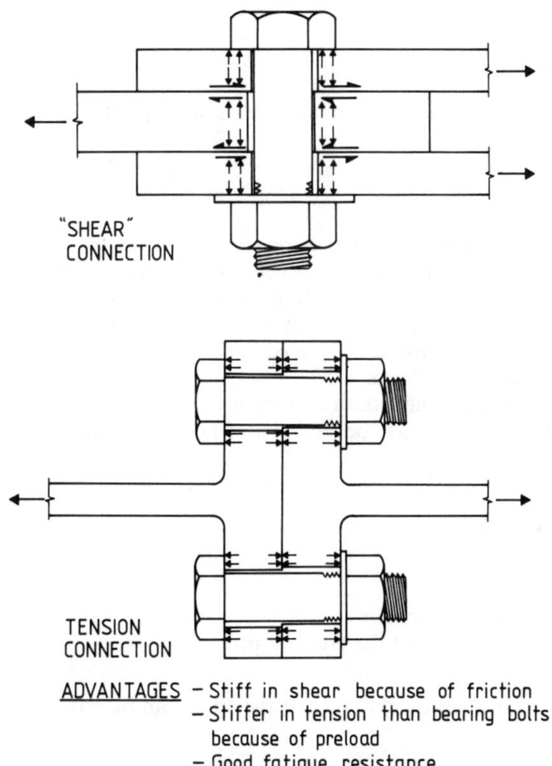

Figure 30.4 High-strength friction-grip bolts

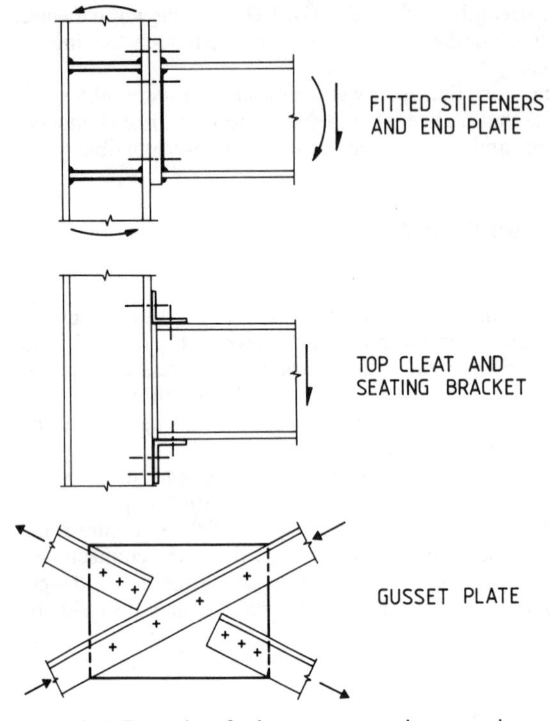

Figure 30.5 Examples of other components in connections

supported, and that the associated end-rotations will not impair the shear capacity of the end-connections.

30.2.2 Economy of the structure

Connection design is also of paramount importance to the economy of the structure. Steel sections and plate cost £300–£400 per tonne. Conventional steel structures are £600–£1200 per tonne, while complex or specialized steel structures may be much more. Except for corrosion prevention and transport, all this added value is related to connection design. The cost of the connectors themselves, the work in the fabrication shop and the erection on site may together cost considerably more than the material for the elements.

Designers of connections can have considerable influence on economy. As they carry out their design and detail the connections they must always be thinking of how these are to be made:

1. Maintaining ease of access for welding;
2. Optimizing the use of automatic equipment;
3. Minimizing precise fitting;
4. Achieving repetition of standard details, etc.

are all important factors in achieving economic fabrication.

In a similar way designers must always have in mind the erection process:

1. Ease of access for site bolting;
2. Provision for supporting the element self-weight quickly so that the crane can be released;
3. Ease of adjustment for alignment;
4. Simplicity and repetition, etc.

are important factors in achieving economic erection.

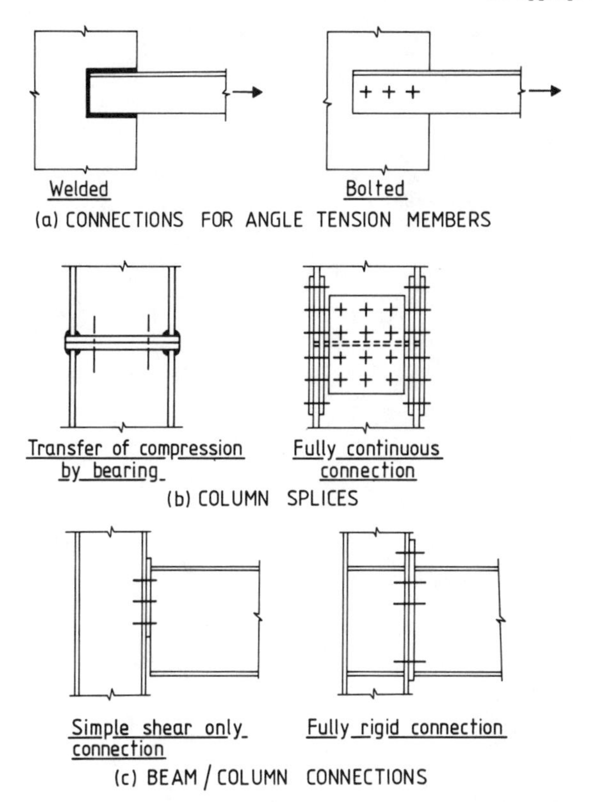

Figure 30.6 Examples of partial and full-strength connections

30.2.3 Economy of design

Connection design is a significant part of total design cost for the structure. Most connection design, especially for standard connections, should be a simple, routine matter that can be carried out by technician engineers or designer/detailers. Both sides of the design team need to develop an understanding of the safe limits to such procedures. Only when the situation is outside these limits, because of unusual arrangement of members, unusual loading or some other requirement, should it be necessary to design from first principles in the way presented in the second part of this chapter. An appreciation of these limits can come either from experience or from an understanding of first principles.

With the connection shown in Figure 30.7(a) a simple calculation to determine the minimum number of bolts, on the basis of their shear and bearing capacity in the presence of the eccentricity shown, may be judged to be sufficient because of experience and past practice. Alternatively, the student who wishes to achieve a deeper understanding may consider the full behaviour of the connection in the way outlined later in this chapter and shown in Figure 30.7(b). He or she will then understand that the simple procedures are appropriate, for connections of normal proportions, because the other modes of behaviour will not govern strength in such circumstances.

30.3 An appropriate design approach for connections

Connection behaviour is both more complex and more variable than that of the elements which are being connected. There are several reasons why this is so:

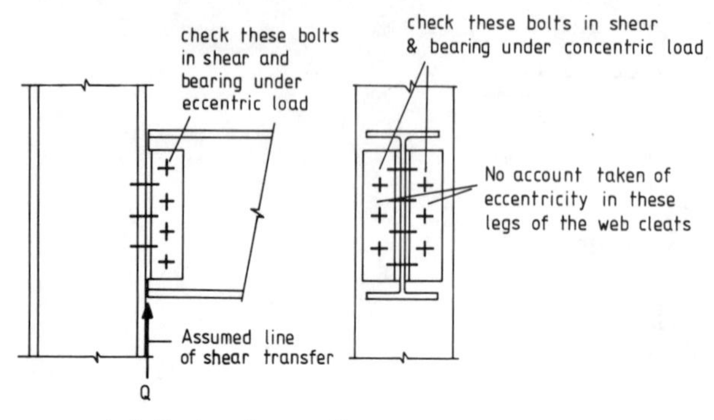

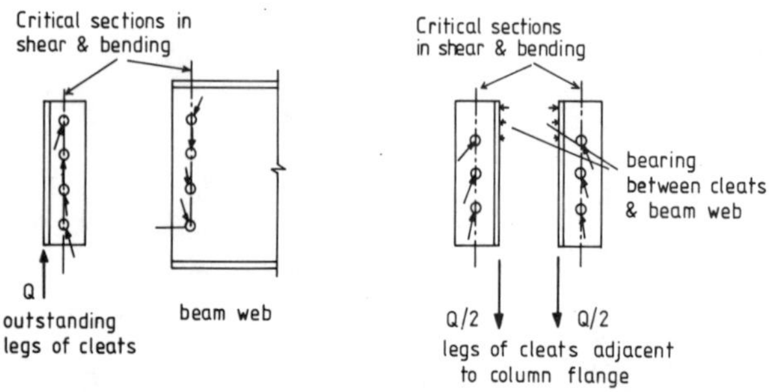

Figure 30.7 Simple web cleat beam/column connections: contrast between design assumptions and detailed behaviour

1. Geometric imperfections arising from welding distortions, cutting and machining imperfections, and permitted misalignment of clearance holes for bolts between plies;
2. Residual stresses and strains due to these lacks of fit, and welding shrinkage;
3. Geometric complexity within the connection.

These complicating factors can, in extreme circumstances, cause the variations in behaviour shown in Figure 30.8.

It is not easy to devise an appropriate design approach. It is tempting to rely solely on empiricism:

This is what to do in such a situation. The method of analysis does not really reflect what happens but it has worked in the past.

Unfortunately this reliance on past practice is little help when faced with a connection of unusual proportions or of an innovatory nature. In such circumstances it is necessary to try to be more scientific:

Wait a minute! Let's think about this unusual case more carefully. We must look at various analytical models and try to determine the likely real behaviour, or else!

There is little to be gained from any attempt to model the complexities of behaviour by using complex analysis. The variability of connection behaviour makes it impossible to justify any complexity.

What is needed, and what follows, is a simple but complete design method that should produce a connection with sufficient robustness and ductility to overcome the uncertainties of behaviour.

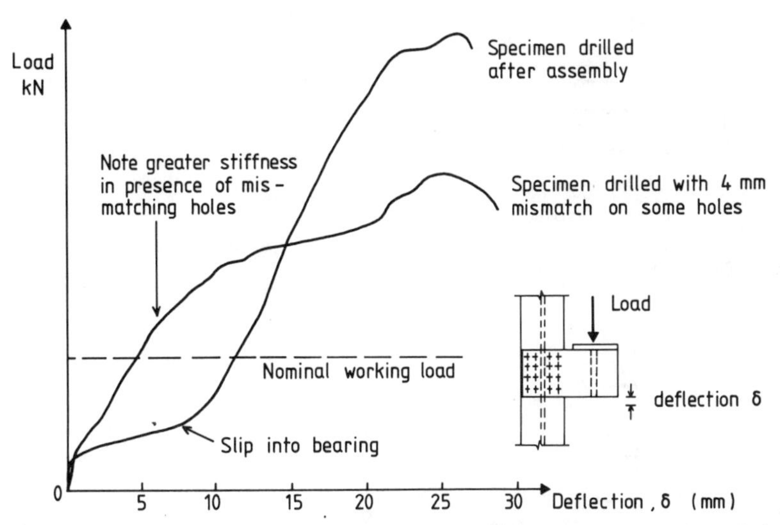

Figure 30.8 Response of bracket connection to eccentric load, 16 M20 Grade 4.6 bolts

(1) *Carry out an appropriate simple analysis to determine a realistic distribution of forces within the connection that is in equilibrium with the applied loading*

This analysis may use one of the methods outlined in Chapter 35 or be based on the concept of 'force paths'. In the latter case the overall loads acting on the connection are replaced by statically equivalent systems of forces which can then be assigned specific paths through the connection.

In carrying out the analysis account should be taken of:

1. The distribution of forces in the elements that are to be connected. For example, if the connection involves an I-beam carrying shear and moment, then, remote from the connection, the shear will be concentrated in the beam web and the flanges will carry most of the moment. A simple and sensible assumption is that the web splice should carry all the shear and the flanges carry all the moment. This is quite satisfactory for building structures provided that it does not lead to an overstress in the flanges.
2. The flexibilities of the components of the connection. It is the most flexible components that will govern the distribution of forces. For example, in the end-plate connections shown in Figure 30.9, if the bolts are of small diameter and the end-plate is thick, it is the bolt flexibility that will govern the distribution of forces – as indicated by conventional analysis. However, if the bolts are stiff compared to the end-plates it is the flexural action of the latter that will primarily govern the distribution of forces in the connection, including the distribution of forces in the bolts.

It is most important to ensure that the analysis is consistent throughout the connection. This may be achieved by carrying out a single analysis, usually of the most critical part of the connection, and using that to determine the distribution of forces in other parts of the connection. It is not uncommon to see designs where serious inconsistencies in analysis have occurred. These most commonly arise when more than one analysis has been used to determine the distribution of forces. For example, in the end-plate connection illustrated in Figure 30.10 it would not be correct to use separate conventional analyses to determine both the distribution of forces in the bolts and in the weld attaching the end-plate to the beam. Such separate analyses would assign different proportions of the forces to different levels in the beam and imply an instantaneous redistribution of forces at the plane of contact between the weld and the end-plate!

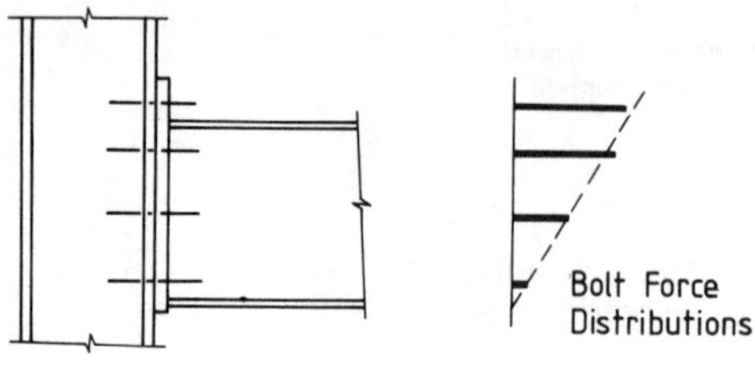

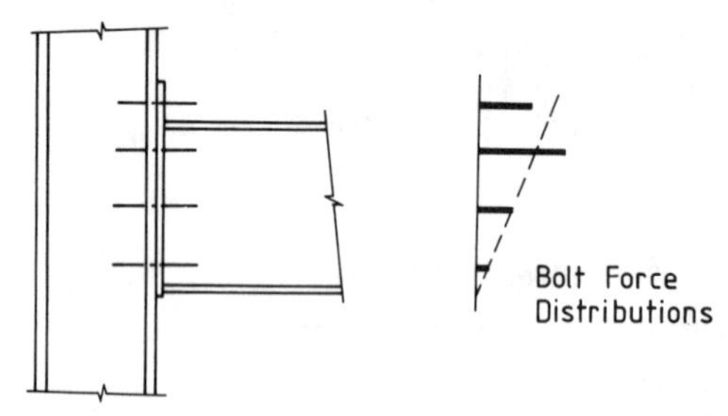

Figure 30.9 Bolt force distributions in beam/column connections with end-plates

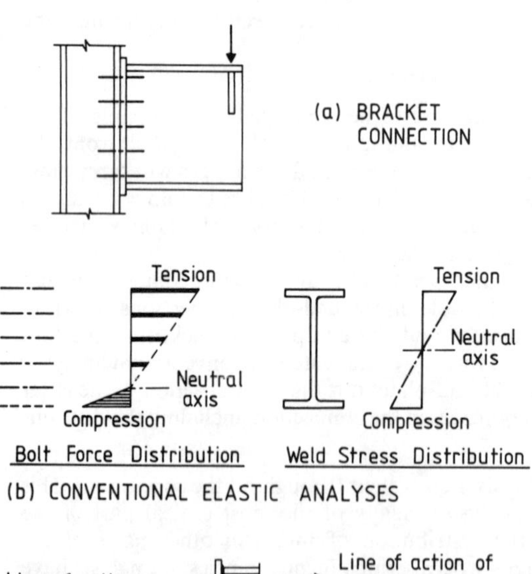

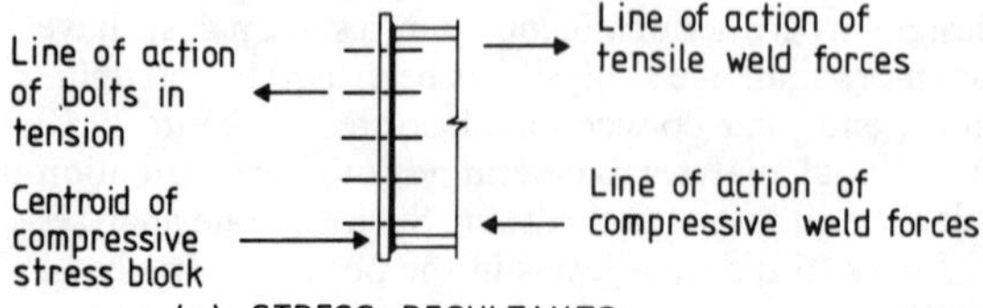

Figure 30.10 Inconsistency in connection analysis

(2) *Ensure that each component of each force path has sufficient strength to transmit the required force*

This is self-evident and yet it is disturbing how frequently designers leave a weak link somewhere in a connection. A major disadvantage of traditional methods of analysis is that they concentrate on the distribution of forces in the connectors. Many codes of practice only give guidance on connector strength. Unwary or inexperienced designers are thus beguiled into thinking that, providing that they have checked the bolts and/or welds, the connection is satisfactory. In reality, more design effort has sometimes to be devoted to the other components than to the connectors themselves.

The only way to be certain that designs are satisfactory is for designers to have a clear understanding of how they wish the connection to behave and for them to ensure that all the components and critical sections have the capacity for this mode of behaviour.

(3) *Recognizing that the above procedure can only give a connection where equilibrium is capable of being achieved but, where compatibility is unlikely to be satisfied, ensure that the connection is capable of ductile behaviour*

The incompatibilities may arise either from simplifications of the analysis or because of some lack of fit; their cause is not important. However, because of their probable presence it is **essential** that the connection is sufficiently ductile for plastic deformation to remove them prior to failure. Providing that this precaution is taken, even if it has not been possible to predict elastic response accurately, the connection will redistribute forces until it is acting in the way that was assumed in design. Fortunately, it is usually a straightforward matter to ensure that the components can achieve the necessary ductility.

(4) *Recognizing that the preceding steps only relate to static ultimate capacity, ensure that the connection will achieve satisfactory serviceability, fatigue resistance, etc.*

For connections in buildings that have been designed by conventional elastic approaches this step may generally be omitted. However, in the following cases further calculations will be necessary:

1. Where either overall analysis or individual component design has been based on simple rigid-plastic methods it will be necessary to ensure that only limited plasticity has occurred at working load levels.
2. Where the connection is subject to significant repeated loading a separate assessment of fatigue resistance should be carried out. This can create considerable difficulties because it requires both detailed consideration of the elastic response of the connection and an evaluation of important stress concentration factors. In extreme circumstances (for example, tubular connections in offshore structures) design for fatigue resistance should govern the overall design procedure and the sequence outlined above should be reversed.

30.4 Application of the design philosophy

Most connection design is very straightforward and satisfies the preceding criteria by implication rather than by specific calculation. For example, conventional design of the simple beam column connection shown in Figure 30.11(a) is carried out in two lines of calculation as a designer/detailer determines the number of bolts necessary to resist the eccentric load shown on the basis of their shear and bearing capacity. Here, the simple analysis is based on the assumption that the eccentricities between the lines of bolts A and A′ may be discounted because of symmetry in a connection of normal proportions with two web cleats. Experience indicates that the only critical checks are the bolts in shear and the cleats and beam web in

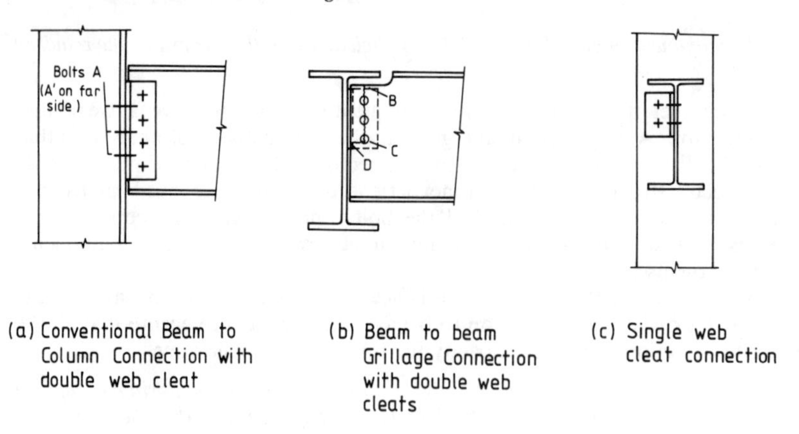

Figure 30.11 'Simple' beam/column and beam/beam connections

bearing. Other 'weak links' are likely to be designed out by minimum pitch criteria and practical edge distance requirements. Experience has also indicated that such connections have sufficient ductility to accommodate both lack of fit within the connection and the beam end-rotation as it takes up its deflected profile.

If all connection design were as straightforward there would be little point in the explicit procedure presented in the preceding section. However, it is only necessary to vary some details in this simple example to illustrate the importance of a sound appreciation of connection design. Figure 30.11(b) shows the beam end-detail that results if the top flange has to be notched. Line BCD becomes a very important critical section to be checked in shear and tension. If a single short web cleat were to be used, as shown in Figure 30.11(c), the same local moments that were reasonably ignored in the double-cleat connection may at least cause unserviceability as the beam twists and could lead to an unacceptable reduction in strength. Similarly, the same simple approach to analysis will clearly lead to an unacceptable reduction in strength in a fillet-welded beam end-connection because of lack of ductility in the weld.

The direct use of the design philosophy outlined in the previous section is well demonstrated on a beam splice with high-strength friction-grip bolts. Figure 30.12(a) shows the straightforward set of forces that can replace the applied moment and shear. (This simplification with its implied redistribution is only applicable for building structures. In bridges such redistribution is not usually permitted, primarily because of general concern over fatigue.) Figure 30.12(b) enumerates the checks that have to be carried out to ensure that there are no weak links within the connections. These checks are as follows:

1. The capacity of the flange to resist the tensile force. The critical section is the vertical net section through the first line of holes, in tension, together with the horizontal web/flange intersection in shear. In addition, the efffective section of the flange through the first line of holes should be checked under the flange stress resultant without any redistribution of bending moment to the flanges. The effective section is defined as the net area times a coefficient greater than unity which recognizes that nominal stresses slightly above yield may occur on a net section without detriment.
2. The frictional capacity of the high-strength friction-grip bolts.
3. The net sections of the splice plates, assuming that the force T is equally divided between the pair of cover plates.

Note that because of the symmetry of the connection these checks also demonstrate the adequacy of most of the compressive flange splice. The only additional checks are:

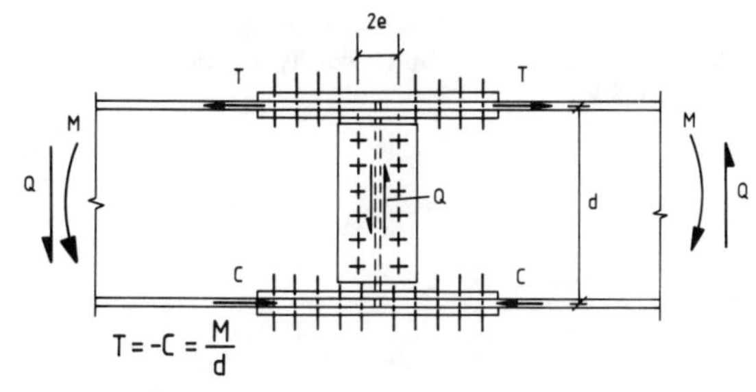

(a) CONVERSION OF APPLIED LOADING TO EQUIVALENT SYSTEM OF FORCES

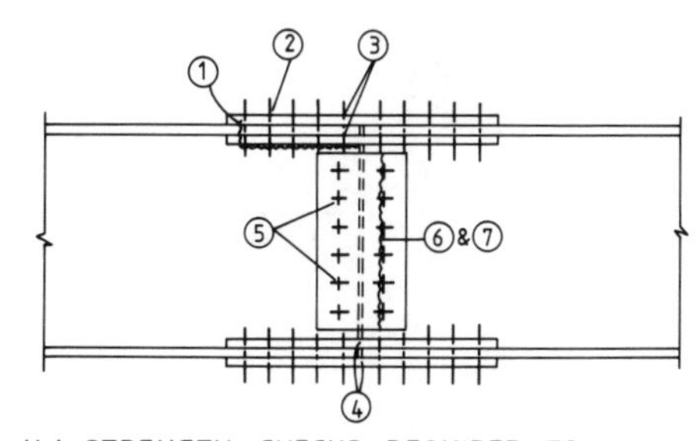

(b) STRENGTH CHECKS REQUIRED TO DEMONSTRATE ADEQUACY OF CONNECTION

Figure 30.12 Analysis and strength assessment of beam splice

4. The compressive capacity of the splice plates, free to buckle vertically. (This is satisfied by maximum pitch criteria.)
5. The frictional capacity of each web bolt group under a load Q at an eccentricity e.

6 and 7. The capacity of the net section of the web and splice plates. These are most unlikely to be critical unless the shear is a very high proportion of the beam capacity.

30.5 Concluding summary

1. Sound connection design is essential for the safety and economy of steel structures.
2. Successful connection design must take account of practicalities of fabrication and erection: how a connection is to be made and inspected, how it is to be assembled and maintained.
3. True connection behaviour may be both complex and variable.
4. Connection design should be based on straightforward analysis, with a clear visualization of how the connection is expected to behave.
5. The load, or force, paths should be traced through the connection and each link on each path checked, at least by implication, for adequacy both of strength and deformation capacity.
6. Special consideration may be necessary for connections where fatigue is a design consideration.

Background reading

FISHER, J.W. and STRUIK, J.H.A. (1974) *Guide to Design Criteria for Bolted and Rivetted Joints*, Wiley, Chichester

MORRIS, L.J. and MANN, A. (1981) *Lack of Fit in Structural Steelwork*, CIRIA Report 87
OWENS, G.W. and CHEAL, B.D. (1988) *Structural Steelwork Connections*, Butterworths, London
PASK, J.W. (1988) *Manual on Connections*, 2nd edn, BCSA

31

Welds: static strength

Objective To review basic weld profiles and layouts and to explain the basis of static strength assessment in welded joint design.

Prior reading Chapter 11 (Principles of welding).

Summary A brief introduction summarizes the basic types of welded joint. Design bases for full and partial penetration welds are presented. Side-fillet and end-fillet welds are presented separately with discussions on constraints on layout and behaviour. A discussion on fillet-weld behaviour under combined loading precedes a review of general design approaches for combined stresses on fillet welds.

31.1 Introduction

The fillet weld and the butt weld account for about 80% and 15%, respectively, of all welding in the construction industry. The remainder comprises plug, slot and spot resistance welds which are not considered here. A brief description of the two major types of weld is given prior to considering their load-carrying characteristics.

By definition, a butt weld is one made within the surface profile of abutting plates, either filling a cavity created by 'edge preparation' or penetrating and fusing an unprepared junction (see Figure 30.2). Full-strength butt welds can be completed from one side of the workpiece by using a backing plate. This may be a steel strip which remains permanently fused to the underside of the weld or a ceramic-coated strip or tile that is removed on completion (see Figure 11.9). When access to both sides of the workpiece is possible full-strength welds may be achieved by gouging the root of the weld on the reverse side and then sealing or filling the cavity. Partial penetration butt welds are generally admissible for static loading, although BS 5400 does not permit their use to transmit tension. They are used in preference to full-penetration butt welds if reduced strength is acceptable. Clearly, a butt weld may be at an edge-to-edge junction or at a tee junction.

In contrast to the above, a fillet weld is applied outside the surface profile of the plates. Thus the joint must be formed either by the overlapping of members or by the use of secondary joining material. Although this implies eccentricity in the flow of force there is the practical benefit that the cutting tolerance on members may be relaxed appreciably. There are three distinct fillet-weld applications, i.e. the lap joint, the tee joint and the corner joint (see Figure 30.1).

31.2 Strength of butt welds

With suitable consumable electrodes the design strength of full or partial penetration butt weld metal may be taken as that of the parent metal. The joint capacity is based on the 'throat' area which, for BS 5400 and BS 5950, effectively means the area of penetration (see Figure 31.1).

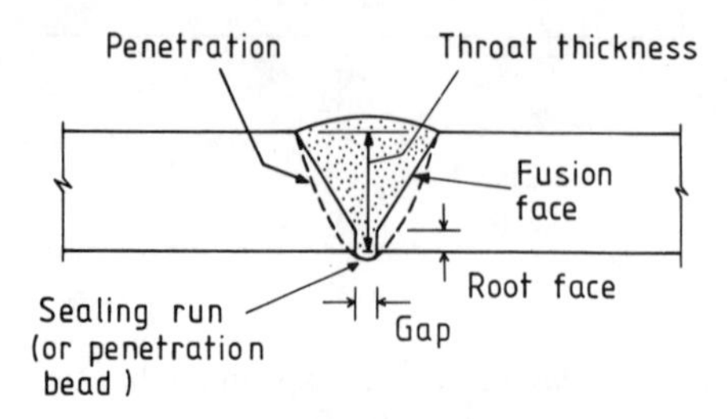

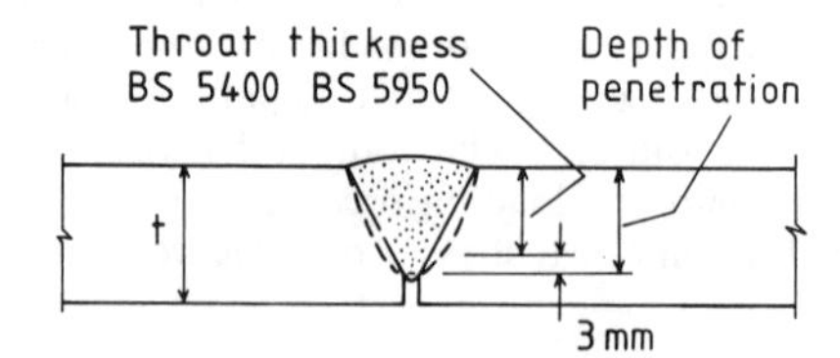

Figure 31.1 Butt welds

Thus in the case of full-strength butt welds the capacity may be taken to be that of the weaker part joined.

With partial penetration butt welds the depth of penetration is assumed to be the depth of preparation, except that a further 3 mm is deducted in the case of single-vee butt welds. BS 5950 also requires that secondary bending stresses, arising from eccentricity of the throat area, are accounted for.

31.3 Strength of fillet welds

31.3.1 Side- and end-fillets

The stress distribution in a fillet weld is complex, and depends on the direction of loading and joint geometry. There are two basic configurations for lap joints, referred to as:

1. Side-fillet (loaded longitudinally, see Figure 31.2(a));
2. End-fillet (loaded transversely, see Figure 31.2(b)).

In practice welds can lie between the above categories, i.e. they may be loaded obliquely. Tee and corner joints differ from lap joints in that they may be loaded in two transverse directions as well as longitudinally. The welds themselves are simply referred to as either transverse or longitudinal fillets.

31.3.2 Side-fillet weld

The side-fillet weld transfers load between connected parts in the direction of its longitudinal axis. In the majority of applications the weld metal will be stronger than the parent metal but, because of the smaller shear area at the weld throat, failure will occur there rather than in the base metal. Given the wide-ranging geometry of fillet welds it is useful to define the throat distance universally as the height of the triangle inscribed within the weld which has its base along the free boundary of the weld (see Figure 31.3).

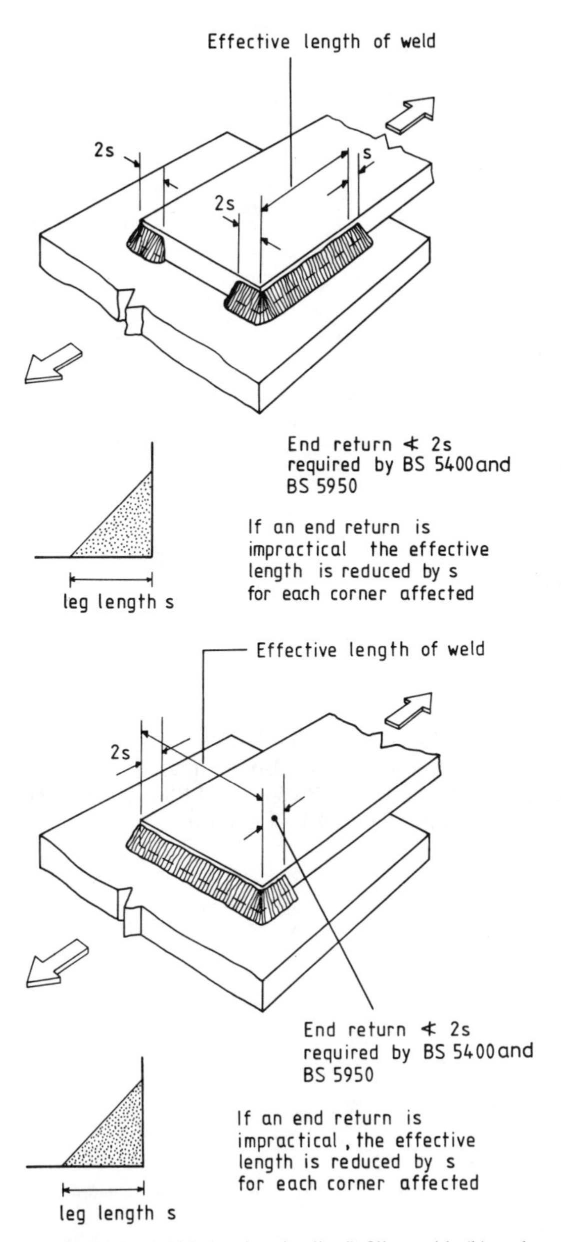

Figure 31.2 (a) Side (or longitudinal) fillet weld; (b) end (or transverse) fillet weld

At low loads the shear stress along the throat plane is affected by differential plate strain similar to that occurring in long-bolted joints. Stresses at the ends of the weld run are significantly higher than at the centre. However, tests on lap joints show that a more uniform stress distribution develops as ultimate load is approached because of yielding of the weld in shear. The stress variation is therefore disregarded by **BS** 5950. **BS** 5400 imposes a reduction in average shear stress for very long welds.

Another characteristic of the side-fillet weld is that large longitudinal forces are concentrated locally on the member cross-section. If the lateral spacing between weld runs is large in

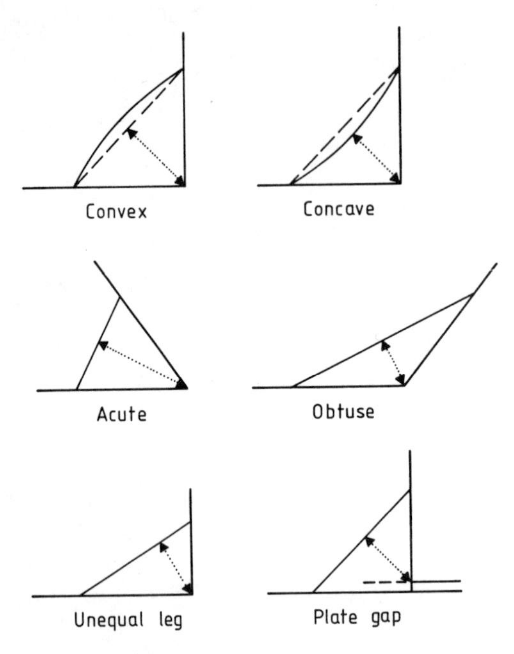

Figure 31.3 Effective throat thickness for various fillet weld profiles

comparison with their length a significant variation in tensile stress will occur across the width of the member, as shown in Figure 31.4. Transverse stresses may also be induced because of lateral strain differences in the plates. To control this variation, BS 5400 and BS 5950 limit the spacing between weld runs to a distance not exceeding the length of one run.

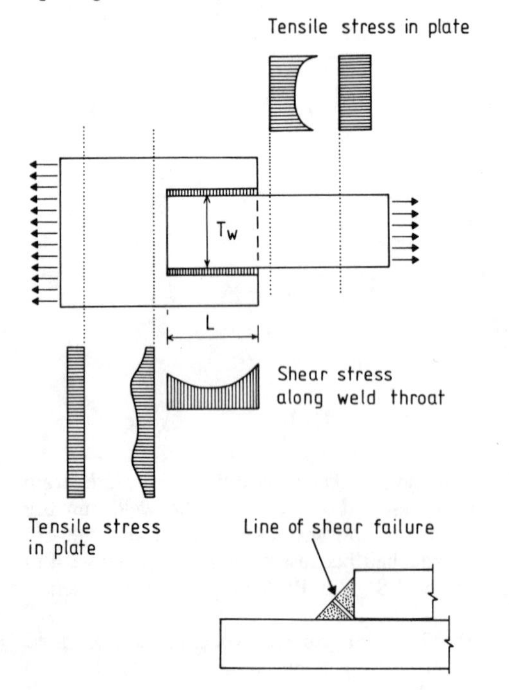

Figure 31.4 Behaviour of side-fillet welds

31.3.3 End-fillet weld

The theoretical elastic stress distribution in an end-fillet weld is shown in Figure 31.5. This result does not take into account shrinkage restraint stresses or plate friction, and merely provides an indication of the likely zones of plasticity in the weld prior to failure. For the case of horizontal load shown failure by shearing across the fillet occurs at a somewhat shallower angle than the throat. Unlike the side-fillet weld concentrations of horizontal and vertical direct stress are present because of the sudden change in the force path, and they influence the line of maximum shear stress, increasing the length of the failure surface to give a significant increase in joint strength.

Tests have confirmed that end-fillet welds are 30–40% stronger than side-fillet welds and this is recognized by BS 5400. However, BS 5950 does not distinguish between side- and end-fillet welds. A common design strength is specified, irrespective of loading direction.

Apart from the large difference in strength there are marked variations in both stiffness and ductility of end- and side-fillet welds (see Figure 31.6). When the two types of weld are mixed a greater share of load is attracted to end-fillets because of their higher stiffness, with the result that side-fillets may not develop their full capacity.

31.3.4 Transverse fillet weld

Although end-fillets in lap joints and transverse fillets in tee joints have comparable strength under uniaxial loading, the possibility of both normal and tangential loading on one fusion face must be considered for transverse welds. Figure 31.7 shows the generalized loading for fillet welds. In the present case only the tangential shear $Q_{\perp}$ and normal force $F_{\perp}$ are applicable. The forces and reactions are not collinear and additional moments must be present to satisfy equilibrium. Except in the case of double-fillet welds (where they cancel), the moments must be reacted by the connected plates. Their precise magnitude may be masked by, among other things, unknown residual moments from cooling; very little confidence can therefore be placed in linear analysis.

Because failure consists of shear yielding at the throat it is simpler to transfer all forces from one or other fusion face to the throat plane and consider a plastic failure criterion instead. In

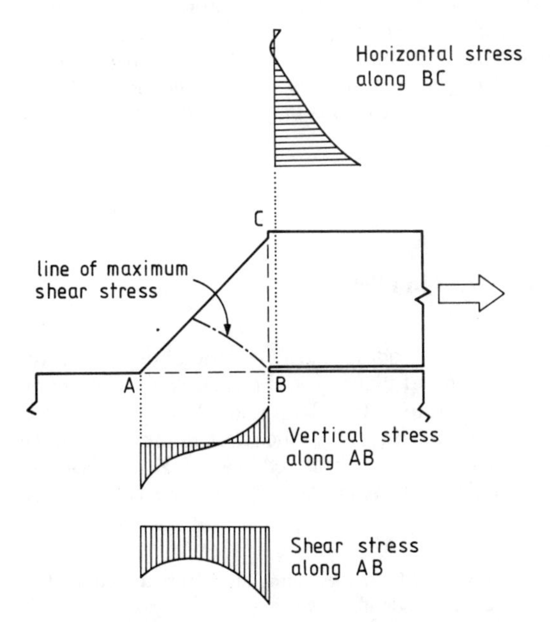

Figure 31.5 Elastic stress distribution in end-fillet weld

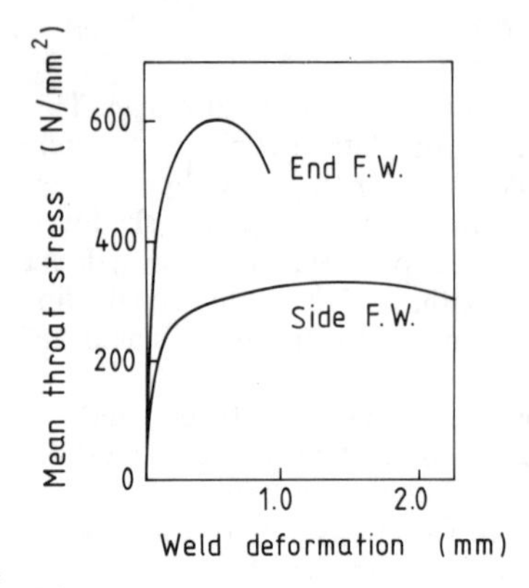

Figure 31.6 Load–deformation curves for 8 mm fillet weld

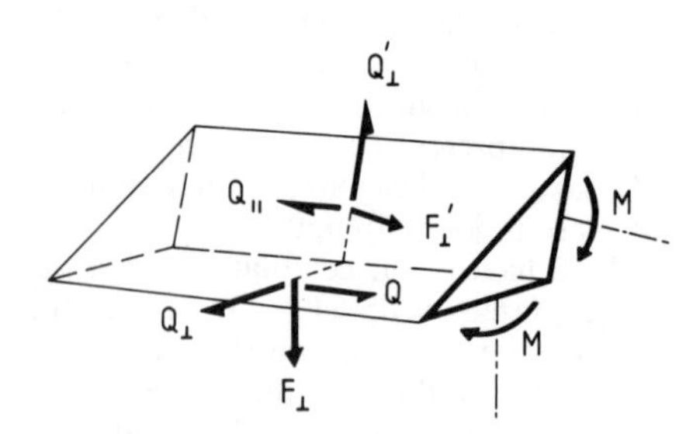

Figure 31.7 Forces acting on the fusion of an obtuse angle fillet weld

this case there are no moments to consider as they will be relieved by plastic deformation at the throat. The stresses acting at the throat plane are shown in Figure 31.8. Stresses $\tau_{\parallel}$ and $\sigma_{\parallel}$ are zero in the present context of transverse welds. Assuming the Huber–von Mises–Hencky yield criterion, the limiting condition for failure is:

$$\sqrt{(\sigma_{\perp}^2 + 3\tau_{\perp}^2)} \leqslant \sigma_y \qquad (31.1)$$

where σ_y is the tensile yield strength of the weld metal.

However, there are two disadvantages to this approach. First, the failure surface for transverse fillets is not well represented by the simple throat plane definition. Second, the need to transform stresses to the throat plane is perhaps unnecessarily complicated for design calculations.

Some results obtained in early tests are shown in Figure 31.9. The shaded area represents the scatter of failure stresses referred to the throat plane. The vertical axis represents tension or compression normal to the throat and the horizontal axis the shear across it. Equation 31.1 is seen to be conservative in relation to these results. A major discrepancy exists between the assumed and measured tensile/shear yield strength relationships of the weld metal. If the superior compression strength at the throat is ignored the results may be safely represented by the ellipse:

$$\sqrt{(\sigma_{\perp}^2 + k_w\, \tau_{\perp}^2)} \leqslant \sigma_y \qquad (31.2)$$

where k_w is approximately 1.8.

31.3.5 The general problem of combined stress in fillet welds

The two UK codes, in common with other national codes, provide a general expression for combined stress which includes shear along the throat plane. The direct stress ($\sigma_{\parallel}$) acting on the weld cross-section is governed by the elastic strain in the plates and, although influential in promoting initial yield, is likely to be relieved by subsequent plastic strain in the weld (unlike the three other basic components of load defined). It is not considered in any UK specification.

The longitudinal shear stress ($\tau_{\perp}$) acts on the same plane as ($\tau_{\parallel}$), and the two may be

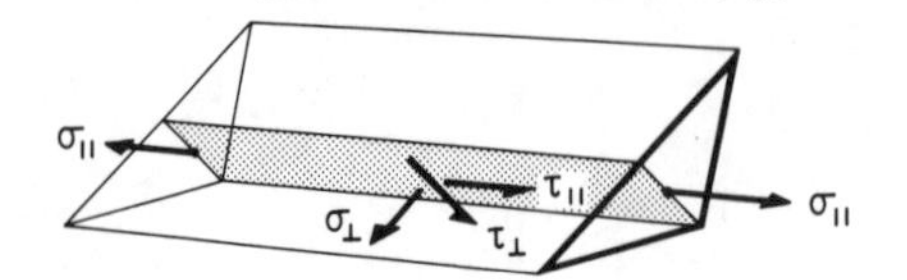

Figure 31.8 Stresses acting at the throat section of an obtuse angle fillet weld

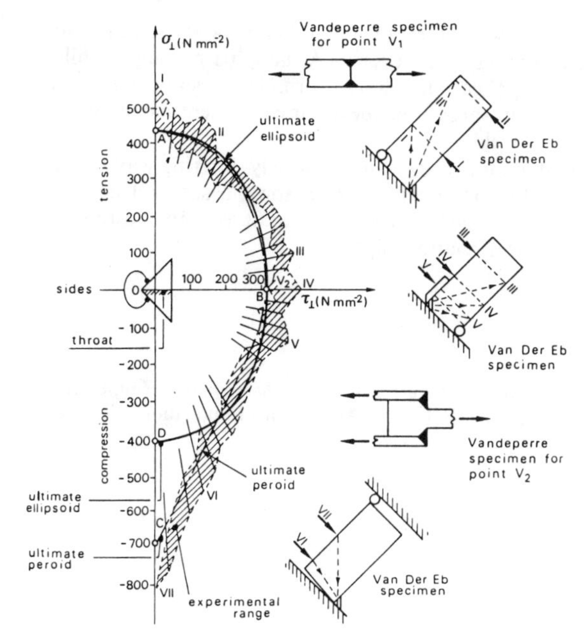

Figure 31.9 Test results for transversely loaded fillet welds

combined vectorially in Equation 31.1 to give:

$$\sqrt{[\sigma_\perp^2 + k_w(\tau_\perp^2 + \tau_\parallel^2)]} \leqslant \sigma_y \qquad (31.3)$$

This is the general form proposed by the IIW-Commission XV. The definition of k_w and σ_y is left to individual national codes. BS 5400 uses this form with $k_w = 3$ and

$$\sigma_y = \frac{(\sigma_y + 455)}{2\,\gamma_m\,\gamma_{f3}}$$

where γ_m and γ_{f3} are partial safety factors (both equal to 1.1 in the present context).

A simple method which considers the vector sum of shear forces acting in the three principal directions tangential to the fusion faces is also specified in BS 5400. The resultant force divided by the throat area should not exceed

$$\frac{k\,(\sigma_y + 455)}{2\,\gamma_m\,\gamma_{f3}\sqrt{3}}$$

where k is a factor distinguishing side and fillet welds.

BS 5950 has redeemed the 'vector sum' method. This procedure requires the resultant of normal and tangential forces on one fusion face (i.e. $Q_\parallel$, $Q_\perp$ and $F_\perp$ in Figure 31.7) to be divided by the throat area and compared with the limiting unidirectional design strength appropriate to parent material and electrode classification. The method is conservative in comparison with BS 5400, being based on a spherical rather than ellipsoidal domain of stress, but it is far simpler to visualize and calculate.

31.4 Concluding summary

1. In checking a butt weld the designer has few, if any, calculations to prepare. On the other hand, he must give serious consideration to the choice of edge preparation, electrode quality and plate properties.

2. The fillet weld poses different problems. Layout is important and must be chosen carefully. Also, design checks on strength are invariably necessary. Although right-angled fillets occur in more than 90% of fillet weld applications, design formulae in standard specifications must cater for the general case. Thus throat thickness is used, rather than leg length, for assessing strength per unit length of weld.
3. Fillet welds subject to multi-directional loading are treated in widely differing ways by the main national codes. Although interaction formulae have in some cases evolved from approximate stress models, they have been extensively adjusted to fit available test results and can only be regarded as semi-empirical relationships.

Background reading

BALLIO, G. and MAZZOLANI, F.M. (1983) *Theory and Design of Steel Structures*, Chapman and Hall, London
BLODGETT, O.W. (1972) *Design of Welded Structures*, James F. Lincoln Arc Welding Foundation, Cleveland, Ohio

32

Welds: fatigue strength

Objective To summarize the main factors affecting the fatigue strength of welded joints and to illustrate the method of carrying out a fatigue check.

Prior reading Chapter 8 (Characteristics of iron–carbon alloys); Chapter 30 (Introduction to connection design); Chapter 11 (Principles of welding).

Summary The chapter begins with an explanation of the mechanism of fatigue and the influence of welding on that mechanism. It summarizes the primary factors affecting fatigue strength and introduces *S–N* curves. The classification of fatigue details is described and the calculation of stress range is summarized. The principal types of fatigue loading and the bases for their calculation are presented with an introduction to damage calculations for mixed-amplitude loading.

32.1 Introduction

32.1.1 How welds fatigue

In welded steel structures fatigue cracks will almost certainly start to grow from welds rather than from other details. This is because:

1. Most welding processes leave minute metallurgical discontinuities from which cracks may grow. This means that the initiation period, which is normally needed to start a crack in plain wrought material, is either very short or non-existent. Cracks therefore spend most of their life propagating, i.e. getting longer.
2. Most structural welds have a rough profile. Sharp changes of direction generally occur at the toes of butt welds and at the toes and roots of fillet welds (see Figure 32.1). These points cause local stress concentrations of the type shown in Figure 32.2. Small discontinuities close to these points will therefore react as though they are in a more highly stressed member and grow faster.

32.1.2 Crack-growth history

The study of fracture mechanics shows that the growth rate of a crack is proportional to the square root of its length, given the stress fluctuation and degree of stress concentration. For this reason, fatigue cracks spend most of their life as very small cracks which are hard to detect. Only in the last stages of life does the crack start to cause a significant loss of cross-sectional area. This poses problems for in-service inspection of structures.

32.2 Fatigue strength

32.2.1 Definition of fatigue strength and fatigue life

The fatigue strength of a welded component is defined as the stress range (σ_r) which, fluctuating

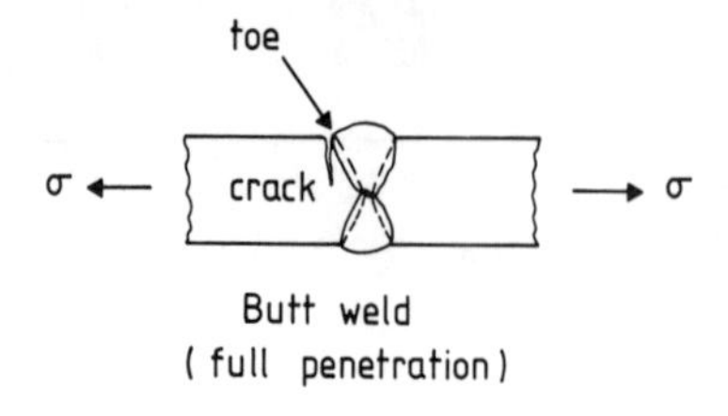

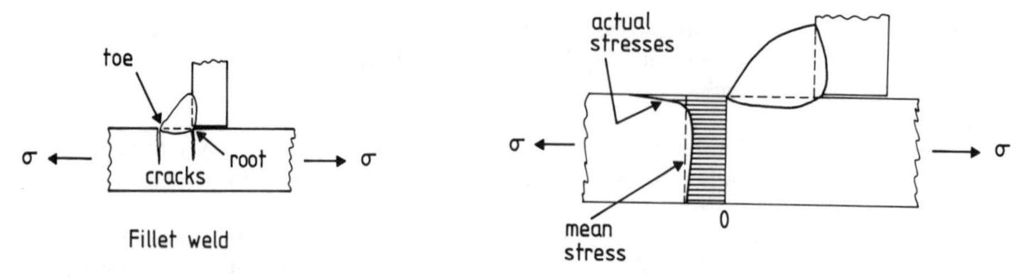

Figure 32.1 Local stress concentrations at welds

Figure 32.2 Typical stress distribution at weld toe

at constant amplitude, causes failure of the component after a specified number of cycles (N). The stress range is the difference between the maximum and minimum points in the cycle (see Figure 32.3). The number of cycles to failure is known as the fatigue life or endurance.

32.2.2 Primary factors affecting fatigue life

For practical design purposes there are two main factors which affect the fatigue life of a detail:

1. The stress range (σ_r) at the location of crack initiation. There are special rules for calculating this.
2. The fatigue strength of the detail. This is primarily a function of its geometry and is defined by the parameter K_0; this varies from joint to joint.

The life (N), or endurance, in number of cycles to failure can be calculated from the expression:

$$N = \frac{K_o}{\sigma_r^m}$$

where m is a constant, which for most welded details is equal to 3. Predictions of life are therefore particularly sensitive to accuracy of stress prediction.

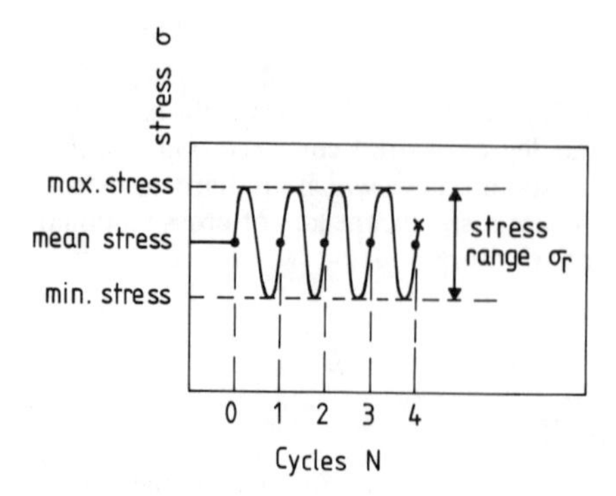

Figure 32.3 Constant-amplitude stress history

32.2.3 *S–N* curve

The expression linking *N* and σ_r^m can be plotted on a logarithmic scale as a straight line and is referred to as an *S–N* curve; an example is shown in Figure 32.4 and the relationship holds for a wide range of endurances. It is limited at the low-endurance end by static failure when the ultimate material strength is exceeded. At endurances exceeding about 10 million cycles the stress ranges are generally too small to permit propagation under constant amplitude loading. This limit is called the non-propagating stress (σ_0). Below this stress range, cracks will not grow.

For design purposes it is usual to use design *S–N* curves which give fatigue strengths about 25% below the mean failure values, as shown in Figure 32.5. K_2 is used instead of K_0 to define these lines.

32.2.4 Effect of mean stress

In non-welded details the endurance is reduced as the mean stress becomes more tensile. In welded details this is not usually the case. This is because the weld shrinkage stresses (or residual stresses), which are locked into the weld regions at fabrication, often attain tensile yield. The crack cannot distinguish between applied and residual stress. Thus for the purposes of design the *S–N* curve always assumes the worst (i.e. that the maximum stress in the cycle is at yield point in tension). It is particularly important to appreciate this point, as it means that fatigue cracks can grow in parts of members which are nominally 'in compression'.

32.2.5 Effect of mechanical strength

The rate of crack growth is not significantly affected by variations in proof stress or ultimate tensile strength within the range of low-alloy steels used for general structural purposes. These properties only affect the initiation period, which, being negligible in welds, results in little influence on fatigue life. This behaviour contrasts with the fatigue of non-welded details, where increased mechanical strength generally results in improved fatigue strength.

32.3 Classification of details

32.3.1 Detail classes

The fatigue strength parameter (K_2) of different welded details varies according to the severity of the stress concentration effect. As there is a wide variety of details in common use those with similar K_2 values are grouped together into a single detail class and given a single K_2 value.

These data have been obtained from constant-amplitude fatigue tests on simple specimens containing different welded detail types. For the most commonly used details it has been found convenient to divide the results into nine main classes. These are given in Table 32.1. As shown in Figure 32.5, they can be plotted as a family of *S–N* curves. The difference in stress range between neighbouring curves is usually between 15% and 20%.

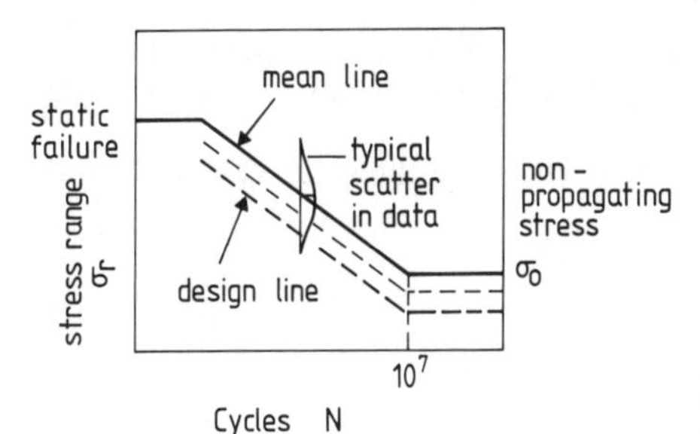

Figure 32.4 Typical *S–N* curve for constant-amplitude tests

Table 32.1 Values of K_2 and *m* for welded details (from BS 5400)

Class	K_2	*m*
B	1.01×10^{15}	4
C	4.23×10^{13}	3.5
D	1.52×10^{12}	3
E	1.04×10^{12}	3
F	0.63×10^{12}	3
F2	0.43×10^{12}	3
G	0.25×10^{12}	3
W	0.16×10^{12}	3
S	2.08×10^{22}	8

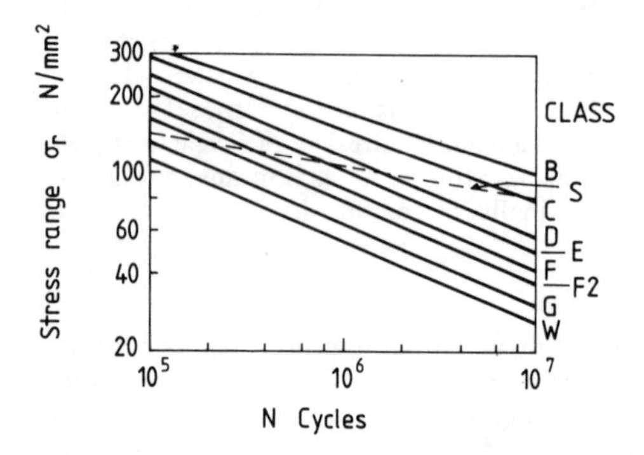

Figure 32.5 Family of design *S–N* curves (from BS 5400: Part 10)

32.3.2 Detail types

There is usually a number of detail types within each class. Each type has a specific description which defines the geometry both microscopically and macroscopically. The main features that affect the detail type, and hence its classification, are:

1. Form of the member (for example, plate, rolled section, reinforcing bar);
2. Location of anticipated crack initiation. This must be defined with respect to the direction of stress fluctuation. Also, a given structural joint may contain more than one potential initiation site, in which case it may fall into two or more detail types;
3. Leading dimensions (for example, weld shape, size of component, proximity of edges, abruptness of change of cross-section);
4. Fabrication requirements (for example, type of weld process, any grinding smooth of particular parts of the joint);
5. Inspection requirements. Special inspection procedures may be required on higher-class details to ensure that detrimental welding defects are not present.

Note that if fatigue is critical in the design the extra controls on fabrication incurred by the last two requirements may increase the total cost significantly above that for purely static strength.

Examples of different types of welded detail and their classes are shown in **Tables 17b** and **17c** of BS 5400.

32.4 Stress parameters for fatigue

32.4.1 Stress area

The stress areas are essentially similar to those used for static design. For a crack starting at a weld toe the cross-section of the member through which propagation occurs is used. For one starting at the root, and propagating through the weld throat, the minimum throat area is employed.

32.4.2 Calculation of stress range σ_r

The force fluctuation in the structure must be calculated elastically. No plastic redistribution is permitted. The stress on the critical cross-section shall be the principal stress at the position of the weld toe (in the case of weld toe cracks). Simple elastic theory shall be used assuming that plane sections remain plane (see Figure 32.6). The effect of the local stress concentration caused by the weld profile shall be ignored, as this is already catered for by the parameter K_2 which determines the weld class. In the case of throat failures the vector sum of the stresses on the weld throat at the position of highest vector stress along the weld shall be used, as in static design.

32.4.3 Geometrical stress concentrations and other effects

Where a member has large changes in cross-section (for example, at access holes), there will be regions of stress concentration due to the change of geometry. In static design the stresses are based on the net area as plastic redistribution will normally smooth out these peaks at ultimate load. With fatigue this is not so, and if there is a welded detail in the area of the geometrical stress raiser the true stress must be used, as shown in Figure 32.7. Similarly, any secondary effects, such as those due to joint fixity in latticed structures, and shear lag and other distortional effects in slender beams, must be allowed for in calculating the stresses.

32.5 Loadings for fatigue

32.5.1 Types of loading

Examples of structures and the loads which can cause fatigue are:

Bridges: Commercial vehicles, goods trains
Cranes: Lifting, rolling and inertial loads
Offshore structures: Waves
Slender towers: Wind gusting

The designer's objective is to anticipate the sequence of service loading throughout the structure's life. The magnitude of the peak load, which is vital for static design purposes, is generally of little concern, as it only represents one cycle in millions. For example, highway bridge girders may experience 100 million significant cycles in their lifetime. The sequence is important because it affects the stress range, particularly if the structure is loaded by more than one independent load system.

For convenience, loadings are usually simplified into a load spectrum, which defines a series

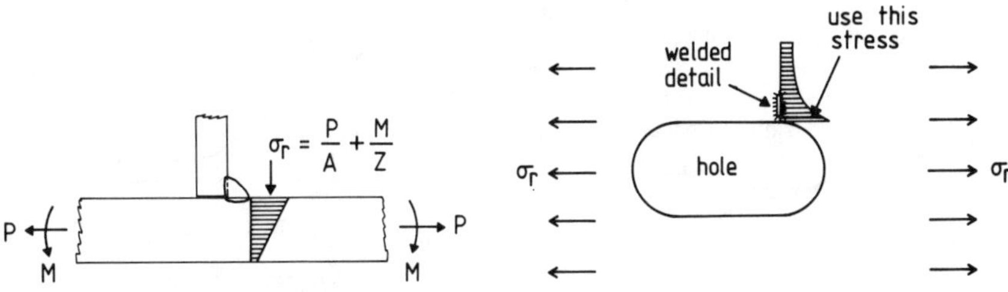

Figure 32.6 Design stress parameter for cracks propagating in parent material

Figure 32.7 Design stress parameter for cracks initiating at geometric stress concentrations

of bands of constant load levels and the number of times that each band is experienced, as shown in Figure 32.8.

Slender structures, with natural frequencies low enough to respond to the loading frequency, may suffer dynamic magnification of stress. This can shorten the life considerably.

Two useful sources of information on fatigue loading are BS 5400: Part 10 for bridges and BS 2573 for cranes.

32.6 Calculation of damage

Under variable amplitude loading the life has to be estimated by calculation of the total damage done by each cycle in the stress spectrum. In practice, the spectrum is simplified into a manageable number of bands, as shown in Figure 32.9. The damage done by each band in the spectrum is defined as n/N, where n is the required number of cycles in the band during the design life and N the endurance under that stress range (see Figure 32.10).

If failure is to be prevented before the end of the specified design life Miner's Rule must be complied with. This states that the damage done by all bands together must not exceed unity, i.e.:

$$\frac{n_1}{N_1} + \frac{n_2}{N_2} + \ldots \frac{n_n}{N_n} \leqslant 1$$

Note that, when variable amplitude loading occurs the bands in the spectrum with σ_r values less than σ_0 may still cause damage. This is because the larger-amplitude cycles may start to propagate the crack. Once it starts to grow, lower cycles become effective. In this case the horizontal cut-off at σ_0, shown in Figure 32.4, is replaced by a sloping line with a log gradient of $1/(m + 2)$.

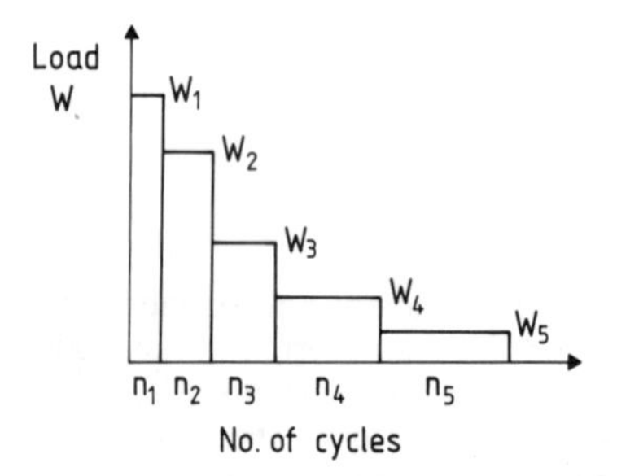

Figure 32.8 Typical load spectrum for design

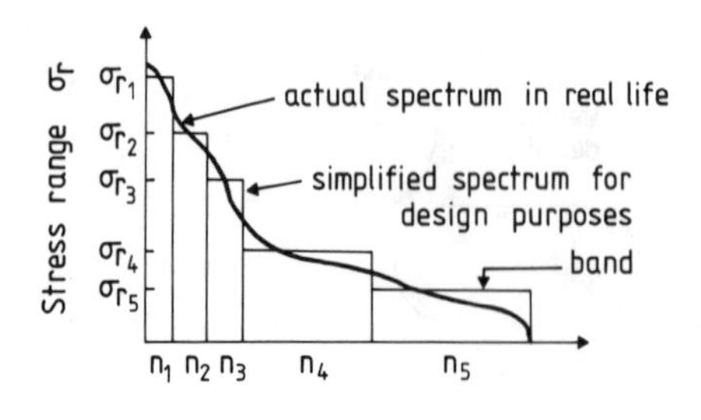

Figure 32.9 Simplification of stress spectrum

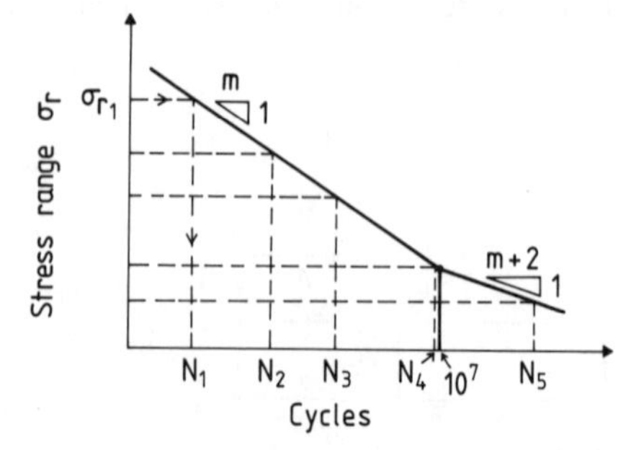

Figure 32.10 Determination of endurance for each band

32.7 Practical implications for design of welded structures

1. Fatigue and static failure (whether by rupture or buckling) are dependent on very different factors, i.e. fatigue:
   (a) Depends on the whole service-loading sequence (not one extreme load event);
   (b) Of welds is not improved by better mechanical properties;
   (c) Is very sensitive to the geometry of details;
   (d) Requires more accurate prediction of elastic stress;
   (e) Makes more demands on workmanship and inspection;
2. It is therefore important to check early in the design whether fatigue is likely to be critical. Acceptable margins of safety against static collapse cannot be relied upon to give adequate safety against it.
3. Areas with a high live/dead stress ratio and low class details should be checked first. The check must cover any welded attachment to a member, however insignificant, and not just the main structural connections. Note that this includes welded additions to the structure in service.
4. If fatigue is critical, then the choice of details will be limited. Simplicity of detail and smoothness of stress path should be sought.
5. Be prepared for fatigue critical structures to cost more.

Background reading

BS 2573: Part 1 (1977) Specification for Permissible Stresses in Cranes and Design Rules, Part 1, Structures, British Standards Institution. Gives loading for cranes. *S–N* data are not up to date.

BS 5400: Part 10 (1980) Steel, Concrete and Composite Bridges, Part 10, Code of Practice for Fatigue, British Standards Institution

BS 6235 (1982) Code of Practice for Fixed Offshore Structures, British Standards Institution. Useful for *S–N* data on tubular joints.

GURNEY, T.R. (1979) *Fatigue of Welded Structures*, 2nd edn, Cambridge University Press, Cambridge

The Welding Institute, *Fatigue Strength of Welded Structures*

Bolts and bolting

Objective

To describe the type of fastener and their installation in structural steelwork. To define their individual behaviour under conditions of simple and combined stresses and show how this is recognized in BS 5950 and BS 5400: Part 3.

Summary

The following aspects of ordinary bolts are presented: their types, installation, spacing requirements and behaviour in shear. The discussion is extended to include bolt tension induced by nut tightening, pretensioned bolts with axial load and prying action. The following section on HSFG bolts covers methods of control of bolt pretension and other aspects of installation, clamping pressure distribution, relaxation of bolt pretension and slip factors. Reference is also made to those aspects of ordinary bolt behaviour that are relevant to HSFG bolting. Performance under fatigue loading is discussed. Guidance is given on usage of HSFG bolts and pretensioned high-strength bolts.

33.1 Introduction

Loads may be transferred from one structural element to another either by adhesion (welding, glueing) or through separate mechanical fasteners (rivets, bolts). Only the latter load transfer mechanism is considered in this chapter. Connections which use mechanical fasteners transmit shear between the connected parts in one of two ways:

1. Load is transmitted into and out of the fasteners by bearing on the connected parts, so that the principal actions in the fastener are bearing (compression) combined with transverse shear (Figure 30.3). Axial tension may also be present in some cases. Bolts which transfer load by this mechanism are referred to as ordinary (or bearing) bolts.
2. The fastener is pretensioned so that a compressive force is exerted on the connected parts (Figure 30.4). This gives rise to high frictional resistance which enables load to be transferred between the connected parts. When the applied load in Figure 33.1(b) exceeds the frictional resistance that can be developed between the plates they will slip relative to each other, allowing the fastener to act in bearing. Bolts which transfer load by this mechanism are known as pretensioned or, more commonly, as high-strength friction-grip (HSFG) bolts. Controlled tightening of HSFG bolts allows the frictional action to be quantified for design.

The behaviour of a connection which uses mechanical fasteners is governed by the properties of both the fasteners and the connected parts. The two properties of connections principally influencing overall structural behaviour are strength and stiffness. Strength is governed by the mechanical properties of the fasteners and the connected parts, stiffness (i.e. the load/deflection response) by elastic modulus, geometry (including clearance) and ductility. Connections which use HSFG bolts effectively act as adhesive connections and are therefore stiffer than those using bearing bolts.

Figure 33.1 compares the effect of using bearing and HSFG bolts to make up a double-cover

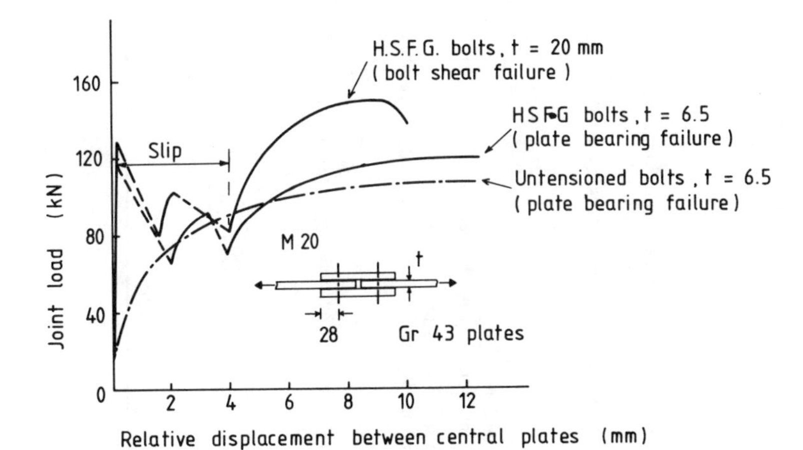

Figure 33.1 Comparison of load-deformation response for ordinary and HSFG bolts

plate butt joint. Until slip occurs the connection which uses HSFG bolts is seen to be much stiffer than that which employs bearing ones. Once slip occurs the HSFG connection progressively becomes a bearing one and, after the hole clearance has been taken up, both connections behave in a similar way.

In general, bearing bolts (which are cheaper to use than HSFG bolts) are used in simple building structures, HSFG bolts being reserved for bridge work, where their increased connection stiffness is necessary. Table 33.1 summarizes the principal types of bolt, clarifying terminology, listing relevant British Standards and indicating primary uses.

The use of rivets as fasteners in structural steelwork has effectively ceased in Europe and North America. They are therefore not considered in detail here. Steel rivets were driven hot and so, contracting during cooling, exerted some (unquantifiable) clamping force on the joined parts. Loose, or missing, rivets in existing structures can be satisfactorily replaced by HSFG bolts.

33.2 Ordinary bolts

33.2.1 Description

Hexagon-headed bolts and nuts are normally available in a range of sizes up to 68 mm diameter and in steels of (minimum) tensile strengths up to 1370 N/mm^2. Washers were traditionally used under the nut but are now frequently omitted. The strength grades most commonly used for structural bolting are 4.6 and 8.8, the former in general applications, the latter where more severe loading is encountered.

33.2.2 Installation

Bolts are normally installed hand-spanner tight in 2 or 3 mm clearance holes for diameters up to or over 24 mm, respectively. Where exact location and prevention of relative movement between the joined parts is required, accurately machined bolts fitted in reamed holes are used. 'Fitted bolts' require precision work and are costly to install. Where rigidity is required pretensioned HSFG bolts are normally used.

33.2.3 Spacing requirements: pitch, edge-distance, end-distance

There are both minimum and maximum limits for the spacing between neighbouring bolt centre lines (pitch) and the distance between a bolt centre line and a plate edge (end-distance when

Table 33.1 Summary of bolt nomenclature specifications and usage

Nomenclature (after BS 5950)	*Other nomenclature*	*Common grades*	*Specifications*	*Guidance on usage*
Ordinary bolts	Bearing bolts Dowel bolts Black bolts (usually 4.6 only)	4.6	BS 4190	Light shear and/or tension connections in buildings. Poor in fatigue; no shear reversal, impact or vibration
	Spanner-tight bolts	8.8	BS 3692	Generally offer economy over 4.6 bolts for main connections in buildings. Poor in fatigue; no shear reversal, impact or vibration other than wind
Pretensioned ordinary bolts (*NB*: not recognized in BS 5950 – would specify General Grade HSFG bolts)		4.6	BS 4190	Only used if required not to work loose under vibration – locking devices are preferred
		8.8	BS 3692	As 8.8 above but pretensioned: – to improve stiffness in tension – to improve fatigue resistance in tension only – to prevent working loose under vibration (locking devices preferred)
High-strength friction-grip bolts	Non-slip bolts Pretensioned bolts	General grade	BS 4395: Part 1 Bolts BS 4604: Part 1 Installation	Bridges and other situations where shear slip is unacceptable Where fatigue is a design consideration. Where high-tensile capacity is required
		Higher grade, parallel shank	BS 4395: Part 2 Bolts BS 4604: Part 2 Installation	Some increase in capacity on general grade but not usually economic
		Higher grade, waisted shank	BS 4395: Part 3 Bolts BS 4604: Part 3 Installation	Expensive and not much stronger than general grade, but may be re-used

measured parallel to the direction of principal loading in a shear or lap splice, otherwise called an edge-distance).

Minimum values were originally defined to ensure access for tightening, but they now provide a useful control on the amount of material surrounding an individual bolt. Maximum values are specified for two reasons. First, they ensure that corrosion cannot develop between plies and, second, they control local buckling in plies that are in compression.

33.2.4 Bolts in shear

The principal action on a bolt in a splice joint of the type shown in Figure 30.3 is shearing on its cross-sectional plane caused by bearing between opposing plates in the joint. The elastic distribution of these bearing stresses and the stresses produced in the bolt are complex. However, for fully developed plastic conditions the distribution of shear stress is effectively uniform, so that the shear strength is the product of the cross-sectional area of the bolt in the shear plane and the shear strength of the material. If threads are excluded from the shear plane, the shank area ($A_s = \pi d^2/4$) may be used. Otherwise, the 'stress area' of the threaded portion should be used: this is the cross-sectional area for tension; its determination is discussed in the section on bolts in tension. In detailing practice it is common to use the smaller area and not to contrive to exclude the threads from the shear plane.

Shearing tests on bolts have shown the shear strength to be about 60% of the tensile strength. The effective shear strength of bolts in joints is reduced by secondary bending actions caused by uneven bearing of the plates and by bending of the bolt due to excessive hole clearance. The reduction increases with the length of the bolt for a given diameter. It is particularly significant in lap joints with a single bolt, where the loading tends to straighten out the joint and rotate the bolt, as shown in Figure 33.2, causing both shear and tension in the bolt and severe local bending stresses in the plates. The reduction in shear capacity of a single fastener is about 10%; increasing the length of the joint (i.e. the number of bolts) reduces the bending and hence the loss of shear capacity.

33.2.5 Long joints

The distribution of load between the bolts in a joint when the hole clearance has been taken up depends upon the joint length, the relative cross-sectional areas of the joined plates, the bolt pitch and the shear deformation capacity of the bolts and their immediately surrounding plate (fastener flexibility).

Figure 33.3 shows the distribution of loads between the bolts in a long joint. The loads transferred through the outer bolts (1 and 9 in the figure) are greater than those through bolts towards the centre of the joint. (If the total area of the cover plates exceeds that of the centre plate the distribution will not be symmetrical, and bolt 1 will transfer more load than any other.)

When the fasteners yield, their flexibility increases, causing a more uniform sharing of the load (the broken line in Figure 33.3). However, for long steelwork joints of normal proportions this behaviour will be insufficient to produce an equal load distribution. Thus the end-bolts will reach their deformation limit and so fail before the remaining ones have been fully loaded. This will result in progressive failure at an average shear value per bolt below the single-bolt shear capacity. Tests have confirmed that joint length, rather than the number of bolts, is the dominant parameter.

33.2.6 Bearing

Yielding due to pressure between the bolt shank and plate material will result in excessive

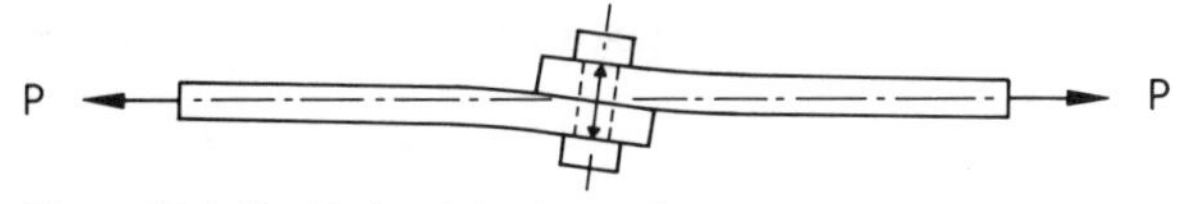

Figure 33.2 Simple lap joint in tension

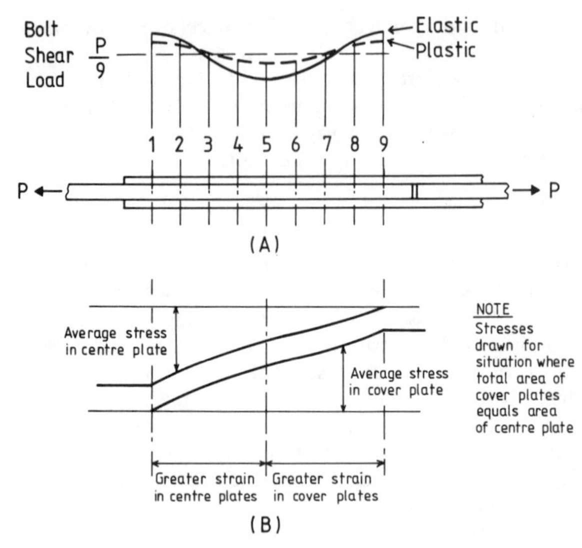

Figure 33.3 Long-joint (A) distribution of bolt shear loads arising from (B) incompatibility of tensile strains in connected elements

deformation of the plate around the bolt hole and possibly some distortion of the bolt. The area resisting the bearing pressure is assumed to be the product of the plate thickness and the nominal bolt diameter. The distance (e) of the bolt from the end of the plate must be sufficient to provide adequate resistance to the shearing-out mode of failure shown in Figure 33.4, which is governed by the area of the shear path. The presence of threads in the grip does not significantly affect the bearing resistance but will increase the deformation.

33.2.7 Bolts in axial tension

The axial tension capacity (P_t) of a bolt is the product of its tension strength (p_t) and the stress area (A_t). The stress area of a bolt is the core area of the threaded portion together with the single thread that intersects all possible failure planes. For bolts with the ISO thread form, i.e. all those listed in Table 33.1:

$$A_t = \frac{\pi}{16}(\text{effective dia.} + \text{minor dia.})^2$$

The extension of a bolt takes place mainly in the threaded portion by elongation of the core at the roots of the threads and by distortion of the threads themselves. The results of axial tension tests on two M20 bolts show, in Figure 33.5, the reduced ductility resulting from failure by thread stripping as compared with failure at the thread root.

33.2.8 Bolts in combined shear and tension

If the tensile and shear stresses at a point in the steel having a yield strength f_k in simple tension are p_t and q, the von Mises–Hencky theory predicts yielding will start when:

$$p_t^2 + 3q^2 = f_k^2$$

Noting that yielding due to shear alone will be caused by a shear stress $q_y = (f_k/\sqrt{3})$, the above relation can be written:

$$\left(\frac{p_t}{f_k}\right)^2 + \left(\frac{q}{q_y}\right)^2 = 1.0$$

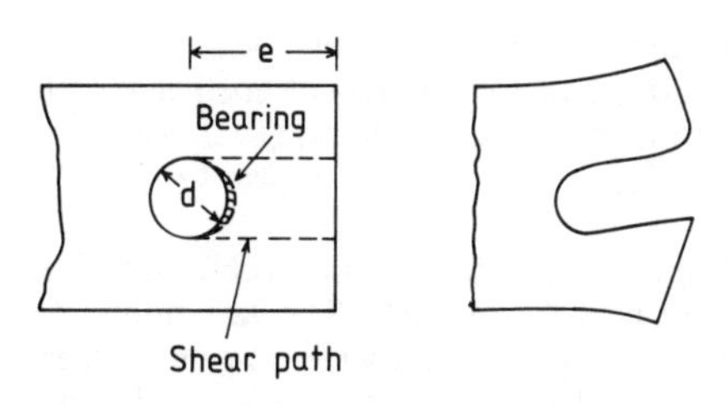

Figure 33.4 Bearing: shear-out failure of plate

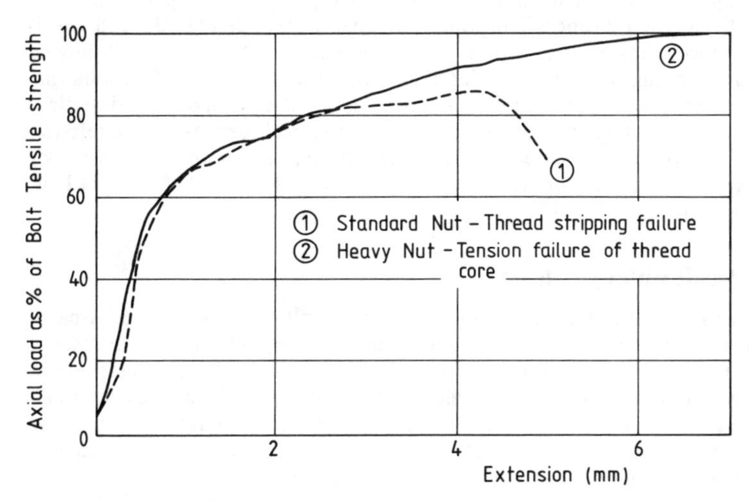

Figure 33.5 Axial tension tests: M20 Grade 4.6 bolt

The yield locus is thus a quarter circle (see Figure 33.6), which has also been shown to represent ultimate combined tension and shear behaviour satisfactorily. In design codes this is often modified to a tri-linear relationship (for example, Figure 33.16).

33.2.9 Bolt tension caused by tightening the nut

The torque applied to the nut is used partly to overcome friction between the nut and the surface against which it rotates and the remainder – approximately half – to drive the threads up the helix, overcoming the friction between the mating screw surfaces and the resolved components of the axial force.

If the geometry of the screw thread and the coefficients of friction between the various mating

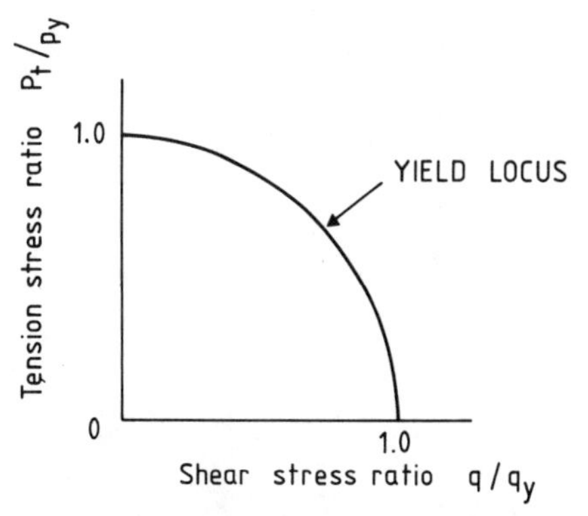

Figure 33.6 Interaction diagram for yielding under combined tension and shear

surfaces were known it would be possible to estimate the tension induced by a given torque. The uncertainties about distribution of contact pressures, and variabilities of coefficients of friction in practice, do not justify the use of anything other than a simple rule such as:

$$P_o = CT/d$$

where T = applied torque (Nm), d = bolt diameter (mm), P_0 = bolt tension (kN), and, for lightly lubricated bolts in good condition, the coefficient $C = 5.5 \pm 20\%$. The variation in C for any batch of bolts, even with threads in good condition, results in a wide range of values of tension induced by a given torque. This simple relation will apply only as long as conditions remain elastic. Part of the applied torque will produce torsional shear stresses in the bolt which will interact with the induced tensile stress. The bolt tension when yielding starts will thus be less than the load corresponding to yielding under axial tension alone. For HSFG bolts this reduction with lubricated threads in good condition may be up to 20%; with damaged or dirty threads the reduction will be much greater. Figure 33.7 shows a typical torque–tension relation for a 16 mm diameter bolt with unlubricated threads, which gave a maximum bolt tension of only 75% of the direct tensile ultimate strength. Another feature was the rapid fall-off of load after the maximum was reached.

33.2.10 Pretensioned bolt with axial load

Let us suppose that two steel plates are clamped together by a bolt of cross-sectional area A_s passing through them, which has been tightened to a tension P_0. For equilibrium, in the absence of external forces, the total compressive force between the plates will be P_0. The distribution of this force over an annular area of the interface between the plates is non-linear, as described later. However, for the purpose of this simple analysis it will be assumed that the compressive force is uniformly distributed over an annulus of area A_p, as shown in Figure 33.8, which is defined as the effective area of the plates participating in the action.

When an external tensile force (T) is applied parallel to the bolt axis, provided that all the parts remain in contact, both bolt and plates will suffer equal strain changes (ε_t), giving equal stress changes ($E\varepsilon_t$), so that:

Reduction in compressive force in plates $= E\varepsilon_t A_p = P_p$
Increase in tensile force in bolt $= E\varepsilon_t A_s = P_b$
For equilibrium $E\varepsilon_t A_p + E\varepsilon_t A_s = T$
giving $\varepsilon_t = T/E(A_p + A_s)$.

Substituting this expression for ε_t gives the forces:

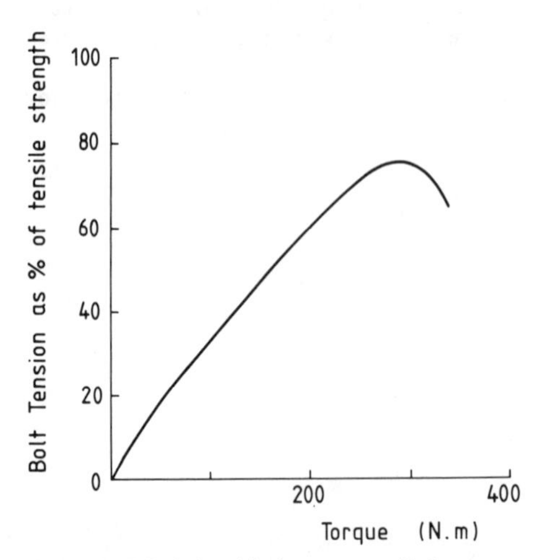

Figure 33.7 Relationship between applied torque and induced tension for a 16 mm diameter Grade 8.8 bolt

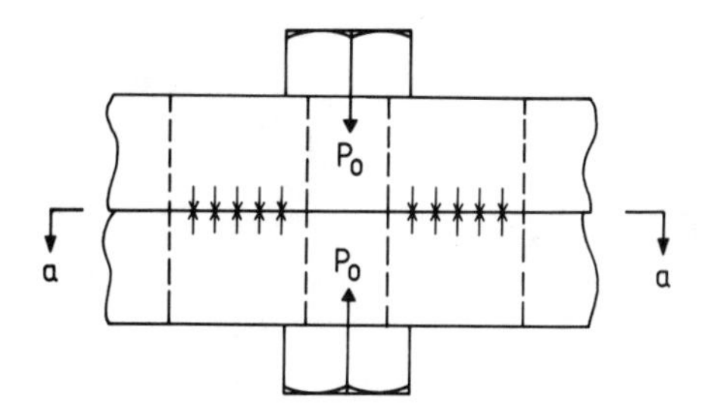

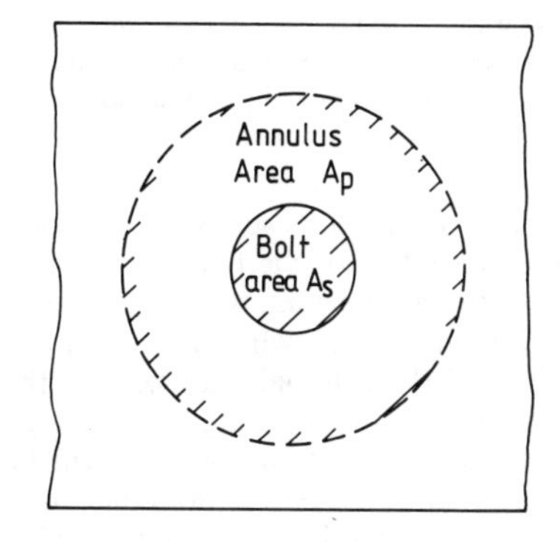

Figure 33.8 Diagrammatic representation of a pretensioned bolt

$$P_p = T\frac{A_p}{A_p + A_s} \quad \text{and} \quad P_b = T\frac{A_s}{A_p + A_s}$$

or, putting $(A_p/A_s) = k$, the axial stiffness ratio:

$$P_p = T\left(\frac{k}{k+1}\right), \quad P_b = T\left(\frac{1}{k+1}\right)$$

Superimposing these forces, induced by the applied load, on the existing forces due to the pretension gives:

Bolt tension $P_b = P_o + T/(k + 1)$

Clamping force between plates $P_p = P_o - T\dfrac{k}{(1 + k)}$

The variation of bolt tension and clamping force for $k = 10$, a typical value for steelwork joints, is shown in Figure 33.9. Note that the increase in bolt tension is only 10% of the applied tension (T) up to the value $T = 1.1\,P_0$. At this load the clamping force between the plates has become zero. Subsequently the bolt takes the whole of the applied load.

Initial imperfections (i.e. lack of flatness) reduce the axial stiffness of the plates and thus the stiffness ratio. The changes in bolt tension caused by an external load are larger than when the plates are flat.

33.2.11 Prying

When the line of action of the applied force is eccentric to the axis of the bolt – as in a beam-to-column end-plate connection – additional tension will be induced in the bolt by the prying action. This is most easily illustrated in terms of the Tee stub, loaded by a tension force

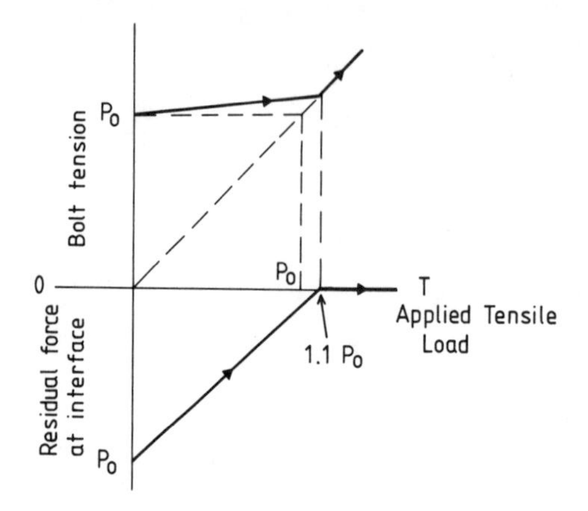

Figure 33.9 Relationship between bolt tension, applied tension load and residual force at the interface for a pretensioned bolt under axial load

$2T$, shown in Figure 33.10. In the bending of the flanges of the Tee the bolts act as a pivot point so that there is a compressive reaction (H) between the outer edges of the flanges, which is defined as the prying force. The tension induced in the bolts, for equilibrium, is thus ($T + H$).

Figure 33.11 shows the results of tests on a Tee joint. Note that when the bolts are pretensioned the full effect of the prying force is not felt until the external load exceeds the preload. If the bolt load due to the combined effects of applied load and prying force is large enough to cause even a small amount of plastic extension in the bolts its preload will be significantly reduced when the external tension is removed. (This would be a major concern in an HSFG connection subject to combined shear and tension – prying action is equally possible in such a connection.)

The prying ratio, H/T, depends upon the geometry and stiffness of the connected parts and the bolt stiffness. An estimate of the prying ratio, taking full account of all the parameters, requires a sophisticated analysis. Generally, the prying ratio is reduced by: increasing the plate thickness (t) (Figure 33.10); decreasing the eccentricity (b) of the load (i.e. by placing the bolts as near as possible to the web); and increasing the bolt flexibility by ensuring an adequate thread length in the grip. The edge distance (a) is significant only when the other parameters produce high prying forces.

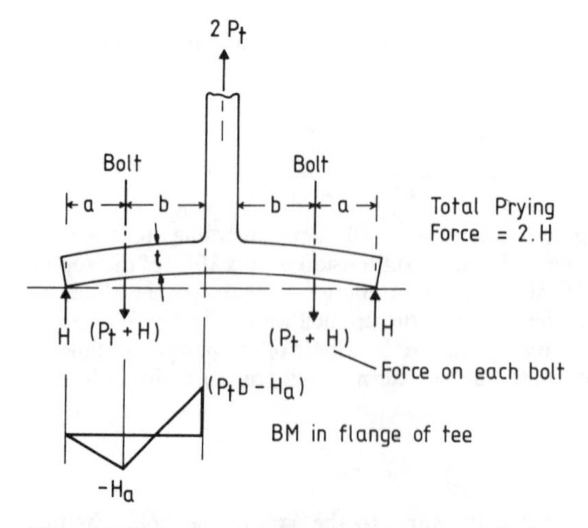

Figure 33.10 Prying and other forces acting on a Tee-stub connection under external tension

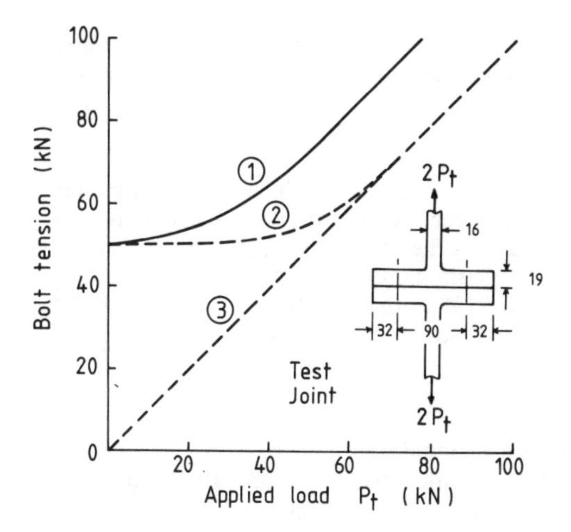

Figure 33.11 Relationships between bolt tensions and external tension on a Tee-stub joint

33.3 High-strength friction-grip (HSFG) bolting

33.3.1 General

In HSFG bolted joints the shear load is transferred between the connected parts by friction, as shown in Figure 30.4. The friction force is provided by the clamping action of the bolts, which are tightened in a controlled way to provide a specific shank tension. The bolts are installed in clearance holes and thus there may be no bearing action in transferring the load.

These joints are characterized by their rigidity, which is almost equal to that of welded joints; they are therefore to be used where the connection is subject to impact, vibration or reversal of stress (reversals of stress due to wind need not be provided for in this way according to BS 5950), or where it is essential to preserve geometrical integrity.

In order to make practical use of the friction effect it is necessary to use high-tensile steel bolts so that an adequate clamping force can be obtained with reasonably sized bolts. The stress induced in the bolts by the pretensioning is at, or near, the proof stress.

There are restrictions on the minimum thickness of outer plies in HSFG bolted joints which have the effect of limiting the bearing stresses at the interfaces. They also ensure that such a connection will have a reserve of strength after it has slipped into bearing.

Two strength grades of parallel shank bolts are available, the General Grade (equivalent to 8.8) and the Higher Grade (10.9). A waisted shank bolt (10.9) is also obtainable. Nuts are designed to develop the full strength of the bolt. Hardened washers are used under the element which is to be rotated during tightening.

33.3.2 Control of bolt pretension

In order to quantify the friction effect it is necessary to be able to obtain, with acceptable accuracy, the desired bolt pretension. This may be done either by controlled tightening of the nuts, using the torque-control or the part-turn methods, or by the use of proprietary devices.

The latter may be special bolts or fasteners, or separate devices such as load-indicating washers. Despite the use of these techniques, the actual tension in a particular bolt may vary significantly from its intended value.

33.3.2.1 The torque-control method

For this method of tightening a calibrated torque wrench is required which may be either hand operated or, for larger bolt diameters, power operated. A wide scatter of results is to be expected and estimates of the result of tightening with the objective of achieving a minimum shank tension of 80% of the specified bolt UTS have shown that about 3% of bolts would be undertightened, 6% would break and the remaining 91% would be tightened satisfactorily. The estimate was based on the normal distribution of bolt strength and variation of the coefficient C in the torque–tension formula.

33.3.2.2 Part-turn method

This method makes use of the ductility of the bolt material by rotating the nut sufficiently to take the bolt well into the plastic state in which the shank tension is comparatively insensitive to further nut rotation (see Figure 33.12). The maximum shank tension that can be obtained is equal to the maximum torqued-tension strength of the bolt under the friction conditions occurring at the time of tightening. The success of this method rests upon having adequate extensibility of the bolt. This depends not only on the ductility of the bolt material but also on the length of the bolt in the grip over which yielding can take place. With parallel shank bolts, yielding takes place mainly in the threads, so that the longer the thread in the grip length, the greater the extensibility and rotational capacity before fracture. With waisted shank bolts, plastic elongation can take place in the whole length of the reduced section, whose cross-section is less than that of the core of the thread. Care must always be taken with short bolts and with parallel shank bolts which have only a small amount of thread in the grip (three threads is a minimum). The part-turn method is not allowed with Higher Grade (parallel shank) bolts, which have a lower ductility than those of the General Grade; it is not recommended for use with M12 bolts.

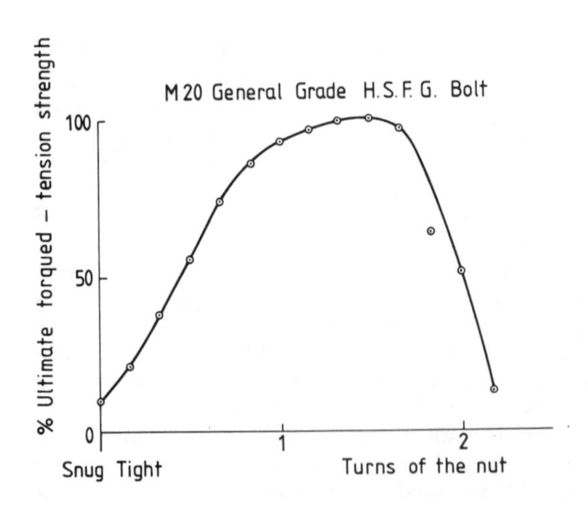

THIS TEST WAS CARRIED OUT IN A BOLT LOAD METER. A WELL FITTING SET OF PLIES WOULD SHOW GREATER STIFFNESS SO THAT A BOLT OF NORMAL PROPORTIONS WOULD BE ABOVE PROOF LOAD WHEN ITS NUT HAD BEEN SUBJECTED TO HALF A TURN, FROM A 'SNUG' TIGHT POSITION

Figure 33.12 Relationship between induced bolt load and turn of nut for an M20 general grade HSFG bolt

33.3.2.3 Load-indicating devices

Any means of load control is permitted provided that it gives the specified shank tension. A variety of special load-indicating bolts has been produced, some of which are complex and so unsuitable for general structural steelwork use. A comparatively simple device is the load-indicating washer, which has a number of protruding nibs on one surface. This is normally fitted under the bolt head with its protrusions in contact with the underside of the head (see Figure 33.13). As the nut is tightened the protrusions are crushed. When the gap between the load indicator and the bolt head has reached a prescribed value, measured by a feeler gauge, the required shank tension will have been achieved. The device is simple to use and makes for easy inspection after tightening.

33.3.3 Installation

HSFG bolts are installed in drilled holes with 2 mm clearance for bolts under 24 mm diameter and 3 mm clearance for those over 24 mm. The holes must be sufficiently well aligned so that the bolts can be inserted freely. Oversize and slotted holes are allowed but higher load factors are required since slip would be serious. A hardened steel washer is used under the nut or head, whichever is to be rotated. When the bolt axis is not normal to the contact surface the appropriate taper washer must be used. Where there are a number of bolts in a joint they should be tightened incrementally in a staggered pattern. Failure to do this may cause increases or decreases of tension in bolts which have been completely tightened at the beginning of a sequence. This effect is aggravated by initial imperfections of the joint and is reduced by tightening the bolts in two stages, the first when a 'bedding torque' is applied which should bring the plies (plates) into tight contact and the second stage when the bolts are finally tightened to the specified pretension.

Successful achievement of the specified shank tension depends on the threads being in good condition. Bolts and nuts must therefore be stored and handled in a way which ensures that the threads are not damaged or contaminated. For a fastener to be in a usable condition the nut must run freely on the bolt thread.

33.3.4 Clamping pressure distribution

The normal pressure between the plies caused by the bolt pretension is distributed over a limited area around the bolt hole. If the plates are flat the area is a circular annulus of a diameter depending on the ratio of the bolt diameter to the grip length. For practical sizes the grip length is generally between about 1.8 and 3.5 times the bolt diameter. The pressure is greatest at the edge of the hole, falling away to zero at the outer periphery of the annulus, as shown in Figure 33.14.

The plates outside the pressure annulus suffer an outward curvature, causing a separation

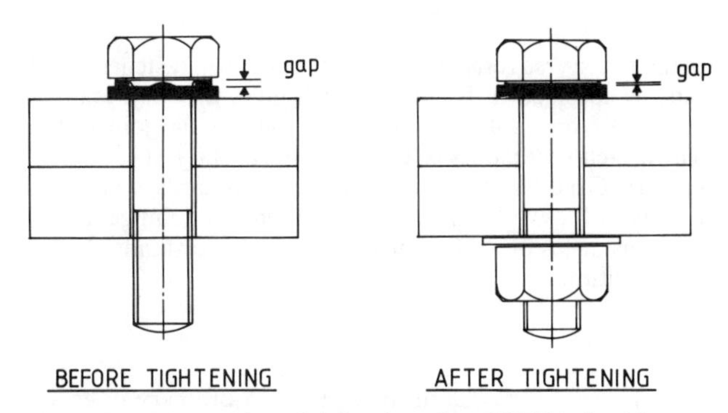

Figure 33.13 Assembly and tightening of an HSFG bolt with a load-indicating washer, where the nut is rotated

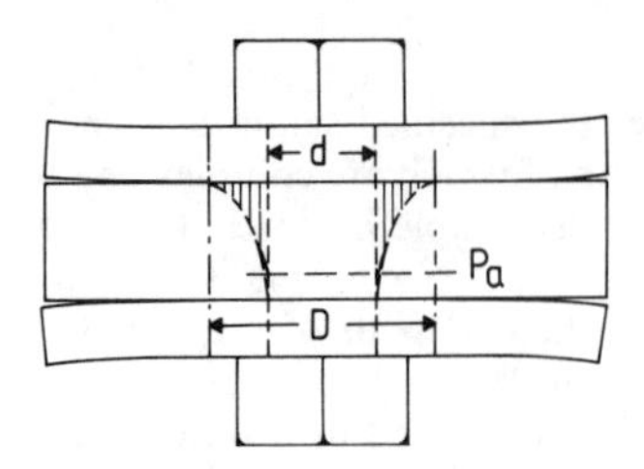

D = dia of pressure contact area
P_a = maximum interface pressure

Figure 33.14 Interface pressure distribution with a tightened HSFG bolt

which increases towards the edges. This should be noted when considering corrosion protection. A knowledge of this pressure distribution is also essential to an understanding of the fatigue behaviour of HSFG joints.

The regular nature of the pressure distribution is upset by lack of fit of the plates and also by lack of flatness and surface irregularities, such as would be caused by severe corrosion. In these cases the slip resistance can be adversely affected.

33.3.5 Relaxation of bolt pretension

Very high local stresses occurring in the threads and at other contact surfaces result in creep and consequent loss in bolt pretension with time. Most of the relaxation takes place within the first 24 h after tightening. It is least with the smoothest surface (approximately 5% for grit-blasted surfaces), and is increased when a surface coating is used, becoming larger as the thickness of the coating is increased. Tests of joints after a year's weathering have shown relaxation for surfaces coated with zinc silicate primer of 10.8% and for zinc metal spray of 22.8%.

Relaxation of bolt tension is also caused by the load applied to the joint. In-plane tensile stress in the plates results in a reduction in thickness (the Poisson's ratio effect) and thus of the pretension strain in the bolt. A further reduction is associated with the surface damage of the contact surfaces in the pressure area. The plate thinning is elastic and the relaxation due to this effect is recoverable. Relaxation caused by surface damage is irrecoverable and, after slip, can be as much as 25%.

The comments earlier concerning possible relaxation in the presence of prying for pretensioned bolts subject to combined shear and tension should also be noted. It is clearly important to make a realistic estimate of prying forces in such cases and thus ensure that the bolt is not overstrained.

33.3.6 Slip factor

The slip factor, μ, which determines the transverse resistance of the friction grip fastener, must be obtained from the results of tests, as specified in BS 4604, which gives μ = slip load ÷ (n × specified minimum shank tension of one bolt × number of bolts in the joint). The number of interfaces (n) through which each bolt passes is 2 for the typical slip test specimen.

The values of μ, appropriate to General Grade Bolts, that are given in Standards are a guide for conventional situations. Actual values of μ will be affected by bolt tension, joint geometry and surface treatment. If unusual circumstances prevail, μ must be found from standard tests representative of practical assembly conditions.

33.3.7 Other considerations

The following aspects of behaviour that were discussed in the section on ordinary bolts are equally applicable to HSFG bolts:

1. Spacing requirements: pitch, edge-distance, end-distance;
2. Long joints in shear;
3. Bolt tension caused by tightening the nut;
4. Pretensioned bolt with axial load;
5. Prying.

In addition, some design codes permit the design strength of HSFG bolted connections to be based on their post-slip reserve of strength at the ultimate limit state. In such circumstances the bolts are considered as ordinary bolts at the ULS.

33.4 Fatigue

33.4.1 Introduction

The ranges of stress caused by fluctuation of loading on steel building structures are normally insufficiently large to cause fatigue damage in members of uniform cross-section. However, the presence of geometrical discontinuities (holes, changes of section, welds) cause stress concentrations which increase the stresses locally to values which may be large enough to cause fatigue damage. Stress concentrations will occur in bolts at the thread roots, thread run-out and at the radius under the head. Fatigue failures in bolts in fluctuating tension commonly occur at the latter location or in the first thread under the nut.

Fatigue damage increases with the range of stress fluctuation ($\sigma_r = \sigma_{max} - \sigma_{min}$) and with the stress ratio ($R = \sigma_{min}/\sigma_{max}$). Bolt fatigue strength is frequently found to be independent of R. This is probably due to the yielding which takes place at the high stress concentrations (thread roots, etc.) giving, in effect, a constant upper stress in the cycle equal to the yield stress.

Threads may be formed by cutting or rolling. The latter process, which is more common for quantity manufacture, gives an improved fatigue strength because of the residual compressive stress caused at the thread roots. This beneficial effect will be lost, however, if the bolts are subsequently heat-treated. As a rough guide the endurance limits for cut and rolled threads are approximately 0.18 and 0.30 $(\sigma_{max} - \sigma_{min})$/UTS, respectively.

If a bolt is pretensioned the increase in tension caused by an external load less than the preload is very small. The effect of pretensioning is thus to reduce the effective stress range (see Figure 33.10) and so improve the fatigue life.

33.4.2 Shear connections

In lapped or spliced connections fastened with pretensioned HSFG bolts failure of the main plate will occur from cracks initiated in front of the bolt hole. These cracks will be initiated by fretting (caused by the relative movement between the plates), provided that the upper load in the cycle is less than about 75% of the slip load. The fatigue design stress for connections with HSFG bolts which have been designed for no slip at the ultimate load may therefore be calculated using the gross cross-section of plate; otherwise failure may be expected at the hole and the net cross-section of plate must be used. For all other fasteners failure occurs at the hole, the fatigue resistance being inferior to that of HSFG bolts.

Fatigue life is improved by making the bolt group compact. However, note that in many circumstances requirements for static strength demand that the bolts be widely dispersed. Examples might be where torsion is applied and also where local overstressing can occur (for example, in the web at end-connections).

33.4.3 Corrosion fatigue

Fatigue damage is accelerated by the presence of a corrosive environment. The two effects are mutually aggravating, each being more severe in the presence of the other. The threshold stress, at which cracks cease to propagate, is lowered in a corrosive environment.

33.5 The use of HSFG bolts and pretensioned high-strength bolts

The installation of fasteners to a specified pretension requires close control of the tightening process and subsequent inspection to ensure that the specified conditions have been attained. Controlled preparation of the mating surfaces of the plates is also required for HSFG bolts. Because of their considerable extra cost, these fasteners should only be used where necessary, for example:

1. Where movement (slip) in the connections would cause unacceptable deformation of the structure under working loads, non-slip friction grip bolted joints are required.
2. Where a connection is subjected to impact or vibration, pretensioned fasteners (or bolts with locking washers) should be used. If the loading causes shear on the joint, HSFG bolted connections should be specified.
3. The fatigue resistance of bolts loaded in axial tension is greatly improved by pretensioning to a load in excess of the highest tension experienced by the bolt.
4. The fatigue resistance of shear connection is considerably improved by using HSFG bolted connections.
5. The specified tension capacity of a Grade 8.8 bolt in BS 5950 is from 71% for M12 to 82% of the specified tension capacity of HSFG bolts. In exceptional circumstances it may be helpful to use HSFG bolts to take advantage of this increase in design strength. (Note that this increase in strength is not due to any anomaly in the specifications: it arises partly because of the stronger nuts for HSFG bolts and partly because each fastener is proof tested during assembly.)

33.6 Specifications

33.6.1 Ordinary bolts

Bolt strengths are specified as grades, by the use of two numbers separated by a decimal point. The first number indicates the strength of the steel expressed as 0.1 × minimum tensile strength in kgf/mm^2 (1 kgf/mm^2 = 9.8 N/mm^2). The second describes the capacity of the material for plastic deformation expressed as the ratio:

$$\frac{\text{Stress at yield or permanent set limit } (R_{0.2}) \times 10}{\text{Minimum tensile strength}}$$

Black bolts and nuts of strength grades 4.6 and 8.8 are described in BS 4190. The 4.6 grade are made of hot-formed mild steel and the 8.8 of low-alloy steel. HSFG bolts to BS 4395: Part 1 may be used as 8.8 bolts, i.e. in the untensioned condition, provided that the special hardened washer is omitted, indicating that the bolt was not pretensioned at the installation stage. BS 3692 specifies strength grades up to 14.9. The grades above 6.6 use alloy steels and the required strength is obtained by heat treatment. This gives a smooth dull black surface, and stock may be supplied in this form or, if required, with a bright finish.

Apart from variations in strength, the principal differences between the two standards are in the dimensional tolerances, which are larger in BS 4190. For example, the shank diameter specified for M30 bolts in BS 4190 is 30 mm ± 0.84 mm, whereas in BS 3692 the corresponding values are 30 mm ± 0.33 mm.

Various protective coatings may be obtained, such as galvanizing or sheradizing (BS 729) and cadmium, chromium or nickel plating (BS 3382). It must be remembered when specifying protective coatings that the thread tolerance has to be increased to allow for the thickness of the coating.

33.6.2 HSFG bolts

These fasteners are specified in BS 4395, 'High Strength Friction Grip Bolts and Associated Nuts and Washers for Structural Engineering', Parts 1–3. The fasteners in Part 1 are of the

General Grade, i.e. approximately 8.8 strength grade. Part 2 describes the Higher Grade (parallel shank) bolts which are strength grade 10.9 (nuts grade 12). Because of their limited ductility the use in tension of this grade of bolt is not allowed. Part 3 describes the Higher Grade (waisted shank) bolts, in which the diameter of the unthreaded shank is reduced so that its maximum area is 90% of the core area of the thread. The purpose of this is to increase the axial flexibility of the fastener, as compared with the Higher Grade (parallel shank) type, permitting their use in tension.

Other types of friction grip fasteners may be used provided that they have mechanical properties which are not inferior to bolts complying with BS 4395 and can be reliably tightened to the minimum shank tensions specified in BS 4604.

33.7 Design to BS 5950

33.7.1 Ordinary bolts

33.7.1.1 Bolts in shear

The shear capacity of a bolt is:

$$P_s = p_s A_s$$

where p_s is the shear strength of the bolt material, as given in **Table 32** of BS 5950. The shear area of the bolt (A_s) is either the shank area or, when threads occur in the shear plane, the (smaller) stress area (A_t), where:

$$A_t = \frac{\pi}{16}(\text{effective dia.} + \text{minor dia.})^2$$

A_t is specified in both BS 3692 and BS 4190. Where, for non-standard bolts, A_t is not specified the core area at the bottom of the thread should be used.

The bearing capacity is the lesser of the bearing capacity of the bolt ($P_{bb} = p_b dt$) or of the ply ($P_{bs} = p_s dt$), where p_b and p_s are the bearing strengths of the bolt and ply respectively; d is the nominal diameter of the bolt and t the ply thickness.

The distance of the bolt from the end of the plate (e) must be sufficient to provide adequate resistance to shearing out (Figure 33.4), which is governed by the area of the shear path ($2et$). For convenience, this is also expressed in terms of bearing stress, so that a further condition on the bearing strength of the connected ply is:

$$P_{bs} \nless \tfrac{1}{2} e t p_{bs}$$

33.7.1.2 Long joints

If the distance L_j between the end-rows of a splice or end connection exceeds 500 mm the shear capacity of the bolts should be reduced by a factor $(5500 - L_j)/5000$.

33.7.1.3 Bolts in axial tension

The axial tension capacities of both ordinary and HSFG bolts are given by:

$$P_t = p_t\,(\text{tensile strength}) \times A_t\,(\text{stress area})$$

Values are compared in Figure 33.15.

33.7.1.4 Prying

BS 5950 requires no calculation of prying forces provided that the bolt stresses are limited to the values specified in the Standard. Though not stated in the code, non-standard situations where prying forces are likely to exceed conventional values (30% or so of applied tension) should be treated rigorously, from first principles.

33.7.1.5 Bolts in combined shear and tension

The effect of combined stress is allowed for by the linearized interaction formula:

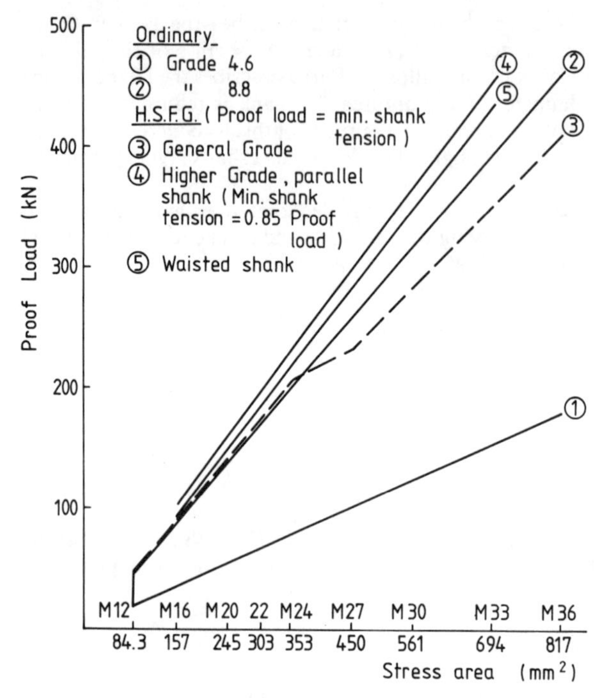

Figure 33.15 Proof or yield loads of bolts

$$\frac{F_s}{P_s} + \frac{F_t}{P_t} \not> 1.4$$

noting that $\dfrac{F_s}{P_s}$ and $\dfrac{F_t}{P_t} \not> 1.0$

(Also note that F_t is *external* tension, not preload – if any.)

This allows 40% of the tension capacity to be used before a reduction of the shear capacity is incurred (and vice versa). It is noted that the specified tension and shear capacities are, generally, only $0.77Y_f$ and $0.64Y_f$, respectively. The combined tension/shear capacities of 4.6 bolts are shown in Figure 33.16.

33.7.2 HSFG bolting

33.7.2.1 Parallel shank fasteners

The slip resistance of a parallel shank fastener is based on a serviceability criterion in BS 5950 but, for ease of calculation, has been presented in a form which may be checked against factored load. The result of this is that the connection will slip into bearing between working and failure load, and so bearing failure and long joint capacity must be considered.

For each effective interface, defined as the common contact surface between two load-transmitting plies (excluding packing pieces) through which the bolt passes, the slip resistance (P_{sL}) is given by:

$$P_{sL} = 1.1\ K_s \mu P_0$$

where P_0 is the minimum shank tension specified in BS 4604. $K_s = 1.0$ for fasteners in clearance holes, and 0.85 or 0.7 for oversize and short-slotted or long slotted holes, respectively.

The value for the slip factor (μ) must be determined from the test specified in BS 4604; the maximum allowable value is 0.55. However, for general grade fasteners, in connections where

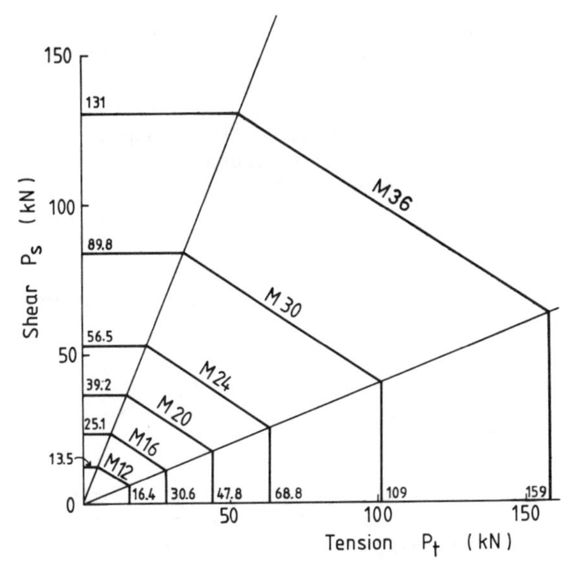

Figure 33.16 Interaction diagram for design strength to BS 5950 of 4.6 bolts under combined shear and tension

the contact surfaces comply with the **BS 4604** requirements for untreated surfaces (i.e. free of paint, oil, dirt, loose scale or rust, and burrs), a value of $\mu = 0.45$ may be assumed without testing. However, although **BS 4604** states that 'tight mill scale is not detrimental', tests have shown that μ is likely to be nearer 0.25 than 0.45 in this case.

Bearing and shear-out capacities are specified by the relation:

$$P_{bg} = dtp_{bg} \not> \tfrac{1}{3}etp_{bg}$$

where p_{bg} is the bearing strength of the parts connected (**Table 34**). At failure by shearing-out of the fastener to the end of the plate, major slip will have taken place. With lower-strength plate material, local plasticity and surface damage will have caused substantial loss of bolt-tension so that the contribution of friction to the ultimate strength becomes small. The effect is most severe with joints containing a single row of bolts.

33.7.2.2 Long joints

As with untensioned bolts, a reduction in shear capacity must be made for joints where the distance between the end-fasteners (L_j) exceeds 500 mm. In such cases the slip resistance for parallel shank fasteners (P_{sL}) is given by:

$$P_{sL} = 0.6\ P_0(5500 - L_j)/5000$$

provided that $P_{sL} \not> 1.1\ K_s\mu P_0$ or the bearing resistance.

33.7.2.3 Waisted shank fasteners

Slip of a connection with waisted shank friction grip fasteners involves major deformation and is regarded as failure. The slip resistance is therefore the failure criterion. As the connection will not slip into bearing until the factored load has been reached, bearing and long joint capacities need not be considered. The slip resistance per interface provided by a waisted shank fastener is to be taken as:

$$P_{sL} = 0.9\ K_s\mu P_0$$

33.7.2.4 External tension

HSFG bolts required to carry external tension must be either general grade or waisted shank types; the use of the higher grade, parallel shank type is not permitted. The tension capacity (P_t) of an HSFG bolt is $0.9P_0$, where P_0 is the minimum shank tension specified in BS 4604.

33.7.2.5 Combined shear and tension

The application of external axial tension to a pretensioned bolt reduces the clamping pressure between the plates through which the bolt passes, thus reducing the transverse slip resistance. This is dealt with in BS 5950 by the simple interaction formula:

$$\frac{F_s}{P_{sL}} + 0.8\frac{F_t}{P_t} \not> 1$$

Note that when the maximum value of F_t (i.e. $P_t = 0.9P_0$) is reached the slip resistance has been reduced to 20% of its value in the absence of external tension.

33.8 Design to BS 5400

33.8.1 General

The bolting clauses in BS 5400: Part 3 are written in the partial safety factor format. The partial load factor γ_{f3} has the value of 1.1 for the ultimate limit state and 1.0 for the serviceability limit state throughout. The values to be taken for the partial material factor γ_m are:

(a) Ultimate limit state
 Fasteners in tension 1.20
 Fasteners in shear 1.10
 Friction capacity of HSFG bolts 1.30
(b) Serviceability limit state
 Friction capacity of HSFG bolts 1.20

A serviceability check is required only in the case of connections made with HSFG bolts when the ultimate limit state capacity is based on the shear or bearing capacity, and this exceeds the friction capacity.

33.8.2 Ordinary bolts

33.8.2.1 General

Black bolts are not permitted in permanent main structural connections in highway or railway bridges. For fasteners, other than HSFG bolts acting in friction, the ultimate limit state of a connection is reached when the design load on any fastener equals its ultimate capacity in axial tension, shear or bearing, whether singly or in combination.

33.8.2.2 Axial tension

For bolts subjected to axial tension the tensile stress:

$$\sigma = \left(\frac{P_t + H}{A_{et}}\right) \leqslant \frac{\sigma_t}{\gamma_m \gamma_{f3}}$$

where σ_t is the lesser of $0.7\sigma_u$ and σ_y (or 0.2% proof stress).

33.8.2.3 Shear

For bolts subjected to shear only, where there are n shear planes, the average shear stress:

$$\tau = \frac{V}{nA} \leqslant \frac{\sigma_q}{\gamma_m \gamma_{f3}\sqrt{2}}$$

where $A = A_{eq}$ or A_{et} if the shear plane passes through the unthreaded or threaded part of the shank, respectively.

Where bolts are subject to combined shear stress (τ) and axial tensile stress (σ) the interaction condition to be satisfied is:

$$\sqrt{\left[\left(\frac{\sigma}{\sigma_t}\right)^2 + 2\left(\frac{\tau}{\sigma_q}\right)^2\right]} \leqslant \frac{1}{\gamma_m \gamma_{f3}}$$

33.8.2.4 Bearing

The area of the bolt resisting bearing (A_{eb}) is taken to be its shank diameter times the thickness of each connected part loaded in the same direction, whether or not this includes the threaded part. The bearing pressure,

$$\sigma_b = \frac{V}{A_{eb}} \leqslant \frac{k_1 k_2 k_3 k_4}{\gamma_m \gamma_{f3}} \sigma_y$$

The factors k_1 and k_3 account for variation in the effectiveness of the bearing resistance. For black bolts $k_1 = 0.85$; otherwise $k_1 = 1.0$. When the part being checked is enclosed on both faces with the fastener acting in double shear, so that full bearing resistance is developed, $k_3 = 1.2$, otherwise $k_3 = 0.95$. For HSFG bolts acting in friction, $k_4 = 1.5$, to account for the additional resistance provided by the frictional action; otherwise $k_4 = 1.0$.

For bearing resistance $k_2 = 2.5$. When the force transmitted by the fastener is towards the edge, shear-out is allowed for by taking the following values for k_2:

Edge-distance $\geqslant 3d$ $\quad k_2 = 2.5$

Edge-distance $\geqslant 1.5d$ $\quad k_2 = 1.7$

Edge-distance $\geqslant 1.2d$ $\quad k_2 = 1.2$

33.8.3 HSFG bolts acting in friction

33.8.3.1 General

The design ultimate capacity of HSFG bolts in normal clearance holes, acting in friction, is the greater of either the friction capacity or the lesser of the shear and bearing capacities. Slip between the connected parts in such a connection is a serviceability limit state, and is deemed to occur when the load applied to any bolt reaches its friction capacity. Slip of HSFG bolts in oversize or slotted holes results in gross deformation of the structure and constitutes the ultimate limit state.

33.8.3.2 Friction capacity (P_D)

$$P_D = \frac{k_h F_v \mu N}{\gamma_m \gamma_{f3}}$$

The prestress load (F_v) is taken as $F_0 - F_t$, which allows for the effect of external tension in reducing the clamping force. Values of the slip factor (μ) for various friction surfaces are quoted. For other surfaces the slip factor must be found from tests. When the higher grade (BS 4395: Part 2) bolts are used the quoted slip factors should be reduced by 10% to allow for the effect of the higher local stress around the bolt lowering the slip resistance.

The factor k_h provides for the reduction in slip resistance resulting from the use of oversize or slotted holes, when the bolt pressure is distributed over a smaller area of the joint interface. For oversized or short-slotted holes $k_h = 0.85$ and for long-slotted ones $k_h = 0.70$; for normal clearance holes $k_h = 1.0$. The sizes of oversized and slotted holes are given in **Table 12** of BS 5400.

33.8.3.3 Long connections

The effect of the unequal distribution of load between the bolts in a long joint ($L \geqslant 15d$) is accounted for by reducing the strength of all fasteners by the multiplying factor:

$$k_5 = 1 - \left(\frac{L - 15d}{200d}\right) \text{ but not less than } 0.75$$

where L is the distance between the centres of the end-fasteners.

33.8.3.4 Fatigue design: connected parts

Structural details are rated in BS 5400: Part 10: 1980 according to their resistance to fatigue and described in ten classes. A design $\sigma_r - N$ relationship is given for each class based on a 2.3% probability of failure, N being the number of repetitions to failure of the stress range σ_r.

33.8.3.5 Fatigue design: bolts in tension

For bolts to BS 3692, BS 4395 and those to BS 4190 which have been machined as specified in BS 5400: Part 10 the design value of stress in the bolt for fatigue is to be taken as:

$$\frac{F}{\sigma_u} \times \sigma_B$$

where $F = 1.7\,\text{kN/mm}^2$ for diameters up to 25 mm, or $2.1\,\text{kN/mm}^2$ for diameters over 25 mm, σ_B = stress range on the core area of the bolt (minor diameter) and σ_u is the ultimate strength of the bolt material in kN/mm^2. The fatigue life N (number of repetitions to failure at stress range in bolt σ_B) is obtained by entering this calculated stress value in the design $\sigma_r - N$ relationship for Class B details.

33.9 Concluding summary

1. Bolts for structural steelwork are described in two categories: ordinary and high-strength friction-grip. The clamping action of the latter causes transfer of shear loads by friction between the connected parts, the former by bearing between the bolts and the connected parts.
2. The capacity of an ordinary bolt may be determined by its own shear, bearing or axial tension capacity, or by the bearing capacity of the joint parts. Reduction in capacity arises in long joints and where the grip length of the bolt is long compared with its diameter. The presence of axial tension reduces the shear capacity and vice versa.
3. The shear resistance of a connection with high-strength friction-grip fasteners is determined by the compressive forces between the plies, caused by the pretension in the fasteners, and the slip coefficient of the mating surfaces of the plates. The presence of an external tension reduces shear capacity and vice versa.
4. Pretensioned friction-grip bolts provide slip-resistant connections which have a superior fatigue resistance. Because of their much higher costs, pretensioned and high-strength friction-grip fasteners should be used only when they serve an essential structural function.

Background reading

FISHER, J.W. and STRUIK, J.H.A. (1987) *Guide to Design Criteria for Bolted and Riveted Joints,* 2nd edn, Wiley, Chichester

Lack of Fit in Steel Structures, CIRIA Report 87

Manual on Connections (1988) 2nd edn, BCSA, London

Steel, Concrete and Composite Bridges, BS 5400: Part 10: 1980, Code of Practice for Fatigue. Contains a concise summary of information on fatigue resistance of bolts

34

Local elements in connections

Objective To review behaviour and the bases for design for local elements in connections, i.e. those components other than the bolts and welds.

Prior reading Chapter 30 (Introduction to connection design).

Summary The chapter commences with a brief summary of the different types of local elements that occur in connections. It then examines the situations where effective width and effective area concepts may usefully be used in design, i.e. web yielding or buckling under local in-plane loads and local bending of plates and flanges. The design of stanchion base plates under moment and/or axial load is considered. Finally, stiffening for panels in shear is reviewed.

34.1 Introduction

This chapter considers the behaviour of connection material, other than bolts and welds, and introduces the principal design checks specified in BS 5400 and BS 5950. The connection material under review includes components such as cleats, welded end-plates and stiffeners, together with those parts of structural members which participate in the connection. Typical examples of connection material are shown in Figure 34.1.

As in the case of bolts and welds, provided that fatigue loading can be discounted, plastic redistribution of stress is feasible and design methods based on ultimate strength are appropriate. In some instances it is necessary to limit deformation, either because of poor ductility in the connection material or to prevent unsightliness, corrosion, etc. This can be achieved by

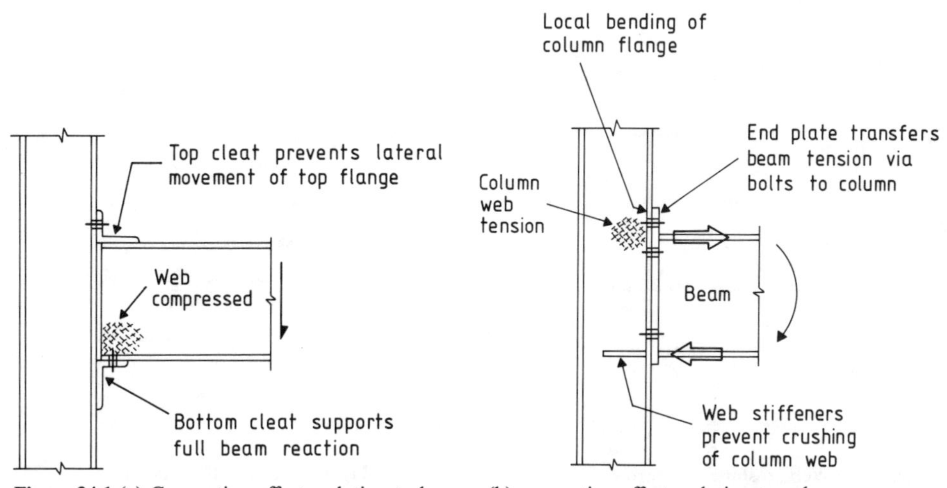

Figure 34.1 (a) Connection effects relating to beams; (b) connection effects relating to columns

controlling stresses at working load. Usually, the stress distribution is complex, and it is expedient to consider representative mean stresses, based on effective widths or areas. Adequate margins against unwanted deformation are built into the recommended limiting stresses.

Effects relevant to connection design which are treated by use of effective widths or areas include:

1. Web yielding at points of concentrated load or reaction;
2. Web buckling at points of concentrated load or reaction;
3. Local bending of plates or flanges;
4. Eccentric connection to tie members;
5. Shear in panels at member junctions.

34.2 Web yielding at points of concentrated load or reaction

Concentrated load at bearings and intermediate points on girder flanges may produce compressive yielding in the adjacent web. To check for this the extent of concentration must be assessed. This is dependent on the effective stiff length of the bearing surface but another major influence is the depth of material between the seating and the girder web (see Figure 34.2).

The stiff portion of a bearing is defined as 'that length which cannot deform appreciably in bending'. BS 5950 stipulates that this should be determined from an assumed 45° dispersion of load through solid seating material (Figure 34.3). Thus the stiff bearing length defines the extent of a load applied to the girder flange which is then tempered by dispersion through the flange material before affecting the girder web. In the latter case, however, the dispersion is greater because of continuity of the flange material; BS 5950 stipulates an angle of 1:2.5 (approx. 22°), BS 5400 specifies 30° (see Figure 34.4). The resulting length of web at a level just above the flange fillet defines the effective bearing area. BS 5950 specifies an effective bearing stress at the ultimate limit state equal to the full design strength of the web material.

If the limiting stress is exceeded, bearing stiffeners must be provided. These are either fitted in close contact with, or attached to, the flanges and must carry the excess load, based on their nominal compression area. In addition, the connection to flanges and web must be adequate to transfer the force from the stiffener.

Tensile yielding of webs opposite points of concentrated loading produces two situations which are associated with welded attachments (Figure 34.5(a)) and bolted end-plate or tee

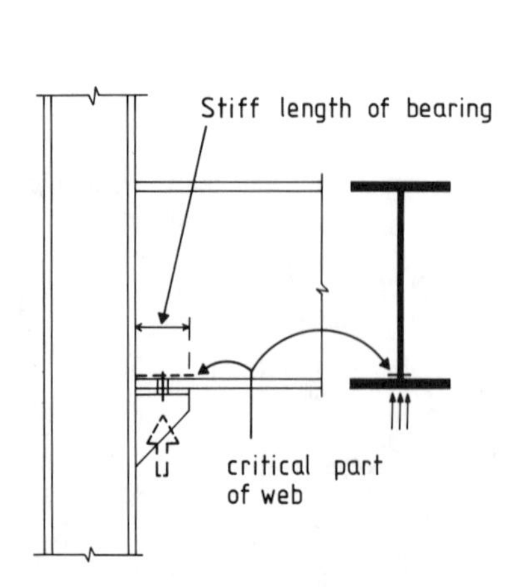

Figure 34.2 Web crushing or bearing

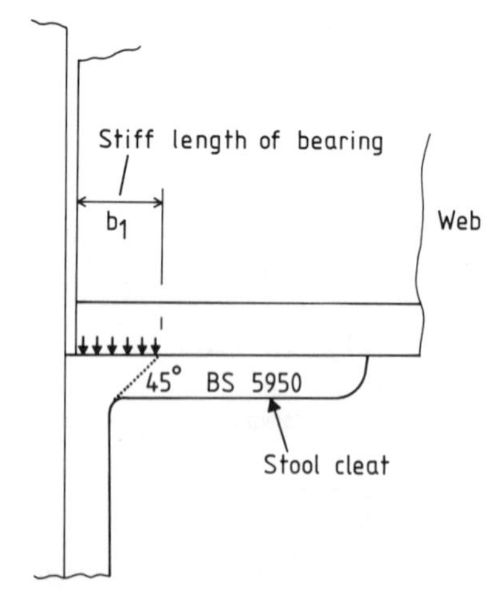

Figure 34.3 Length of bearing

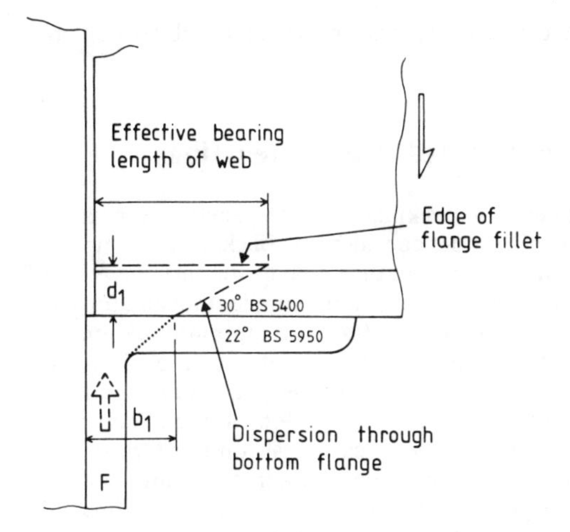

Figure 34.4 Effective bearing zone in beam web

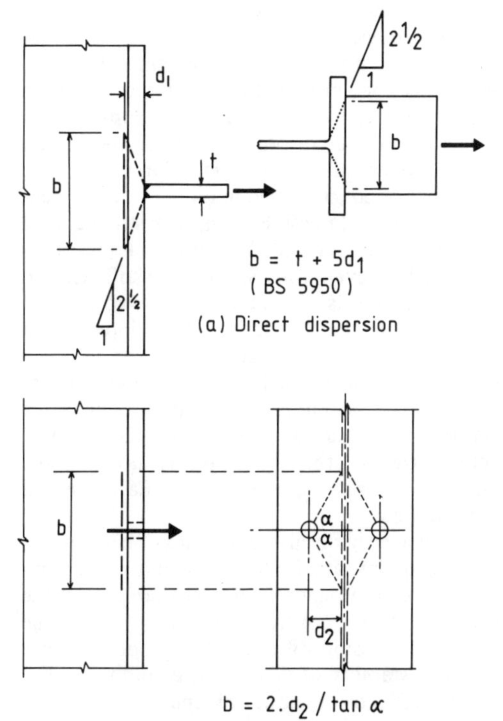

Figure 34.5 Effective tensile length

attachments (Figures 34.1(b) and 34.5(b)). The first is treated in BS 5950 in an identical way to compressive load, using the same angles of dispersion, etc. The second involves flexural dispersion along the column flange of tensile forces at bolt positions, leading to an effective length of column web resisting tension. This form of dispersion is considered only in BS 5400 where, strictly speaking, the context is flange bending resistance rather than web tension, and is discussed later under that heading.

34.3 Web buckling at points of concentrated load or reaction

Although there is a general analytical treatment for buckling of plate panels under various conditions of edge-loading and restraint the British codes consider web buckling over supports and at points of concentrated load as a special case and are thus able to simplify the form of checks to be made. The type of support envisaged corresponds to the simple stool and top cleat connection between a beam and column, shown in Figure 34.1(a). An important feature here is that the web is restrained against sidesway at upper and lower flange level. It is also rotationally restrained at lower flange level by the bearing. One further point is that loading is not applied on opposite edges of the web; it enters normal to the bottom edge and departs as a vertical shear acting over the full web depth. In the light of early tests on this type of joint it was accepted that the web may be represented by an equivalent pin-jointed column of height $0.5d$ and width $B + 0.5d$. This had the advantage that the procedure for column design could be adapted to web buckling. As a further simplification, the web slenderness ratio was specified directly as $\sqrt{(3d/t)}$ rather than stating an effective height of $0.5d$. This follows from the relationship $r_y = \sqrt{(I_y/A)}$. In BS 5950 a revised value of $2.5d/t$ has been introduced, implying an effective height of $0.7d$ approximately, to take account of more recent experience.

The above approach was extended to checking of webs over intermediate supports and at points of concentrated load by adopting an effective height $0.5d$ and width $B + d$. Again, it is assumed that the upper and lower flanges may not displace laterally relative to one another. This is particularly important when load is applied on both flanges, as is sometimes the case, for example, with foundation grillages and continuous floor girders. Somewhat casual design in this respect of steel I-beams used in temporary falsework for steel or concrete construction has led to several notable failures in steel construction. A code of practice dealing specifically with falsework design has been introduced, partly in consequence of these failures.

Web buckling may be prevented at points of support or concentrated loading by attachment of 'load-carrying' stiffeners. Two modes of buckling failure are possible when such stiffeners are used. First, the outer fibres of the stiffener may buckle locally. Limiting breadth/thickness ratios for the stiffener section are given in the two UK codes to prevent this. Second, the stiffener and associated web region may buckle as a whole. To evaluate this effect, a length of web on each side of the stiffener of $20t$ (if available) is assumed to participate. An effective pin-jointed height of $0.7d$ is assumed for the compound section, using basic column design information to check for buckling as before. BS 5950 allows an optional effective height of $1.0d$ if rotational restraint is not available at the loaded end of the stiffener. This standard imposes a further requirement that the bearing capacity, based on the stiffener area in contact with the loaded flange, must not be less than 80% of the applied load or reaction. Among other things, this ensures that the web will develop its contribution without excessive deformation in the stiffener. Early yield in the stiffener is undesirable from a buckling point of view.

Although web-crushing and web-buckling effects have been discussed in terms of beam webs, similar considerations apply to column webs loaded by horizontal forces from connecting beam flanges. Usually, the force concentration in the compression region is greater than that in the tension region, and stiffeners will often be necessary at that position. Necessary or not, the stiffness of the connection is improved markedly by compression stiffeners. The proportions are such that it should not be necessary to check overall stability of a stiffened column web but merely to consider local buckling of the stiffener in terms of outstand/thickness ratio, and to check the provision of 80% minimum bearing capacity, as discussed earlier.

34.4 Local bending of plates and flanges

Some common forms of connection involve local bending of plate elements. At the simplest level we may be concerned with a short plate element subject to knife-edge loading and support, which may be treated as a wide beam or cantilever. In more complex situations a plate may be cantilevered from two adjacent edges, or may be a cantilever of indefinite width with locally concentrated loading.

This second category of problem may be transferred into simple cantilever terms by allocating an effective width of bending. In the case of the elastic plate shown in Figure 34.6 it may be seen that the maximum bending stress produced by point load P approximates to that induced in a simple cantilever of width $B_{eff} = 3.5a - 1.5a^2/L$. At the same time we may imagine that the point load disperses linearly into a knife-edge load of length B_{eff} so far as the root of the cantilever is concerned. It is, in fact, very convenient to consider the problem in terms of dispersion. For typical load positions, say a/L in the range 0.3–0.6, the dispersion angle α varies only from 52° to 57°.

BS 5400 adopts this principle for local bending checks on flanges. A dispersion angle of 60° is specified, allowing for the relieving effect of double-curvature caused by contact between the outer edges of the flanges. In joints consisting of two flanges pulling against one another (Figure 34.7(a)) significant prying forces may be generated. An indication of the maximum value may be obtained by assuming a plastic distribution of bending moments in the plate, as shown in Figure 34.7(b), taking a fulcrum force near the edge of the plate. If the strength-reducing effect of the hole is ignored it follows that the prying force $Q = F.c/2n$. In fact, due to the presence of the bolt hole and because of bolt extension, the force will usually be less than this.

BS 5950 implies that prying action is allowed for in the tabulated strengths for bolts in tension. BS 5400 recommends that it should be assumed to be at least 10% of the applied bolt tension. Clearly, a high value will relieve the bending moment at the flange/web junction but will increase the required bolt capacity. BS 5400 allows a free choice, provided that equilibrium is satisfied and the resulting moments at the bolt line and flange root do not exceed the respective net moments of resistance.

Yield line theory has been used as an alternative to the effective-width approach for flange-strength assessment. Two simple mechanisms for the collapse of a column flange are shown in

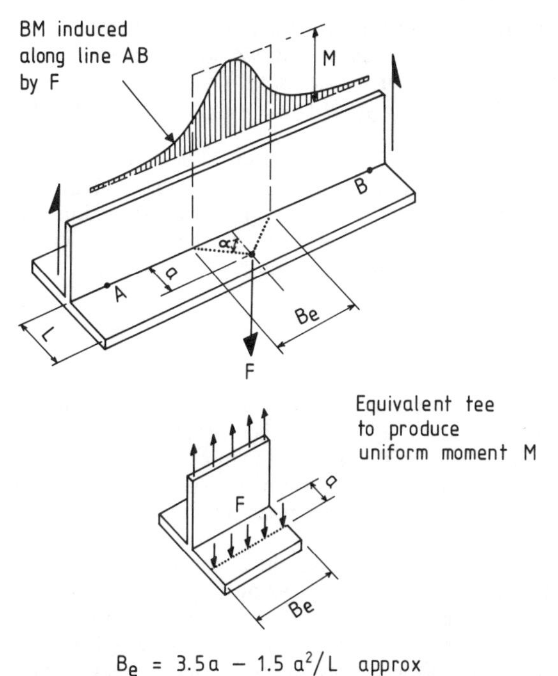

Figure 34.6 Equivalence of plate-bending effects

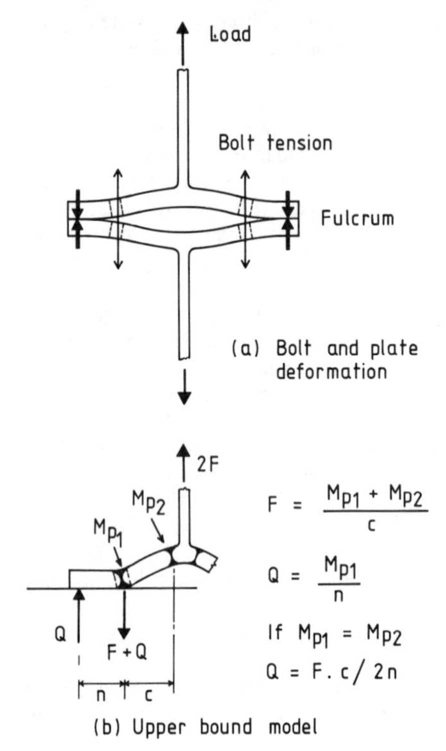

Figure 34.7 Bolt prying action

Figure 34.8. In Figure 34.8(a) bolt yielding has permitted a comparatively weak mode of collapse in the flange, in effect cancelling prying action. Stronger bolts would force a mechanism of the type shown in Figure 34.8(b). Many variations of mechanism (b) may be found in the literature (see Figure 34.9(a), for example) but the analytical solutions give broadly similar results.

An approximate mechanism for a stiffened flange is shown in Figure 34.9(b). The effect of stiffening is shown in Figure 34.10, where improvement in strength is seen to be marginal. The results are presented in relation to performance of a beam/column end-plate connection and stress the need for adequate stiffness in the column flange. Nevertheless, the improvement in stiffness for the stiffened case is somewhat better than that in strength.

34.5 Stanchion baseplates

For purely axial load a plain square steel plate or slab lightly attached to the column is adequate. If uplift or overturning forces are present more positive attachment is necessary. This invariably means welding, and a practical limit is thus imposed on the maximum thickness of plate possible, which will normally be in the region of 40–50 mm. If plate at this limit is insufficient to develop the applied bending moment, or if thinner plate is preferred, some form of stiffening must be adopted.

In high-moment bases the length of the baseplate determines the anchorage force on the tension side. Ideally, for practical purposes, a total of four holding-down bolts should be used (one in each corner). Occasionally, up to four bolts may be necessary in tension and bolt diameters up to 80 mm may be adopted. More typically, diameters in the range 20–30 mm are used. Base dimensions can be determined by trial and error, as outlined in many texts on steel design. A direct method is indicated briefly in Figures 34.11 and 34.12.

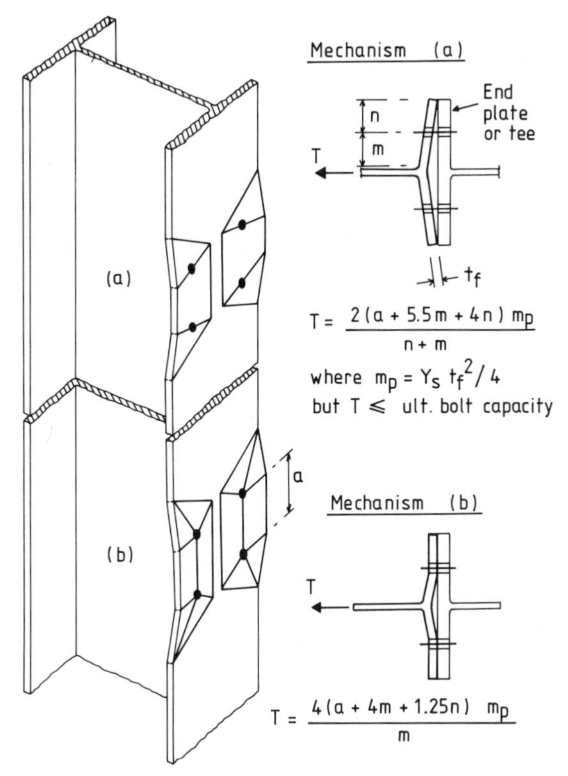

Figure 34.8 Yield line collapse models for a column flange

Two basic models are in common use for moment-resisting bases. The rectangular pressure distribution in Figure 34.12 is more convenient to analyse and for a given limiting pressure it permits thinner baseplates. On the other hand, the associated lever arm is lower and larger bolt areas may be required. It is worth noting that concrete compression failure is only a remote possibility for practical foundations. In contrast to a reinforced concrete beam or concrete test cube, the compression zone in a foundation block is confined by a ring of unloaded concrete which inhibits conical failure and can produce compressive strengths in excess of ten times normal cube-crushing strength. This is not considered in any of the UK steelwork codes but a significant enhancement of compressive strength is now provided for in Eurocode 3.

Thus if the equivalent eccentric load lies within the base length there is a every likelihood that it will pass directly into the concrete with little or no tension induced in the bolts. To provide an upper limit for eccentricity under these conditions an arbitrary minimum length to the compression block of $L/5$ has been assumed.

When the eccentricity exceeds $0.4L$, bolt tension is assumed to occur, as in Figure 34.12. In this case the compression zone must be confined within a length which would ensure strain compatibility with the bolts up to the point of collapse, as provided for in RC beam limit design theory. An arbitrary maximum length of $L/2$ has been taken for this.

In Figures 34.11 and 34.12 baseplate width is assumed to be a fixed proportion of length. If there are practical grounds for fixing the baseplate width, the basic equilibrium equations are simplified, replacing αL with B. Baseplate thickness may be calculated from the simple cantilever model or, in the case of axially loaded bases, from the formula given in BS 5950, which provides for the bending moment at the root of the larger plate projection to be reduced by an amount proportionate to the bending moment at the root of the smaller projection. Two-way bending will stiffen the baseplate and inhibit yield in the region of the column corners

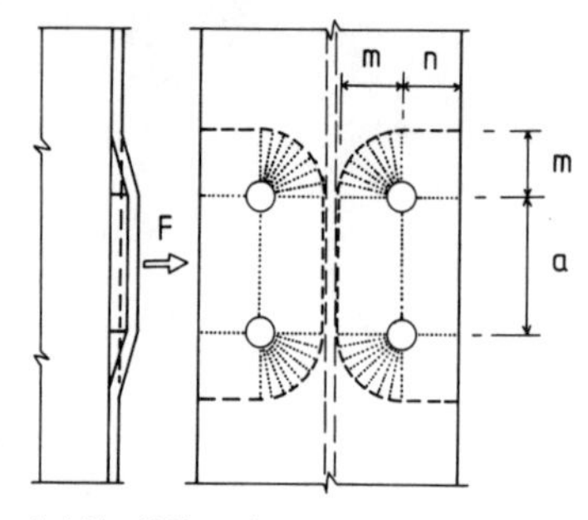

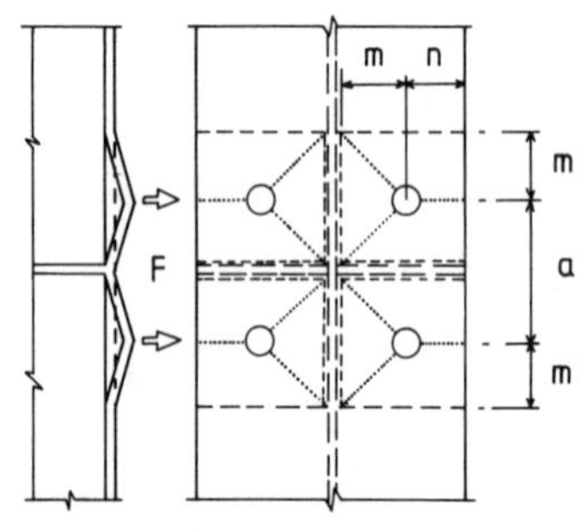

Figure 34.9 Stiffened and unstiffened flanges

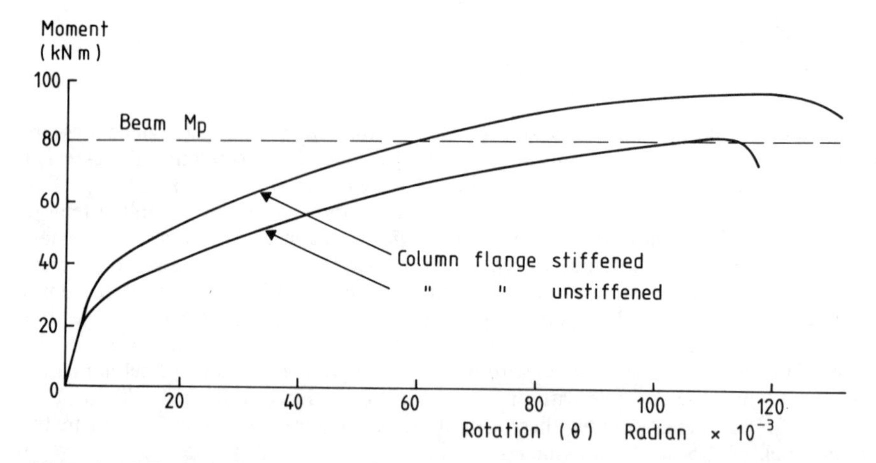

Figure 34.10 Typical test comparison of stiffened and unstiffened weak column flanges

but the bending moment elsewhere at the column boundary will depend on the shape of the column section and is unlikely to be well represented by the above formula. However, the considerable reserve of compressive strength in the concrete, referred to earlier, means that only an inner core of bearing contact is really necessary and bending deformation in the plate will not weaken the base significantly. In the case of moment connections the plate must nevertheless be of sufficient thickness to resist bending moment from the tensile force in the holding-down bolts.

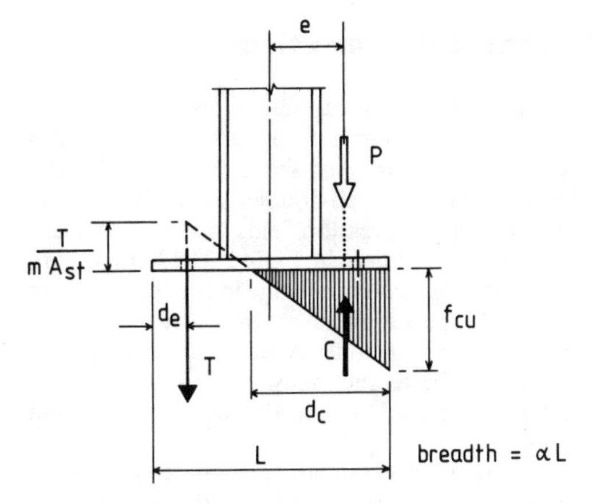

Compatibility : $$\frac{d_c}{f_{cu}} = \frac{L - d_e}{f_{cu} + \dfrac{T}{m A_{st}}} \quad \cdots\cdots(1)$$

Equilibrium : $$C = P + T = \alpha L d_c f_{cu}/2 \quad \cdots\cdots(2)$$

$$\left(L - d_e - \frac{d_c}{3}\right) C = \left(e + \frac{L}{2} - d_e\right) P \quad \cdots\cdots(3)$$

May be solved for L, given f_{cu}, A_{st}, m, d_e, P, e and α

Figure 34.11 Elastic models for baseplate forces: uplift on bolts [$e > 1/6(L + 2d_e)$]

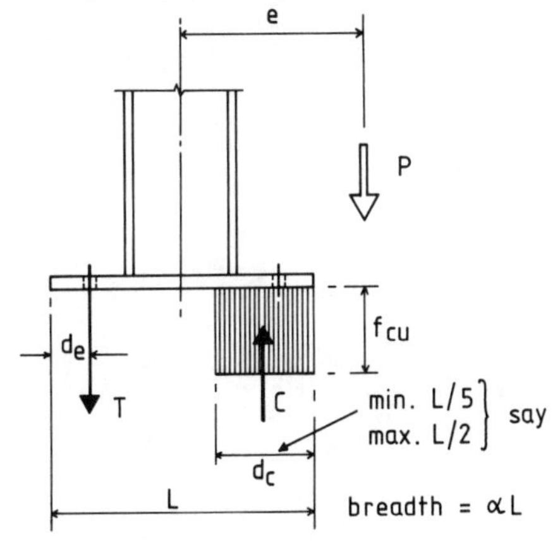

Equilibrium only:

$$C = T + P = \alpha L d_c f_{cu} \quad \cdots\cdots(1)$$

$$T\left(\frac{L}{2} - d_e\right) + C\left(\frac{L}{2} - \frac{d_c}{2}\right) = Pe \quad \cdots\cdots(2)$$

$$\therefore L = \frac{e + d_e T/P}{1 + 2\,T/P}\left[1 + \sqrt{1 + \frac{(1 + T/P)^2}{(e + d_e T/P)^2 \alpha f_{cu}}}\right]$$

Figure 34.12 Plastic model for baseplate forces: uplift on bolts ($e > L/4$)

34.6 Shear panels and shear reinforcement in connections

It is sometimes necessary to reinforce column webs against shear failure in the region of connecting beams, particularly when one side only of the column is connected. The shear is generated by opposing tension and compression forces transferred from the beam flanges. Forms of reinforcement include diagonal and K stiffeners and web doubler plates. The action of stiffeners is better understood in terms of tension and compression fields, rather than shear flow. Two examples are shown for a portal haunch connection in Figure 34.13. In Figure 34.13(a) any supplementary stiffening must be welded to the column flange in the region of the tension bolts to complete the force path. Curved or J type stiffeners are therefore essential in order to provide adequate clearance for inserting and fixing bolts. A different arrangement (shown in Figure 34.13(b)) uses a cap plate to transfer the haunch tension force to the outer corner of the column and thence down a stiff path formed by the diagonal compression stiffener. Bolt-fixing problems are avoided in this case.

The effective contribution of the web in resisting diagonal forces may be estimated intuitively in terms of path width. An alternative approach is to calculate the capacity of the web in terms of shear stress and regard stiffening material as making up the difference to full design load. The latter method should normally lead to lower stiffener costs.

34.7 Concluding summary

1. The stress distribution in connection material is difficult to analyse due to variable conditions in the real structure and because of the complexity of the physical model.
2. Simplified models which incorporate important variables without necessarily reflecting real behaviour are used to assess strength. One common example is the replacement of variable stress distributions by uniform distributions assumed to act over notional effective lengths or widths.

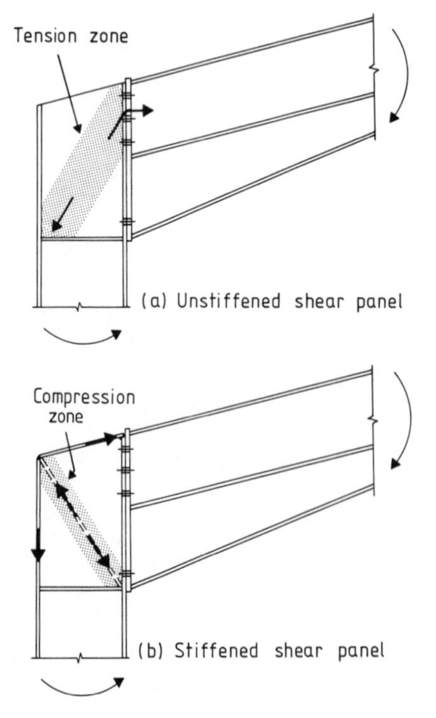

Figure 34.13 Force paths in shear panels

3. As an aid to derivation of effective lengths for checking direct stresses, or widths for bending stresses, the concept of load dispersion has been introduced. This assumes that concentrated loads disperse into plate and slab material at a constant rate (or constant angle to the direction of loading). It should be appreciated that this assumption rarely corresponds with the elastic stress distribution, and therefore presumes either that yielding is acceptable or an adequate margin for discrepancy has been allowed in the model.
4. It is also important to appreciate whether a design check aims to control the extent of yielding, amount of displacement, onset of embrittlement or some other aspect. Codes of practice could, to some advantage, be more explicit in these matters when dealing with aspects of connection design.

Background reading

Eurocode No. 3: *Common Unified Rules for Steel Structures*, 2nd draft (1988) Commission of the European Communities, Brussels

HOGAN, T.J. and THOMAS, I.R. (1978) *Standardized Structural Connections*, Australian Institute of Steel Construction, Sydney

OWENS, G.W. and CHEAL, B.D. (1988) *Structural Steelwork Connections*, Butterworths, London

PASK, J.W. (1988) *Manual on Connections for Beam and Column Construction*, 2nd edn, BCSA, London

35

Analysis of connections

Objective To present the basic methods of analysis for connections.

Prior reading Simple concepts of elasticity and kinematics.

Summary The chapter commences with a brief review of the simplifications that are inherent in the practical analysis of connections. A discussion of the behaviour of concentrically loaded shear joints is followed by the analysis of eccentrically loaded shear joints with illustrations of applications. Bolted connections which resist moment by tensile forces in the bolts are discussed. The simple analysis of fillet-welded joints is presented with a discussion of eccentrically loaded lap joints and welded moment connections.

35.1 Introduction

Accurate analysis of connection behaviour is impractical and certainly not essential for satisfactory design. For most applications it is only necessary to consider simple force paths through the connection which accord with external forces and then to ensure that components are capable of transmitting the resultant system of internal forces. Precise deformational compatibility is set aside.

While this may seem a major omission it so happens that the methods of analysis in common use do not usually involve large incompatibilities of strain. There is a simple reason for this. In many connections the deformation is concentrated mainly at the connectors (i.e. the bolts and welds) rather than in the connected material. In such cases it is reasonable to assume that:

1. Simple displacement patterns are imposed on the connectors by relative displacement of rigid connected parts;
2. There is a simple relationship between force and displacement in the connectors. Usually it is assumed that force is proportional to displacement and thus the internal force model corresponds exactly with the simple displacement pattern.

In other cases, however, the stiffness of connectors and associated material may vary widely within the connection and further assumptions are necessary.

Suitable applications for simple modelling will now be explored. Attention will be focused on analysis of connector forces as these invariably point to the general force pattern within the connection. Limit state codes do not preclude 'fully elastic' and other conservative assumptions at the ultimate limit state. However, it should be noted that serviceability requirements (for example, avoidance of slip, fatigue, etc.) may demand special care in the choice of model.

35.2 Simple analysis of bolted joints

35.2.1 Concentrically loaded shear joints

A tension splice is shown in Figure 35.1. In effect, this consists of two identical connections, one on each side of the junction, but we need only concern ourselves with one half of the splice,

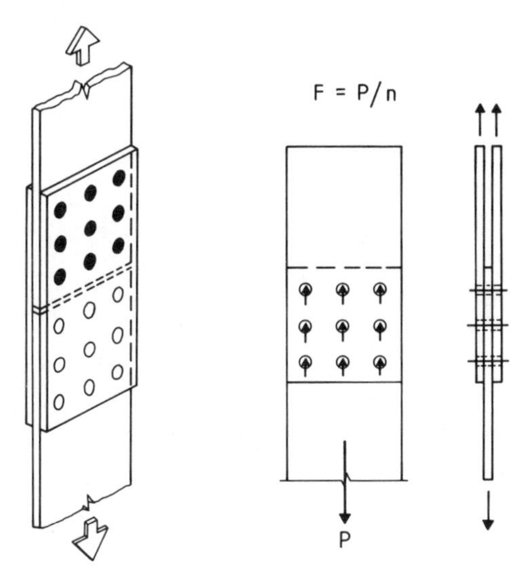

Figure 35.1 Simple tension splice

say the connection between the lower tie member and the cover plates. If the bolts were symmetrical about the line of action of the tie force we would expect equal shear displacements at each bolt position; that is, a pure translation of the cover plates relative to the tie member. In this circumstance the bolt force/displacement relationship is immaterial provided that it is the same for each bolt. The effect is that the tie force is shared equally between the bolts. Thus the force carried by each bolt acting in double shear is:

$$F = P/n$$

where P is the tie force, n is the number of bolts.

For the rare case of differing bolt diameters the force would be shared according to bolt stiffness, if we were concerned with service loading, or bolt strength, if we were considering ultimate behaviour. In either case it is commonly assumed that bolt shear force is proportional to the gross area of the bolt. Thus for a pure translation of the cover plates relative to the tie member it would be necessary for the line of action of the tie force to coincide with the centre of area of the bolt group. Given this condition, we could assume that the force in bolt i is:

$$F_i = P\,A_i/\Sigma A$$

where A_i is the area of bolt i and ΣA is the total area of the bolts.

35.2.2 Eccentrically loaded shear joints

When the line of action of applied force does not pass through the centre of area of the group equlibrium can only be satisfied by a non-parallel system of internal and external forces. As the plates are assumed to be rigid we may consider this to be the result of translation parallel to the external force plus rotation about the centre of area of the group, as shown in Figure 35.2. If the bolts have a linear force/displacement characteristic we may treat these two effects separately and apply the principle of superposition to combine them. That is, we can replace the eccentric force (P) by a concentric force (P) and a pure moment (Pe), where e is the eccentricity of the external force with respect to the centre of area of the bolt group.

The effect of a pure moment is analysed as follows. Referring to Figure 35.3, if the plates are rigid, reactive bolt shear forces (F_i) resulting from external moment (M) must conform to an instantaneous centre of displacement IC. The magnitude of F_i will be proportional to the distance r_i for each bolt. Thus F_{max} will be associated with r_{max} and, in general,

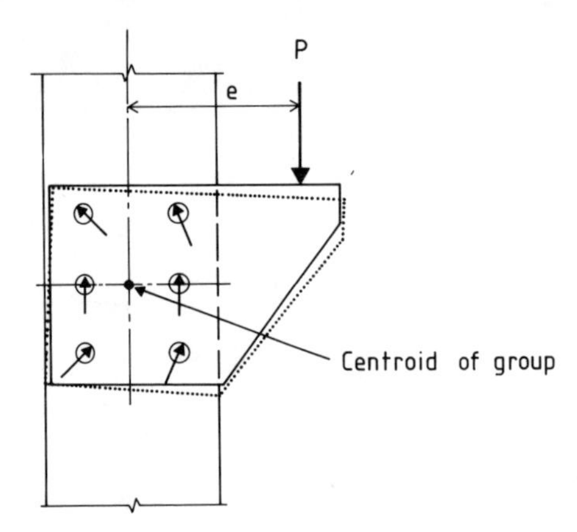

Figure 35.2 Eccentrically loaded shear joint

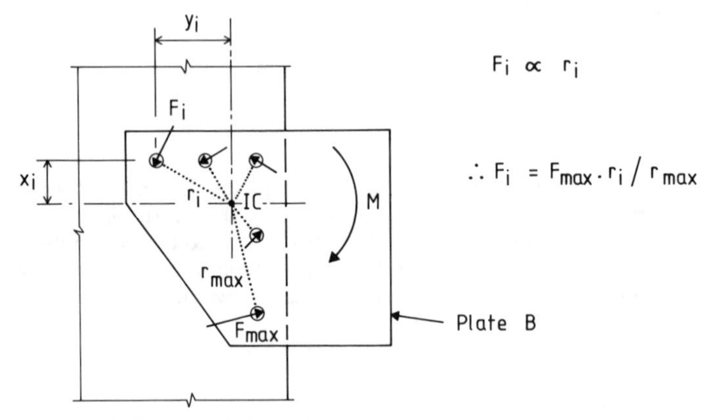

Figure 35.3 Elastic analysis of shear joint under pure moment

$$F_i = F_{max} r_i / r_{max} \tag{35.1}$$

Considering horizontal, vertical and rotational equlibrium between external forces and reactions on plate B, we have:

$$\Sigma F_i x_i / r_i = 0 \tag{35.2}$$

$$\Sigma F_i y_i / r_i = 0 \tag{35.3}$$

$$\Sigma F_i r_i = M \tag{35.4}$$

Consequently, from Equations 35.1–3:

$$\Sigma x_i = 0 \quad \text{and} \quad \Sigma y_i = 0$$

This confirms the assumption that the centre of displacement is at the centroid of the bolt group. Also, for Equations 35.1 and 35.4:

$$F_{max} \sum r_i^2 / r_{max} = M$$

i.e.

$$F_{max} = M/Z$$

where

$$Z = \Sigma r_i^2 / r_{max}$$

Z is referred to as the modulus of the bolt group and is dependent solely on bolt geometry. While this is a comparatively simple calculation, tables of group moduli prepared for standard bolt configurations may be used to advantage.

The analysis of eccentric external force is illustrated in Figure 35.4. Although we may simply combine the effects of direct force and pure moment vectorially, it is worth noting that the resulting forces must also conform to another (different) instantaneous centre of displacement because of the rigid plate assumption. The location of this centre is calculable but generally need not be determined. It is only introduced here notionally as a means of identifying the most highly loaded bolt. If we consider a line normal to the applied external force and passing through the centroid of the group the instantaneous centre of displacement will be on this line, but located on the opposite side of the centroid to the applied force. As in the case of pure moment the bolt furthest from the instantaneous centre will be the most highly loaded one. Usually, any approximate location of IC within the above definition is sufficient to indicate the critical bolt. Direct and rotational shears may then be computed for the bolt in question and combined vectorially, as shown in Figure 35.5.

Analysis of commonly used bolt patterns under eccentric loading is also facilitated in practice by use of standard tables. These usually provide coefficients to relate maximum bolt force directly to external load for various configurations and load eccentricities.

Two distinct forms of eccentric shear connection occur in practice. The bracket connection shown in Figures 32.2–5 is the simpler of these. It connects two parts only, and is fully

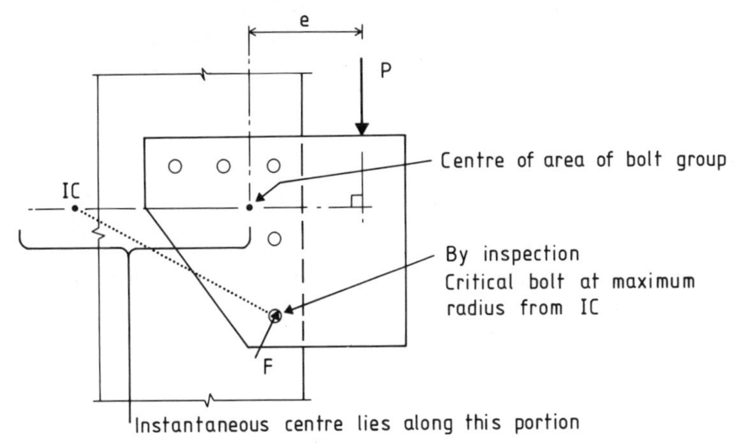

Figure 35.4 Elastic analysis of shear joint under eccentric load: identification of critical bolt

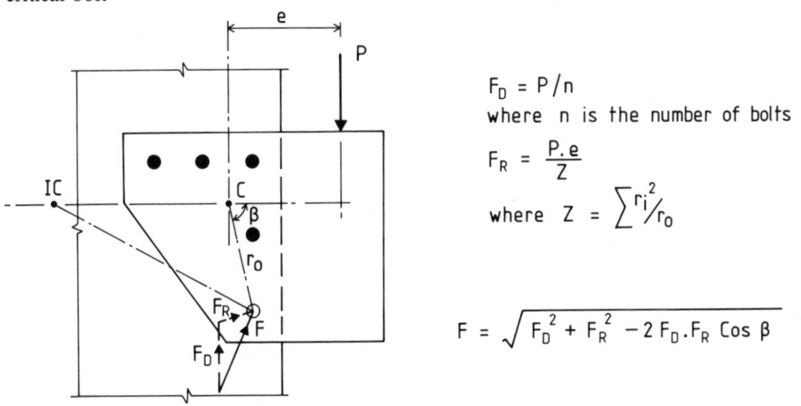

Figure 35.5 Elastic analysis of shear joint under eccentric load: force vectors on critical bolt

determinate. The other, typified either by the web splice shown in Figure 35.6 or the simple end-cleated connection shown in Figure 35.7, may be regarded as two eccentric connections acting in opposition with common shear force (Q). In this case, simplifying assumptions must be made before the elastic model can be applied.

For the web splice it is usual to assume that the beam slope is identical on each side of the joint (i.e. the joint is at a point of contraflexure) and thus the two halves of the connection may be solved by virtue of their skew-symmetry. In other words, Q is the only external force acting on each half of the splice.

The end-cleat connection can be simplified by assuming either that the primary beam is free to rotate or that it is prevented from rotating. The former assumption would be appropriate for a one-sided connection to a primary beam of low torsional stiffness (see Figures 35.7(a) and 35.7(b)) and also for double-sided cleats that were sufficiently flexible to permit rotation under shear loading. The latter assumption could be applied to a two sided beam-to-beam connection, as shown in Figure 35.7(c), or a beam-to-column connection, provided that the cleats were sufficiently stiff not to rotate under the applied shear loading. In this case we would assume that the secondary beam end-rotation was small in comparison with the vertical movement at failure. In other words, bolts connecting the cleats to the web of the secondary beam would ultimately have to resist a vertical load Q only. With the former (single-sided) case they would be required to resist a direct force Q plus a moment Qe.

In the complementary portion of the connection, i.e. between the end-cleats and primary beam as shown in Figure 35.8, this action is reversed. For the single-sided case, or where the cleats are flexible in the double-sided one, the bolts are required to resist a vertical shear force (Q) only, whereas in the two-sided case, with stiff cleats, tension will be induced in the upper bolts under the action of a bending moment not less than Qe.

35.2.3 Moment connections with bolts in shear

The examples discussed so far concern transfer of direct force, perhaps with small accompanying bending moment as a result of eccentricity. Figure 35.9 shows two connections designed primarily to transmit bending moment. This is achieved very efficiently by concentrating the transmission forces at the extremity of the tension and compression zones. The beam splice in

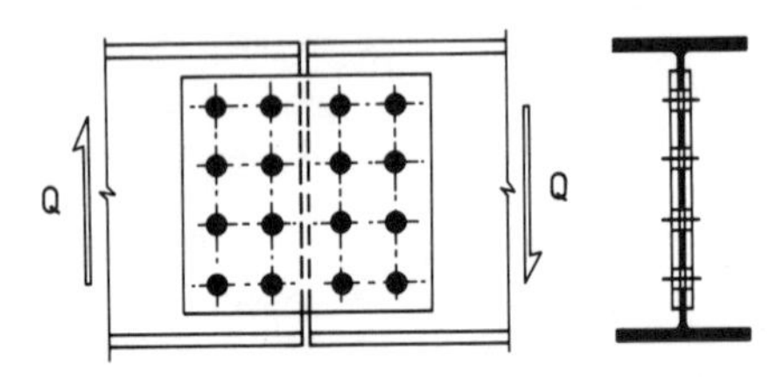

(a) Overall splice

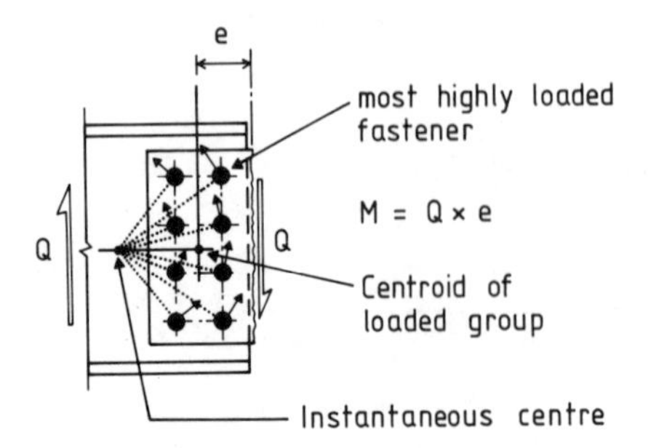

(b) Forces acting on left splice

Figure 35.6 Elastic analysis of beam web splice

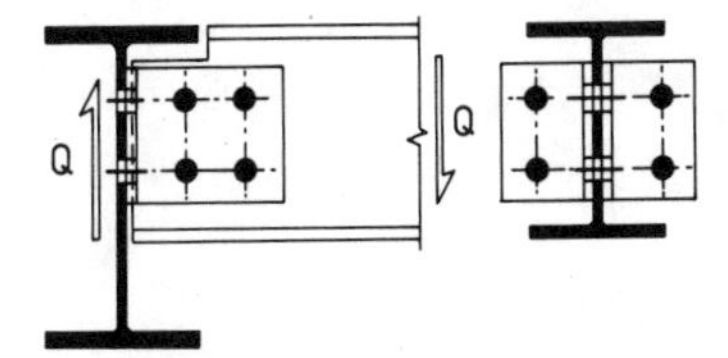

(a) Single sided and double sided connections

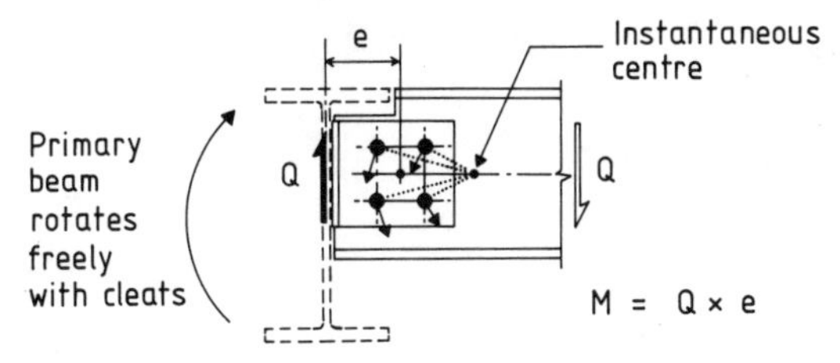

(b) Forces acting on single sided connection

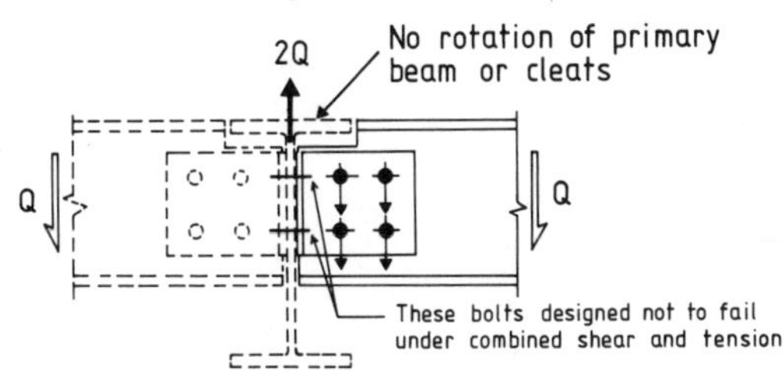

(c) Forces acting on double sided connection with stiff cleats

Figure 35.7 End-cleated beam-to-beam connections

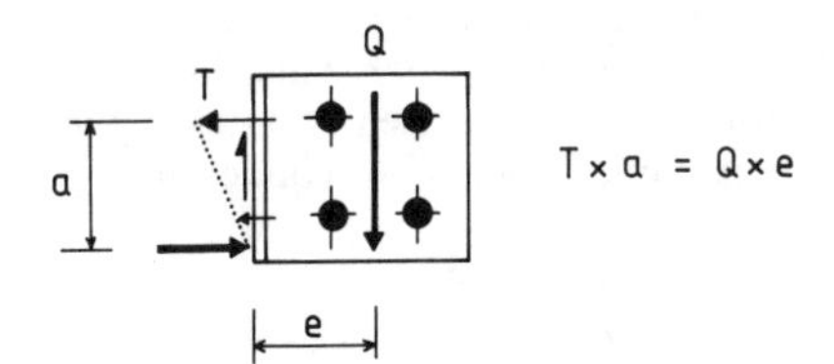

Figure 35.8 Forces in bolts attaching cleats to primary beams in Figure 35.7

Figure 35.9(a) mobilizes the double-shear strength of the flange bolts. The contribution of the web to bending resistance is small, except in the case of deep girders, and they may be used instead to resist vertical shear, possibly under a different load case from that associated with maximum bending moment. Thus:

$$M = nP_s d_f$$

where n = number of bottom flange bolts on one side of splice, P_s = double shear or bearing strength of one bolt, d_f = depth between flange centres. The corner connection in Figure 35.9(b) differs from the beam splice in that the compression force is transmitted by direct bearing of connection material. Horizontal deformation in the compression zone will be of a lower order than in the tension zone and the connection will effectively pivot about the compression flange. Provided that the beam end-plate is flexible (say, 8–12 mm thick) the tensile force induced in the column flange bolts will be negligible and the latter may be designed for vertical shear only. Thus the analysis for the bolts on the beam tension is similar to the previous case except that all bolts act in single shear.

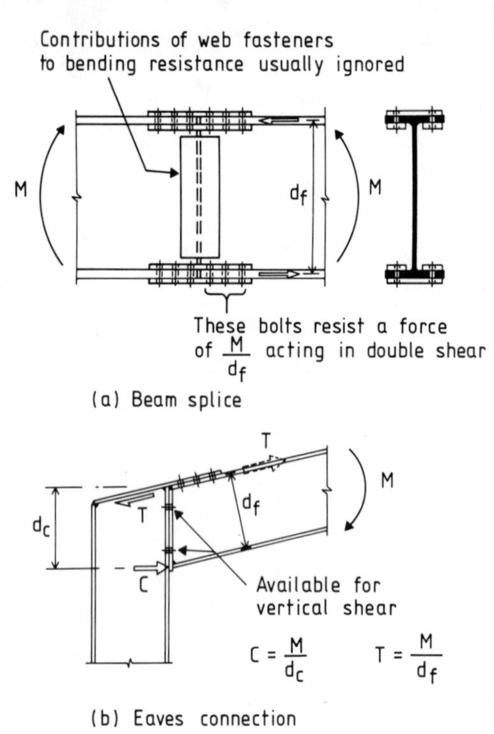

Figure 35.9 Moment connections with bolts in shear

35.2.4 Moment connections with bolts in tension

Shear bolts present a moderately stiff path once they have slipped into bearing, whereas tension bolts, though stiff in themselves, induce plate flexure. This can lead to quite large displacements in a connection and may be accompanied by prying action on the bolts. Clearly, thicker plies are better in this respect but they may not be feasible. For example, if one of the connected plies

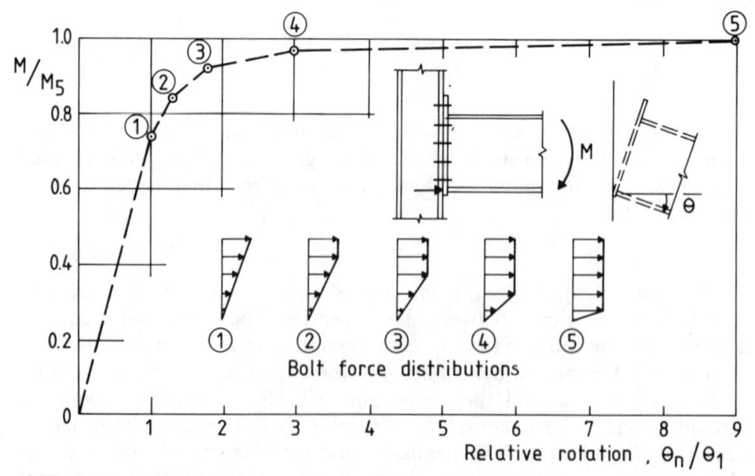

Figure 35.10 Moment-rotation characteristics for beam-to-column connection with end-plate and bolts in tension

is a column flange it may be uneconomic to choose a much heavier section for the column simply to stiffen the connection. There are, nonetheless, many applications for moment connections which use bolts in tension.

Considering a typical connection, the transition from fully elastic to fully plastic bending resistance is shown in Figure 35.10. This is based on a linear pattern of bolt strain pivotal about the bottom flange, which would be modified in reality by plate flexure and prying. Nevertheless, it shows the high order of rotation necessary to develop the full capacity of the connection and, more importantly, the small gain in strength after the point at which the upper bolts have yielded. As a result, two simple analytical models are in common use as shown in Figure 35.11.

There is some disagreement among designers about methods for considering shear resistance for these connections. The more conservative approach is to distribute any shear equally between all the bolts and check the top bolts under combined shear and tension. Other designers simply determine the residual shear strength of all the bolts within the group in the presence of the tensions shown in either Figure 35.11(a) or 35.11(b). It is then checked that this shear capacity is greater than the applied shear. With the assumptions shown in Figure 35.11(b), this has the effect of assigning all the shear to the lower two pairs of bolts.

35.3 Simple analysis of fillet-welded joints

35.3.1 Introduction

The basic mechanisms of force transfer occurring in welded lap and tee joints are similar to those described for bolts in shear and tension, respectively. We can regard the lap joint as stiff, almost co-planar, plate material connected by comparatively extensible fillet welds (see Figure 35.12). The high deformation at the weld arises from shear strain in the fillet and adjoining plate, mainly as a result of the sudden lateral discontinuity. In the case of tee joints the force distribution along the weld is dependent on the way in which the 'transverse' plate is supported on the side remote from the weld. A typical instance is shown in Figure 35.13. Clearly, stress concentrations are relieved if thicker intervening plates are used, but the only way to ensure a uniformly stressed tee joint is to arrange stiffening material on the reverse side.

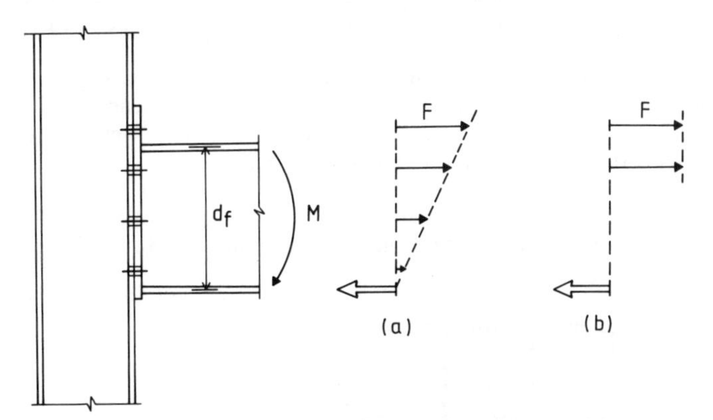

Figure 35.11 Alternative bolt tension distributions assumed in design for moment connections

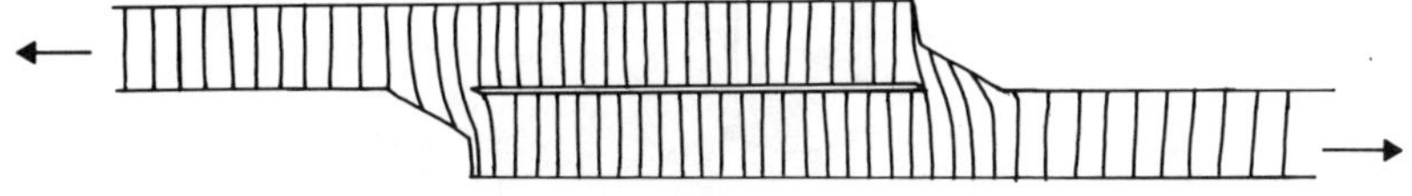

Figure 35.12 Deformations in welded lap joint

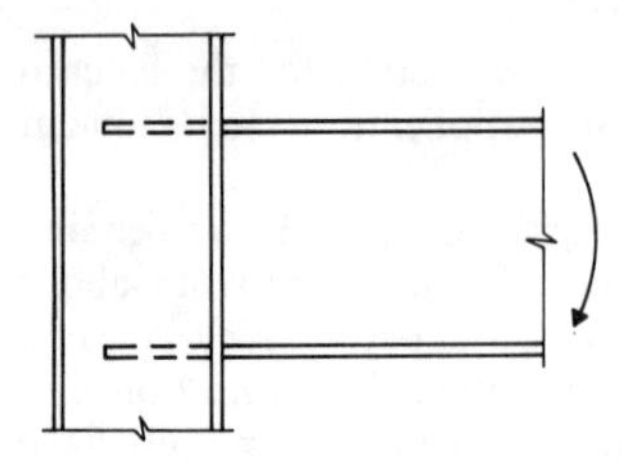

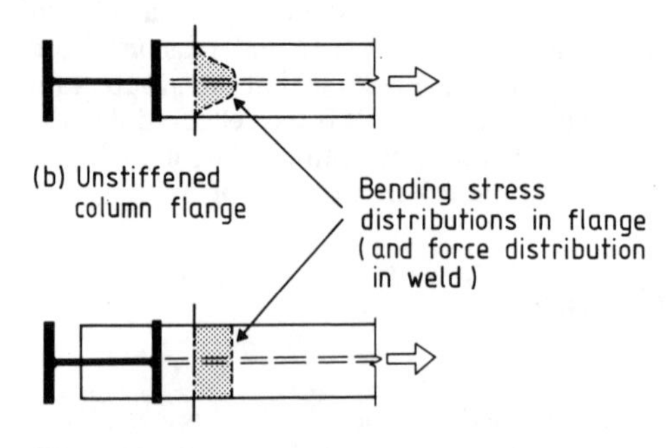

Figure 35.13 Influence of connection geometry on distribution of forces in a welded joint

35.3.2 Eccentrically loaded lap joints

The theory and techniques described for bolt groups subject to eccentric shear loading can be applied readily to welded lap joints of the type shown in Figure 35.14. We can regard the weld as a continuous zone of connection, unlike the discrete bolts previously considered. The connected elements are assumed to be rigid and the weld response is assumed to be linearly elastic.

In the weld group of Figure 35.15 a pure torsion, i.e. moment in the plane of the weld group, will produce a rotation about the centroid 0. Linearly elastic weld response will lead to weld stresses that are a function of distance from the centroid. If F_{max} is the maximum weld stress, at a radius of r_{max}, then the weld stress (F), at some radius (r), is given by:

$$F = F_{max}\, r/r_{max}$$

An element of weld, area dA, at distance r from the centroid will contribute $F r\, dA$ to resist the applied torque. Thus:

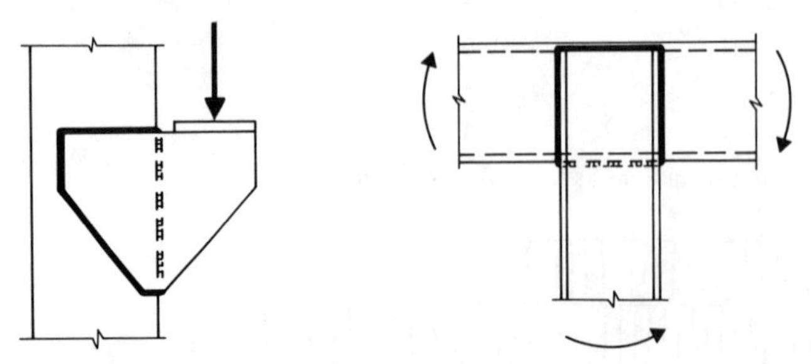

Figure 35.14 Types of welded lap joint

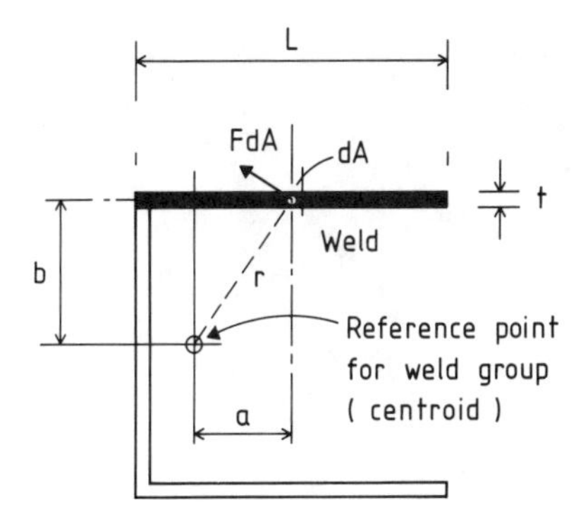

Polar second moment of shaded area

$$I_0 = L.t.\left[a^2 + b^2 + L^2/12\right]$$

$$(t << L)$$

where t is the throat thickness

Figure 35.15 Analysis of weld group under pure torsion

$$M = \int_{\substack{\text{Weld}\\\text{group}}} Fr\,dA$$

$$= \frac{F_{max}}{r_{max}} \int_{\substack{\text{Weld}\\\text{group}}} r^2\,dA = \frac{F_{max}}{r_{max}} I_0$$

where I_0 is the polar moment of throat area of the weld group about its centroid. Hence:

$$F_{max} = \frac{M r_{max}}{I_0}$$

Figure 35.15 also indicates how I_0 may be evaluated for most weld groups.

Where the torsion is acting in conjunction with a shear, elastic superposition may be used. The shear is assumed to be equally distributed over the weld group. A vector summation of the two stresses is carried out at the critical point in the weld, which is determined by inspection.

35.3.3 Welded moment connections

Moment connections may be formed using lap joints, as considered above, or by means of tee joints. General considerations are similar to those for bolted moment connections. For maximum economy welds should be disposed at as great a lever arm as possible. The assumed force distribution in the weld should take account of local conditions such as the absence of stiffening material on the reverse side of the tee joint. Welded joints are often used to attach secondary material to members in site-bolted connections and may be subject to local concentrations of force at bolt positions.

In fully continuous stiffened connections a linear force distribution consistent with plane bending theory is frequently adopted, as shown in Figure 35.16, Method 1. Alternatively, partitioning of web and flange welds to resist vertical shear and bending, respectively, as shown in Method 2, will give a comparable result. If stiffening is not provided on the remote side of the tee joint, an effective width of weld less than B might be taken to allow for stress concentration.

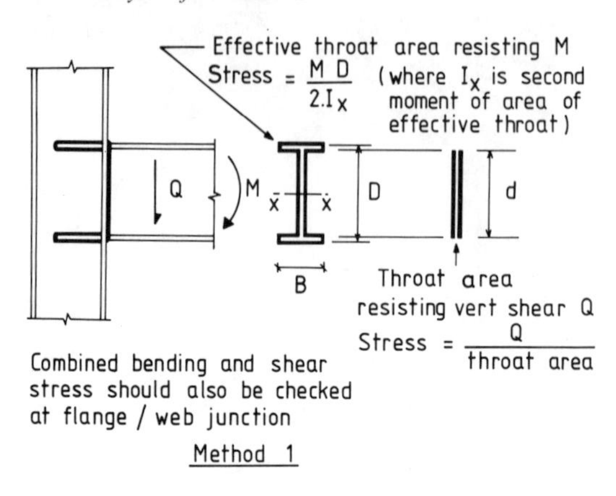

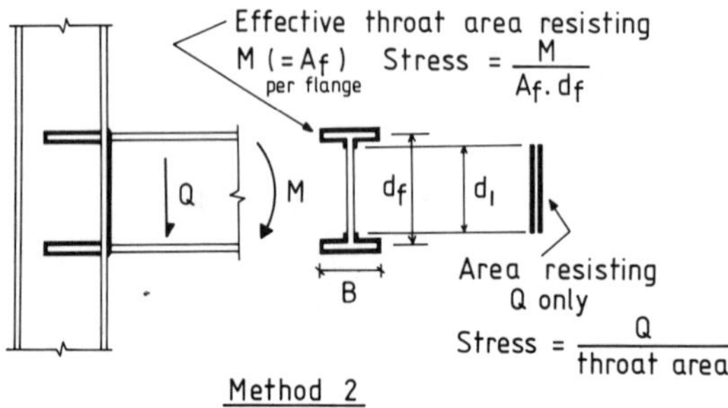

Figure 35.16 Alternative analyses for weld groups under shear and moment

35.4 Concluding summary

1. Analysis of steelwork connections may be based on either:
   (a) A simple overall compatibility model, assuming rigid connected parts and linear fastener behaviour; or
   (b) Any reasonable fastener force distribution, taking account of possible non-linear fastener behaviour and stiff/weak force paths.
2. It is important to appreciate what constitutes a stiff or a weak force path. Ideally, direct forces should be transmitted by in-plane or membrane action. Alternatively, selected forces may be relieved by plate flexure. This enables the flow of force through a connection to be controlled to best effect and often simplifies the analysis.

Background reading

Eurocode No. 3, *Common Unified Rules for Steel Structures* (1984) Commission of the European Communities, Brussels

HOGAN, T.J. and THOMAS, I.R (1978) *Standardized Structural Connections*, Australian Institute of Steel Construction, Sydney

Metric Practice for Structural Steelwork, 2nd edn, (1982) BCSA, London

PASK, J.W. (1988) *Manual on Connections for Beam and Column Construction*, 2nd edn, BCSA, London

36

Multi-storey frame connections

Objective To consider typical connections for multi-storey frames, examine their behaviour and show how they may be designed.

Prior reading Chapter 6 (Design of multi-storey Buildings); Chapter 30 (Introduction to connection design); Chapter 31 (Welds: static strength); Chapter 33 (Bolts and bolting); Chapter 34 (Local elements in connections); Chapter 35 (Analysis of connections).

Summary The principal types of the following connections are discussed: beam-to-beam connections, beam-to-column connections, column splices and column bases.

36.1 Introduction

Multi-storey frame connections may be conveniently classified into five types:

1. Beam-to-beam connections;
2. Beam-to-column connections;
3. Column splices;
4. Column bases;
5. Bracing connections.

Fasteners can be ordinary bolts, pretensioned high-strength friction-grip bolts or welds. The choice of bolts is usually between ordinary Grade 4.6 and Grade 8.8, in clearance holes. So far as practicable, butt welds are avoided because of the high cost of preparation. Fillet welds are preferred, of such a size that they can be put down in a single run.

36.2 Beam-to-beam connections: flexible connections

Figures 36.1(a)-(c) show three different flexible web-to-web connections and Figures 36.1(d) and 36.1(e) the behaviour of the beams at these. In Figure 36.1(d), when beam 1 rotates, beam 2 resists by virtue of its torsional stiffness, and a small restraint moment $H.d$ develops. In Figure 36.1(e) beam 2 does not rotate but a small restraint moment still develops because of the continuity provided by the bolts joining beams 1 together. These factors are ignored in practice, on the grounds that the connecting elements are so flexible that any restraint moments developed are negligible. Instead, the connecting elements are regarded as extensions of the web of beam 1. The fasteners connecting to the web face of beam 2 are designed for a shear force (Q), and the fasteners connecting to the web face of beam 1 are designed for the combined effect of a shear (Q) and a torsional moment ($Q.e$).

In Figure 36.1(a) the bolted cleats are self-positioning. In Figure 36.1(b), if the end-plate is short as shown, the connection is more flexible, but the plate is difficult to position for welding. Fabricators prefer it to run full depth so that it is squared up automatically by the bottom flange. The detail in Figure 36.1(c) has the virtue that workmanship on beam 2 is restricted solely to welding, while that on beam 1 is restricted solely to drilling, so that double handling of elements is avoided.

36.3 Beam-to-column connections: erection-stiff connections

The erection-stiff connections shown in Figure 36.2 impart some temporary sway stiffness to parts of the multi-storey frame during erection, before permanent stiffening is provided by floors, walls and bracings. They do this because they possess in-built (but uncalculated) rotational stiffness. However, apart from the bottom cleats the connecting elements are made fairly thin, and the connections are not regarded as capable of developing appreciable end-restraint moments when the beams carry their full design loads.

In Figures 36.2(a) and 36.2(b) the bottom cleat is designed to carry the shear force Q; the bolts/welds on the vertical leg are designed for shear only, and the leg is made sufficiently thick to limit bearing and buckling stresses in the beam, as shown in Figure 36.2(d). Where the maximum thickness of standard angles is insufficient, a fabricated seating may be used (Figure 36.2(c)). The reaction Q can also be regarded as acting at mid-length of the horizontal leg of the angle, which is designed for the resulting bending moment. The bolts/welds on the vertical leg are also designed for this bending moment. Top cleats provide torsional restraint for the end of the beam. Figure 36.2(b) is arranged to limit workmanship on the columns to welding only and on the beams to drilling only.

Connections shown in Figures 36.2(e)–(h) are designed like the beam-to-beam connections. Erection-stiffness is achieved by positioning the top and bottom bolts close to the flanges as

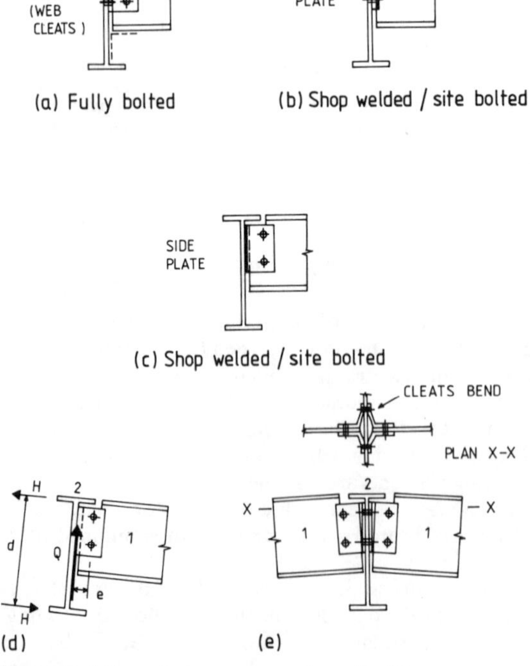

Figure 36.1 Beam-to-beam connections: flexible

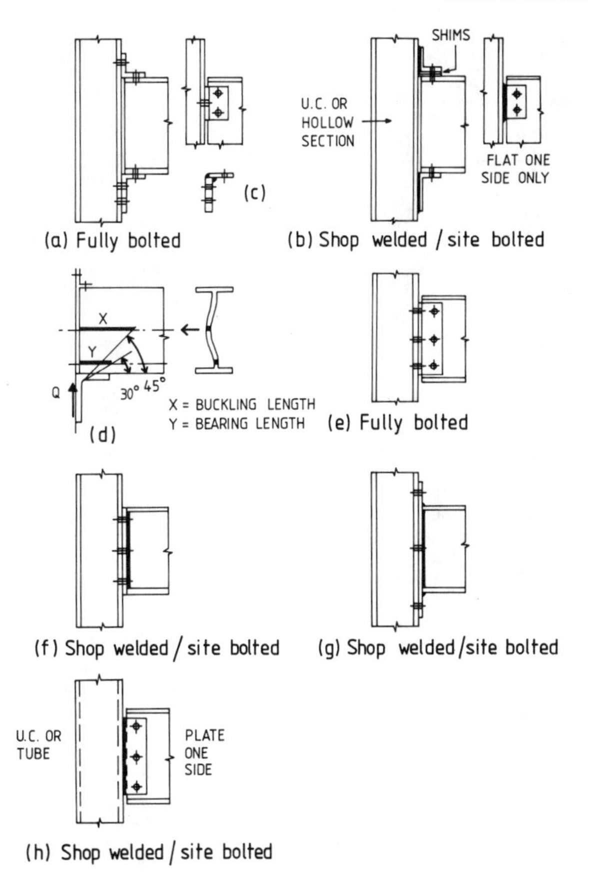

Figure 36.2 Beam-to-column connections: erection-stiff

shown. Practice varies considerably, but some designers prefer to limit the use of these types of connection to beams not exceeding 600 mm deep.

36.4 Beam-to-column connections: fully rigid connections

Figures 36.3(a)–(c) show three different fully rigid connections, and Figure 36.3(d) illustrates the forces to be transmitted by them. The beam moment (M) and axial force (F) are resolved into tensile and compressive forces (PT) and (PC), which induce a high local shear in the column web, which may need to be stiffened, as shown in Figures 36.3(e) or 36.3(f). The beam reaction (Q) passes through the bolts/welds in shear into the column flange.

Fully rigid connections may be required to develop the full moment of resistance of the beam. For this purpose, Figure 36.3(b) is least effective as the connection lever arm (centre of compression flange to centroid of tension bolt group) is *less* than the beam lever arm (distance between centres of flanges). Figure 36.3(a) does not have this disadvantage and is effective where:

1. Sufficient bolts can be clustered round the tension flange of the beam to transfer the full force from the flange; and

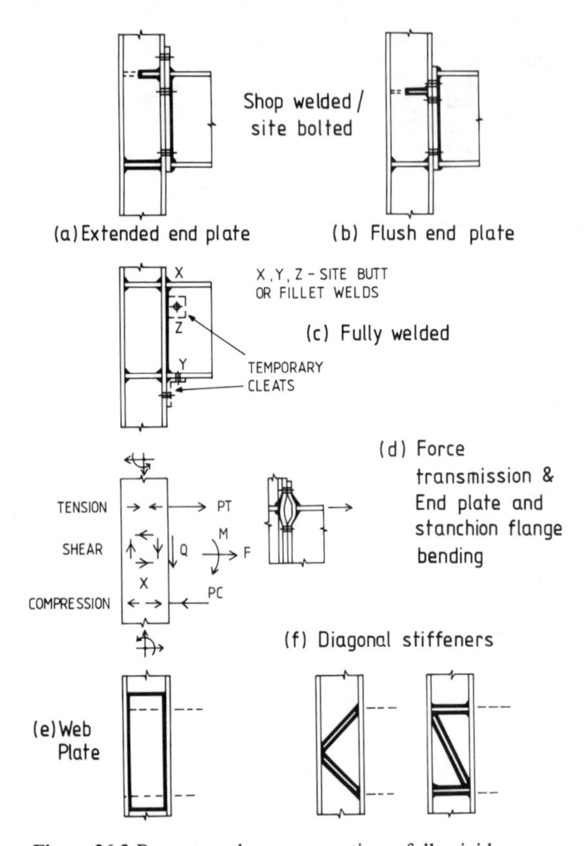

Figure 36.3 Beam-to-column connections: fully rigid

2. The bending resistance of the column flange is also sufficient to transfer this force.

Figure 36.3(c) is the most effective, as:

1. Welds of sufficient strength can be accommodated easily at the end of the tension flange; and
2. Connection strength is independent of column flange bending strength.

The disadvantage of Figure 36.3(c), compared with Figures 36.3(a) and 36.3(b), is that it is site welded; this tends to be expensive.

36.5 Column splices

Figures 36.4(a)-(c) show three different site-bolted splices for columns and Figure 36.4(d) illustrates the force transmission system. The column ends are machined and axial force is assumed to be transmitted in direct bearing. Where different serial sizes are joined, a division plate is used of sufficient thickness to transfer the flange stresses in bearing, at an angle not less than 45° to the horizontal, as shown in Figure 36.4(b). Cover plates and fasteners are sized initially by simple empirical rules to give the splices a certain minimum robustness, to preserve continuity in the finished structure and to ensure erection-stiffness; they are then checked for strength. Any bending moment about the major axis is resolved into couple forces (tension and compression), which are transmitted through cover plates/bolts/welds, which are not designed to transmit axial load. Any shear force is taken through the web covers. Where net tension

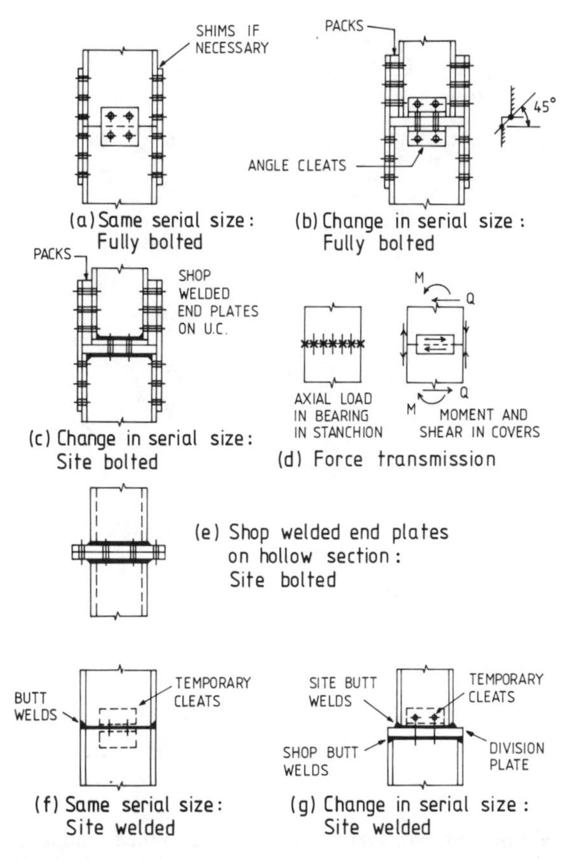

Figure 36.4 Column splices

develops, some designers use pretensioned bolts in the flanges to avoid bolt slip. Figures 36.4(a) and 36.4(b) allow plenty of play for adjustment when bolting up. Figure 36.4(c) is suited to a structure in which the beams also have welded end-plates, since the method of fabrication is the same.

Figure 36.4(e) shows a site-bolted splice which can be buried in the floor thickness. It is normally used in conjunction with hollow sections, but can also be employed with Universal Column sections.

Figures 36.4(f) and 36.4(g) show site-welded splices. Comments made on site welding in the previous section also apply to these splices. They are not erection-stiff, and the upper column lengths must be temporarily guyed or propped until site welding is completed. In Figure 36.4(g) it is essential to check the division plate for laminations during fabrication, as it will have little moment resistance if they are present.

36.6 Column bases

Slab bases are used for multi-storey columns, as normally the axial forces are large and the bending moments are small. Alternative types are shown in Figures 36.5; both types are erection-stiff.

The column ends are machined, so is the upper surface of the slab if it is very thick; otherwise it is just flattened. In either case it is assumed that axial loads is transmitted by direct bearing,

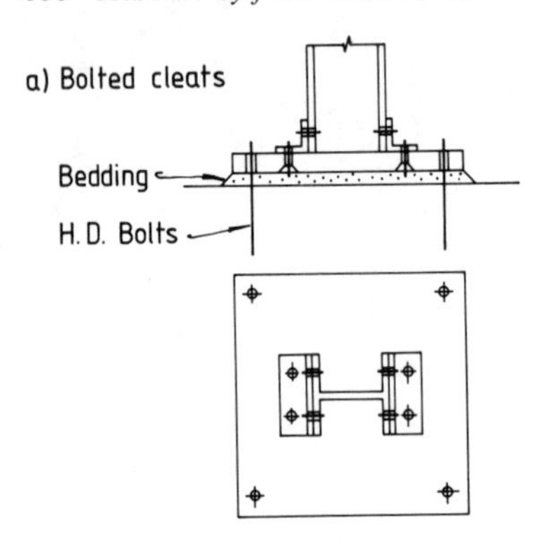

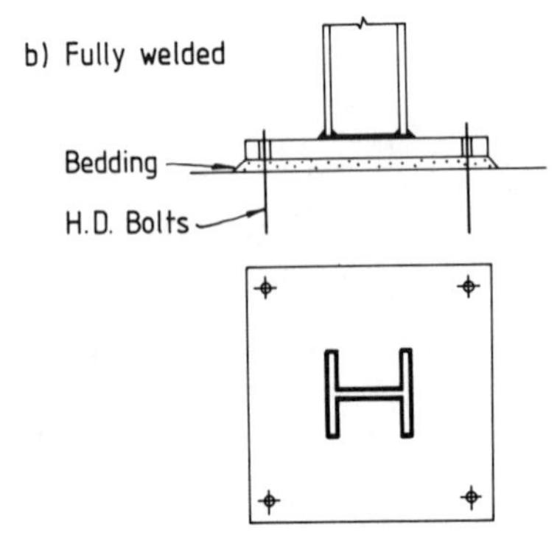

Figure 36.5 Column bases

from column to slab and from slab to bedding material. The slab is anchored to the foundation by suitably proportioned holding-down bolts, set in the concrete in clearance holes, which are later grouted up. Any bending moment about the major axis is resolved into couple forces (tension and compression) which are transmitted through the flange cleats/bolts/welds, which are not designed to take axial load.

Slabs are often large and heavy, and bolted construction enables them to be transported separately and attached on site. This is not possible when the slabs are shop-welded. Some designers advocate welding all round the Universal Column section. Others suggest that only the flanges and part of the web need by welded. Provided that strength requirements are met, either method is acceptable.

36.7 Concluding summary

1. The stability of multi-storey steel frames depends upon the nature of their connections. Stability may be achieved by introducing diagonal bracings, in which case flexible (or erection-stiff) connections may be used. Alternatively, it may be obtained by the use of fully rigid connections.
2. Fasteners are normally Grade 4.6 and 8.8 bolts and fillet welds.
3. Connection configurations are variable, but functional requirements are constant.

Background reading

HOGAN, T.J. and FERKINS, A. (1981) *Standardised Structural Connections*, Australian Institute of Steel Construction, Sydney
OWENS, G.W. and CHEAL, B.D. (1988) *Structural Steelwork Connections*, Butterworths, London
PASK, J.W. (1988) *Manual on Connections for Beam and Column Construction*, 2nd edn, BCSA, London
THE STEEL CONSTRUCTION INSTITUTE (1989) *The Steel Designers Manual*, 6th edn, Blackwell, Oxford

37

Single-storey frame connections

Objective To consider typical connections for single-storey frames, examine their behaviour and show how they may be designed.

Prior reading Chapter 4 (Design of Industrial buildings); Chapter 30 (Introduction to connection design); Chapter 31 (Welds: static strength); Chapter 33 (Bolts and bolting); Chapter 34 (Local elements in connections); Chapter 35 (Analysis of connections).

Summary The principal types of the following connections are discussed: portal knees, portal apexes, portal bases and column bases.

37.1 Introduction

Single-storey frame connections may be conveniently classified into seven types:

1. Portal knees;
2. Portal apexes;
3. Portal bases and column bases;
4. Column caps;
5. Purlin and side-rail brackets;
6. Compression flange restraints;
7. Bracing connections.

Fasteners can be ordinary bolts, pretensioned high-strength friction-grip bolts, or welds. The choice of bolts is usually between ordinary Grade 4.6 and 8.8, in 2 mm clearance holes. So far as practicable, butt welds are avoided because of the high cost of preparation. Fillet welds are preferred, of such a size that they can be put down in a single run.

37.2 Unhaunched portal knees

Figures 37.1(a), 37.1(b), 37.1(d)-(f) show five different, unhaunched, fully rigid, portal knee connections and Figures 37.1(c) and 37.1(g) the forces to be transmitted by them. In Figure 37.1(a) the behaviour of this unhaunched knee is seen to be similar to the behaviour of the haunched knees described in the next section. Figure 37.1(b) illustrates how member forces are resolved, as shown in Figure 37.1(c). The vertical shear force in the rafter web passes through the web welds into the end-plate. From this it passes through the connecting bolts (in shear) into the column flange. The tensile force in the top flange of the rafter passes (in shear) through the connecting bolts into the cover plate. The designer may consider the use of pretensioned high-strength friction-grip bolts here to avoid bolt slip. Tensile force in the cover plate passes (in shear) through the connecting welds into the column web. Portion x of the web will be highly stressed in shear and may need to be stiffened. The compressive force in the bottom flange of the rafter passes into the flat stiffeners on the column and through the connecting welds (in shear) into the column web.

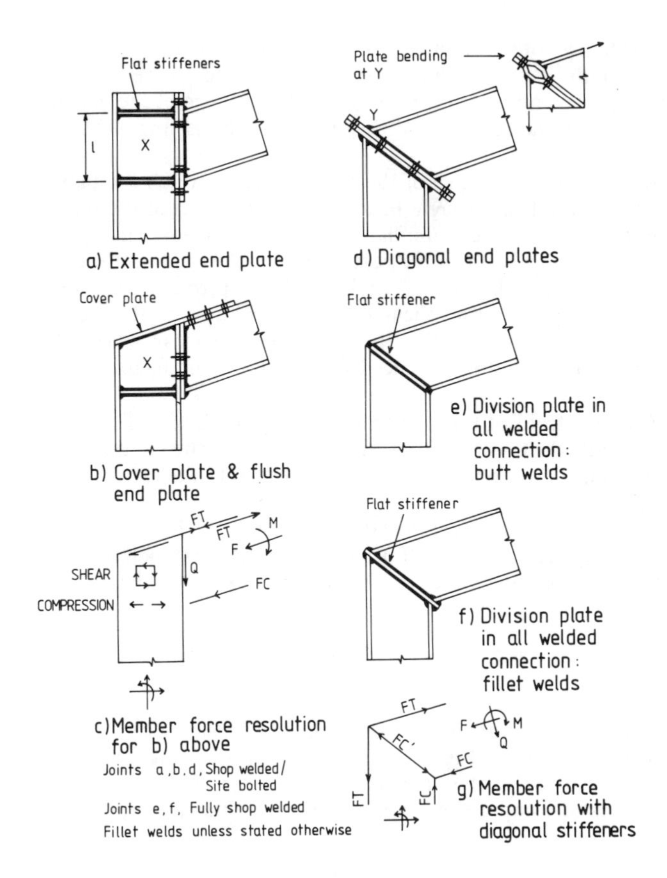

Figure 37.1 Portal knees: unhaunched

In Figures 37.1(d)–(f) the member forces are resolved as shown in Figure 37.1(g). The welds are designed to transmit the flange forces and web forces into the end-plates or flat stiffeners. These elements are designed as struts to take the compression force shown in Figure 37.1(g). In addition, the end-plates in Figure 37.1(d) are checked for local bending near the tension flange. The bolts in this knee are designed to transmit the bending moment at the knee, and the axial and shear forces that act perpendicular and parallel to the plane of the end-plates. In Figures 37.1(e) and 37.1(f) it is essential to check the flat stiffeners for laminations during fabrication, as they will have little moment resistance if laminations are present.

It is possible to compare the merits of these unhaunched portal knees. The fully welded joints result in frames, or parts of frames, which are of awkward shape, and difficult to handle, transport and erect. They will usually be limited to frames of small overall dimensions. The shop-welded/site-bolted joints are suitable when it is possible to obtain sufficient bolts of a practical size to take the moment. When this is not possible, the knees will need to be deeper, to increase the moment of resistance of the bolt group, and it will be preferable to change to one of the haunched knees described in the following section.

37.3 Haunched portal knees

Figures 37.2(a)–(c) show three different haunched, fully rigid, portal knee connections and Figure 37.2(d) illustrates how the member forces are resolved. The vertical shear force in the rafter and haunch webs passes through the web welds into the end-plate, and from this it passes through the connecting bolts (in shear) into the column flange. Some designers share the force equally between all the bolts in the flange; others assume that it is carried solely by the group of bolts near the haunch compression flange, so that no shear force is carried by the group of bolts near the rafter tension flange. The compression force in the bottom flange of the haunch passes into the compression stiffeners on the column and through their connecting welds (in shear) into the column web.

In Figure 37.2(a) the tensile force in the top flange of the rafter passes through the flange welds into the end-plate, which bends, and then through the tension bolts (arranged symmetrically around the end of the rafter flange) into the column flange, which also bends. This flange may need to be stiffened as shown, in which case the force passes through the welds at the ends of the tension stiffeners, into the stiffeners themselves, and (in shear) through the welds connecting them to the column web. The resulting shear force in portion x of the web is dependent upon the haunch depth (1), and by adjusting this depth the shear can usually be made less than the safe web shear, thus avoiding the need for flat plate or diagonal web stiffening.

In Figures 37.2(b) and 37.2(c) the centroid of the tension bolt group lies below the end of the top flange of the rafter. Consequently the tension force tends to be concentrated in the rafter web (and web welds) opposite the bolts, and probably only finds its way into the rafter flange at some distance from the column face.

It is possible to compare the merits of the haunched portal knees: Figure 37.2(a) provides a smoother tension load path than Figures 37.2(b) and 37.2(c), with the top of the stanchion skew sawn and plated, presents a more pleasing appearance but is more expensive to fabricate.

37.4 Portal apexes

For fully rigid apexes suitable details are the unhaunched ones already considered in Figures 37.2(d)–(f), and the haunched details shown in Figures 37.3(a) and 37.3(b). Figure 37.3(c) shows how the member forces are resolved in the haunched connections, with the bottom flanges in tension and the top flanges in compression. Apart from this difference, fully rigid

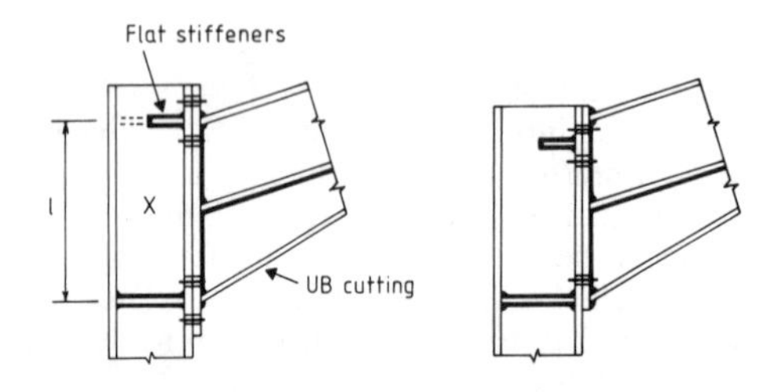

(a) Extended end plate (b) Flush end plate

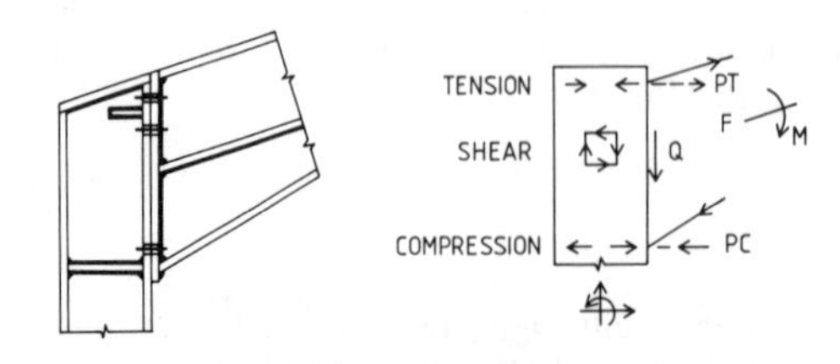

(c) Flush end plate (d) Member force resolution

Figure 37.2 Portal knees: haunched

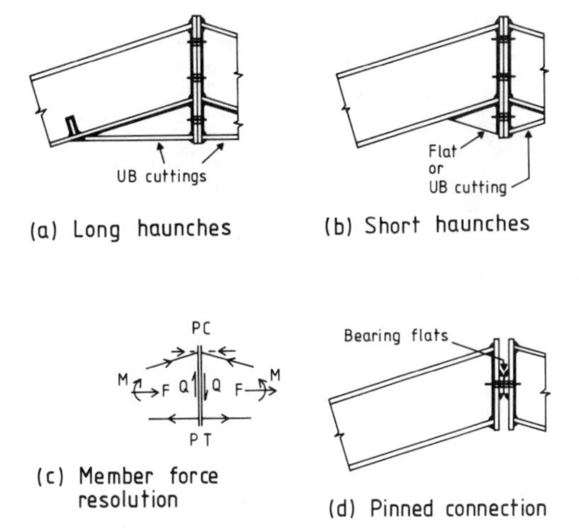

Figure 37.3 Portal apexes

apexes and knees behave in a similar way. Figure 37.3(d) shows a pinned apex connection, which allows fairly free rotation, and is suitable for a three-pin portal.

37.5 Portal bases and column bases

37.5.1 Pinned bases

These are shown in Figure 37.4. 'Pinned' bases should somehow meet the following (incompatible) requirements:

1. They should be free to rotate about the major axis; yet
2. They should be 'erection-stiff' so that the columns can be free-standing, with the minimum of temporary guys and props, before the portal rafters or trusses are erected; and

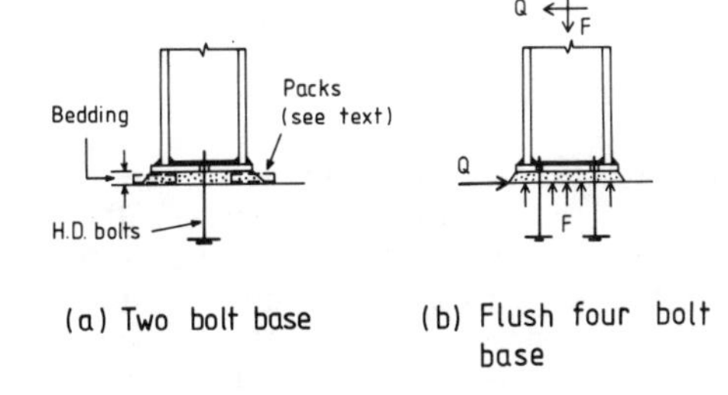

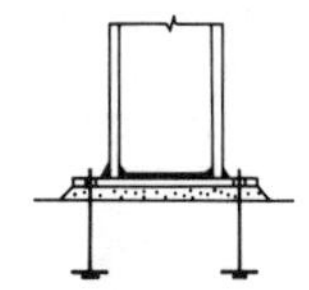

Figure 37.4 'Pinned' bases

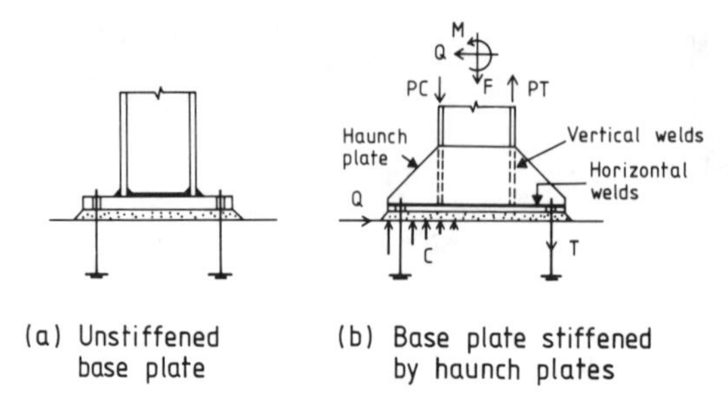

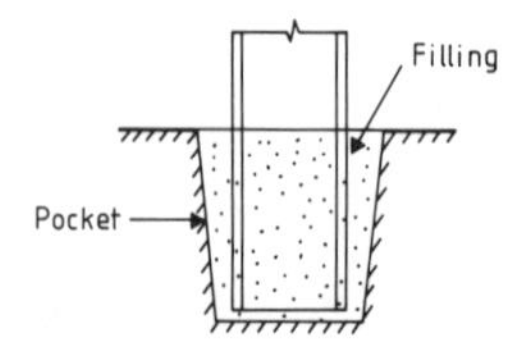

(c) Pocket base

Figure 37.5 Fixed bases

3. Where the columns form part of a fire-enclosing wall, the Building Regulations require the bases to have partial fixity so that the wall will remain standing, even when the roof has collapsed.

The first of these requirements is generally ignored, as partial fixity is not regarded as detrimental to the performance of the steel frame. The bases in Figures 37.4(b) and 37.4(c) are inherently erection-stiff by virtue of having four well-spaced-out holding-down bolts. The base in Figure 37.4(a) can be made erection-stiff by sitting it upon steel packs, positioned as shown dotted. All these bases have some degree of partial fixity in service, and can, where appropriate, be designed to meet the requirements for boundary conditions. Baseplates are usually fairly thin.

37.5.2 Fixed bases

Figure 37.5(a) shows a welded slab base, usually restricted to smaller columns where the bending moment does not require the bolts to have a large lever arm, and Figure 37.5(b) a built-up base, which is used when the bending moment does require a large lever arm for the bolts. The bending moment and axial force are resolved into tensile and compressive forces, which pass from the column flanges, through the vertical connecting welds, into the haunch plates. From the haunch plates they pass through the horizontal connecting welds into the base plate, producing pressure on the bedding material on the compression side and tension in the holding-down bolts on the tension side. It is recommended that shear force should be taken by friction between the base and the bedding, by mechanical shear keys in the foundation, by embedding the base in concrete, or by a combination of these means. Nevertheless, many designers do take shear on the holding-down bolts.

Figure 37.5(c) shows a pocket base, where fixity is achieved by concreting the plain end of the column into a preformed pocket in the foundations.

37.6 Concluding summary

1. Portal knee and apex connections are usually fully rigid, shop welded and site bolted; bases are pinned or fixed.
2. Fasteners are normally Grade 4.6 and 8.8 bolts, and fillet welds.
3. Connection configurations are variable, but functional requirements are constant.

Background reading

Behaviour of Steel Portal Frames in Boundary Conditions (1982) The Steel Construction Institute.
HOGAN, T.J. and FERKINS, A. (1981) *Standardised Structural Connections*, Australia Institute of Steel Construction, Sydney
OWENS, G.W. and CHEAL, B.D. (1988) *Structural Steelwork Connections*, Butterworths, London
PASK, J.W. (1988) *Manual on Connections*, 2nd edn, BCSA, London
THE STEEL CONSTRUCTION INSTITUTE (1989) *The Steel Designers Manual*, 6th edn, Blackwell, Oxford

Truss and girder connections

Objective To describe the types of connections used in trusses and lattice girders, discuss selection and set out their behaviour and design.

Prior reading Chapter 29 (Trusses and lattice girders).
The relevant parts of:
Chapter 4 (Design of industrial buildings); Chapter 6 (Design of multi-storey buildings); Chapter 30 (Introduction to connection design); Chapter 31 (Welds: static strength); Chapter 33 (Bolts and bolting); Chapter 34 (Local elements in connections).

Summary A description of the main types of connections used in trusses and lattice girders and for bracing in building is given. Joint eccentricity is reviewed and the design of bolted and welded joints and gusset plates is described. For worked examples see Appendix, page 387.

38.1 Introduction

Connections are needed to join the individual members together to form a complete truss or lattice girder. External joints are needed to connect the truss to the building frame. The joints actually used in truss construction do not reflect the idealization of pinned joints assumed for simple manual analysis. It would be expensive to make a truss with truly pinned joints, and experience has shown that traditional joints give satisfactory service.

38.2 Types of connections

Joints may be classified as follows:

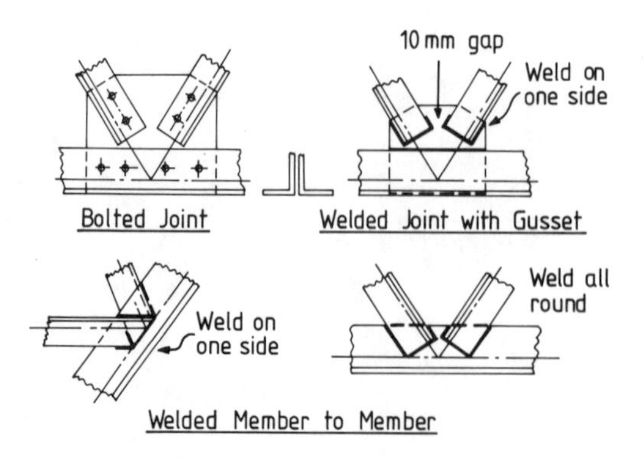

Figure 38.1 Internal joints

1. Internal joints. These connect the individual truss members together and usually join discontinuous web members to a continuous chord. Some typical examples are shown in Figure 38.1.
2. Site splices. Large trusses must be subdivided for transport to site where they are assembled into complete units through the site splices. Some typical examples are shown in Figure 38.2.
3. External cap or face joints are required to connect the truss to a column or another truss. Typical examples are shown in Figure 38.3.
4. Bracing connections are required to fix the diagonal members between building columns, portal frame members and adjacent trusses. Examples are shown in Figure 38.4.

38.3 Economic considerations

Guidance can be given on the selection of joint type to give the most economic solution:

1. In general, shop joints should be welded and site joints bolted. In special cases site-welded joints are used.
2. The choice between welded and bolted connections depends on the fabricator's equipment. Alternative prices can be requested or the decision left to the fabricator. If a large number of trusses are to be made, welded joints are usually more economical than bolted ones. Welded joints give a better appearance and maintenance costs are less than with bolted joints.
3. With welded joints, gussets can be eliminated where members are joined directly. The type of member used is a controlling factor in this case; for example, double-angle members generally require gusseted joints whereas hollow sections have to be welded together directly.
4. Standard joints should be used with as much repetition of member shapes and sizes, end-preparation and fabrication operations as possible. This can be achieved readily with parallel chord lattice girders.

38.4 Joint behaviour

BS 5950 states that members meeting at a joint should be so arranged that either their centroidal axes or, in the case of angles with bolted connections, the bolt centre lines meet at a point. If this is not so the members and connections must be designed to resist the moment arising from the eccentricity. In the case of gusseted joints making member axes meet at a point can lead to a need for large gusset plates. These can be made smaller if the members are nested as shown in Figure 38.5. The moment due to the eccentricity is distributed between the members meeting

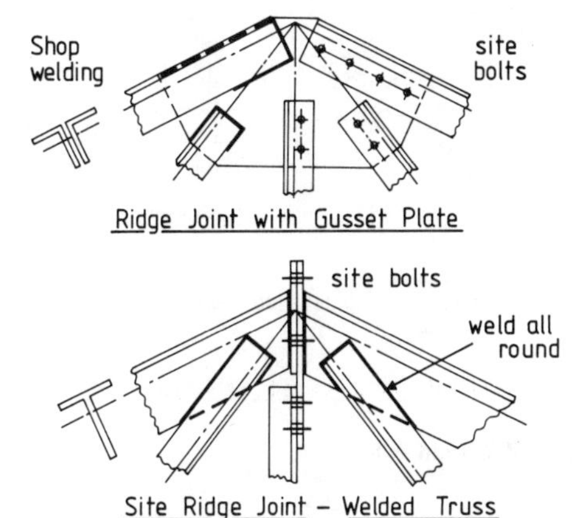

Figure 38.2 Site splices

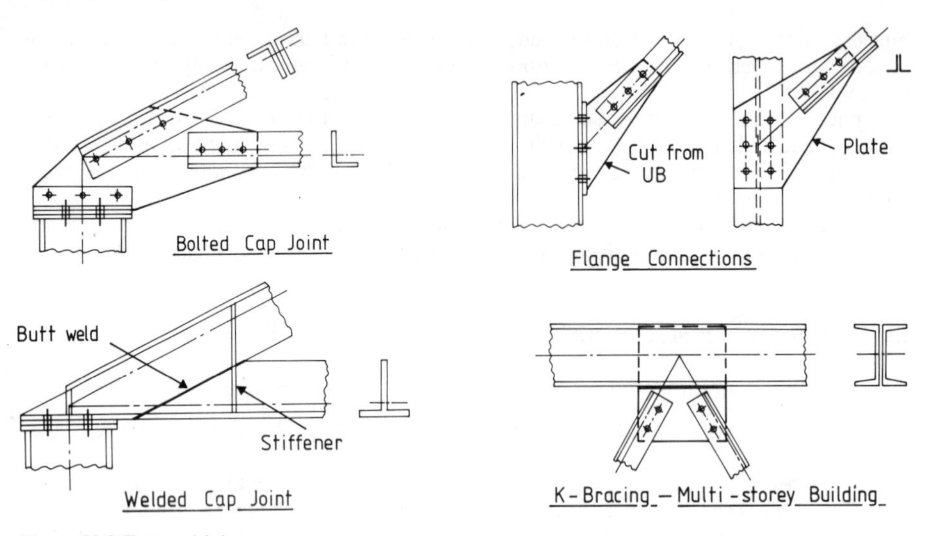

Figure 38.3 External joints

Figure 38.4 Bracing connections

at the joint and the connections in proportion to their stiffnesses. The effect of small eccentricities is usually negligible.

The real behaviour of a joint is complicated and far removed from the idealization taken for analysis and design. Referring to Figure 38.6, even when the centroidal axes of the members meet in the plane of the joint, the force path goes through the connected legs of the angle, the bolts and the gusset plate to transmit the load from one member to the other. The secondary stresses from the out-of-plane eccentricities are ignored in conventional design. Construction in structural hollow sections does give direct transmission of forces; but again there are stress concentrations at the joint.

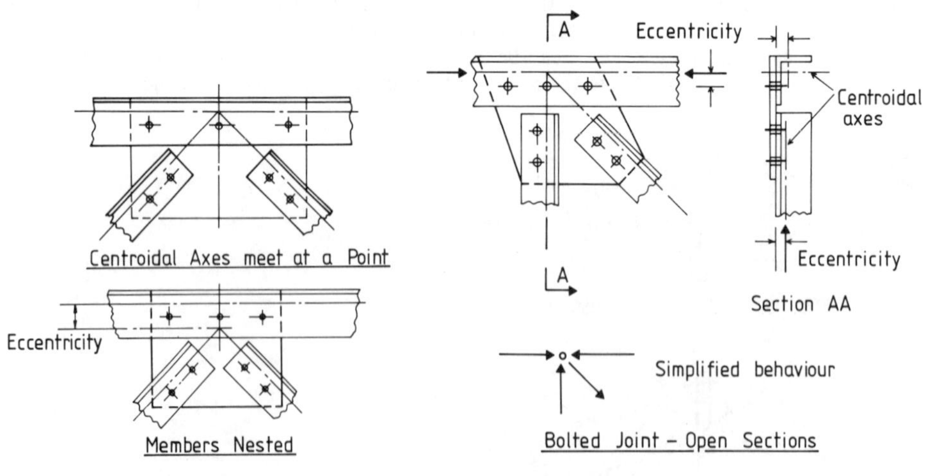

Figure 38.5 Joint eccentricity

Figure 38.6 Joint behaviour

38.5 Joint design

This consists of design of the fasteners and/or welds and the gusset plate if provided. The main points of joint design for truss and girder and bracing connections are discussed briefly. Standard connection design procedures are used:

1. Bolted joints (Figure 38.7). Where the centroidal axes of the members meet at a point the bolts are designed to resist direct load. If eccentricity occurs in the plane of the joint this should be included in the design. Eccentricity between the centroidal axis and the bolt gauge line is ignored.

Bolted joints are of two types – ordinary bolts in clearance holes and friction-grip fasteners. In nearly all the joints in trusses and lattice girders the bolts are arranged to resist load in shear. They may be in single or double shear, with the load assumed to be equally distributed between all bolts in the group. Ordinary bolts are checked for shear and bearing. Friction-grip bolts are designed for slip resistance, and shear and bearing where appropriate. The smallest size of bolt recommended is 16 mm for both types of bolts.

In bracing connections the bolt group that connects the diagonal to the building frame is in tension and shear as shown in Figure 38.7. This joint should be arranged so that the line of action of the load passes through the centroid of the bolt group, otherwise the normal component of the load will put a moment on the group in addition to direct tension.

2. Welded joints (Figure 38.8). Welded joints are of two types, butt welds and fillet welds, with fillet welds being used in most cases. The members may be connected together directly or through a gusset. In a lap joint (for example, between an angle and a gusset) the joint may be welded all round or the weld placed on one side only. Welding all round seals the joint against corrosion but the unit must be turned over to complete welding, and this increases fabrication costs. The weld group is eccentrically loaded in this case.

A satisfactory joint can be obtained with a weld on one side only. The weld can be balanced,

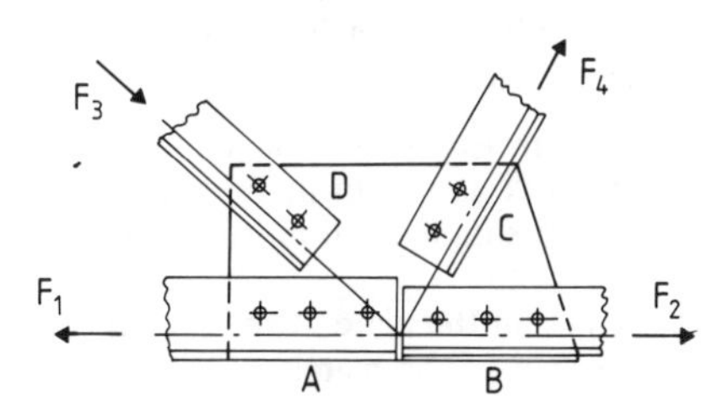

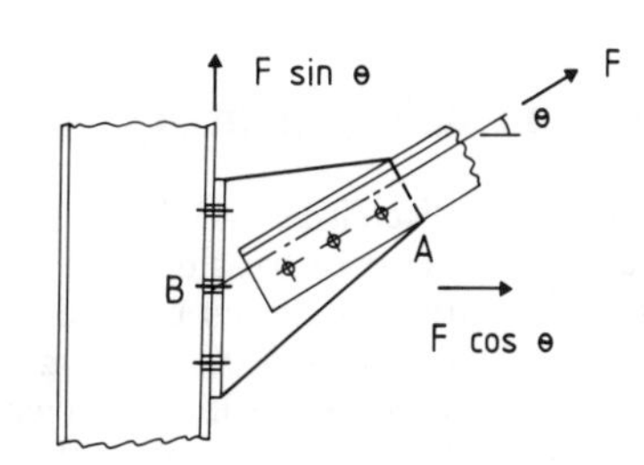

Figure 38.7 Bolt design

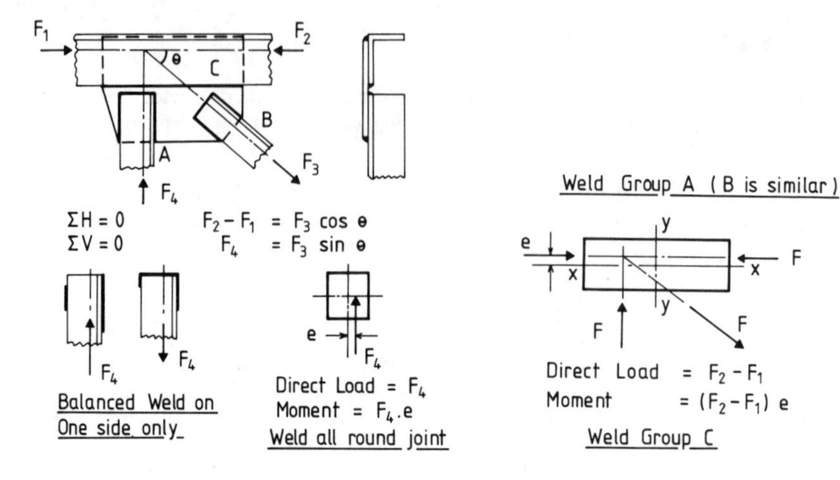

Figure 38.8 Weld design

i.e. arranged so that its centre of gravity coincides with the line of action of the force through the centroid of the angle. A proprietary filler/sealant can be used to seal the joint to prevent corrosion. The weld is considered to be directly loaded in shear only in this case.

Directly loaded welds can also be achieved in the joints of hollow section members. Figure 38.8 shows the analyses for the various groups in a welded joint.

The smallest fillet weld to be used should be 5 mm. Butt welds are also used in some joints, such as in the eaves joint shown in Figure 38.3.

3. Gusset plates (Figure 38.9). The thickness of gusset plates is usually selected from experience but should be such that it is equal to or slightly larger than that of the thickest part to be connected. The numbers of bolts or weld lengths are determined for each member and the gusset plate is made large enough to accommodate the connections.

It has then been accepted practice to check the plate as a beam section in axial load, bending and shear on various cross-sections using simple elastic theory, as shown in Figure 38.9(a). However, the proportions of many gusset plates are such that simple bending theory gives a poor estimate of the maximum stresses in the plate.

An alternative method of design is to check the direct stress in the plate at the end of each member. The load is assumed to disperse at 30° on either side of the member, as shown in Figure 38.9(b).

38.6 Concluding summary

1. A truss or lattice girder is only as strong as its joints. The selection and design of the joints is an integral part of truss design.
2. Truss and girder joints can be made in a variety of ways. Joints may be bolted or welded, with the members connected through gusset plates or directly together.
3. Moments due to eccentricities from member centre lines not meeting in the plane of the joint should be taken into account in the design. Out-of-plane eccentricities inherent in angle joints are ignored.
4. Standard simplified methods of design for bolts and welds are satisfactory. All joints should have a minimum of two bolts. Gusset plates are usually selected from experience but can be checked by beam theory.

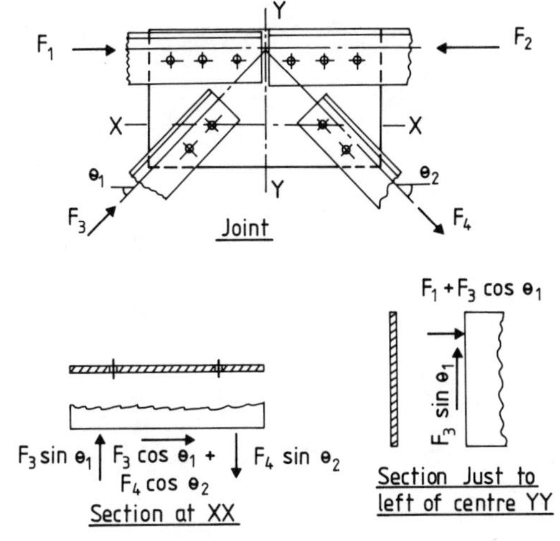

a) Checking Gusset Plate in Bending and Shear

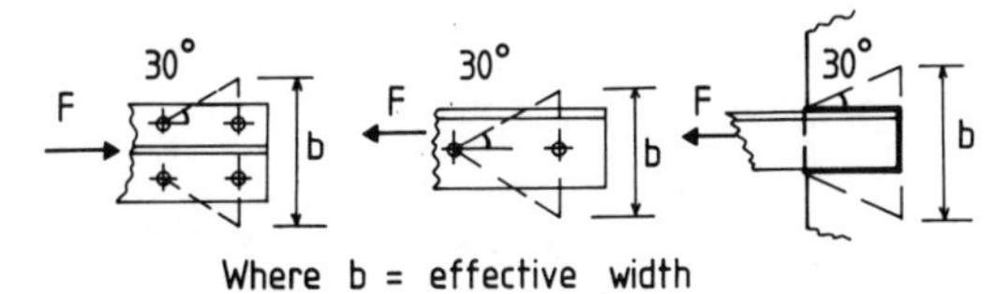

Where b = effective width

b) Checking Gusset Plate Stresses on Effective Width

Figure 38.9 Gusset plate design

Background reading

FISHER, J.W. and STRUIK, J.H.A. (1987) *Guide to Design Criteria for Bolted and Riveted Joints*, Wiley, Chichester. Pages 143–152: Truss type connections – behaviour and design recommendations; Pages 242–249: Method of analysis and experimental work on gusset plates.

LEECH, L.V. (1987) *Structural Steelwork for Students*, 2nd edn, Butterworths, London

OWENS, G.W. and CHEAL, B.D. (1988) *Structural Steelwork Connections*, Butterworths, London

THE STEEL CONSTRUCTION INSTITUTE (1989) *Steel Designers' Manual*, 6th edn, Blackwell, Oxford

The Steel Construction Institute
Silwood Park Ascot Berks SL5 7QN
Telephone: (0990) 23345
Fax: (0990) 22944 Telex: 846843

CALCULATION SHEET

Job No.	Sheet of	Rev.
Job Title: APPENDIX		
Client	Contract No.	
Made by	Date	
Checked by	Date	

Worked Examples :~

Chapter 17 ~ Tension members

Chapter 20 ~ Restrained Compact Beams

Chapter 22 ~ Column design ~ pin ended axially loaded column

Chapter 23 ~ Laterally unrestrained beams

Chapter 24 ~ Column design ~ non sway intermediate column in a building frame with flexible joints

Chapter 25 ~ Plate girder design

Chapter 38 ~ Truss and girder connections

Further design examples can be obtained from:~

1). Steelwork Design Guide to BS.5950: Part 1: 1985

Volume 2 ~ Worked Examples
The Steel Construction Institute

2). Structural Steel Design

A comprehensive guide to undergraduate teaching
The Steel Construction Institute

3). Introduction to Steelwork design to BS.5950: Part 1
The Steel Construction Institute

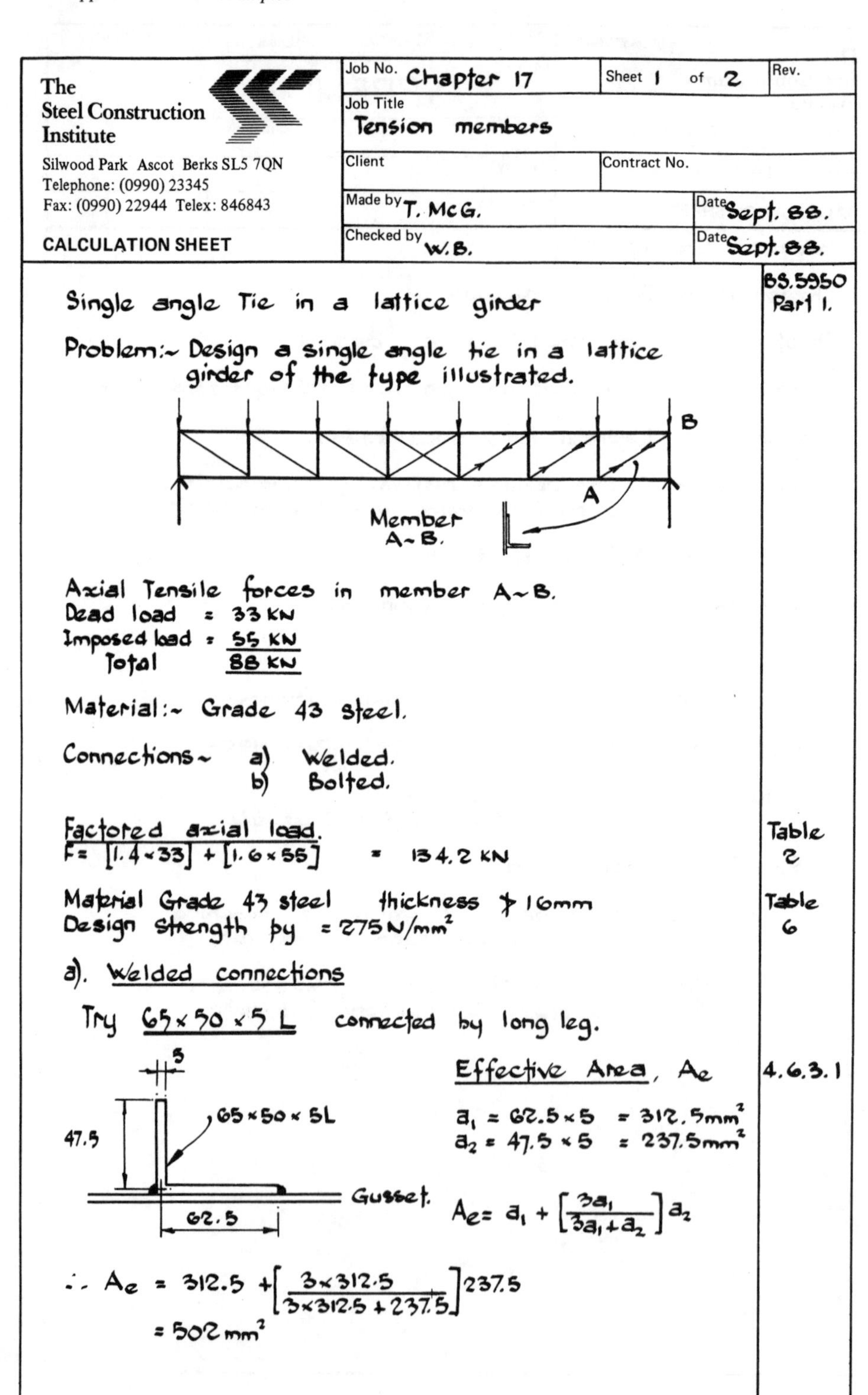

The Steel Construction Institute
Silwood Park Ascot Berks SL5 7QN
Telephone: (0990) 23345
Fax: (0990) 22944 Telex: 846843

CALCULATION SHEET

Job No. Chapter 17	Sheet 1 of 2	Rev.
Job Title Tension members		
Client	Contract No.	
Made by T. McG.	Date Sept. 88.	
Checked by W.B.	Date Sept. 88.	

BS.5950 Part 1.

Single angle Tie in a lattice girder

Problem:– Design a single angle tie in a lattice girder of the type illustrated.

Axial Tensile forces in member A–B.
Dead load = 33 kN
Imposed load = 55 kN
Total 88 kN

Material:– Grade 43 steel.

Connections – a) Welded. b) Bolted.

Factored axial load. (Table 2)

$F = [1.4 \times 33] + [1.6 \times 55] = 134.2$ kN

Material Grade 43 steel thickness ≯ 16mm (Table 6)
Design strength $p_y = 275$ N/mm²

a). Welded connections

Try 65 × 50 × 5 L connected by long leg.

Effective Area, A_e (4.6.3.1)

$a_1 = 62.5 \times 5 = 312.5 \text{mm}^2$
$a_2 = 47.5 \times 5 = 237.5 \text{mm}^2$

$$A_e = a_1 + \left[\frac{3a_1}{3a_1 + a_2}\right] a_2$$

$$\therefore A_e = 312.5 + \left[\frac{3 \times 312.5}{3 \times 312.5 + 237.5}\right] 237.5$$

$= 502 \text{ mm}^2$

The Steel Construction Institute
Silwood Park Ascot Berks SL5 7QN
Telephone: (0990) 23345
Fax: (0990) 22944 Telex: 846843

CALCULATION SHEET

Job No. Chapter 17	Sheet 2 of 2	Rev.
Job Title Tension members		
Client	Contract No.	
Made by T. McG.	Date Sept. 88.	
Checked by W.B.	Date Sept. 88.	

BS.5950 Part 1

Tension capacity $P_t = A_e . p_y$

Then $P_t = \frac{502 \times 275}{10^3} = 138\text{kN} > 134.2\text{kN}$ OK — 4.6.1

∴ Use 65 × 50 × 5 L ~ Grade 43.

b). Bolted connections.

Try 65 × 50 × 8L connected by long leg.

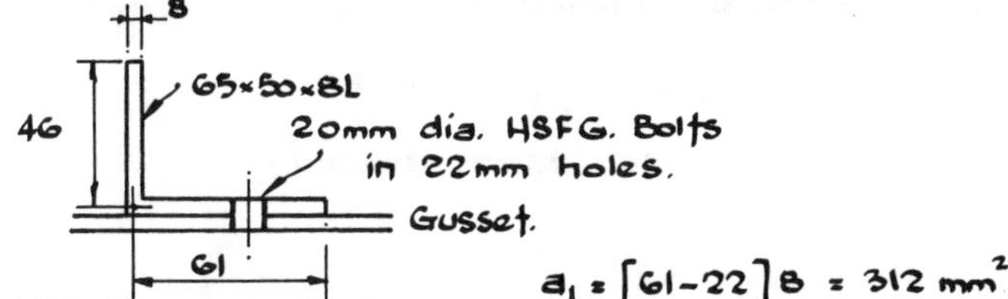

$a_1 = [61 - 22]8 = 312\text{ mm}^2$

Effective Area, A_e — 4.6.3.1

$A_e = a_1 + \left[\frac{3a_1}{3a_1 + a_2}\right]a_2$ $\qquad a_2 = 46 \times 8 = 368\text{mm}^2$

$= 312 + \left[\frac{3 \times 312}{3 \times 312 + 368}\right]368 = 576\text{mm}^2$

Note:- Effective area factor K_e for sections with holes (Clause 3.3.3) does not apply in this case

Tension capacity $P_t = \frac{576 \times 275}{10^3} = 158.4\text{kN} > 134.2\text{kN}$ OK. — 4.6.1

Section has good margin but 65×50×6L too small.
∴ Use 65×50×8L - Grade 43.

Efficiency of single Angle Tie.

Calculate the efficiency of 65×50×8L used as a single angle tension member. Gross area = 8.6cm²

a) Bolted through long leg with bolts in 22dia. holes
From above calculations Net area = 576mm²

Then Efficiency $= \left[\frac{576}{8.6 \times 10^2}\right]100 = \underline{67\%}$.

b). Welded through long leg $\qquad a_1 = 61 \times 8 = 488\text{mm}^2$
Effective area $\qquad a_2 = 46 \times 8 = 368\text{mm}^2$

$A_e = 488 + \left[\frac{3 \times 488}{3 \times 488 + 368}\right]368 = 782\text{ mm}^2$

Then Efficiency $= \left[\frac{782}{8.6 \times 10^2}\right]100 = \underline{91\%}$

The Steel Construction Institute
Silwood Park Ascot Berks SL5 7QN
Telephone: (0990) 23345
Fax: (0990) 22944 Telex: 846843

CALCULATION SHEET

Job No Chapter 20	Sheet 1 of 5	Rev.
Job Title Restrained Compact Beams		
Client	Contract No.	
Made by D.N.	Date Sept. 88	
Checked by W.B.	Date Sept. 88.	

BS.5950 Part 1.

Problem:~ To select a UB. Section to carry a factored imposed load of 400 kN uniformly distributed over a simply supported span of 6m
UB Section ~ Grade 43. [T<16mm $p_y = 275 N/mm^2$]

Factored imposed load W = 400 kN

200 kN — L = 6.0m. — 200 kN

The load factor of 1.6 has been taken assuming all imposed load.

For uniformly distributed load

$$M_{max.} = \frac{WL}{8} = \frac{400 \times 6}{8} = 300 \text{ kN m}$$

Assume use of compact section.

$$S_x \text{ req'd.} = \frac{300 \times 10^3}{275} = 1091 \text{ cm}^3$$

From Structural Steelwork Handbook — S.C.I

457×152×52 UB. ~ $S_x = 1094 cm^3 > 1091 cm^3$

Section Properties:~

Overall Depth D = 449.8mm — Flange thickness T = 10.9mm
Overall Breadth B = 152.4mm — Web thickness t = 7.6mm
Depth between fillets d = 407.7mm — Inertia $I_{xx} = 21345 cm^4$
Elastic modulus $Z_{xx} = 949 cm^3$

Check section to see if Compact. — 3.5.2 Table 7 Fig. 3

$$\varepsilon = \sqrt{\frac{275}{275}} = 1.0$$

$$b = \frac{B}{2} = \frac{152.4}{2} = 76.2\text{mm}$$

$$\frac{b}{T} = \frac{76.2}{10.9} = 7.0 < [9.5 \times 1.0 = 9.5] \text{ ok}$$

$$\frac{d}{t} = \frac{407.7}{7.6} = 53.6 < [98 \times 1.0 = 98] \text{ ok.}$$

∴ Section Compact ~ ok in Bending.

The Steel Construction Institute
Silwood Park Ascot Berks SL5 7QN
Telephone: (0990) 23345
Fax: (0990) 22944 Telex: 846843

CALCULATION SHEET

Job No. Chapter 20	Sheet 2 of 5	Rev.
Job Title Restrained Compact Beams		
Client	Contract No.	
Made by D.N.	Date Sept. 88.	
Checked by W.B.	Date Sept. 88.	

BS.5950 Part 1.

Check for shear

Shear force at supports $= \frac{400}{2} = 200\text{kN} = F_v$

Shear capacity $P_v = 0.6 p_y A_v$ where $A_v = Dt$. — 4.2.3

$\therefore P_v = \frac{0.6 \times 275 \times 449.8 \times 7.6}{10^3} = 564\text{ kN} > 200\text{kN}$ OK.

$\therefore$ Section OK in Shear.

Moment capacity with low shear load — 4.2.5
Where $F_v \leq 0.6 P_v$

$0.6 P_v = 0.6 \times 564 = 338\text{kN} > 200\text{kN}$ $\therefore$ low shear

Moment capacity M_c for plastic or compact sections

$M_c = p_y \cdot S_x$ but $\leq 1.2 p_y Z_x$

$= \frac{275 \times 1094}{10^3} = 300.9\text{kN m} \leq \frac{1.2 \times 275 \times 949}{10^3} = 313.2\text{kN m}$

$\therefore$ Moment capacity $M_{cx} = 300.9\text{kN m} > 300\text{kN m}$ OK.

Check for Deflection [Load factor 1.6 on full load].

Then actual load $= \frac{400}{1.6} = 250\text{kN}$

$\delta_{max} = \frac{5WL^3}{384EI_x}$ where $E = 205\text{ N/mm}^2$

$= \frac{5 \times 250 \times [6 \times 10^3]^3}{384 \times 205 \times 21345 \times 10^4} = 16.06\text{mm}$

$\delta_{max} \not> \frac{L}{200} = \frac{6 \times 10^3}{200} = 30.0\text{mm} > 16.06\text{mm}$ OK — 2.5.1 Table 5

$\therefore$ Section is Satisfactory.

Use 457×152×52 UB. ~ Grade 43.

The Steel Construction Institute
Silwood Park Ascot Berks SL5 7QN
Telephone: (0990) 23345
Fax: (0990) 22944 Telex: 846843

CALCULATION SHEET

Job No. Chapter 20	Sheet 3 of 5	Rev.
Job Title Restrained Compact Beams		
Client	Contract No.	
Made by D. N.	Date Sept. 88.	
Checked by W. B.	Date Sept. 88.	

BS.5950 Part 1.

Problem:~ To select a suitable R.H.S. to act as a cantilever beam of 1.8m projection carrying a tip load of 10KN. vertical and 3KN. horizontal ~ all factored imposed loads. Grade 43 Steel ~ design strength $p_y = 275 N/mm^2$

Built-in or equal

10KN

1.8m. projection.

10KN

3KN

End View

The load factor of 1.6 has been used for imposed load. — Table 2.

Shear parallel to webs = 10 KN

Moment $M_x = 10 \times 1.8 = 18$ KN m

Shear parallel to flanges = 3 KN

Moment $M_y = 3 \times 1.8 = 5.4$ KN m

Guess and check approach required since interaction equation must eventually be satisfied.

Try ~ 120×80×6.3 RHS Grade 43

From Structural steelwork Handbook — S.C.I.

Section Properties:~

$S_x = 92.3 cm^3$, $Z_x = 74.6 cm^3$, $1.2Z_x = 89.5 cm^3 < S_x$

$S_y = 69.1 cm^3$, $Z_y = 58.4 cm^3$, $1.2Z_y = 70.1 cm^3 > S_y$

$D = 120 mm$, $B = 80 mm$

$t = 6.3 mm$, $A = 23.4 cm^2$

$I_x = 447 cm^4$

Section can be considered as compact — 3.5.2

Shear capacity $P_v = 0.6 p_y . A_v$ — 4.2.3

where $A_v = \left[\frac{D}{D+B}\right] A$

$= \left[\frac{120}{120+80}\right] A = 0.6A$

$\therefore A_v = 0.6 \times 23.4 = 14.04 cm^2$

$P_v = \frac{0.6 \times 275 \times 14.04}{10} = 232$ KN

Actual shear F_v = <u>10KN</u> < <u>0.6×232 KN</u> OK.

The Steel Construction Institute
Silwood Park Ascot Berks SL5 7QN
Telephone: (0990) 23345
Fax: (0990) 22944 Telex: 846843

CALCULATION SHEET

Job No. Chapter 20	Sheet 4 of 5	Rev.
Job Title Restrained Compact Beams		
Client	Contract No.	
Made by D.N.	Date Sept. 88.	
Checked by W.B.	Date Sept. 88.	

BS.5950 Part 1.

Then for compact sections, moment capacity M_c with low shear load

4.2.5

$$M_{cx} = p_y . S_x = \frac{275 \times 92.3}{10^3} = 25.38 \text{ kN m}$$

$$\text{or} = 1.6 p_y . Z_x = \frac{1.6 \times 275 \times 74.6}{10^3} = 32.82 \text{ kN m}$$

$$M_{cy} = p_y . S_y = \frac{275 \times 69.1}{10^3} = 19.00 \text{ kN m}$$

$$\text{or} = 1.2 p_y Z_y = \frac{1.2 \times 275 \times 58.4}{10^3} = 19.27 \text{ kN m}$$

$\therefore$ $M_{cx} = 25.38$ kN m
$M_{cy} = 19.00$ kN m

4.8.3

$$\frac{F}{A_e . p_y} + \frac{M_x}{M_{cx}} + \frac{M_y}{M_{cy}} \leq 1.0$$

where $F = 0$

4.9

$$\text{then} \quad \frac{M_x}{M_{cx}} + \frac{M_y}{M_{cy}} \leq 1.0$$

$$\frac{18.0}{25.38} + \frac{5.4}{19.0} = 0.71 + 0.28 = \underline{0.99} < \underline{1.0 \text{ ok.}}$$

$\therefore$ <u>Section satisfactory for strength</u>

<u>Deflection</u>

2.5.1

To gain an idea of serviceability calculate tip deflection under vertical load only.

3.5.1

Actual $W = 10/1.6 = 6.25$ kN $E = 205$ kN/mm^2

$$\delta = \frac{WL^3}{3EI_x} = \frac{6.25 \times (1.8 \times 10^3)^3}{3 \times 205 \times 447 \times 10^4} = 13.26 \text{mm}$$

Deflection of cantilevers should not exceed

Table 5

$$\frac{L}{180} = \frac{1.8 \times 10^3}{180} = 10.0 \text{ mm}$$

The Steel Construction Institute
Silwood Park Ascot Berks SL5 7QN
Telephone: (0990) 23345
Fax: (0990) 22944 Telex: 846843

CALCULATION SHEET

Job No. Chapter 20	Sheet 5 of 5	Rev.	
Job Title Restrained Compact Beams			
Client	Contract No.		
Made by D.N.		Date Sept. 88.	
Checked by W.B.		Date Sept. 88.	

BS.5950 Part 1.

The deflection exceeds this amount. However, this is an advisory limit and the section could be acceptable depending on exact operational requirements. The important factor is that the deflection should not impair the strength or efficiency of the structure ie. should not cause damage to cladding etc. Since, however, there will also be some horizontal deflection at the tip, it would be advisable to increase the section.

Provide ~ <u>120×80×8 RHS. Grade 43.</u>

The Steel Construction Institute
Silwood Park Ascot Berks SL5 7QN
Telephone: (0990) 23345
Fax: (0990) 22944 Telex: 846843

CALCULATION SHEET

Job No. Chapter 22	Sheet 1 of 2	Rev.
Job Title: Column design		
Client	Contract No.	
Made by A.G.	Date Sept. 88.	
Checked by W.B.	Date Sept. 88.	

BS.5950 Part 1.

Problem:-

A pin ended axially loaded column 3.5m. high has to support a working load of 1040 kN.
Design a suitable UC. Section column in Grade 50 steel.

Height of pin ended column = 3.5m
Working load = 1040 kN
Load factor γ_f = 1.4 for dead load. = 1.6 for imposed load. (2.4.1.1 Table 2)
Take average value 1.5.

Then factored load P = 1.5 × 1040 = 1560 kN
For grade 50 steel:- (3.1.1)
p_y = 340 N/mm² [Thickness over 16mm under 63mm] or
p_y = 355 N/mm² [Thickness under or equal to 16mm] (Table 6)

Assuming that buckling reduces p_c to 200 N/mm² [an arbitrary guess].
Cross~sectional area required

$$A = \frac{P}{200} = \frac{1560 \times 10^3}{200} = 7800\,mm^2\ [78.0\,cm^2]$$

Try 203 × 203 × 60 kg/m UC
A = 75.8 cm² ≈ 78.0 cm²
r_y = 5.19 cm T = 14.2 mm < 16.0 mm

From Table 25 (4.7.5 Table 25)
Rolled H Section, thickness < 40mm and buckling about y-y axis requires strut curve 27(c).

$$\text{Slenderness } \lambda = \ell/r_y = \frac{3.5 \times 10^3}{5.19 \times 10} = 67.4$$

From Table 27(c). and p_y = 355 N/mm² (Table 27(c).)

p_c = 223 N/mm²

Then column compressive resistance, P_c
$P_c = A_g p_c$
= 75.8 × 10² × 223 / 10³ = 1690 kN > 1560 kN OK.

Section Adequate but try lighter section.

The Steel Construction Institute
Silwood Park Ascot Berks SL5 7QN
Telephone: (0990) 23345
Fax: (0990) 22944 Telex: 846843

CALCULATION SHEET

Job No. Chapter 22	Sheet 2 of 2	Rev.
Job Title: Column design		
Client	Contract No.	
Made by A.G.	Date Sept. 88.	
Checked by W.B.	Date Sept. 88.	

BS.5950 Part 1.

Try 203×203× 52 kg/m UC.

Area = 66.4 cm^2
r_y = 5.16 cm

Slenderness $\lambda = \ell/r_y = \frac{3.5\times10^3}{5.16\times10}$
$= 67.8$

From Table 27(c). and p_y = 355 N/mm^2 — Table 27(c).

p_c = 223 N/mm^2

Then column compressive resistance P_c

$P_c = A.\,p_c$

$= 66.4\times10^2\times223/10^3$

$=$ 1480 kN < 1560 kN OVER

Section is Inadequate

∴ Use 203×203×60 kg/m UC. Grade 50.

The Steel Construction Institute
Silwood Park Ascot Berks SL5 7QN
Telephone: (0990) 23345
Fax: (0990) 22944 Telex: 846843

CALCULATION SHEET

Job No. Chapter 23	Sheet 1 of 3	Rev.
Job Title: Laterally unrestrained Beams		
Client	Contract No.	
Made by D. A.N	Date Sept. 88	
Checked by D.L.M.	Date Sept. 88.	

BS. 5950 Part 1

Problem:~ Use BS.5950: Part 1 to check the ability of a 457×152×60 UB in Grade 43 steel to carry a factored uniformly distributed load of 24 kN/m over a span of 6m.
Both ends of the beam are attached to the flanges of columns by double web cleats.

w = 24kN/m [factored]
457×152×60 UB
6m

Appraisal:~ For this form of end condition the beam may reasonably be assumed to act as if simply supported in the vertical plane and to be fully restraind against lateral deflection and twist with no rotational restraint in plan at its ends. — Table 9 Condition (3).

Section Classification for 457×152×60 UB, Grade 43.
Section Properties taken from Steelwork Design Guide ~ Volume 1. — S.C.I.

Depth	D = 454.7mm
Width	B = 152.9mm
Web thickness	t = 8.0mm
Flange thickness	T = 13.3mm
Depth between fillets	d = 407.7mm
Radius of gyration about minor axes	r_y = 3.23cm
Plastic modulus about major axes	S_x = 1280 cm^3

As flange thickness T = 13.3mm < 16mm from Table 6, Design strength $p_y = 275\ N/mm^2$ for grade 43 steel. — 3.1.1 Table 6

Outstand of flange — 3.5.3 Fig. 3

$b = B/2 = 152.9/2 = 76.5mm$

Limit for Compact section $b/T \leq 9.5\epsilon$ — Table 7

where $\epsilon = \left[\frac{275}{p_y}\right]^{1/2}$, $p_y = 275N/mm^2$ $\therefore \epsilon = 1.0$ — Table 7 Note 3

$b/T = 76.5/13.3 = \underline{5.75} < \underline{9.5 \times 1.0}$ OK.

The Steel Construction Institute
Silwood Park Ascot Berks SL5 7QN
Telephone: (0990) 23345
Fax: (0990) 22944 Telex: 846843

CALCULATION SHEET

Job No. Chapter 23	Sheet 2 of 3	Rev.
Job Title Laterally unrestrained Beams		
Client	Contract No.	
Made by D.A.N.	Date Sept. 88	
Checked by D.L.M.	Date Sept. 88.	

	BS 5950 Part 1
Web [neutral axis at mid-depth]. Limit for Compact Section $d/t \leq 98\varepsilon$	3.5.3 Table 7
$d/t = 407.7/8.0 = 51 < 98.0 \times 1.0$ OK.	
Section meets both limits and may therefore be considered as COMPACT.	
Since, $d/t \leq 63\varepsilon$ no check for shear buckling is required.	Table 7 Note 2
Lateral torsional buckling [Conservative design].	4.3
Determine λ_{LT}, equivalent slenderness.	
$\lambda_{LT} = n.u.v.\lambda$	4.3.7.5
where n, slenderness correction factor for type of loading at its conservative value of 1.0.	4.3.7.6
u, buckling parameter taken at its conservative value of 0.9	4.3.7.5
λ, minor axis slenderness $= L_E/r_y$ for a beam with these end conditions take $L_E = L = 1.0$	4.3.7.5 4.3.5 Table 9
$\therefore \lambda = \dfrac{6 \times 10^3}{3.23 \times 10} = 186$	
v, slenderness factor. For beams with equal flanges, $N = 0.5$ and using x, torsional index equal to D/T	4.3.7.5
$x = \dfrac{D}{T} = \dfrac{454.7}{13.3} = 34.2$	
$\dfrac{\lambda}{x} = \dfrac{186}{34.2} = 5.44$	
From Table 14 $v = 0.79$	
$\therefore \lambda_{LT} = 1.0 \times 0.9 \times 0.79 \times 186 = 132$	
For $p_y = 275$ N/mm², bending strength p_b from Table 11 $= 82$ N/mm²	4.3.7.4

The Steel Construction Institute
Silwood Park Ascot Berks SL5 7QN
Telephone: (0990) 23345
Fax: (0990) 22944 Telex: 846843

CALCULATION SHEET

Job No. Chapter 23	Sheet 3 of 3	Rev.
Job Title Laterally unrestrained Beams		
Client	Contract No.	
Made by D.A.N.	Date Sept. 88	
Checked by D.L.M.	Date Sept. 88.	

BS.5950 Part 1

Buckling resistance moment, $M_b = S_x . p_b$

$$M_b = \frac{1280 \times 82}{10^3} = 105 \text{ kN m}$$ 4.3.7.3

For a simply supported beam 6m. span with a load of 24 kN/m.

$$M_{max} = \frac{wL^2}{8} = \frac{24 \times 6^2}{8} = \underline{108 \text{ kN m}} > \underline{105 \text{ kN m OVER.}}$$

M_{max}. exceeds buckling resistance moment M_b, check beam using refined analysis to see if buckling resistance moment M_b can be increased enough to exceed the actual moment M_{max}.

<u>Refined calculation.</u>

Calculate slenderness correction factor n using Table 16 — 4.3.7.6

$\gamma = \frac{M}{M_o}$ Say = 0 then n = 0.94.

Using buckling parameter u and torsional index x from Steelwork design guide ✓ Volume 1 — S.C.I

u = 0.869 , x = 37.5 — 4.3.7.5

$$\frac{\lambda}{x} = \frac{186}{37.5} = 4.96$$

from Table 14 v = 0.82

$\lambda_{LT} = n.u.v.\lambda$ — 4.3.7.5

$= 0.94 \times 0.869 \times 0.82 \times 186 = 125$

∴ using $p_y = 275 \text{ N/mm}^2$ and $\lambda_{LT} = 125$ — 4.3.7.4

from Table 11 $p_b = 90 \text{ N/mm}^2$

Buckling resistance moment, M_b — 4.3.7.3

$$= \frac{90 \times 1280}{10^3} = \underline{115.2 \text{ kN m}} > \underline{108 \text{ kN m}} \text{ OK.}$$

∴ Use <u>457 × 152 × 60 UB.</u> <u>Grade 43</u>

The Steel Construction Institute
Silwood Park Ascot Berks SL5 7QN
Telephone: (0990) 23345
Fax: (0990) 22944 Telex: 846843

CALCULATION SHEET

Job No. Chapter 24	Sheet 1 of 3	Rev.
Job Title Column Design		
Client	Contract No.	
Made by A.R.G.	Date Sept. 88.	
Checked by W. B.	Date Sept. 88.	

BS.5950 Part 1.

Problem:-

A non-sway intermediate column in a building frame with flexible joints is 4m. high and carries factored loads as shown in sketch below. Grade 43 steel is to be used and the design is to be to BS.5950: Part 1 using the simple method.

230KN.
Floor Level
170KN
50KN 35KN.
Floor Level
170KN 250KN
60KN. 40KN
℄ Column
℄ Column
View X~X
View Y~Y

Column axial load
From above = 230KN
From beams at top level = 255KN
485 KN

Eccentricity moments (4.7.6)

From clause 4.7.6. beam loads assumed to act at 100mm. from column face.

Try 152×152×37kg/m UC. in Grade 43 steel

from section tables:-

$D = 161.8mm$; $B = 154.4mm$; $t = 8.1mm$

$A = 47.4cm^2$; $r_{yy} = 3.87cm$; $T = 11.5mm$

$Z_{xx} = 274.2cm^3$; $Z_{yy} = 91.78cm^3$

$S_{xx} = 310.1cm^3$; $S_{yy} = 140.1cm^3$

Thickness of web and flanges < 16mm (Table 6)

∴ From Table 6 - Design strength $p_y = 275 N/mm^2$

The Steel Construction Institute
Silwood Park Ascot Berks SL5 7QN
Telephone: (0990) 23345
Fax: (0990) 22944 Telex: 846843

CALCULATION SHEET

Job No. Chapter 24	Sheet 2 of 3	Rev.
Job Title Column design		
Client	Contract No.	
Made by A.R.G.	Date Sept. 88.	
Checked by W.B.	Date Sept. 88.	

BS.5950 Part 1.

Beam moments.

At top. (4.7.6.(a).)

$M_x = 170[D/2 + 100] \times 10^{-3}$ kN m
$= 170 \times 180.9 \times 10^{-3}$ $= 30.75$ kN m

$M_y = (50-35)(t/2 + 100) \times 10^{-3}$ kN m
$= 15 \times 104 \times 10^{-3}$ $= 1.56$ kN m

At Bottom.

$M_x = (250-170)(180.9) \times 10^{-3}$ kN m
$= 80 \times 180.9 \times 10^{-3}$ $= 14.47$ kN m

$M_y = (60-40)(104) \times 10^{-3}$ kN m
$= 20 \times 104 \times 10^{-3}$ $= 2.08$ kN m

Dividing these moments between columns above and below gives:-

At top.
$M_x = $ 15.38 kN m ; $M_y = $ 0.78 kN m

At bottom.
$M_x = $ 7.24 kN m ; $M_y = $ 1.04 kN m

These moments are applied thus:-

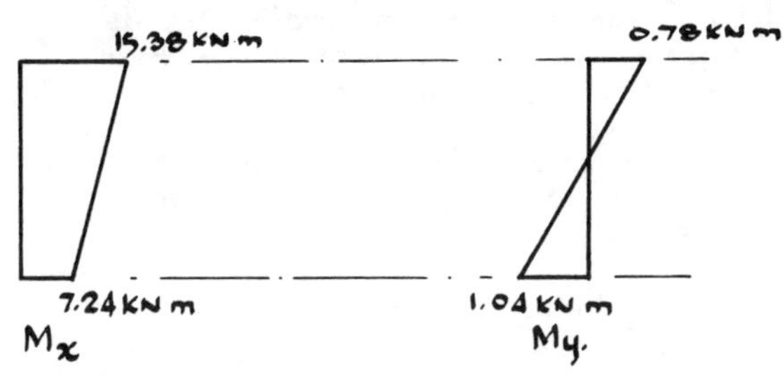

From clause 4.7.7. the column should satisfy clause 4.8.3.3 simplified approach, taking uniform moment factor m = 1.0. ie. the maximum moments anywhere in the column length are used to check the buckling which therefore also checks local capacity (4.8.3.2). (4.7.7.)

Overall buckling check. (4.8.3.3)

$$\frac{F}{A_g . p_c} + \frac{m M_x}{M_b} + \frac{m M_y}{p_y . Z_y} \leq 1.0$$ (4.8.3.3.1)

$F = 485$ kN ; $M_x = 15.38$ kN m ; $M_y = 1.04$ kN m

The Steel Construction Institute
Silwood Park Ascot Berks SL5 7QN
Telephone: (0990) 23345
Fax: (0990) 22944 Telex: 846843

CALCULATION SHEET

Job No. Chapter 24	Sheet 3 of 3	Rev.
Job Title Column design		
Client	Contract No.	
Made by A.R.G.	Date Sept. 88.	
Checked by W.B.	Date Sept. 88.	

BS.5950 Part 1.

From clause 4.7.5 calculate the value of p_c.

From clause 4.7.2 and Table 24 — Table 24
Effective length of column $L_E = 0.85L$

$\therefore L_E = 0.85 \times 4 \times 10^3 = 3400$ mm.

Slenderness $\lambda = L_E / r_{yy} = \frac{3400}{3.87 \times 10} = 87.9$

From Table 25. — Table 25
For Y axis buckling of H section, column curve (c) should be used.
From Table 27 curve (c) for $\lambda = 88$ and $p_y = 275\text{N/mm}^2$ — Table 27 (c).

$p_c = 146\text{N/mm}^2$

From clause 4.3.7.3 the buckling resistance M_b is given by – — 4.3.7.3

$M_b = S_{xx} \cdot p_b$.

$S_{xx} = 310.1\text{cm}^3$; p_b = bending strength.

From clause 4.3.7.4 the bending strength p_b is related to the equivalent slenderness λ_{LT}, the design strength of the material and the member type. — 4.3.7.4

From clause 4.7.7. in calculating M_b only the equivalent slenderness λ_{LT} between restraints should be taken as — 4.7.7.

$$\lambda_{LT} = \frac{0.5L}{r_{yy}} = \frac{0.5 \times 4 \times 10^3}{3.87 \times 10} = 51.7$$

Then bending strength p_b from Table 11 = 233N/mm^2
and resistance moment $M_b = 310.1 \times 233 \times 10^{-3}$
$= 72.25$ KN m

Use these values to resolve the expression from clause 4.8.3.3 — 4.8.3.3

$$\frac{F}{A_g \cdot p_c} + \frac{mM_x}{M_b} + \frac{mM_y}{p_y \cdot Z_y} \leq 1.0 \quad \text{where } m = 1.0$$

$$= \frac{485 \times 10^3}{47.4 \times 10^2 \times 146} + \frac{15.38}{72.25} + \frac{1.04 \times 10^6}{275 \times 91.78 \times 10^3}$$

$$= 0.70 + 0.21 + 0.04 = \underline{0.95 < 1.0 \text{ ok.}}$$

$\therefore$ <u>Section satisfactory</u>

Use <u>152×152×37Kg./m UC Grade 43.</u>

The Steel Construction Institute
Silwood Park Ascot Berks SL5 7QN
Telephone: (0990) 23345
Fax: (0990) 22944 Telex: 846843

CALCULATION SHEET

Job No. Chapter 25	Sheet 1 of 6	Rev.
Job Title: Plate Girder Design		
Client	Contract No.	
Made by R.E.	Date Sept. 88	
Checked by W.B.	Date Sept. 88	

BS.5950 Part 1.

Problem:– To design a plate girder to withstand a vertical shear force of 300kN. and a co-existent bending moment of 3000 KN m. The overall depth of the girder must not exceed 950mm.
Design the plate girder in accordance with BS.5950: Part 1.
Grade 43 steel with py as appropriate will be used, and the compression flange is provided with effective lateral restraints.
It will be assumed that the loads given are factored loads, and alternative approaches will be made for different types of web stiffening.
For purposes of comparison, theoretical plate thicknesses will be used but in practice available thicknesses would have to be adopted. These will be specified where necessary.

1). Flange design.

The girder will be designed in accordance with clause 4.4.4.2(a) for thin webs, where it can be assumed that the flanges alone carry the bending moment [each flange being subject to a uniform stress py] and the web designed for shear only.

Taking the maximum allowable depth of 950 mm.

$$\text{Flange force} = \frac{\text{Moment}}{\text{girder depth}} = \frac{3000 \times 10^3}{950} = 3158 \text{ KN}$$

Flange will be designed to be semi-compact.
It is probable that the thickness of the flange will be more than 16mm. but less than 40mm
$\therefore p_y = 265 \text{ N/mm}^2$ — Table 6

$$\varepsilon = \sqrt{\frac{275}{265}} = 1.02$$

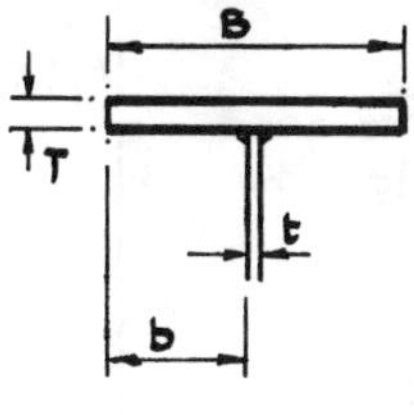

From Table 7– $\frac{b}{T} \leq 13\varepsilon = 13.3\text{mm}$ for semi-compact section. — Table 7, Fig. 3

Assuming flange stress = design strength $= p_y = 265\text{N/mm}^2$

$$\text{Required flange area} = \frac{3158 \times 10^3}{265}$$

$= \underline{11917 \text{mm}^2}$

The Steel Construction Institute
Silwood Park Ascot Berks SL5 7QN
Telephone: (0990) 23345
Fax: (0990) 22944 Telex: 846843

CALCULATION SHEET

Job No. Chapter 25	Sheet 2 of 6	Rev.
Job Title: Plate Girder Design		
Client	Contract No.	
Made by R.E.	Date Sept. 88.	
Checked by W.B.	Date Sept. 88.	

BS.5950 Part 1.

Taking maximum permissible outstand at 13.3T and ignoring web thickness.

Flange area = $2 \times 13.3T \times T = 26.6T^2$

Then required flange thickness T

$= \sqrt{\frac{11917}{26.6}} = 21.2$ mm

Using 25mm. thick plate, minimum width required to give an area of 11917mm² is 476mm

Use Flange plate <u>600mm × 25mm</u>

Check $b/T = \frac{300}{25} = 12 < 13$ ok.

Actual flange stress $= \frac{3000 \times 10^6}{925 \times 25 \times 600} = 216 \text{N/mm}^2 < 265 \text{N/mm}^2$ OK.

The flange not being fully stressed will help the web in shear.

2). <u>Web design</u>
Web depth d = 950 − 2×25 = 900mm
Several options are available

a). <u>Try and design a stocky web to avoid buckling.</u>
To avoid web buckling

$\frac{d}{t} \not> 63\varepsilon = 63 \times 1.02 = 64.3$ — 4.2.3

∴ Minimum $t = \frac{900}{64.3} = 14.0$ mm

From clause 4.2.3:–
Shear capacity $= 0.6 p_y . A_v$
$= 0.6 \times 265 \times 900 \times 14 / 10^3$
$= 2003$ kN.

This is far greater than the required shear capacity of 300kN, so a stocky web will not be economical.

b). <u>Try to avoid using transverse stiffeners</u> — 4.4.5.3

A slender web, which is unstiffened is designed assuming the shear capacity to be equal to the shear buckling resistance V_{cr}.

From clause 4.4.5.3:–

$V_{cr} = q_{cr} . d . t.$

where q_{cr} is the critical shear strength from Table 21a

The Steel Construction Institute Silwood Park Ascot Berks SL5 7QN Telephone: (0990) 23345 Fax: (0990) 22944 Telex: 846843 CALCULATION SHEET	Job No. Chapter 25	Sheet 3 of 6	Rev.
	Job Title Plate Girder Design		
	Client	Contract No.	
	Made by R.E.	Date Sept. 88	
	Checked by W.B.	Date Sept. 88.	

BS.5950 Part 1. Table 21a

Assumed d/t	Corresponding t mm.	q_{cr} N/mm² * from Table 21a	Shear Capacity kN.		Actual Shear kN.
200	4.5	25	101	<	300
150	6.0	44	238	<	300
120	7.5	69	465	>	300
140	6.4	51	294	≈	300

* assuming an infinite (∞) stiffener spacing

∴ For an unstiffened web required plate thickness = 6.5mm [7mm]

Check $d/t = \frac{900}{6.5} = 138.5$; $q_{cr} = 52$

Then $V_{cr} = \frac{52 \times 900 \times 6.5}{10^3} = 304\text{kN} > 300\text{kN. OK.}$

c). Try using transverse stiffeners.

The more stiffeners that are used the greater will be the strength of the girder. However, the introduction of stiffeners will increase fabrication costs so the number of stiffeners will be limited so that the stiffener spacing/web depth ratio (a/d) is 1.4.

BS.5950 allows stiffened webs to be designed with or without tension field action. Both alternatives will be considered.

ci) Stiffened web with no tension field action.

As for the unstiffened web in (b), the shear capacity of a stiffened web is assumed to be equal to the shear buckling resistance (cl. 4.4.5.3). when tension field action is neglected. However, the stiffener spacing is now finite ie. $a/d = 1.4$ Refering to table 21a. — 4.4.5.3

Table 21a.

Assumed d/t	Corresponding t mm	q_{cr} N/mm²	Shear Capacity kN		Actual Shear kN
150	6.0	61	329	>	300
160	5.6	54	273	<	300
155	5.8	58	303	≈	300

∴ For a stiffened web, neglecting tension field action. required plate thickness = 5.8mm [6.0mm].

The Steel Construction Institute
Silwood Park Ascot Berks SL5 7QN
Telephone: (0990) 23345
Fax: (0990) 22944 Telex: 846843

CALCULATION SHEET

Job No. Chapter 25	Sheet 4 of 6	Rev.
Job Title Plate Girder Design		
Client	Contract No.	
Made by R.E.	Date Sept. 88	
Checked by W.B.	Date Sept. 88.	

BS.5950 Part 1.

cii). Stiffened web with Tension field action.
From clause 4.4.5.4:~
Shear buckling resistance

$$V_b = \underbrace{q_b\, d.t.}_{\text{web contribution}} + \underbrace{q_f \sqrt{K_f}.\, dt}_{\text{flange dependent contribution}} \quad \text{but} \leq 0.6 p_y.\, dt.$$

4.4.5.4

Since the flange stress is high due to the bending moment, [see sheet 2/6], the flange contribution to the shear resistance will be low. This is shown by the high value of the second term in the bracket in the expression for K_f.

$$K_f = \frac{M_{pf}}{4 M_{pw}}\left[1 - \frac{f}{p_{yf}}\right]$$

where:~

M_{pf} = plastic moment capacity of flange
$= 0.25(2b)\, T^2.\, p_{yf}$
$= 0.25 \times 600 \times 25^2 \times 265$
$= 24.84 \times 10^6$ N mm

M_{pw} = plastic moment capacity of web
$= 0.25(t)\, d^2\, p_y$
$= 0.25(t) \times 900^2 \times 265$
$= 53.66t \times 10^6$ N mm

f = flange stress = 216 N/mm² from sheet 2/6

$$\therefore K_f = \frac{24.84 \times 10^6}{4 \times 53.66t \times 10^6}\left[1 - \frac{216}{265}\right] = \frac{0.0214}{t}$$

Assmd. d/t	Corr. t	q_b (N/mm²) from Table 22a.	web $q_b.d.t$ KN	$\sqrt{K_f}$	q_f (N/mm²) from Table 23a	Flange $q_f.\sqrt{K_f}.d.t$ KN	V_b KN.	Actual
200	4.5	71	288	0.069	313	87	375	>300
225	4.0	65	234	0.073	319	84	318	>300
235	3.8	64	219	0.075	320	82	301	≮300

Tables 22a. and 23a.

∴ For a stiffened web, allowing for Tension field action required plate thickness = 3.8 mm [5mm plate]

Note:-
In each of the cases the most practical thickness above the theoretical is shown in brackets.

The Steel Construction Institute Silwood Park Ascot Berks SL5 7QN Telephone: (0990) 23345 Fax: (0990) 22944 Telex: 846843 CALCULATION SHEET	Job No. Chapter 25	Sheet 5 of 6	Rev.
	Job Title Plate Girder Design		
	Client	Contract No.	
	Made by R.E.	Date Sept. 88	
	Checked by W.B.	Date Sept. 88.	

BS.5950 Part 1.

Summary of design to BS.5950: Part 1.

a) Flange dimensions = 600 × 25 mm

b) Web dimensions
(i) Stocky web = 900 × 14 mm d/t = 67
or for practical sizes = 900 × 15 mm d/t = 60

(ii) Thin web without stiffeners:–
= 900 × 6.5mm d/t = 139
or for practical sizes = 900 × 7.0 mm d/t = 129

(iii) Thin web with transverse stiffeners, no tension field action
= 900 × 5.8mm d/t = 155
or for practical sizes = 900 × 6.0mm d/t = 150

(iv) Thin web with transvers stiffeners, and tension field action
= 900 × 3.8mm d/t = 235
or for practical sizes = 900 × 5.0mm d/t = 180

These dimensions show that for an efficient design, transverse web stiffeners should be used and the tension field action should be taken into account in the design calculations.
From this summary a plate girder with 600×25mm flanges and 900×5mm stiffened web will be adopted.

The Steel Construction Institute
Silwood Park Ascot Berks SL5 7QN
Telephone: (0990) 23345
Fax: (0990) 22944 Telex: 846843

CALCULATION SHEET

Job No. Chapter 25	Sheet 6 of 6	Rev.
Job Title Plate Girder Design		
Client	Contract No.	
Made by R.E.	Date Sept. 88	
Checked by W.B.	Date Sept. 88.	

<u>Final design checks on selected girder.</u> — BS.5950 Part 1.

950 | 900 = d | h_s = 925 | 25 | t = 5 | 25 | 600 B

All dimensions in mm.

Stiffener spacing $a = 1260$ mm
Web depth $d = 900$ mm
Flange centres $h_s = 925$ mm
Flange width $B = 600$ mm.
Flange area A_f
$= 600 \times 25 = 15000 \text{ mm}^2$

Shear area $A_v = 900 \times 5 = 4500 \text{ mm}^2$

<u>Shear capacity</u>, P_v — 4.2.3

$P_v = 0.6 \cdot p_y \cdot A_v$

$= \dfrac{0.6 \times 265 \times 4500}{10^3}$

$= \underline{715 \text{ KN}} > \underline{300 \text{ KN}}$ OK.

<u>Moment Capacity</u>, M_c — 4.4.4.2 (a).

$M_c = p_y \cdot A_f \cdot h_s$

$= \dfrac{265 \times 15000 \times 925}{10^6} = \underline{3677 \text{ KN.m}} > \underline{3000 \text{ KN m}}$ OK.

<u>Minimum web thickness for serviceability</u> — 4.4.2.2.

b) With transverse stiffeners

(i) where stiffener spacing $a > d$: $t \geq \dfrac{d}{250}$
note $a/d = 1.4$.

$\therefore a = 1260\text{mm} > d = 900\text{mm}$: $t = 5\text{mm} > \dfrac{900}{250} = \underline{3.6\text{mm}}$

<u>Minimum web thickness to avoid flange buckling</u> — 4.4.2.3

b). With intermediate transverse stiffeners:

(2) where stiffener spacing $a \leq 1.5d$: $t \geq \dfrac{d}{250}\left(\dfrac{p_{yf}}{455}\right)^{1/2}$

$\therefore a = 1260\text{mm} < 1.5d = 1.5 \times 900 = 1350\text{mm}$: $t = 5\text{mm} > \dfrac{900}{250}\left(\dfrac{265}{455}\right)^{1/2}$

$= \underline{2.75\text{mm}}$ OK.

$\therefore$ <u>Section</u> <u>Satisfactory</u>.

The Steel Construction Institute
Silwood Park Ascot Berks SL5 7QN
Telephone: (0990) 23345
Fax: (0990) 22944 Telex: 846843

CALCULATION SHEET

Job No. Chapter 38	Sheet 1 of 6	Rev.
Job Title: Truss & Girder Connections		
Client	Contract No.	
Made by T. Mc G	Date Sept. 88.	
Checked by W. B.	Date Sept. 88.	

BS.5950 Part 1

Connection Design.

Summary of Contents.

1). Internal truss joint ~ shop welded and bolted site joints

2). Face joint for Lattice Girder ~ members welded directly to each other ~ end plate bolted to column.

3). Wind bracing connection ~ high strength friction grip bolts ~ Group in double shear ~ Group in shear and tension.

1) Internal Truss Joint.

Problem:~ Design the bottom chord centre joint in the roof truss given

The data is as follows:~

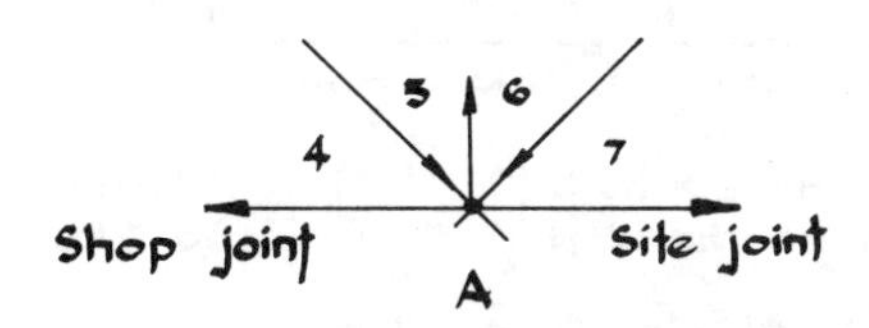

Member	Force in kN [unfactored]		Section.
	Dead	Imposed	
A4, A7	23.5	41.1	90×90×6L
4-5, 6-7	- 6.86	-12.0	80×80×6L
5-6	10.3	18.0	65×50×6L

Members ~ A-4, 4-5, 5-6 welded shop joints
~ A-7 6-7 bolted site joints.

Material ~ Grade 43 Steel.

The Steel Construction Institute
Silwood Park Ascot Berks SL5 7QN
Telephone: (0990) 23345
Fax: (0990) 22944 Telex: 846843

CALCULATION SHEET

Job No. Chapter 38	Sheet 2 of 6	Rev.
Job Title: Truss & Girder Connections		
Client	Contract No.	
Made by T. McG.	Date Sept. 88.	
Checked by W.B.	Date Sept. 88.	

BS. 5950 Part 1, Table 2

Factored Loads

A-4, A-7 $F = [1.4 \times 23.5] + [1.6 \times 41.1] = 98.7$ KN

4-5 , 6-7 $F = [1.4 \times 6.86] + [1.6 \times 12] = 28.8$ KN

5-6 $F = [1.4 \times 10.3] + [1.6 \times 18] = 43.2$ KN

Ordinary bolts ~ Grade 4.6 ~ Steel Grade 43

Shear capacity $P_s = p_s . A_s$.
where $p_s = 160$ N/mm² (Table 32)
A_s = tensile stress area. (6.3.2)
Bearing capacity $P_{bb} = d.t.\ p_{bb}$ for bolt.
$p_{bb} = 435$ N/mm² (Table 32)
or $P_{bs} = d.t.p_{bs} \leq \frac{1}{2} e.t.\ p_{bs}$
$p_{bs} = 460$ N/mm² (Table 33)
e = end distance (6.2.3)

Dia.	Shear Capacity $P_s = p_s . A_s$			Bearing Capacity $P_{bb} = d.t.p_{bb}$				$\frac{1}{2}$ e.t. p_{bs}			
mm	p_s N/mm²	A_s mm²	P_s KN	d mm	t mm	p_{bb} N/mm²	P_{bb} KN	e mm	t mm	p_{bs} N/mm²	P_{bb} KN
20	160	245	39.2	20	6	435	52.2	40	6	460	55.2
16	160	157	25.1	16	6	435	41.8	30	6	460	41.4

A_s, obtained from Structural Steelwork Handbook (S.C.I.)

Member A-7   3-No. 20mm dia. bolts $[3 \times 39.2] = 117.6$ KN
Member 6-7   2-No. 16mm dia. bolts $[2 \times 25.1] = 50.2$ KN

Use 20mm. dia. ~ Ordinary bolts Grade 4.6.

Weld ~ Grade 43 ~ Design Strength $p_w = 215$ N/mm² (6.6.5, Table 36)

Provide 5mm fillet weld
Weld strength $= \frac{0.7 \times 5 \times 215}{10^3} = 0.75$ KN/mm

Member A-4 Length Req'd. $= 98.7/0.75 = 132$mm
" 4-5 " " $= 28.8/0.75 = 38$mm
" 5-6 " " $= 43.2/0.75 = 58$mm

Detailed check may be carried out for joints with weld all round or weld one side only.

Gusset Plate

Moment capacity may be checked. (4.2.5.)
Direct stress can also be checked.

The Steel Construction Institute
Silwood Park Ascot Berks SL5 7QN
Telephone: (0990) 23345
Fax: (0990) 22944 Telex: 846843

CALCULATION SHEET

Job No. Chapter 38	Sheet 3 of 6	Rev.
Job Title Truss & Girder Connections		
Client	Contract No.	
Made by T.McG.	Date Sept. 88.	
Checked by WB	Date Sept. 88.	

BS.5950 Part 1.

2). Face joint for Lattice Girder

Problem:- Design the face joint for a lattice girder for the forces shown in the table.

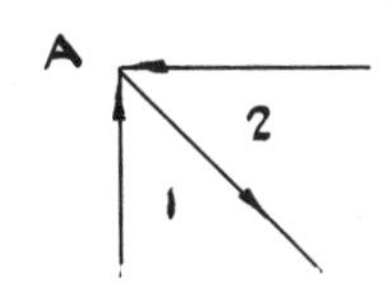

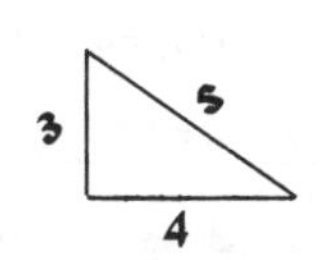

Member	Force KN.		Section.
	Dead	Imposed	
A-1	19.8	33	152×152×30 UC
A-2	26.4	44	102×152×13 T
1-2	33	55	65×50×6 L

Members A-2 and 1-2 are welded directly together.
The lattice girder is bolted to the column.
Material ~ Grade 43 steel.

Factored loads - for members at joint. (Table 2)

A-1 F = [1.4×19.8] + [1.6×33] = 80.5 KN
A-2 F = [1.4×26.4] + [1.6×44] = 111.4 KN
1-2 F = [1.4×33] + [1.6×55] = 134.2 KN

Bolted Joint ~ Lattice Girder to Column
Factored shear = 80.5 KN.

Provide 4 No. 20mm. dia. ordinary bolts. - Grade 4.6
Shear capacity = 4×39.2 = 156.8 KN > 80.5KN OK.

4 No. 16mm. dia. bolts are strong enough but are considered to be too light and not sufficiently durable for the joint.

Welded Joint - Member 1-2 to A-2

Provide 5mm. fillet weld
Strength = 0.75KN/mm

a). Weld all round joint
Factored load = 134.2 KN

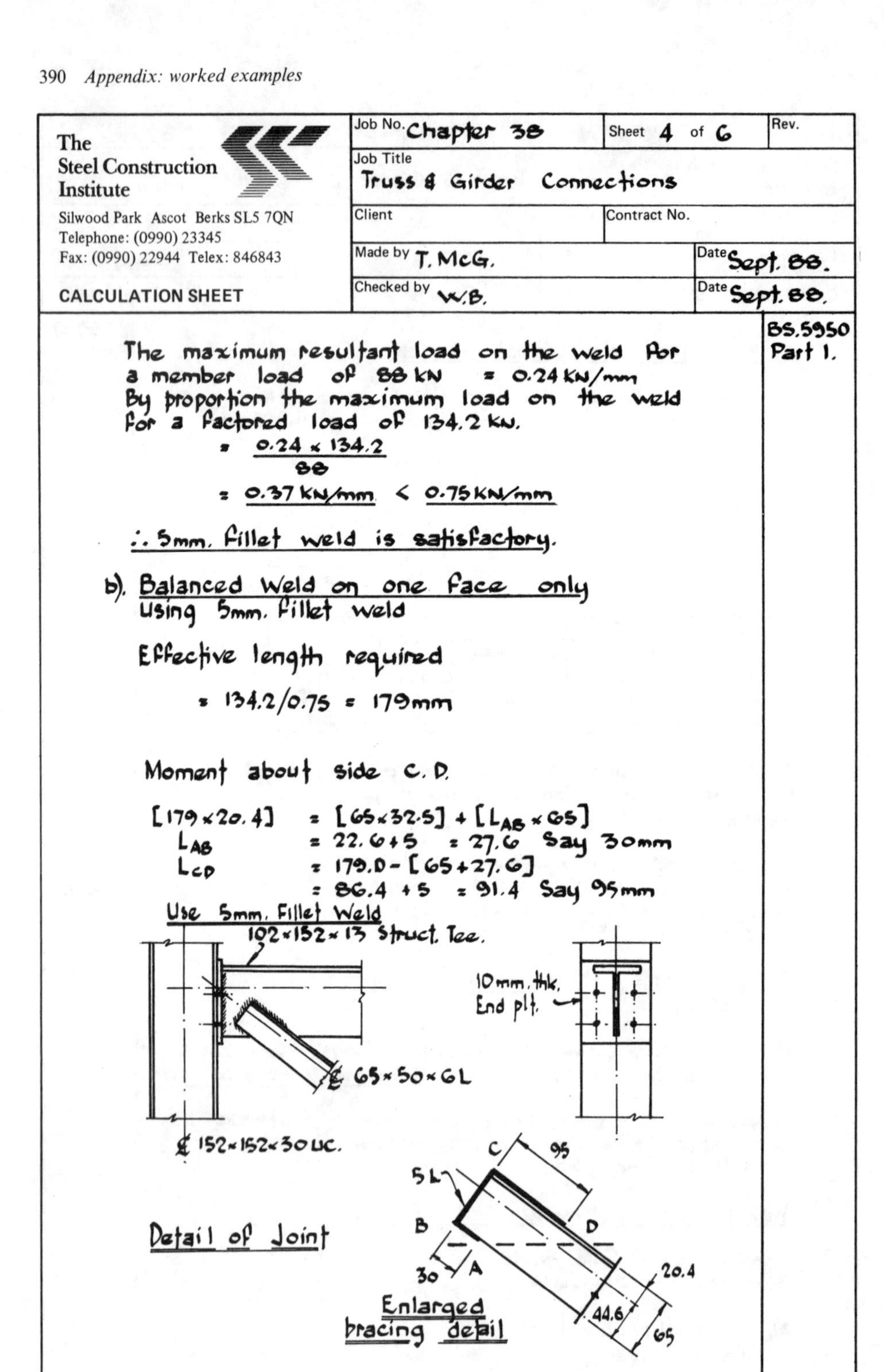

The Steel Construction Institute Silwood Park Ascot Berks SL5 7QN Telephone: (0990) 23345 Fax: (0990) 22944 Telex: 846843 CALCULATION SHEET	Job No. Chapter 38	Sheet 4 of 6	Rev.
	Job Title Truss & Girder Connections		
	Client	Contract No.	
	Made by T. McG.	Date Sept. 88.	
	Checked by W.B.	Date Sept. 88.	

BS.5950 Part 1.

The maximum resultant load on the weld for a member load of 88 kN = 0.24 kN/mm

By proportion the maximum load on the weld for a factored load of 134.2 kN.

$= \frac{0.24 \times 134.2}{88}$

$= 0.37$ kN/mm < 0.75 kN/mm

∴ 5mm. fillet weld is satisfactory.

b). Balanced Weld on one face only using 5mm. fillet weld

Effective length required

$= 134.2/0.75 = 179$ mm

Moment about side C.D.

$[179 \times 20.4] = [65 \times 32.5] + [L_{AB} \times 65]$

$L_{AB} = 22.6 + 5 = 27.6$ Say 30mm

$L_{CD} = 179.0 - [65 + 27.6]$

$= 86.4 + 5 = 91.4$ Say 95mm

Use 5mm. Fillet Weld

The Steel Construction Institute Silwood Park Ascot Berks SL5 7QN Telephone: (0990) 23345 Fax: (0990) 22944 Telex: 846843 CALCULATION SHEET	Job No. Chapter 38	Sheet 5 of 6	Rev.
	Job Title Truss & Girder Connections		
	Client	Contract No.	
	Made by T. McG.	Date Sept. 88.	
	Checked by W.B.	Date Sept. 88.	

3). Wind bracing connection — BS.5950 Part 1.

Problem:-

Design a connection for a bracing member which carries a load due to wind of 315 kN

The member consists of 2 No. angles 80 × 60 × 8 ⅃L and is inclined at an angle of 45° to the column.

Use Grade 43 Steel and High strength Friction Grip Bolts.

315 kN.
Cut from 457×191×89 UB
℄ 2 No. 80×60×8 ⅃L
Shear 222.7 kN
Group A - 3 bolts.
100
100
100
40
80
80
40
110
Tension = 222.7 kN
Group B - 8 bolts.
℄ 203×203×71 UC

Detail of Connection

H.S.F.G Bolts ~ 20mm. diameter

Proof Load = 144 kN

From Structural Steelwork Handbook [obtained from BS. 4604]. — SCI.

Load due to wind = 315 kN

Factored Load = 1.4×315 = 441 kN — Table 2

Try 20mm. dia. parallel shank friction grip fasteners. — 6.4.1

Slip resistance $P_{SL} = 1.1\, K_s \mu P_o$ — 6.4.2.1

P_o = minimum shank tension = 144 kN — BS.4604.

k_s = 1.0 for clearance holes

μ = slip factor taken as 0.45

Then $P_{SL} = 1.1 \times 1.0 \times 0.45 \times 144$ = 71.28 kN

The Steel Construction Institute
Silwood Park Ascot Berks SL5 7QN
Telephone: (0990) 23345
Fax: (0990) 22944 Telex: 846843

CALCULATION SHEET

Job No. Chapter 38	Sheet 6 of 6	Rev.
Job Title: Truss & Girder Connection		
Client	Contract No.	
Made by T.McG.	Date Sept. 88.	
Checked by W.B	Date Sept. 88.	

BS.5950 Part 1.

Bearing resistance

$P_{bg} = d.t.\,p_{bg} \leq \frac{1}{3}\,e.t.\,p_{bg}$ — 6.4.2.2

p_{bg} for Grade 43 Steel $= 825\,N/mm^2$ — Table 34

$d = 20$; $t = 8$

e = edge distance $= 40\,mm$

Then $P_{bg} = \frac{20 \times 8 \times 825}{10^3} = 132\,kN$

but $\frac{1}{3}[40 \times 8 \times 825/10^3] = 88\,kN$

Group A ~ 3 bolts in double shear

Capacity $= 3 \times 2 \times 71.28 = \underline{427.7\,kN} < \underline{441\,kN}$ OVER.

[for the purposes of this design example accept 3% over]

Group B ~ 8 bolts in shear and tension.

$P_{sL} = 71.28\,kN.$

$P_t = 0.9 \times P_o$ — 6.4.4.2

$= 0.9 \times 144 = 129.6\,kN$

Tension $F_t = \frac{441 \times 0.707}{8} = 39.0\,kN$

Shear $F_s = 39.0\,kN$

Combined shear and tension — 6.4.5

$$\frac{F_s}{P_{sL}} + \frac{0.8\,F_t}{P_t} \leq 1.0$$

Then $\frac{39}{71.28} + \frac{0.8 \times 39}{129.6} = \underline{0.79 < 1.0 \text{ OK.}}$

Web of UB. cutting

Tension capacity can be checked.

Index